Probiotic Dairy Products

Society of Dairy Technology Series

Series Editor: Adnan Y. Tamime

The Society of Dairy Technology has joined with Wiley-Blackwell to produce a series of technical dairy-related handbooks providing an invaluable resource for all those involved in the dairy industry; from practitioners to technologists working in both traditional and modern large-scale dairy operations.

Probiotic Dairy Products, 2nd Edition, ISBN 9781119214106
by Adnan Y. Tamime and Linda V. Thomas (Editors)

Microbial Toxins in Dairy Products, ISBN 9781118756430
by Adnan Y. Tamime (Editor)

Biofilms in the Dairy Industry, ISBN 9781118876213
by Koon Hoong Teh, Steve Flint, John Brooks, and Geoff Knight (Editors)

Milk and Dairy Products as Functional Foods, ISBN 9781444336832
by Ara Kanekanian (Editor)

Membrane Processing: Dairy and Beverage Applications, ISBN 9781444333374
by Adnan Y. Tamime (Editor)

Processed Cheese and Analogues, ISBN 9781405186421
by Adnan Y. Tamime (Editor)

Technology of Cheesemaking, 2nd Edition, ISBN 9781405182980
by Barry A. Law and Adnan Y. Tamime (Editors)

Dairy Fats and Related Products, ISBN 9781405150903
by Adnan Y. Tamime (Editor)

Dairy Powders and Concentrated Products, ISBN 9781405157643
by Adnan Y. Tamime (Editor)

Milk Processing and Quality Management, ISBN 9781405145305
by Adnan Y. Tamime (Editor)

Cleaning-in-Place: Dairy, Food and Beverage Operations, 3rd Edition, ISBN 9781405155038
by Adnan Y. Tamime (Editor)

Structure of Dairy Products, ISBN 9781405129756
by Adnan Y. Tamime (Editor)

Brined Cheeses, ISBN 9781405124607
by Adnan Y. Tamime (Editor)

Fermented Milks, ISBN 9780632064588
by Adnan Y. Tamime (Editor)

Probiotic Dairy Products, ISBN 9781405121248
by Adnan Y. Tamime (Editor)

Probiotic Dairy Products

Second Edition

Edited by

Adnan Y. Tamime
Consultant in Dairy Science and Technology
Ayr
Scotland
United Kingdom

Linda V. Thomas
Editor, *International Journal of Dairy Technology*
Dorchester
England
United Kingdom

WILEY Blackwell

SDT Society of Dairy Technology

This edition first published 2018

Edition History
John Wiley & Sons Ltd (1e, 2006)

Registered Offices
John Wiley & Sons, Inc., 111 River Street, Hoboken, NJ 07030, USA
John Wiley & Sons Ltd, The Atrium, Southern Gate, Chichester, West Sussex, PO19 8SQ, UK

Editorial Office
The Atrium, Southern Gate, Chichester, West Sussex, PO19 8SQ, UK

For details of our global editorial offices, customer services, and more information about Wiley products visit us at www.wiley.com.

Wiley also publishes its books in a variety of electronic formats and by print-on-demand. Some content that appears in standard print versions of this book may not be available in other formats.

Library of Congress Cataloging-in-Publication Data

Names: Tamime, A. Y., editor. | Thomas, Linda V., 1955– editor.
Title: Probiotic dairy products / edited by Adnan Y. Tamime, Scotland, United Kingdom, Linda V. Thomas, editor, International Journal of Dairy Technology, England, United Kingdom.
Description: Second edition. | Hoboken, NJ : Wiley, 2017. | Series: Society of dairy technology | Includes bibliographical references and index. | Identifiers: LCCN 2017029408 (print) | LCCN 2017029768 (ebook) | ISBN 9781119214113 (pdf) | ISBN 9781119214120 (epub) | ISBN 9781119214106 (hardback)
Subjects: LCSH: Dairy microbiology. | Dairy products in human nutrition. | Probiotics. | BISAC: TECHNOLOGY & ENGINEERING / Food Science.
Classification: LCC QR121 (ebook) | LCC QR121 .P76 2017 (print) | DDC 641.3/7–dc23
LC record available at https://lccn.loc.gov/2017029408

Cover Design: Wiley
Cover Image: © Chr. Hansen

Set in 10/12.5pt Times by SPi Global, Pondicherry, India
Printed and bound in Malaysia by Vivar Printing Sdn Bhd

10 9 8 7 6 5 4 3 2 1

Contents

List of Contributors

Editors

Dr Adnan Y. Tamime
Dairy Science & Technology Consultant
24 Queens Terrace
Ayr KA7 1DX
Scotland – United Kingdom
Tel: +44 (0)1292 265498
Fax: +44 (0)1292 265498
Mobile: +44 (0)7980 278950
E-mail: draytamime@gmail.com

Dr Linda V. Thomas
57 Queen's Avenue
Dorchester DT1 2EP
Dorset
England – United Kingdom
Mobile: +44 (0)7484 602729
E-mail: drlvthomas@gmail.com

Contributors

Dr Giovanna E. Felis
University of Verona
Department of Biotechnology
via della Pieve 70
37029 S. Floriano di S. Pietro in Cariano (VR)
Italy
Tel: +39 045 6835627
Fax: +39 045 6835631
E-mail: giovanna.felis@univr.it

Professor Gerald Fitzgerald
University College Cork
Department of Microbiology
Cork
Ireland
Tel: +353 (0)21 490 2730
E-mail: g.fitzgerald@ucc.ie

Dr Ana Belén Flórez
IPLA-CSIC
Paseo Río Linares s/n
33300-Villaviciosa
Spain
Tel: +34985892131
Fax: +34985892233
E-mail: abflorez@ipla.csic.es

Dr Hamid B. Ghoddusi
London Metropolitan University
Faculty of Life Sciences and Computing
School of Human Sciences
Head of Microbiology Research Unit (MRU)
166–220 Holloway Road
London N7 8DB
England – United Kingdom
Tel: +44 (0)20 7133 4196
E-mail: H.Ghoddusi@londonmet.ac.uk

Mr Michael Hickey
Derryreigh
Creggane
Charleville
Cork
Ireland
Tel: +353 (0)63 89392
Mobile: +353 (0)87 2385653
E-mail: mfhickey@oceanfree.net

Professor Colin Hill
University College Cork
School of Microbiology
Cork
Ireland
Tel: +353 (0)21490 3000
E-mail: c.hill@ucc.ie

Dr Ron Levin
Haydonhill House
Bushey
Herts WD23 1DU
England – United Kingdom
Tel: +44 (0)208 950 5463
E-mail: ron@ronlevin.co.uk and rlevin@talktalk.net

Dr Daniel M. Linares
Food Biosciences Department
Teagasc Food Research Centre
Moorepark
Fermoy
Cork
Ireland
Tel: +353 (0)2542 273
Fax: +353 (0)2542 340
E-mail: Daniel.Linares@teagasc.ie

Dr Petra Mohar Lorbeg
University of Ljubljana
Biotechnical Faculty
Institute of Dairy Science and Probiotics
Groblje 3
1230 Domžale
Slovenija
Tel: +386 1 3203 844
E-mail: Petra.Mohar@bf.uni-lj.si

Dr Andreja Čanžek Majhenič
University of Ljubljana
Biotechnical Faculty
Chair of Dairy Science
Groblje 3
1230 Domžale
Slovenija
Tel: +386 1 3203 844
E-mail: Andreja.Canzek@bf.uni-lj.si

Dr Baltasar Mayo
IPLA-CSIC
Paseo Río Linares s/n
33300-Villaviciosa
Spain
Tel: +34985892131
Fax: +34985892233
E-mail: baltasar.mayo@ipla.csic.es

Professor Robert A. Rastall
The University of Reading
Department of Food and Nutritional Sciences
PO Box 226
Whiteknights
Reading RG6 6AP
England – United Kingdom
Tel: +44 (0)118 378 6726
Fax: +44 (0)118 931 0080
E-mail: r.a.rastall@reading.ac.uk

Professor Paul Ross
University College Cork
College of Science Engineering and Food Science
Cork
Ireland
Tel: +353 (0)21490 3760
E-mail: p.ross@ucc.ie

Dr Maria Saarela
Industrial Microbiology
Business Development
VTT Biotechnology and Food Research
Box 1501
FIN-02044-VTT
Finland
Tel: +358 40 5760913
E-mail: Maria.Saarela@vtt.fi

Professor Nagendra P. Shah
University of Hong Kong
6N-08, Kadoorie Biological Sciences Building
Dairy and Probiotic Unit
Food and Nutritional Science Programme
The Pokfulam Road
Hong Kong
Tel: +852 (0)2299 0836
Fax: +852 (0)2559 9114
E-mail: npshah@hku.hk

Professor Catherine Stanton
Teagasc Moorepark
Food Research Centre

Fermoy
Cork
Ireland
Tel: +353 (0)2542 606
Fax: +353 (0)2542 340
E-mail: Catherine.Stanton@teagasc.ie

Professor Sandra Torriani
University of Verona
Department of Biotechnology
Strada Le Grazie 15
37134 Verona
Italy
Tel: +39 045 8027051
Fax: +39 045 8027928
E-mail: sandra.torriani@univr.it

Dr Primož Treven
University of Ljubljana
Biotechnical Faculty
Chair of Dairy Science
Groblje 3
1230 Domžale
Slovenija
Tel: +386 1 3203 909
E-mail: Primoz.Treven@bf.uni-lj.si

Professor Effie Tsakalidou
Agricultural University of Athens
Department of Food Science and Human Nutrition
Iera Odos 75
11855 Athens
Greece
Tel: +30 (0)210 5294661
Fax: +30 (0)210 5294672
E-mail: et@aua.gr

Ms Xuedan Wang
The University of Reading
Department of Food and Nutritional Sciences
PO Box 226
Whiteknights
Reading RG6 6AP
England – United Kingdom
Tel: 0118 378 8718
Fax: 0118 378 7708
E-mail: X.Wang6@pgr.reading.ac.uk

Dr Monika Wszolek
Animal Products Technology Department
University of Agriculture in Krakow
Faculty of Food Technology
ul. Balicka 122
30–149 Krakow
Poland
Tel: +48 (0)12 662 4788
Fax: +48 (0) 12 662 4810
E-mail: rtwszole@cyf-kr.edu.pl

Dr Georgia Zoumpopoulou
Agricultural University of Athens
Department of Food Science and Human Nutrition
Laboratory of Dairy Research
Iera Odos 75
11855 Athens
Greece
Tel: +30 (0)210 5294628
Fax: +30 (0)210 5294672
E-mail: gz@aua.gr

Preface to the Technical Series, Second Edition

For more than 70 years, the Society of Dairy Technology (SDT) has sought to provide education and training in the dairy field, disseminating knowledge and fostering personal development through symposia, conferences, residential courses, publications, and its journal, the *International Journal of Dairy Technology* (previously known as *Journal of the Society of Dairy Technology*).

Through this time, there have been major advances in our understanding of milk systems, probably the most complex natural food available to man. Improvements in process technology have been accompanied by massive changes in the scale and efficiency of many milk and dairy processing operations, accompanied by an ever widening range of sophisticated dairy and other related products.

In 2005, the Society embarked on a project to produce a Technical Series of dairy-related books, to provide an invaluable source of information for practicing dairy scientists and technologists, covering the range from traditional to modern large-scale operations. The 2nd edition of 'Probiotic Dairy Products', under the editorship of Drs Adnan Tamime and Linda Thomas, provides a timely update on the advances that have been made in the understanding of the human gut microbiota, the characterisation, enumeration and production of probiotics together with their relationship with prebiotics and the commercial implications for dairy and other products within the legislative constraints.

Andrew Wilbey
Chairman of the Publications Committee, SDT
October 2016

Preface to the Technical Series, First Edition

For more than 60 years, the Society of Dairy Technology (SDT) has sought to provide education and training in the dairy field, disseminating knowledge and fostering personal development through symposia, conferences, residential courses, publications, and its journal, the International Journal of Dairy Technology (previously known as Journal of the Society of Dairy Technology).

In recent years, there have been significant advances in our understanding of milk systems, probably the most complex natural food available to man. Improvements in process technology have been accompanied by massive changes in the scale of many milk/dairy processing operations, and the manufacture of a wide range of dairy and other related products.

The Society has now embarked on a project with Blackwell Publishing to produce a Technical Series of dairy-related books to provide an invaluable source of information for practising dairy scientists and technologists, covering the range from traditional to modern large-scale operations. This, the first volume in the series, on 'Probiotic Dairy Products', under the editorship of Dr Adnan Tamime, complements the second volume on 'Fermented Milks' in providing a wide-ranging review of this group of micro-organisms, which are increasingly recognised as playing a vital role in the maintenance of our health while also contributing to the microbiology of many fermented dairy products.

Andrew Wilbey
President, SDT
February 2005

Preface to the Second Edition

Since the publication of the first edition of this book in 2005, we have witnessed incredible advances in our knowledge and understanding of the human microbiota, mainly due to the development and use of new molecular analysis techniques. One example is the new 'omic' technologies that have been used to detect and analyse all the genes, proteins and metabolites of individuals' gut microbiota. Studies investigating different population groups in various states of health that have used such methods have given a better overall picture of the composition and functions of the gut microbiota. This new edition of 'Probiotic Dairy Products' reflects this scientific interest by incorporating a new chapter on the human gut microbiota (see Chapter 1), which reviews current knowledge.

The vast amount of research that has been conducted in this field, which has included several multi-national projects, has resulted in numerous high-profile scientific papers that have helped to drive medical and consumer interest in probiotics, because of their influences on the gut, its microbiota and overall health. Another new chapter for this edition describes the history of probiotics (see Chapter 2), reminding us of the origins of these products and the early pioneers in this field. It is generally acknowledged that the probiotic concept started with Metchnikoff's idea that a long healthy life could be promoted by increasing numbers of lactic acid bacteria in the colon at the expense of 'putrefying' bacteria that were injurious to health. In the twenty-first century, probiotic benefits have been reported for an extraordinary range of health and disease areas (see Chapter 8), and it is important to note that clinical studies have been conducted not just with tablets or powders but also with probiotic dairy products, in the form of fermented milk drinks and yoghurts. One great advantage of dairy products over pharmaceuticals is that the former can be incorporated readily into one's daily diet, and thus can quite easily be part of a proactive strategy for health maintenance.

It is an absolute requirement that manufacturers can assure product quality and safety. Probiotic products must contain adequate numbers of live microbial strains, and other chapters in this book provide valuable updates on genomic analysis of probiotic strains (Chapter 3) and aspects of probiotic products' production and quality control (Chapter 4). The new molecular technologies can now be applied for the identification and enumeration of the live probiotic strains in dairy products, although culture methods remain important. These methods are reviewed in Chapter 6.

Since the first edition of the book, the sale and marketing of probiotics have expanded to around the world, which has led to regulatory changes to ensure that, among other things, probiotic health claims are substantiated by scientific evidence. This is reviewed in Chapter 5. Probiotics are sometimes combined with prebiotics to make synbiotic products, and the research behind prebiotics is discussed in Chapter 7, whilst Chapter 9 gives an overview of the different metabolites that can be produced by probiotic strains that have potential health benefits. Finally, Chapter 10 speculates on the future for probiotic dairy products, and the current barriers to progress.

A.Y. Tamime and Linda V. Thomas
December 2016

Preface to the First Edition

Fermented foods, including milk and dairy products, have played important roles in the diet of humans worldwide for thousands of years. Since the mid-1950s, there has been increasing knowledge of the benefits of certain micro-organisms, such as lactic acid bacteria (LAB) and probiotic gut flora, and their impact on human biological processes and, at the same time, of the identity of certain dairy and non-dairy components of fermented milks and their role in human health and body function. The purpose of this book, which is written by a team of international scientists, is to review the latest scientific developments in these fields with regard to the 'functional' aspects of fermented milk products and their ingredients.

Some scientific aspects reviewed in this publication are: (a) the latest knowledge regarding the gut microflora (e.g. identifying the beneficial microbiota in terms of probiotic and health aspects); (b) the use of a wide range of probiotic micro-organisms during the manufacture of different dairy products that have dominated the global markets for the past decades and are used as vehicles to increase the probiotic gut flora of humans; (c) the genomic sequences of certain strains of LAB; and (d) the use of prebiotic ingredients, such as galacto- and fructo-oligosaccharides, to enhance the viable count of probiotic microflora in humans.

Furthermore, numerous related topics – for example, the current statutory regulations (national and international), analytical methods to enumerate these beneficial organisms, sensory profiling to improve the quality of the product and enhance consumer acceptability, bioactive components produced by the probiotic microflora, and the treatment of certain human diseases – are also reviewed. It is of interest to note that the current research work on probiotic dairy products, which aims to understand the role of the intestinal microbiota, will underpin new strategies to improve the health status of consumers, and will contribute to a reduction in healthcare costs, particularly in ageing populations.

A.Y. Tamime
February 2005

1 Microbiota of the Human Gut[1]

H.B. Ghoddusi and L.V. Thomas

1.1 Background

The human gastrointestinal (GI) tract has been the subject of intense research over the past decade, since the publication of the first edition of this book. Notably, the Human Microbiome Project in the United States of America (USA) (http://hmpdacc.org) (Turnbaugh *et al.*, 2007) and the Metagenomics of the Human Intestinal Tract consortium in Europe (MetaHIT; www.metahit.eu) (Qin *et al.*, 2010) have been two major initiatives, but very many other research groups have published their findings. Scientists can get qualitative and quantitative information about all the microbes present in the gut (the gut microbiota) in the context of their habitat, genomes and surrounding environment (the gut microbiome), as well as cataloguing all the metabolites in the gut (metabolomics) and getting an overview of microbial functions in the gut based on analysis of all their genes (metagenomics), the genes' activity (transcriptomics) and proteins present (metaproteomics) (Marchesi *et al.*, 2016). Such work has amassed a vast amount of data and helped improve our understanding of microbial communities in the human body. Although the main target of this research has been the human intestinal tract, other body parts, including the skin and the nasal, oral and urogenital tracts, have not been overlooked. Apart from finding an answer to the 'What is there?' question, the main purpose of this research has been to look for associations between any observed changes in the microbiome and the prevalence of certain diseases (Korecka & Arulampalam, 2012). One clear outcome, however, has been the confirmation of the key influence of the human gut microbiota on health, not just of the gut but of the whole body, because of the gut microbiota's influence on different systems in the body (Rooks & Garrett, 2016). In fact, many scientists and medics are now of the opinion that the gut microbiota should be considered equivalent to a body organ (Marchesi *et al.*, 2016).

The highly specialised ecosystem that is the human gut microbiota has evolved to achieve a symbiotic homeostatic relationship with the host (Bäckhed *et al.*, 2005; Flint *et al.*, 2012). The GI tract and its microbiota cannot be really considered as separate

[1] In the book's first edition, this chapter was authored by Dr B. O'Grady and Professor Glenn Gibson of the University of Reading. The current chapter constitutes a major update of that work to reflect the significant advances in this field since 2005.

Probiotic Dairy Products, Second Edition. Edited by Adnan Y. Tamime and Linda V. Thomas.

entities because together they represent a dynamic biological system that has developed together from birth. The human GI tract is composed of highly adapted regions for mediation of its diverse functions, many of which impact markedly upon host health and welfare. Physiological considerations in each unique region influence the degree and type of colonisation, and initial colonisers also modify the physiological conditions therein. This results in the development of distinct microhabitats along the length of the GI tract, which influence metabolism, protection and immune stimulation (Flint *et al.*, 2012; Thomas *et al.*, 2014; Honda & Littman, 2016). Such effects are both local and systemic, as the GI tract is connected to the vascular, lymphatic and nervous systems. The ability of the gut to sustain a microbiota that is supportive of health is critical for host health and reduction of disease risk.

1.2 The human GI tract and its microbiota

It has long been thought that colonisation of the GI tract begins immediately after birth (Castanys-Muñoz *et al.*, 2016), but although this is certainly when the primary colonisation process occurs, recent studies have reported the detection of micro-organisms in meconium, placenta, umbilical cord and amniotic fluid (Thomas, 2016). Micro-organisms have also been detected in breast milk (Fernández *et al.*, 2013).

Microbial colonisation of the neonate mainly occurs during the delivery process. The inoculum may be largely derived either from the mother's vaginal and faecal microbiota (in a conventional birth) or from the environment (in a Caesarean delivery); hence, the micro-organisms that colonise the new-born tract are primarily acquired postnatally. The delivery method is key, as new-borns delivered by Caesarean section are exposed to a different microbiota compared to that found in the vagina. In a recent pilot study, Dominguez-Bello *et al.* (2016) demonstrated that by exposing infants delivered by Caesarean section to maternal vaginal fluids at birth, not only the gut but also the oral and skin bacterial communities of these new-borns were partially altered to become more like those of a naturally delivered infant during the first 30 d of their life. The potential long-term health effects of Caesarean delivery remain unclear, although microbial differences may last for at least one year (Rutayisire *et al.*, 2016), and links to health risks such as childhood obesity (Blustein *et al.*, 2013) and allergic disease (Brandão *et al.*, 2016) have been reported.

Bacterial populations in the gut develop progressively during the first few days of life; facultative anaerobes predominate initially and create a reduced environment that allows for the growth of strict anaerobes (Rodríguez *et al.*, 2015). The choice of diet for the new-born is also of importance as the microbiota of breast-fed infants is predominated by bifidobacteria, whereas formula-fed infants have a more complex microbiota that resembles the adult gut, in that *Bacteroides*, clostridia, bifidobacteria, lactobacilli, Gram-positive cocci, coliforms and other groups are all represented in fairly equal proportions (Lozupone *et al.*, 2012; Ghoddusi & Tamime, 2014). Breastfeeding promotes a more beneficial microbiota; the presence of certain oligosaccharides in human breast milk, for instance, promotes the growth of beneficial bifidobacteria (Smilowitz *et al.*, 2014). During weaning, the microbiota becomes more complex, and the ecosystem is thought to become fairly stable at around two years of age. The prevalence of

Table 1.1 The change in the gut microbiota through life.

Stage of life	Intestinal microbiota profile
Foetus	Usually sterile
Baby	Immediately after birth, there is rapid colonisation of the gut with micro-organisms from the immediate surroundings; the gut microbiota composition is influenced by mode of delivery and type of feeding: • *Breast-fed*: low diversity, dominated by bifidobacteria. • *Formula-fed*: a more diverse microbiota with more Bacteroidetes and fewer bifidobacteria.
Child	The gut microbiota becomes more stable and complex over the first three years (particularly after weaning), so that it becomes much more diverse in its composition and more like that of an adult.
Adults	A diverse composition; dominant phyla are Firmicutes, Bacteroidetes and Actinobacteria.
Old age	The microbiota changes to become less diverse and resilient; there are fewer Firmicutes and bifidobacteria and more Bacteroidetes and Proteobacteria.

bifidobacteria in breast-fed infants is thought to confer protection by improving the colonisation resistance of the gut; among other mechanisms, bifidobacteria exert directly antagonistic activities against gut pathogens. New-borns are susceptible to intestinal infections and atopic diseases as their immune system and GI tract develop. The mode of delivery and subsequent diet, therefore, have important implications, both at birth and later in life, as the initial colonisation process has a strong influence on the development of the GI tract and its microbiota, and in the maturation of the immune system. During the first few years of life and after weaning, the infant microbiota normalises to a composition that remains relatively stable throughout most of adult life (Thomas, 2016). Table 1.1 summarises how the intestinal microbiota develops with age.

In recent years, the development of next-generation sequencing (NGS) techniques has played a major role in revealing that the human body harbours more than 1000 phylotypes, although intestinal bacteria mainly belong to just a few phyla (Tojo *et al.*, 2014). Most of this work comes from analysis of faecal samples; these best represent the distal portion of the gut. Due to the difficulties in obtaining samples higher in the gut, it has proved more difficult to get a true picture of the microbial communities in the small and proximal large intestines (Li *et al.*, 2015; Marchesi *et al.*, 2016).

The GI tract begins with the oral cavity (the mouth, nose and throat), where a complex microbiota exists that comprises viruses, bacteria, archaea and protozoa. Bacterial species cause dental caries and periodontal species, but many bacteria in the oral microbiome remain uncultured (Wade, 2013). Bacteria are found on the posterior and anterior tongue, sub- and supra-gingival plaque, buccal mucosa and vestibular mucosa (Willis *et al.*, 1999). These include members of the *Prevotella*, *Porphyromonas*, *Peptostreptococcus*, *Bacteroides*, *Fusobacterium*, *Eubacterium* and *Desulfovibrio* genera. Bacterial numbers drop dramatically to $<10^3$ colony forming units (cfu) mL^{-1} of gastric contents as they encounter the stomach, which provides a highly effective barrier against invading micro-organisms, both pathogenic and benign. Few micro-organisms, with the exception of acid-tolerant lactobacilli, yeasts and notably *Helicobacter pylori*, can survive the harsh, strongly acidic and peristaltic nature of the stomach.

There is a high degree of variability between the stomach, small intestine and colon in terms of numbers and bacterial population types, due predominantly to different transit times, secretions and nutrient availability (Lambert & Hull, 1996; Guilliams, 1999). Micro-organisms themselves are also determinants because they interact with and influence their surroundings to ensure their survival against competitors. This is achieved through many mechanisms, such as increasing aerobic conditions in the gut or producing inhibitory compounds, such as bacteriocins or short-chain fatty acids (which also lower the pH of the gut milieu). Such compounds may also affect the host with positive or negative consequences (Fooks & Gibson, 2002; Fuller & Perdigón, 2003).

The rapid transit time, low pH and presence of bile associated with the small intestine do not provide an environment that encourages the growth of bacteria. The duodenum also has low microbial numbers due to its short transit time and the secretion of intestinal fluids, which create a hostile environment (Sanford, 1992); however, there is a progressive increase in both numbers and species along the jejunum and ileum. The small intestine harbours enterococci, enterobacteria, lactobacilli, *Bacteroides* and clostridia. These rapidly increase in numbers from 10^4–10^6 cfu mL^{-1} in the small intestine to 10^{11}–10^{12} cfu mL^{-1} in the large intestine, as the flow of intestinal chyme slows upon entry into the colon (Salminen *et al.*, 1998).

The large gut is favourable for bacterial growth with its slow transit time, ready availability of nutrients and more favourable pH. Several hundred culturable species may be present here, although a significant proportion is not cultivable by conventional methods. The proximal colon is the site of saccharolytic fermentation, due to its high substrate availability (Scott *et al.*, 2012; Russell *et al.*, 2013; Shanahan, 2013). Organic acids produced from fermentation result in a lower pH (of 5.5–6.0) compared to the more neutral pH found in the distal colon. Transit in the distal colon is slower and nutrient availability is minimised, producing slower growing populations that tend towards more proteolytic fermentations.

An intriguing question about the human microbiota is the relevance of microbial variations in healthy and diseased individuals, and whether microbial mapping could help predict specific conditions (Knights *et al.*, 2014). Despite the diverse range of micro-organisms found in the human digestive tract, it has been suggested that just five or six genera and two phyla shape the mainstream biomass. Numerically dominant genera include *Bacteroides*, *Bifidobacterium* and *Eubacterium* and, to a lesser extent, although still important, *Clostridium*, *Enterobacteriaceae* and *Streptococcus* (Gibson & Roberfroid, 1995; Salminen *et al.*, 1998). Five bacterial phyla represent the bulk of the bacteria in the gut, with the two major phyla being the Gram-positive Firmicutes and the Gram-negative Bacteroidetes (LePage *et al.*, 2013), which have relatively similar proportions in different individuals (Jeffery *et al.*, 2012). In 2011, three different profiles for the human gut microbiota were proposed, termed 'enterotypes', that were dominated by *Bacteroides*, *Prevotella* or *Ruminococcus* (Arumugam *et al.*, 2011). The situation, however, may be more complex than this, and further research is also needed to elucidate the health implications of such enterotypes (Gibson *et al.*, 2016).

Table 1.2 illustrates the representation of the microbiota of the GI tract, highlighting some of the common bacteria and their abundance in different parts of the human digestive system. Yeasts, including the opportunistic pathogen *Candida albicans*, are also

Table 1.2 Representative bacteria in the gastrointestinal (GI) tract.

Bacterial family or genus	GI tract region	Microbial count (colony forming units (cfu) mL^{-1}	Function of the GI tract region
Lactobacillus *Streptococcus* *Helicobacter* *Peptostreptococcus*	Stomach	$1–10^2$	• Hydrochloric acid secretion • Macromolecule digestion • pH 2
Streptococcus *Lactobacillus*	Duodenum Jejunum Ileum	$10^1–10^3$ $10^3–10^4$ $10^7–10^9$	• Main digestion • Absorption of monosaccharides, amino acids, fatty acids and water • pH 4–5
Bacteroides *Clostridium* *Streptococcus* Actinomycineae	Caecum	NR[1]	• Absorption of fluids and salts • Mixing of the lumen contents with mucus • pH 5.7
Bacteroides *Clostridium* *Bifidobacterium* Enterobacteriaceae *Eubacterium*	Colon	$10^{11}–10^{12}$	• Microbial production of secondary bile acids and vitamin B_{12} • Water absorption • pH 7
NR	Rectum	NR	• Storage of faeces before evacuation • pH 6.7

NR = Not reported.
Adapted from Korecka and Arulampalam (2012).

present in the gut microbiota, although in healthy individuals its counts do not exceed 10^4 cfu g^{-1} in faeces (Bernhardt *et al.*, 1995; Bernhardt & Knoke, 1997). The vast majority (>90%) of the total cells in the body are present as bacteria in the colon. It is thought that over 60% of the faecal mass exists as prokaryotic cells. As well as the different microhabitats along the length of the GI tract, there are other microhabitats, such as the surface of the gut epithelia, the gut lumen, the colonic mucus layers and the ileum/caecum and colon (Donaldson *et al.*, 2016).

The classification of the microbiota as autochthonous or allochthonous complements the distinction between these different habitats of the GI tract (Savage *et al.*, 1968). Autochthonous micro-organisms are indigenous and colonise the GI tract, whereas allochthonous micro-organisms are transient and will predictably be found in the lumen. The slow transit time of the large intestine allows multiplication of the luminal microbiota; allochthonous micro-organisms exert equally important effects on the GI tract as their autochthonous counterparts.

1.3 Functions of the GI microbiota

The GI tract along with its microbiota comprise one of the most metabolically active organs in the human body. The intestinal microbiota is involved in the fermentation of endogenous and exogenous microbial growth substrates. The metabolic end products of carbohydrate fermentation are benign or even advantageous to human health (Macfarlane

& Gibson, 1994; Flint *et al.*, 2012; Rooks *et al.*, 2016). Major substrates available for the colonic fermentation are starches that, for various reasons, are resistant to the action of pancreatic amylases but can be degraded by bacterial enzymes, as well as dietary fibres, such as pectins and xylans. Other carbohydrate sources available for fermentation in lower concentrations include oligosaccharides and a variety of sugars and non-absorbable sugar alcohols. Saccharolysis results in the production of short-chain fatty acids (SCFAs), such as butyrate, acetate, propionate and lactate that contribute towards the energy metabolism of the large intestinal mucosa and colonic cell growth; they can also be metabolised by host tissues, such as the liver, muscle and brain. The production of SCFAs concomitantly results in a lower pH that can protect against invading micro-organisms and also reduces the transformation of primary bile acids into secondary pro-carcinogenic bile acids (Cummings & Macfarlane, 1997; Marchesi *et al.*, 2016). This is one of the mechanisms utilised by beneficial bacteria in the gut that results in protection for the host.

Proteins and amino acids can be effective growth substrates for colonic bacteria, whilst bacterial secretions, lysis products, sloughed epithelial cells and mucins may also make a contribution. However, diet provides, by far, the predominant source of nutrients, with around 70–100 g d^{-1} of dietary residues available for the colonic microbiota. These materials are degraded by a wide range of bacterial polysaccharidases, glycosidases, proteases and amino-peptidases to smaller oligomers and their component sugars and amino acids (Macfarlane & Gibson, 1994).

The gut profile of each adult represents a population of microbes that has evolved since birth and that can best cope with the physiological and microbiological pressure encountered within this ecosystem. This stability provides resistance for the host, also known as the 'barrier effect', against invading micro-organisms, both pathogenic and benign. The indigenous gut microbiota is better adapted to compete for nutrients and attachment sites than any incoming micro-organism, which it may also inhibit through the production of compounds (Alderbeth *et al.*, 2000). The role of the intestinal microbiota in challenging invading micro-organisms and preventing disease through competitive exclusion is best demonstrated by the studies showing that germ-free animals are more susceptible to infection (Baba *et al.*, 1991). This demonstrates the individual role of beneficial micro-organisms in preventing infection through colonisation resistance.

Another important function of the gut microbiota is the production of vitamins B and K; this is best demonstrated by studies where germ-free animals required a 30% increase in their diet to maintain their body weight, and supplementation with vitamins B and K as compared to animals with a microbiota (Hooper *et al.*, 2002).

The ability of the gut microbiota, however, to utilise biologically available compounds can have negative outcomes. *Helicobacter pylori* can affect the absorption of vitamin C and important micronutrients for host health (Annibale *et al.*, 2002). Moreover, the fermentation of proteins and amino acids in the distal colon can lead to the production of toxic substances such as ammonia, phenols and amines that are undesirable for host health (Mykkanen *et al.*, 1998; Kim *et al.*, 2013). This highlights the importance of ensuring a balance of beneficial bacteria to prevent the multiplication of pathogens or bacteria whose growth and metabolism may increase disease risk.

The GI tract is in more contact with the external environment than our skin, which exposes ~2 m^2, whereas the GI tract exposes a surface area of ~200 m^2 (Guilliams, 1999). The microbiota of the GI tract is therefore heavily involved in gut maturation. As mentioned in this chapter, exposure to the intestinal microbiota after birth plays a critical role in stimulating local and systemic responses and supporting the maturation of the immune system. The intestinal microbiota also provides a source for non-inflammatory immune stimulation, throughout life, by stimulating the production of secretory IgA, which neutralises foreign bacteria and viruses (Moreau, 2000; Mathias *et al.*, 2014). The immune system–microbiota alliance provides a dynamic environment by defending the host from pathogens as well as maintaining a balanced and controlled tolerance to harmless antigens. Many factors can play a role in destabilising this coalition and disturbing this symbiotic relationship, including changes in diet and overuse of antibiotics, which in turn could allow the proliferation of a microbiota lacking in diversity or the resilience and tolerance needed for a well-functioning immune system. The rise in autoimmune diseases and inflammatory disorders has been suggested to be partly the result of this troubled reciprocal relationship. Overall, the ability of the GI tract to perform its functions of nutrient uptake in conjunction with the exclusion of foreign antigens or micro-organisms is a complex and difficult process. The interplay between the host immune response and the GI microbiota is critical to health; loss of tolerance may become clinically manifest through disorders, such as inflammatory bowel disease (IBD) (Malloy & Powrie, 2011).

The gut microbiota and host health has found a new clinical frontier in recent years, the so-called gut–brain axis (El Aidy *et al.*, 2015), which is described as a two-way communication between the central and the enteric nervous systems, in which the emotional, intuitive, decision-making and cognitive centres of the brain are linked with peripheral intestinal functions (Mayer, 2011). This bidirectional interaction is believed to include signal exchange between gut microbiota and the brain through neural, endocrine, immune and humoral links (Carabotti *et al.*, 2015; Kountouras *et al.*, 2015). To provide evidence of these interactions, studies on germ-free animal models, probiotics, antibiotics and infection have been carried out. At a clinical level, studies have focused on central nervous disorders such as autism, anxiety-depressive behaviours and GI disorders, such as (typically) irritable bowel syndrome. It is hoped that such investigations lead to new therapeutic strategies (Distrutti *et al.*, 2016).

1.4 Influences on the GI tract and its microbiota

The profile of the intestinal microbiota that develops in each individual is a result of their host genetics (as shown in twin studies in the UK) (Goodrich *et al.*, 2014), environmental factors and microbiological influences. These factors result in a stable community of micro-organisms that is more unique than an individual's own fingerprint; even homozygotic twins develop distinct microbial profiles (Zoetendal *et al.*, 2001). Notwithstanding this, the overall metabolism of a healthy gut ecosystem varies little from one individual to another, as evinced by the ratios of major metabolic end products. Modern living presents numerous challenges to the human GI tract, particularly in the developed world, with often stressful lifestyles and unhealthy intake of processed foods. Antibiotics and other medications, however, can cause immediate serious disruption of the gut microbiota, and the resulting dysbiosis may be long term (Jernberg *et al.*, 2010; Francino, 2015). Disturbances of the microbiota can have serious implications, and this fragility merits careful consideration of the external influences on the GI tract and how they may disrupt host health (O'Sullivan *et al.*, 2013). The numerous factors which act upon the intestinal microbiota are briefly outlined in Table 1.3; some of the more relevant influences are discussed here.

The influence of diet on the neonatal intestinal microbiota has already been outlined (do Rosario *et al.*, 2016; Ojeda *et al.*, 2016). The GI tract of healthy humans remains relatively stable throughout life apart from later life, when a significant decrease of beneficial bifidobacteria and loss of microbial diversity have been reported. Such changes have also been linked to indications of increased risk of disease and frailty (van Tongeren *et al.*, 2005; Claessen *et al.*, 2012; Jackson *et al.*, 2016). Diet is an effective and rapid modulator of the microbial composition and metabolic activity of the human gut, which in turn can impact health (Claesson *et al.*, 2012; Conlon & Bird, 2015) with temporary and/or lasting effects. For example, the ELDERMET study in Ireland has shown clear differences between the core microbiota in older people compared to younger ones. Furthermore, clear differences were observed in the gut microbiota that correlated to these older persons' place of residence: long-term residential care, rehabilitation hospital care for less than six months, attending hospital outpatients or living in the community (Claessen *et al.*, 2012). The profile of the microbiota of those living at home was the one most similar to that of healthy younger adults, whereas the gut microbiota of the older people living in long-term care was significantly different and much less diverse. These microbiota differences correlated with the different diets eaten at home or in residential care; the latter had a much lower intake of fruit, vegetables and fibre, and a higher intake of fatty, starchy and sugary foods. Whilst long-term diet clearly influences the composition of gut microbiota, even short-term dietary modifications lead to significant and relatively swift changes in the composition of the microbiota, but

Table 1.3 Influences on the composition of the gastrointestinal microbiota.

- Type of feeding
- Amount, chemical composition and availability of growth substrate
- Availability of colonisation sites
- Immunological interactions
- Individual fermentation strategies by the bacteria
- Intestinal transit time
- Gut pH
- Redox potential
- Availability of inorganic electron acceptors
- Production of bacterial metabolites
- Presence of antimicrobial compounds
- Xenobiotic compounds
- Age of the host
- Peristalsis

Adapted from Fooks *et al.* (1999).

these would not be expected to cause a lasting shift in microbiota composition or affect the core profile. Data indicate that such changes may be at genus and species level, but not at phylum level (Wu *et al.*, 2011).

Type of dietary intake has consequences in the colon as carbohydrate fermentations usually result in benign end products (Wong *et al.*, 2006; do Rosario *et al.*, 2016). However, when carbohydrate levels become diminished, proteolytic fermentation in more distal regions produces toxic compounds that can predispose to diseases such as colorectal cancer or ulcerative colitis (Nyangale *et al.*, 2012); thus, protein-based diets such as the Atkins diet could potentially have serious long-term repercussions for gut health (Russell *et al.*, 2011). High intakes of processed food and other dietary aspects will reduce levels of fibre in the diet, which is of concern as dietary fibre influences stool volume, colon motility, water absorption and faecal transit time (Dhingra *et al.*, 2012).

Chronic illness, immune suppression and the use of broad-spectrum antibiotics can severely compromise the crucial balance between beneficial and harmful micro-organisms in the gut microbiota. The loss of any beneficial genera sensitive to antibiotic therapy, such as lactobacilli and bifidobacteria, has implications for GI health, as opportunistic pathogens can overgrow the gut, and the host will have increased risk for iatrogenic disease. For example, the serious concerns about the risks of antibiotic-associated diarrhoea, including that caused by *Clostridium difficile*, are well documented (Burke & Lamont, 2014; Elseviers *et al.*, 2015).

The increase in antibiotic resistance, the lack of progress in developing new antibiotics, concerns over (possibly long-term) adverse effects associated with antibiotic use (such as increased risk of obesity) (Reid, 2006; Langdon *et al.*, 2016; Ouwehand *et al.*, 2016) plus consumer interest in dietary supplements to maintain GI health have fuelled scientific research into alternative strategies. The potential for preventing dysbiosis, increasing the resilience of the gut microbiota or otherwise fortifying the GI tract through modulation of the intestinal microbiota has been widely explored. The principle of using harmless bacteria to prevent disease dates back to the suggestion of Metchnikoff at the turn of the twentieth century that ingested bacteria could promote longevity and well-being (Metchnikoff, 1907; see Chapter 2 for details). Micro-organisms associated with health benefits *in vivo* include many members of the *Lactobacillus* and *Bifidobacterium* genera, although *Escherichia coli*, streptococci, enterococci, lactococci, bacilli and yeasts, such as *Saccharomyces cerevisiae* var. *boulardii*, have also been used (Table 1.4). Such strains have been researched for their probiotic potential, and many strains (including those marketed commercially) are the focus of intense research (see Chapter 8 for further details).

1.5 Conclusions

A number of disease states have been linked to dysbiosis and/or low diversity of the gut microbiota, suggesting that its manipulation at any stage of life but particularly in infancy could have beneficial consequences in reducing the risk of both short-term and long-term disease (Thomas *et al.*, 2014; Carding *et al.*, 2015; Prosberg *et al.*, 2016). Differences in the ratio of Firmicutes to Bacteroidetes have also been observed between

Table 1.4 Examples of microbial species that contain probiotic strains.

Microbial genus or group	Species
Bifidobacterium	*Bifidobacterium bifidum*
	Bifidobacterium longum subsp. *longum*
	Bifidobacterium breve
	Bifidobacterium adolescentis
	Bifidobacterium longum subsp. *infantis*
Enterococcus	*Enterococcus faecalis*
	Enterococcus faecium
Lactococcus	*Lactococcus lactis* subsp. *lactis*
Lactobacillus	*Lactobacillus acidophilus*
	Lactobacillus rhamnosus
	Lactobacillus reuteri
	Lactobacillus casei
	Lactobacillus gasseri
	Lactobacillus plantarum
Yeast	*Saccharomyces cerevisiae* var. *boulardii*

individuals and patient groups. Other examples include IBD, where low counts of *Faecalibacterium prausnitzii* have been associated with increased risk of ulcerative colitis (Sokol *et al.*, 2009), and several species have been implicated in colorectal cancer, including *Streptococcus gallolyticus*, *Enterococcus faecalis* and *Bacteroides fragilis* (Wu *et al.*, 2009; Boleij & Tjalsma, 2013; Wang *et al.*, 2015).

A key question in gut microbiota research, however, is whether such microbial changes are the *cause* of the disease or are the *result* of disease (Zhang, 2013). One tactic to explore this 'correlation/causality' microbial conundrum is to conduct clinical trials in patients or people at risk of disease, investigating the health effects of modulating the microbiota. Faecal microbiota transplantation, for example, has shown strong efficacy for treatment of *C. difficile* infection (Borody *et al.*, 2015). Probiotics work through multiple mechanisms of activity, including the modulation of the gut microbiota, and evidence of probiotic benefit for a broad range of disorders has accumulated. This is discussed further in Chapter 8.

References

Alderbeth, I., Cerquetti, M., Poilane, I., Wold, A.E. & Collignon, A. (2000) Mechanisms of colonisation and colonisation resistance of the digestive tract. *Microbial Ecolology in Health and Disease*, **12**, 223–239.

Annibale, B., Capurso, G. & Delle Fave, G. (2002) Consequences of *Helicobacter pylori* infection on the absorption of micronutrients. *Digestive and Liver Disease*, **34**, S72–S77.

Arumugam, M., Raes, J., Pelletier, E., Le Paslier, D., Yamada, T., Mende, D.R., Fernandes, G.R., Tap, J., Bruls, T., Batto, J.M., Bertalan, M., Borruel, N., Casellas, F., Fernandez, L., Gautier, L.,

Hansen, T., Hattori, M., Hayashi, T, Kleerebezem, M., Kurokawa, K., Leclerc, M., Levenez, F., Manichanh, C., Nielsen, H.B., Nielsen, T., Pons, N., Poulain, J., Qin, J., Sicheritz-Ponten, T., Tims, S., Torrents, D., Ugarte, E., Zoetendal, E.G., Wang, J., Guarner. F., Pedersen, O., de Vos, W.M., Brunak, S., Doré, J., MetaHIT Consortium, Antolín, M., Artiguenave, F., Blottiere. H.M., Almeida, M., Breczhot, C., Cara, C., Chervaux, C., Cultrone, A., Delorme, C., Denariaz, G., Dervyn, R., Foerstner, K.U., Friss, C., van de Guchte, M., Guedon, E., Haimet, F., Huber, W., van Hylckama-Vlieg, J., Jamet, A., Juste, C., Kaci, G., Knol, J., Lakhdari, O, Layec, S., Le Roux, K., Maguin, E., Mérieux, A., Melo Minardi, R., M'rini, C., Muller, J., Oozeer, R., Parkhill, J., Renault, P., Rescigno, M., Sanchez, N., Sunagawa, S., Torrejon, A., Turner, K., Vandemeulebrouck, G., Varela, E., Winogradsky, Y., Zeller, G., Weissenbach, J., Ehrlich, S.D. & Bork, P. (2011) Enterotypes of the human gut microbiome. *Nature*, **473**, 174–180.

Baba, E., Nagaishi, S., Fukata, T. & Arakawa, A. (1991) The role of intestinal microflora on the prevention of salmonella colonization in gnotobiotic chickens. *Poultry Science*, **70**, 1902–1907.

Bäckhed, F., Ley, R.E., Sonneburg, J.L., Peterson, D.A. & Gordon, J.I. (2005) Host bacterial mutualism in the human intestine. *Science*, **307**, 1915–1920.

Bernhardt, H. & Knoke, M. (1997) Mycological aspects of gastrointestinal microflora. *Scandinavian Journal of Gastroenterology*, **32**, 102–106.

Bernhardt, H., Wellmer, A., Zimmerman, K. & Knoke, M. (1995) Growth of *Candida albicans* in normal and altered faecal flora in the model of continuous flow culture. *Mycoses*, **38**, 265–270.

Blustein, J., Attina T., Ryan, A.M., Cox, L.M., Blaser, M.J. & Trasande, L. (2013) Association of caesarean delivery with child adiposity from age 6 weeks to 15 years. *International Journal of Obesity (London)*, **37**, 900–906.

Boleij, A. & Tjalsma, H. (2013) The itinerary of *Streptococcs gallolyticus* infection in patients with colonic malignant disease. *Lancet Infectious Diseases*, **13**, 719–724.

Borody, T., Fischer, M., Mitchell, S. & Campbell, J. (2015) Fecal microbiota transplantation in gastrointestinal disease: 2015 update and the road ahead. *Expert Reviews in Gastroenterology & Hepatology*, **9**, 1379–1399.

Brandão, H.V., Vieira, G.O., de Oliveira Vieira, T., Camargos, P.A., de Souza Teles, C.A., Guimarães, C., Cruz, A.A. & Cruz, C.M. (2016) Increased risk of allergic rhinitis among children delivered by cesarean section: a cross-sectional study nested in a birth cohort. *BMC Pediatrics*, **16**, 57. DOI: 10.1186/s12887-016-0594-x

Burke, K.E. & Lamont, J.T. (2014) *Clostridium difficile* infection: a worldwide disease. *Gut Liver*, **8**, 1–6.

Carabotti, M., Scirocco, A., Maselli, M.A. & Severia, C. (2015) The gut-brain axis: interactions between enteric microbiota, central and enteric nervous systems. *Annals of Gastroenterology*, **28**, 203–209.

Carding, S., Verbeke, K., Vipon, D.T., Corfe, B.M. & Owen, L.J. (2015) Dysbiosis of the gut microbiota in disease. *Microbial Ecology in Health & Disease*, **26**, 26191.

Castanys-Muñoz, E., Martin, M.J. & Vazquez, E. (2016) Building a beneficial microbiome from birth. *Advances in Nutrition*, **7**, 323–330.

Claesson, M.J., Jeffery, I.B., Conde, S., Power, S.E., O'Connor, E.M., Cusack, S., Harris, H.M., Coakley, M., Lakshminarayanan, B., O'Sullivan, O., Fitzgerald, G.F., Deane, J., O'Connor, M., Harnedy, N., O'Connor, K., O'Mahony, D., van Sinderen, D., Wallace, M., Brennan, L., Stanton, C., Marchesi, J.R., Fitzgerald, A.P., Shanahan, F., Hill, C., Ross, R.P. & O'Toole, P.W. (2012) Gut microbiota composition correlates with diet and health in the elderly. *Nature*, **488**, 178–184.

Conlon, M.A. & Bird, A.R. (2015) The impact of diet and lifestyle on gut microbiota and human health. *Nutrients*, **7**, 17–44.

Cummings, J.H. & Macfarlane, G.T. (1997) Role of intestinal bacteria in nutrient metabolism. *Journal of Parenteral and Enteral Nutrition*, **21**, 357–365.

Dhingra, D., Michael, M., Rajput, H. & Patil, R.T. (2012) Dietary fibre in foods: a review. *Journal of Food Science & Technology*, **49**, 255–266.
Distrutti, E., Monaldi, L., Ricci, P. & Fiorucci, S. (2016) Gut microbiota role in irritable bowel syndrome: new therapeutic strategies. *World Journal of Gastroenterology*, **22**, 2219–2241.
Dominguez-Bello, M.D., De Jesus-Laboy, K.M., Shen, N., Cox, L.M., Amir, A., Gonzalez, A., Bokulich, N.A., Song, S.J., Hoashi, M., Rivera-Vinas, J.I., Mendez, K., Knight, R. & Clemente, J.C. (2016) Partial restoration of the microbiota of cesarean-born infants via vaginal microbial transfer. *Nature Medicine*, **22**, 250–253.
Donaldson, G.P., Lee, S.M. & Mazmanian, S.K. (2016) Gut biogeography of the bacterial microbiota. *Nature Reviews in Microbiology*, **14**, 20–32.
El Aidy, S.E., Dinan, T.G. & Cryan, J.F. (2015) Gut microbiota: the conductor in the orchestra of immune-neuroendocrine communication. *Clinical Therapy*, **37**, 954–967.
Elseviers, M.E., Van Camp, Y., Nayaert, S., Duré, K., Annemans, L., Tanghe, A. & Vermeersch, S. (2015) Prevalence and management of antibiotic associated diarrhea in general hospitals. *BMC Infectious Disease*, **15**, 129. DOI: 10.1186/s12879-015-0869-0
Fernández, L., Langa, S., Martín, V., Maldonado, A., Jiménez, E., Martín, R. & Rodríguez, J.M. (2013) The human milk microbiota: origin and potential roles in health and disease. *Pharmacological Research*, **69**, 1–10.
Flint, H.J., Scott, K.P., Louis, P. & Duncan, S.H. (2012) The role of the gut microbiota in nutrition and health. *Nature Reviews Gastroenterology and Hepatology*, **9**, 577–589.
Fooks, L.J., Fuller, R. & Gibson, G.R. (1999) Prebiotics, probiotics and human gut microbiology. *International Dairy Journal*, **9**, 53–61.
Fooks, L.J. & Gibson, G.R. (2002) Probiotics as modulators of the gut flora. *British Journal of Nutrition*, **88**, S39–S49.
Francino, M.P. (2015) Antibiotics and the human gut microbiome: dysbioses and accumulation of resistances. *Frontiers in Microbiology*, **6**, 1543. https://doi.org/10.3389/fmicb.2015.01543
Fuller, R. & Perdigón, G. (eds.) (2003) *Gut Flora, Nutrition, Immunity and Health*. Blackwell Publishing, Oxford.
Ghoddusi, H.B. & Tamime, A.Y. (2014) Microflora of the intestine: biology of bifidobacteria. In *Encyclopedia of Food Microbiology* (eds. C.A. Batt and M.L. Tortorello), Vol. **2**., 639–645. Elsevier Ltd and Academic Press, London.
Gibson, G.R. & Roberfroid, M.B. (1995) Dietary modulation of the human colonic microbiota – introducing the concept of prebiotics. *Journal of Nutrition*, **125**, 1401–1412.
Gibson, T.E., Bashan, A., Cao, H.T., Weiss, S.T. & Liu, Y.Y. (2016) On the origins and control of community types in the human microbiome. *PLoS Computational Biology*, **21**, e1004688.
Goodrich, J.K., Waters, J.L., Poole, A.C., Sutter, J.L., Koren, O., Blekhman, R., Beaumont, M., Van Treuren, W., Knight, R., Bell, J.T., Spector, T.D., Clark, A.G. & Ley, R.E. (2014) Human genetics shape the gut microbiome. *Cell*, **159**, 789–799.
Guilliams, T.G. (1999) Healthy microbial organisms. *The Standard*, **2**, 1–8.
Honda, K. & Littman, D.R. (2016) The microbiota in adaptive immune homeostasis and disease. *Nature*, **535**, 75–84.
Hooper, L.V., Midtvedt, T. & Gordon, J.L. (2002) How host-microbial interactions shape the nutrient environment of the mammalian intestine. *Annual Review of Nutrition*, **22**, 283–307.
Jackson, M.A., Jeffery, I.B., Beaumont, M., Bell, J.T., Clark, A.G., Ley, R.E., O'Toole, P.W., Spector, T.D. & Steves, C.J. (2016) Signatures of early frailty in the gut microbiota. *Genome Medicine*, **8**, 8. doi:10.1186/s13073-016-0262-7
Jeffery, I.B., Claesson, M.J., O'Toole, P.W. & Shanahan, F. (2012) Categorization of the gut microbiota: enterotypes or gradients? *Nature Reviews of Microbiology*, **10**, 591–592.
Jernberg, C., Löfmark, S., Edlund, C. & Jansson, J.K. (2010) Long-term impacts of antibiotic exposure on the human intestinal microbiota. *Microbiology*, **156**, 3216–3223.

Kim, E., Coelho, D. & Blachier, F. (2013) Review of the association between meat consumption and risk of colorectal cancer. *Nutrition Research*, **33**, 983–994.
Knights, D., Ward, T.L., McKinlay, C.E., Miller, H., Gonzalez, A., McDonald, D. & Knight, R. (2014) Rethinking "enterotypes." *Cell Host & Microbe*, **16**, 433–437.
Korecka, A. & Arulampalam, V. (2012) The gut microbiome: scourge, sentinel or spectator? *Journal of Oral Microbiology*, **4**, 9367–9381.
Kountouras, J., Zavos, C., Polyzos, S.A. & Deretzi, G. (2015) The gut-brain axis: interactions between *Helicobacter pylori* and enteric and central nervous systems. *Annals of Gastroenterology: Quarterly Publication of the Hellenic Society of Gastroenterology*, **28**, 506–510.
Lambert, J. & Hull, R. (1996) Upper gastrointestinal disease and probiotics. *Asia Pacific Journal of Clinical Nutrition*, **5**, 31–35.
Langdon, A., Crook, N. & Dantas, G. (2016) The effects of antibiotics on the microbiome throughout development and alternative approaches for therapeutic modulation. *Genome Medicine*, **8**, 39. doi:10.1186/s13073-016-0294-z
Lepage, P., Leclerc, M.C., Joossens, M., Mondot, S., Blottière, H.M., Raes, J., Ehrlich, D. & Doré, J. (2013) A metagenomic insight into our gut's microbiome. *Gut*, **62**, 146–158.
Li, G., Yang, M., Zhou, K., Zhang, L., Tian, L., Lv, S., Jin, Y., Qian, W., Xiong, H., Lin, R., Fu, Y. & Hou, X. (2015) Diversity of duodenal and rectal microbiota in biopsy tissues and luminal contents in healthy volunteers. *Journal of Microbiology & Biotechnology*, **25**, 1136–1145.
Lozupone, C.A., Stombaugh, J.I., Gordon, J.I., Jansson, J.K. & Knight, R. (2012) Diversity, stability and resilience of the human gut microbiota. *Nature*, **489**, 220–230.
Macfarlane, G.T. & Gibson, G.R. (1994) Metabolic activities of the normal colonic flora. In *Human Health – The Contribution of Microorganims* (ed. S.A. Gibson), 17–52. Springer-Verlag, London.
Malloy, K.J. & Powrie, F. (2011) Intestinal homeostasis and its breakdown in inflammatory bowel disease. *Nature*, **474**, 298–306.
Marchesi, J.R., Adams, D.H., Fava, F., Hermes, G.D.A., Hirschfield, G.M., Hold, G., Quraishi, M.N., Kinross, J., Smidt, H., Tuohy, K.M., Thomas, L.V., Zoetendal, E.G. & Hart, A (2016). The gut microbiota and host health: a new clinical frontier. *Gut*, **65**, 330–339.
Mathias, A., Pais, B., Favre, L., Benyacoub, J. & Corthésy, B. (2014) Role of secretory IgA in the mucosal sensing of commensal bacteria. *Gut Microbes*, **5**, 688–695.
Mayer, E.A. (2011) Gut feelings: the emerging biology of gut–brain communication. *Nature Reviews Neuroscience*, **12**, 453–466.
Metchnikoff, E. (1907) *The Prolongation of Human Life*. Heinemann, London.
Moreau, M.C. (2000) Flore intestinale, prébiotique et effets sur la réponse immunitaire intestinale à IgA (in French). *Archives de Pediatre*, **2000**, 247–248.
Mykkanen, H., Laiho, K. & Salminen, S. (1998) Variation in faecal bacterial enzyme activities and associations with bowel function and diet in elderly subjects. *Journal of Applied Microbiology*, **85**, 37–41.
Nyangale, E.P., Mottram, D.S. & Gibson, G.R. (2012) Gut microbial activity, implications for health and disease: the potential role of metabolite analysis. *Journal of Proteome Research*, **11**, 5573–5585.
Ojeda, P., Bobe, A., Dolan, K., Leone, V. & Martinez, K. (2016) Nutritional modulation of gut microbiota – the impact on metabolic disease pathophysiology. *Journal of Nutritional Biochemistry*, **28**, 191–200.
O'Sullivan, O., Coakley, M., Lakshminarayanan, B., Conde, S., Claesson, M.J., Cusack, S., Fitzgerald, A.P., O'Toole, P.W., Stanton, C., Ross, R.P. & the ELDERMET Consortium (2013) Alterations in intestinal microbiota of elderly Irish subjects post-antibiotic therapy. *Journal of Antimicrobial Chemotherapy*, **68**, 214–221.
Ouwehand, A.C., Forssten, S., Hibberd, A.A., Lyra, A. & Stahl, B. (2016) Probiotic approach to prevent antibiotic resistance. *Annals of Medicine*, **48**, 246–255.

Prosberg, M., Bendtsen, F., Vind, I., Petersen, A.M. & Gluud, L.L. (2016) The association between the gut microbiota and the inflammatory bowel disease activity: a systematic review and meta-analysis. *Scandinavian Journal of Gastroenterology*, **51**, 1407–1415.

Qin, J., Li, R., Raes, J., Arumugam, M., Burgdorf, K.S., Manichanh, C., Nielsen, T., Pons, N., Levenez, F., Yamada, F., Yamada, T., Mende, D.R., Li, J., Xu, J., Li, S., Li, D., Cao, J., Wang, B., Liang, H., Zheng, H., Xie, Y., Tap, J., Lepage, P., Bertalan, M., Batto, J-M., Hansen, T., Le Paslier, Linnebberg, A., Nielseon, H.B, Pelletier, E., Renault, P., Sicheritz-Ponten, T., Turner, K., Zhu, H., Yu, C., Li, S., Jian, M., Zhou, Y., Li, Y., Zhang, X., Li, S., Qin, N., Yang, H., Wang, J., Brunak, S., Doré, J., Guarner, F., Kristiansen, K., Pedersen, O., Parkhill, J., Weissenbach, J., MetaHIT Consortium, Bork, P., Dusko Ehrlich, S. & Wang, J. (2010) A human gut microbial gene catalogue established by metagenomic sequencing. *Nature*, **464**, 59–65.

Reid, G. (2006) Probiotics to prevent the need for, and augment the use of, antibiotics. *Canadian Journal of Infectious Disease & Medical Microbiology*, **17**, 291–295.

Rodríguez, J.M., Murphy, K., Stanton, C., Ross, R.P., Kober, O.I., Juge, N., Avershina, E., Rudi, K., Narbad, A., Jenmalm, M.C., Marchesi, J.R. & Collado, M.C. (2015) The composition of the gut microbiota throughout life, with an emphasis on early life. *Microbial Ecology in Health and Disease*, **26**, 26050. http://dx.doi.org/10.3402/mehd.v26.26050

Rooks, M.G. & Garrett, W.S. (2016) Gut microbiota, metabolites and host immunity. *Nature Reviews Immunology*, **16**, 341–352.

Do Rosario, V.A., Fernandes, R. & Trindade, E.B. (2016) Vegetarian diets and gut microbiota: important shifts in markers of metabolism and cardiovascular disease. *Nutrition Reviews*, **74**, 444–454.

Russell, W.R., Gratz, S.W., Duncan, S.H., Holtrop, G., Ince, J., Scobbie, L., Duncan, G., Johnstone, A.M., Lobley, G.E., Wallace, R.J., Duthie, G.G. & Flint, H.J. (2011) High-protein, reduced-carbohydrate weight-loss diets promote metabolite profiles likely to be detrimental to colonic health. *American Journal of Clinical Nutrition*, **93**, 1062–1072.

Russell, W.R., Hoyles, L., Flint, H.J. & Dumas, M.E. (2013) Colonic bacterial metabolites and human health. *Current Opinions in Microbiology*, **16**, 246–254.

Rutayisire, E., Huang, K., Liu, Y. & Tao, F. (2016) The mode of delivery affects the diversity and colonization pattern of the gut microbiota during the first year of infants' life: a systematic review. *BMC Gastroenterology* **16**, 86. doi:10.1186/s12876-016-0498-0

Salminen, S., Bouley, C., Boutron-Ruault, M.C., Cummings, J.H., Franck, A., Gibson, G.R., Isolauri, E., Moreau, M.C., Roberfroid, M.B. & Rowland, I.R. (1998) Functional food science and gastrointestinal physiology and function. *British Journal of Nutrition*, **80**, S147–S171.

Sanford, P.A. (1992) *Digestive System Physiology*, 2nd ed. *Physiological Principles in Medicine* series (series eds. M. Hobsley, K.B. Saunders & J.T. Fitzsimons). Edward Arnold, London.

Savage, D.C., Dubos, R. & Schaedler, R.W. (1968) The gastrointestinal epithelium and its autochthonous bacterial flora. *Journal of Experimental Medicine*, **127**, 67–76.

Scott, K.P., Gratz, S.W., Sheridan, P.O., Flint, H.J. & Duncan, S.H. (2012) The influence of diet on the gut microbiota. *Pharmacological Research*, **69**, 52–60.

Shanahan, F. (2013) The colonic microbiota in health and disease. *Current Opinion in Gastroenterology*, **29**, 49–54.

Smilowitz, J.T., Lebrilla, C.B., Mills, D.A., German, J.B. & Freeman, S.L. (2014) Breast milk oligosaccharides: structure-function relationships in the neonate. *Annual Reviews in Nutrition*, **34**, 143–169.

Sokol, H., Seksik, P., Furet, J.P., Firmesse, O., Nion-Larmurier, I., Beaugerie, L., Cosnes, J., Corthier, G., Marteau, P. & Doré, J. (2009) Low counts of *Faecalibacterium prausnitzii* in colitis microbiota. *Inflammatory Bowel Disease*, **15**, 1183–1189.

Thomas, L.V. (2016) The gut microbiota and the role of probiotics in children. In *Probiotics and Children* (eds. M. Manfredi & G. Luigi de'Angelis), 1–29. Nova Science Publishers Inc., New York.

Thomas, L.V., Ockhuizen, T. & Suzuki, K. (2014) Exploring the influence of the gut microbiota and probiotics on health. *British Journal of Nutrition*, **112**, S1–S18.
Tojo, R., Suárez, A., Clemente, M.G., de los Reyes-Gavilán, C.G., Margolles, A., Gueimonde, M. & Ruas-Madiedo, P. (2014) Intestinal microbiota in health and disease: role of bifidobacteria in gut homeostasis. *World Journal of Gastroenterology*, **20**, 15163–15176.
Turnbaugh, P.J., Ley, R.E., Hamady, M., Fraser-Liggett, C.M., Knight, R. & Gordon, J.I. (2007) The Human Microbiome Project. *Nature*, **449**, 804–810. doi:10.1038/nature06244
Van Tongeren, S.P., Slaets, J.P., Harmsen, H.J. & Welling, G.W. (2005) Fecal microbiota composition and frailty. *Applied & Environmental Microbiology*, **71**, 6438–6442.
Wade, W.G. (2013) The oral microbiome in health and disease. *Pharmacological Research*, **69**, 137–143.
Wang, X., Yang, Y. & Huycke, M.M. (2015) Commensal bacteria drive endogenous transformation and tumour stem cell marker expression through a bystander effect. *Gut*, **64**, 459–468.
Willis, C.L., Gibson, G.R., Holt, J., Atherton, S. & Allison, C. (1999) Negative correlation between oral malodour and numbers and activities of sulphate-reducing bacteria in the human mouth. *Archives of Oral Biology*, **44**, 665–670.
Wong, J.M., de Souza, R., Kendall, C.W., Emam, A. & Jenkins, D.J. (2006) Colonic health: fermentation and short chain fatty acids. *Journal of Clinical Gastroenterology*, **40**, 235–243.
Wu, S., Rhee, K.J., Albesiano, E., Rabizadeh, S., Wu, X., Yen, H.R., Huso, D.L., Brancati, F.L., Wick, E., McAllister, F., Housseau, F., Pardoll, D.M. & Sears, C.L. (2009) A human colonic commensal promotes colon tumorigenesis via activation of T helper type 17T cell responses. *Nature Medicine*, **15**, 1016–1022.
Wu, G.D., Chen, J., Hoffmann, C., Bittinger, K., Chen, Y.-Y., Keilbaugh, S.A., Bewtra, M., Knights, D., Walters, W.A., Knight, R., Sinha, R., Gilroy, E., Gupta, K., Baldassano, R., Nessel, L., Li, H., Bushman, F.D. & Lewis, J.D. (2011) Linking long-term dietary patterns with gut microbial enterotypes. *Science*, **334**, 105–108.
Zhang, L. (2013) The gut microbiota and obesity: from correlation to causality. *Nature Reviews Microbiology*, **11**, 639–647.
Zoetendal, E.G., Akkermans, A.D.L., Akkermans van Vliet, W.M., de Visser, R.A.G.M. & de Vos, W.M. (2001) The host genotype affects the bacterial community in the human gastrointestinal tract. *Microbial Ecology in Health and Disease*, **13**, 129–134.

2 Probiotics: The First 10 000 Years

R. Levin

> It can only be a matter of time, we shall obtain exact information on the influence of diets which prevent intestinal putrefaction, prolong life and maintain the body's forces.
>
> Metchnikoff (1907)

2.1 In the beginning

Milk is not only an important food for humans, it is the first food of infants. It is believed humans began domesticating animals somewhere between 8000 BC and 5000 BC. Not long after, it must have been realised that the milk of other animals was just as able as human milk to satisfy the nutritional, energy and fluid needs of both adults and children. Moreover, early humans must have soon discovered that, whereas milk normally has a short life, under certain conditions, it forms curds with an extended shelf-life.

Probiotics in the form of fermented milk products have been in regular and continuous use as a source of nutrition and, unknowingly, for health and well-being, since time immemorial. Indeed, early evidence comes from a sculptured relief found at Tel Ubaid in ancient Babylon that appears to depict the production of cultured milk products for food some 8000 years ago. Milk is also mentioned in the Old Testament several times: when three angels visited Abraham, he asked Sarah to bake bread and he brought curds and milk (Genesis 18:7). This could be the first record of processed foods containing living micro-organisms, but the Vedic Hymns of India, written before 2000 BC, also reveal that Hindu people used fermented milk in their diet (Kroger *et al.*, 1989). Sumerians also crossed expanses of desert with milk carried in bags made from sheep's stomach where bacteria fermented the milk to curd, improving its flavour and keeping qualities. Hippocrates named milk as both a food and a medicine for curing stomach disorders. Plinius, the Roman historian, also recorded that fermented milk was used for stomach disorders. Since earliest times, many Nomadic and semi-Nomadic tribes have produced sour milk because of its improved keeping qualities. The best known of these are Kefir, Leben, Koumiss and Matsun (known also as Mazoor, Mazun, Matsoni or Madzoon). In the eleventh century AD, Yuseuf Has Hajib recorded the use of yoghurt by ancient Turks in his book 'Kutadgu Bilig'.

Probiotic Dairy Products, Second Edition. Edited by Adnan Y. Tamime and Linda V. Thomas.

The earliest of these milk beverages were probably produced because of spontaneous fermentation by miscellaneous bacteria that contaminated the goat skin bags carried by nomadic peoples, such as the Bulgars, who migrated from Asia to Europe in the second century AD, eventually settling in the Balkans. Many of today's traditional fermented drinks came from the Asian nomads, since fermented milks (together with animal meats) comprised their main nutritional and energy source. Nomads of Central Asia produced a variety of fermented milks, influenced by the animals they bred. Milk from at least eight species of domestic mammals (cow, buffalo, sheep, goat, horse, camel, yak and zebu) has been used to make traditional fermented milk products for human consumption. The following are details of nomadic beverages taken from descriptions given by Douglas (1911).

- Kefir has been used in the Caucasus for about as long as Koumiss has been used in the steppes. It differs in that it is prepared from the milk of sheep, goat or cow. The process is started with the addition of kefir grains to the milk contained in 'leathern' bottles. The grains are small solid kernels, kept by families and handed down from one generation to the next. They were described as a 'zoogloea' composed of bacilli and yeast, the latter being *Saccharomyces cerevisiae* (Kern, 1881). After the grains were added to the 'leathern' vessels containing the milk, in summer fermentation would proceed to completion in cool chambers for 1–2 days. During winter, the vessels would be placed in the sunshine at about ~16 to 18 °C. Agitation of the process would be supplied in the form of kicks from passers-by or children at play.
- Koumiss, thought by some to be the greatest of all the fermented milks, is made from mare's milk. It has been celebrated since ancient times as the principal food of the wandering tribes of Bashkirs, Kalmucks and Tartars who inhabit the steppes of European Russia and plains of West and Central Asia. Dr John Grieve, a surgeon in the Russian Army in 1784, sent a description of it to The Royal Society of Edinburgh (of which he was a member), entitled 'Method of Making Wine called by the Tartars Koumiss, with Observations on its Use as a Medicine'. This resulted in the establishment of sanatoria at Samura and elsewhere in Russia, which 'successfully' treated pulmonary consumption.
- Leben is a soured milk product associated with Middle Eastern countries, prepared from the milk of buffalo, cow and goat. It is prepared by adding fermented milk from the previous preparation to boiled fresh milk. The fermentation is rapid, finishing in ~6 h.
- Arka is a strong alcoholic beverage prepared by the Tartar and Burgaten tribes by distillation of fermented milk. It contains 7–8 g 100 mL^{-1} alcohol and also volatile fatty acids.
- Matzun is a drink mainly found in West Asia, prepared from buffalo, goat or cow milk and partly used for butter making. It is prepared in the Caucasus, using a similar procedure as that for Kefir.
- Yoghurt is related to Matzun and Leben. After boiling to concentrate the milk, it is inoculated with a small quantity of an old culture, then allowed to ferment at a comparatively higher temperature.

These practices still continue in some isolated mountain and desert areas of Asia and Africa. It was in the fifteenth century AD that the science behind fermentation began to be elucidated. Girolamo Fracastoro (1478–1553), an Italian physician and professor at the University of Padua, was possibly the first to propose that epidemic diseases are caused by external factors. He conceived the possibility that tiny transferable particles could transmit disease by direct or indirect contact, or even without contact, over long distances, but he did not anticipate that such tiny particles would be living entities. This hypothesis persisted for three centuries until Louis Pasteur revealed their true nature (Pasteur, 1878). Meanwhile, in about 1590, two Dutch spectacle makers discovered that when two lenses were placed in a tube, nearby objects appeared greatly enlarged. One century later, Anton van Leeuwenhoek, also from Holland, while working in a store where magnifying glasses were used to count the threads on cloth, taught himself methods of grinding tiny lenses to great curvature for higher magnification. Two of these, placed in a tube, led to his first microscope and the consequent first visualisation of bacteria, yeasts and blood corpuscles upon which his fame became based. He reported his findings meticulously in more than one hundred letters to the Royal Society in London and the French Academy in Paris.

2.2 The intervention of science

The first major discovery in bacteriology was by French chemist Louis Pasteur (1822–1895) who, using a microscope, revealed that the cause of spoilage in local beer, wine and milk was microbial contamination. He and Claude Bernard went on to invent a process in which milk was heated to kill most of any bacteria and moulds present, completing its first test in April 1862. The process duly became known as pasteurisation. On becoming Professor of Chemistry at the University of Strasbourg in 1849, Pasteur married the daughter of the university's rector and together they had five children, but only two survived to adulthood, the others succumbing to typhoid. These personal losses undoubtedly inspired Pasteur to seek cures for deadly microbial diseases, such as typhoid. Convinced from his contaminated beverage studies that animals and humans could be similarly afflicted by disease causing micro-organisms, he formally presented the evidence for his Germ Theory of Disease in 1878, for which he would subsequently be awarded the Nobel Prize for Medicine. Pasteur is now recognised as one of the founders of preventative medicine.

Fresh milk can turn sour within hours, but fermented milks (e.g. yoghurt) last much longer and, moreover, are characterised by the presence of microbial metabolites that, fortunately, render the product pleasant to taste. The sensory properties of fermented milks (taste, aroma and viscosity) are all the direct result of specific bacterial action. Pasteur's publication of his Germ Theory of Disease prompted and coincided with an intensive period of progress in the scientific study of milk. The dairy industry appeared to have captured the attention of scientific investigators throughout the world, but especially in the Pasteur Institute in Paris. A wholly unexpected and very close relationship between milk, intestinal disease and longevity then began to emerge.

2.3 A remarkable sequence of important discoveries

The sequence of discoveries that ultimately led Metchnikoff to his lactic microbe hypothesis was as follows:

- Senator (1868) declared that the decomposition of protein within the alimentary tract under normal conditions results in the formation of substances toxic to the host.
- Billroth (1874) was credited with being the first to observe that the meconium of the new-born infant is sterile. This was later confirmed by other researchers between 1880 and 1900.
- Bouchard (1884) elaborated the theory of intestinal intoxication in which he claimed that the amount of putrefactive products eliminated in urine was a measure of intestinal putrefaction, calling his measurements 'urotoxic coefficients'.
- Ortweiller (1886) demonstrated that the administration of certain carbohydrates tended to lessen putrefaction in the digestive tract.
- Hirschler (1886) was the first to observe that feeding particular carbohydrates, such as sucrose, lactose, dextrin and starch, as well as alcohol and glycerol, has an inhibitory effect on intestinal putrefaction.
- Escherich (1886) was a pioneer paediatrician, who devoted himself to improving child-care, particularly with regard to infant hygiene and nutrition, and he published his extensive systematic study of the microbes in infants' 'dejecta', in both health and disease states. He noted a predominance of Gram-positive rods, but (surprisingly) failed to isolate the two species that were soon to generate considerable and continuous interest, which were then known as '*Bacillus bidifus*' (presumed to be *Lactobacillus bifidus* and later renamed as *Bifidobacterium bifidum*) and '*Bacillus acidophilus*' (presumed to be *Lactobacillus acidophilus*). Nevertheless, the quality of his study and his monograph on the relationship of intestinal bacteria to the physiology of digestion in the infant established him as the leading bacteriologist in the field of paediatrics. In 1919, *Bacterium coli* was renamed *Escherichia coli*, after its discoverer.
- Poehl (1887) noted that ingestion of soured milk tended to decrease the undesirable products of protein decomposition by bacteria. This was confirmed by other researchers between 1887 and 1903.
- Döderlein (1892) reported that vaginal lactobacilli were much depleted in numbers in women with vaginitis; he was probably the first to suggest a potentially beneficial role for lactic acid bacteria in the treatment of vaginitis.
- Grigoroff (1905), a Bulgarian postgraduate at Geneva University, was aware of the number of centenarians to be found in Bulgaria, a region in which yoghurt, a soured milk, was a staple food. Working with Professor Massol at Geneva University, he isolated several microbes from 'podkvassa' starter used for the production of Bulgarian yoghurt. Among these was a very active lactic acid-producing species that he called '*Lactobacillus bulgaricus*' (presumed to be *Lactobacillus delbrueckii* subsp. *bulgaricus*). Another species he found in the starter, *Streptococcus thermophilus*, received no attention as it was then considered to be a pathogen. Specimens of the lactic acid-producing cultures were sent, at Metchnikoff's request, to the Pasteur Institute, where they were further investigated by Döderlein and Michelson

(Cohendy, 1906a). Further investigations at the Pasteur chemical laboratories were conducted by Bertrand and Weisweiler (1906), who found it to be an extremely active producer of lactic acid, producing 25 g L^{-1} milk. Fortunately, it produced no alcohol or acetone, two other by-products of fermentation. All these qualities made the '*Bulgarian bacillus*' (presumed to be *Lb. delbrueckii* subsp. *bulgaricus*) appear the ideal microbe to colonise the digestive tract for the purpose of 'arresting putrefaction and pernicious fermentations'.

- Solukha (1896) further researched the effects of specific milk components on the reduction of intestinal putrefaction, concluding that lactose inhibited putrefaction when given orally.
- Tissier (1900), who worked at the Pasteur Institute, was the first to note that Y-shaped "bifid" bacteria were predominant in the faeces of new-born breastfed babies.
- Moro (1900) revealed a predominance of a distinctive, highly acidic, gut bacillus in the intestinal tract of breast milk-fed babies, which he accordingly named '*B. acidophilus*' (presumed to be *Lb. acidophilus*).
- Tissier and Martelly (1902) believed that the chief agent responsible for the inhibition of the putrefying bacteria was the lactic acid produced by such bacteria.
- Bienstock (1902) reported that certain microbes, which sour milk by the production of lactic acid, hinder putrefaction of milk.
- Tissier & Gasching (1903) demonstrated that acid-producing bacteria were able, in a sugar-containing medium, to arrest the growth of putrefactive organisms.
- Weiss (1904) demonstrated the presence of large numbers of *B. acidophilus* (presumed to be *Lb. acidophilus*) in the human intestine after milk was consumed.
- Cohendy (1906b) reported treating two men and two women with 250 mL of a milk culture of the '*Bul. bacillus*' (presumed to be *Lb. delbrueckii* subsp. *bulgaricus*). The organisms could be recovered from the faeces in considerable numbers from the third to fifth days, as well as every day thereafter for 24 days. Prior to this administration, no '*Bul. bacillus*' (presumed to be *Lb. delbrueckii* subsp. *bulgaricus*) appeared in any pre-treatment faecal specimens. In a separate study, Cohendy (1906a) administered 250 mL of a 24 h milk culture of the bacteria to himself for 5 months and to 30 others for 7 months. He reported evidence of disinfection of the intestines after seven days, together with stool deodorisation and easier evacuation of stools. Absence of putrefaction persisted 2 weeks after cessation of treatment.

2.4 Could disinfection be the solution?

Having established a relationship between intestinal bacteria and putrefaction, efforts were made to eliminate putrefaction by disinfecting the intestines. Bouchard (1887), who was regarded as the pioneer, administered charcoal, naphthalene or iodoform and observed a reduction in toxicity of the stool and urine. Wassilieff (1882) claimed that calomel reduced the number of intestinal micro-organisms, and Rovighi (1892) employed turpentine, camphor, menthol and boric acid with moderate success. Many more attempts were made until 1912, when the use of intestinal disinfectants was finally accepted to be impractical.

Many scientists tried to quantify the enormous number of bacteria in the intestine of humans and animals. Eberle (1896), Klein (1900) and Hehewerth (1900) tried direct microscopic counting. Winterberg (1898) used the Thoma–Zeiss blood counting chamber for the first time, while Strasburger (1902) resorted to the gravimetric method, which proved efficient, and calculated that a healthy adult daily excretes about 8 g dry weight of dried bacteria or 128 trillion cells. Others used plate culture and other methods to estimate the total number of bacteria in the faeces.

2.5 On the cusp of a major breakthrough

In a remote village near Kharkov, which was within the Russian Empire in 1845 but is now in Ukraine, a child was born into the family of a semi-retired officer of the Russian Imperial Guard and his reputedly beautiful and intelligent wife, the daughter of a Russian Jewish author. The infant, Ilya (later changed to Elie) Metchnikoff, would become an internationally recognised scientist, as well as becoming the principal figure in the probiotic story to date (Figure 2.1). By the age of eight, he had developed an interest in collecting and studying plant and insect life. At the age of 11, he attended the Lycée in nearby Kharkov where he was first introduced to the world of microscopic

Figure 2.1 Elie Metchnikoff (1894–1916) in front of a conventional light microscope. Reproduced with permission of the Wellcome Photo Library.

biology. His time there spanned a period of intense scientific activity, especially in the fields of biological and microbiological science. In 1862, the publication of Darwin's 'Origin of the Species' excited the young Metchnikoff, as well as stimulating much argument among biologists and theologians.

His mother, with whom he maintained a close attachment until her death in 1889, persuaded him against studying medicine. Instead he chose natural sciences and marine zoology, a decision that was to have profound and lasting effects in the worlds of medicine and microbiology. The larvae of the humble starfish became a focus of his investigations, where he first noted the presence of mobile cells that engulfed and apparently digested bacteria, thereby preventing them from threatening the host organism. This led to the Nobel Prize-winning concept of phagocytosis, a concept totally at variance with the belief held by most pathologists at the time (Tauber, 1992).

In 1889, following a visit to Louis Pasteur, Metchnikoff joined the Pasteur Institute in Paris where he initiated his studies with fermented milk. In 1895, following Pasteur's death, he was named Director of the Pasteur Institute. At the turn of the twentieth century and following the general acceptance of his phagocytosis theory, Metchnikoff turned to investigating the mechanisms of ageing, influenced no doubt by Darwin's theory of organ and tissue adaptation. The fact that he himself was approaching 60 may also have played a part. He became convinced not only that the large intestine and its massive bacterial population were obsolete, but also that its population contained vast numbers of proteolytic organisms. These, he believed, were constantly producing toxic metabolites that were absorbed systemically by the host, acting as accumulative poisons and accelerating the ageing process. He proposed to fight these by introducing an opposing force: lactic acid-producing bacilli. This original approach would involve oral administration of living cultures of the bacilli suspended in milk. The bacteria would ferment milk carbohydrates, such as lactose, into lactic acid, thereby causing a fall in pH. The resulting soured, acidic milk would, he believed, be unsuitable for the continued survival and multiplication of any putrefactive organisms present in the intestines.

His first public presentation of this hypothesis was a Wilde Lecture given to an audience in Manchester in 1901, entitled 'Flora and the Human Body'. The hypothesis then played a significant role in two of his books: 'Nature of Man' in 1903 followed by 'Prolongation of Life' (Metchnikoff, 1907). Praise began to be heaped on him when the latter was translated into English and favourably reviewed in Harper magazine. In 1911, The Independent named him one of 'Twelve Major Prophets of Today' (along with H.G. Wells, G.B. Shaw and nine others), and the lead story in the 13 July 1912 issue of Scientific American was entitled 'Professor Elie Metchnikoff: The Most Distinguished of Living Bacteriologists'.

Metchnikoff planned his research along two routes: (a) to investigate the role of the intestinal microbiota as chronic intoxicants, and (b) to study tissues that show prominent changes with advancing age, such as hair and skin. It was his belief that the application of science could extend the normal life span to 100 to 120 years. To achieve this, he proposed to strengthen beneficial cells within the body and transform the 'wild' intestinal bacteria by the introduction of useful and harmless acid-producing bacteria. This belief was backed by correspondence from contacts in Asia and Africa, describing the apparent longevity of 'well preserved natives', who showed few signs of senility and consumed soured milk as a major part of their diet. He was also aware of certain exceptionally long-living populations in Bulgaria and the Russian Steppes, who existed largely on sour milk. In order to test his lactic acid hypothesis, however, he needed lactic acid-producing bacteria that were safe to administer, stable, harmless, accessible and beneficial. Possibly at Metchnikoff's request, his colleague in the Pasteur laboratories, Cohendy (1906a, 1906b), had been conducting a series of feeding studies which involved a milk-based culture of the Bulgarian sourced lactic acid bacteria, after he had ascertained that it passed safely through the digestive tract to reach the colon to live there on a permanent or semi-permanent basis. He found this a very powerful lactic ferment, naming it '*Bul. bacillus*' (presumed to be *Lb. delbrueckii* subsp. *bulgaricus*) and noting evidence of reduced putrefaction and greater ease of evacuation. Tested in the Pasteur chemistry laboratory, it had proved to be an extremely active producer of lactic acid, producing $25\,g\,L^{-1}$ of milk.

All these and other qualities commended the '*Bul. bacillus*' (presumed to be *Lb. delbrueckii* subsp. *bulgaricus*) to Metchnikoff as the most useful method for preventing putrefaction and pernicious fermentation in the gut. In practice, however, he preferred to use another lactic acid-producing microbe known as 'paralactic bacillus' (presumed to be *Lb. delbrueckii* subsp. *bulgaricus*) because, although it produced less acid, this strain prevented the breakup of fats and gave the curdled milk a pleasant flavour. For more than eight years, Metchnikoff experimented on himself by taking it as a regular part of his diet. The soured milk was prepared from boiled milk, which after rapid cooling was inoculated with the mixed lactic bacilli. The resulting fermentation took a few hours, depending on the ambient temperature. Prepared according to his recipe, the milk provided about $10\,g\,L^{-1}$ of lactic acid. He consumed 300–500 mL daily; when his health appeared to benefit, his friends followed his example and soon physicians were prescribing this sour milk for their patients.

In the concluding remarks of his 'Prolongation of Life' thesis, Metchnikoff (1907) commented:

> if it be true that our unhappy and precocious old age is due to poisoning of the tissues, it is clear that agents which arrest intestinal putrefaction must at the same time postpone and ameliorate old age. It can only be in the future, near or remote, we shall obtain exact information upon what is one of the chief problems of humanity. In the meantime, those who wish to preserve their intelligence as long as possible and make their life as complete and normal as possible, must depend on general sobriety and on habits conforming to the rules of rational hygiene.

Demand for his sour milk became so great that the Pasteur Institute began selling cultures. The market was soon flooded with commercial products claiming to contain the '*Bul. bacillus*' (presumed to be *Lb. delbrueckii* subsp. *bulgaricus*), including one by a company named Le Ferment sold under the trade name 'Lactobacilline', the label of which, much to his annoyance, bore the name of Professor Metchnikoff. The outbreak of World War I (WWI) and the German occupation of Paris brought an end to Metchnikoff's research, although he continued working at the Institute on problems related to soldiers' health. He died of heart failure in 1916 at the (then) ripe old age of 71. A scientist to the end, his final words to his friend Dr Salimbeni were to ask him to conduct his post mortem and to look carefully at the intestines.

Mechnikoff's achievements were recognised by numerous honours, including The Royal Society's most prestigious prize, the Copley Medal, in 1906. It is also worth mentioning that, prior to developing his lactic acid hypothesis, Metchnikoff had regarded the large intestine as a disused organ, best surgically removed. This approach appealed to Sir Arbuthnot Lane, a distinguished surgeon, who began performing colectomies. He appears to have continued this procedure in spite of a 50% fatality rate, but it eventually prompted a debate on alimentary toxaemia at the Royal Society of Medicine in 1913, during which Lane's ideas were comprehensively dismissed and from which Lane is reported to have "driven away crushed" (Hamilton-Miller, 2008).

2.6 The urge for progress switches to the USA (1914–1931)

Following receipt of a doctorate degree from the Department of Physiology and Physiological Chemistry at Yale University, in 1902, Leo Fredrick Rettger chose to pursue his career in what was then a comparatively young science: bacteriology. Fifteen years later, as Professor of Bacteriology at Yale, he delivered the Presidential Address to the Society of American Bacteriologists in which he expressed a hope of bacteriology becoming a wholly independent science. Author of the first scientific paper to be published in the Journal of Bacteriology, he lived to see his hope realised.

According to Rettger *et al.* (1935), no aspect of modern bacteriology received more research attention than the bacteriology of the digestive tract. He was undoubtedly the most prolific researcher in the immediate post-WWI era to study bacterial implantation in the intestines for therapeutic purposes. Much of this work, which was almost continuous over three decades, was prompted by the discovery by others that '*B. acidophilus*' (presumed to be *Lb. acidophilus*) was a normal resident in the large intestine and, therefore, more likely to be successfully implanted there (Rotch & Kendall, 1911). During the first decade of the twentieth century, most investigators confirmed Metchnikoff's claim that the '*Bul. bacillus*' (presumed to be *Lb. delbrueckii* subsp. *bulgaricus*) could be implanted in the human intestine; however, in the second decade, Hull & Rettger (1917) and many others all failed to achieve this. Luerssen & Kuhn (1908) and Spiegel (1911) found a decided lack of evidence for the benefits of the '*Bacillus bulgaricus*' (presumed to be *Lb. delbrueckii* subsp. *bulgaricus*) milk and tablets. In addition, Distaso & Schiller (1914) concluded that the implantation of this organism in the intestines was impossible. Rettger & Cheplin (1921a) eventually concluded that the reason why ingestion of milk soured with '*B. bulgaricus*' (presumed to be *Lb. delbrueckii* subsp. *bulgaricus*) appeared to Metchnikoff to have such a marked transforming influence on the intestinal flora (now known as microbiota) was that, on examining the stool, what appeared at first sight to be '*B. bulgaricus*' (presumed to be *Lb. delbrueckii* subsp. *bulgaricus*), was, in fact, the almost indistinguishable, normally resident, '*B. acidophilus*' (presumed to be *Lb. acidophilus*), invigorated by the lactose present in the milk. This conclusion was claimed to substantiate reports of beneficial effects from feeding lactose and milk to typhoid fever patients by Torrey (1915), who also demonstrated that administering a high lactose-containing diet suppressed putrefying bacteria and favoured the growth of '*B. acidophilus*' (presumed to be *Lb. acidophilus*). Cheplin & Rettger (1920) thought the explanation for this was that incompletely digested carbohydrate reaching the large intestine served as a readily available source of energy for these bacteria; therefore, they argued that the fundamental principle of Metchnikoff's sour milk therapy to transform the intestinal microbiota was justified. Hull & Rettger (1917) claimed that lactose and dextrin preferentially favoured the growth of '*B. acidophilus*' (presumed to be *Lb. acidophilus*) more than any other intestinal micro-organisms.

As early as 1915, Cheplin & Rettger (1920) had suggested that '*B. acidophilus*' (presumed to be *Lb. acidophilus*) might work better than '*B. bulgaricus*' (presumed to be *Lb. delbrueckii* subsp. *bulgaricus*) and thereafter until 1932, having developed a method for preparing acidophilus milk; however, Rettger *et al.* (1935) conducted a series of clinical studies with it in a variety of intestinal conditions. In the first study with 40 constipated patients, 32 (77.5%) showed positive effects from the treatment, and of these, 27 (87%) continued to show no symptoms and maintained an intestinal microbiota with apparently high levels of '*B. acidophilus*' (presumed to be *Lb. acidophilus*) for 12–16 weeks following discontinuance of treatment. In a second study involving 17 patients with constipation and biliary symptoms, 13 (76.5%) responded favourably, while four (23.5%) did not. Of the positive cases, nine (69%) became 'implanters'. A third investigation comprised eight patients with chronic ulcerative colitis and eight patients with 'mucous colitis', of whom 75% and 87.5%, respectively, showed positive effects. None of the 13 responders became 'implanters' (Rettger & Cheplin, 1921b).

The results of the clinical and bacteriological studies convinced the investigators that this species, now known as *Lb. acidophilus*, was a significant factor in the successful application of acidophilus milk in these cases. The investigations and methods with detailed clinical results are fully described in a book by Rettger *et al.* (1935), in conclusion of which they claimed:

> on the basis of our clinical investigations, we feel justified in concluding that acidophilus milk of high viability, given in massive quantities over relatively long periods, will, in the majority of cases, be beneficial in relieving patients suffering from simple constipation, constipation accompanied by biliary symptoms and idiopathic ulcerative colitis

They were pleased to report that their acidophilus milk was pleasant to taste and smell, was not very acid and was stable. The curd was described as soft and of creamy consistency, and easily prepared. They advised that pure strains of '*B. acidophilus*' (presumed to be *Lb. acidophilus*) should be used, which should be grown in milk kept at 35–37 °C sufficiently long enough to produce acidity and a soft curd within 24 h.

Also, Rettger *et al.* (1935) lamented the fact that as their acidophilus milk treatment increased in popularity, the principle on which the therapy was based, namely massive doses of the correct culture accompanied by large amounts of lactose in milk as the vehicle, met with considerable abuse in the hands of the producers of acidophilus products and physicians. This undermined confidence in the principle. The market became flooded with numerous 'acidophilus' products that, with few exceptions, belied the labels. Few contained *Lb. acidophilus* or any other aciduric organism in appreciable numbers, if at all, at the time of purchase. Others that boasted of cultures of high viability contained, in place of the recommended intestinal species of *Lactobacillus*, strains resembling common oral and dental species of the same genus. Furthermore, concentrates of the products were prescribed in such small doses that they would be ineffective, even if they contained high numbers of viable bacteria. Such concentrates, however, found willing advocates and buyers because they preferred a treatment in a vial, rather than one supplied in a large volume.

Describing their clinical experiences with acidophilus milk, Griffith & Matt (1932) stated that therapists and bacteriologists agreed that viable cultures of the *Lb. acidophilus* bacteria in several types of vehicles, when administered in proper dosage to animals and humans, were fully capable of implantation in the intestinal tract, with an eventual transformation of the intestinal bacterial community from a proteolytic to an aciduric predominance. Moreover, it was accepted by leading clinical researchers that this benefited the average patient, particularly in the alleviation of constipation, as well as in the treatment of diarrhoeas.

In 1921, Cannon (1921) confirmed the surprising observation by Torrey (1915) that, whereas animal protein encourages putrefaction in the intestinal tract, 'vegetable proteins do not offer the slightest encouragement to the growth of the putrefactive intestinal types of bacteria and, moreover, encourage the overgrowth of non-gas producing acidic flora'.

Having noted from a survey of the then contemporary literature on *Lb. acidophilus* and '*Lb. bulgaricus*' (presumed to be *Lb. delbrueckii* subsp. *bulgaricus*) that the advocates of each were still irreconcilable, Nicholas Kopeloff, a bacteriologist working at the Pasteur Institute, undertook a series of human studies designed to resolve the dispute. His objective was to establish whether either, both or neither was capable of being implanted in the large intestine, and he was careful to ensure that the strain of '*Lb. bulgaricus*' (presumed to be *Lb. delbrueckii* subsp. *bulgaricus*) was Metchnikoff's original culture, while the *Lb. acidophilus* strain was that used by Kulp & Rettger (1924), principal proponents of the argument. The survey checked the effects of both strains for the treatment of constipation, and Kopeloff & Beerman (1925) concluded the following:

- '*Lactobacillus bulgaricus*' (presumed to be *Lb. delbrueckii* subsp. *bulgaricus*) could rarely, if ever, be implanted in the human intestine. In only one instance out of 12 was the strain recovered from faeces 2 weeks after the feeding of large quantities of the bacteria.
- *Lactobacillus acidophilus* was recovered from the faeces of the same patients, in large numbers, during the 2 weeks following feeding.
- Constipation was markedly alleviated by the administration of acidophilus milk, whereas no improvement was recorded with '*Lb. bulgaricus*' (presumed to be *Lb. delbrueckii* subsp. *bulgaricus*).

Kopeloff & Beerman (1924) also investigated the influence of *Lb. acidophilus* on the intestinal microbiota, following their earlier observations, and they concluded the following aspects: (a) there was evidence to indicate that the ingestion of *Lb. acidophilus* in large numbers transformed the intestinal microbiota from Gram-negative to Gram-positive species, (b) the higher the percentage of viable *Lb. acidophilus* recovered in faeces, the higher the number of normal defecations, plus the character of the faeces changed to become larger in quantity, a softer consistency and a lighter colour, (c) it was possible to recover *Lb. acidophilus* from the faeces of patients long after treatment was discontinued, and (d) *Lb. acidophilus* could be administered rectally in situations where oral administration needed to be supplemented or was undesirable.

Fascinated by its potential, Kopeloff (1923) contemplated whether the action of *Lb. acidophilus* was physical, mechanical or bacterial and, therefore, devised a series of studies from which he finally concluded that the bacterium exerted the described effects bacteriologically. His recommended daily dosage for treatment was one litre of acidophilus milk containing approximately 200 million viable *Lb. acidophilus* mL^{-1}, given in two doses and accompanied by 100–300 g lactose.

According to Frost *et al.* (1931), numerous articles and two books appeared each year between 1914 and 1930, mostly from scientists in the USA, all dealing exclusively with '*B. acidophilus*' (presumed to be *Lb. acidophilus*). Among these were 18 articles and one book reporting successful clinical results from such bacterium implantation in the intestines. These led, once again, to a surge of commercial acidophilus milk products being marketed. Concerned about the dubious quality of some of these, James (1927) examined 107 samples purchased from retail stores, mostly pharmacies. Of these, 34 were in broth, 28 in milk cultures, 34 in tablets, three in powdered form and eight semi-solid (bacteria suspended in solidified agar or petrolatum). Of the 107 samples, only 13 contained the species indicated on the label, in reasonably pure form and sufficient numbers; 15 contained pure cultures but in insufficient numbers to have any benefit. All the others were 'worthless'. As a consequence, in 1934, the American Medical Association (AMA) Council of Chemistry & Pharmacy issued a specification requiring that 'Acidophilus Milk' must contain not less than 200 million viable cells mL^{-1} on the day of manufacture and not less than 100 million cells mL^{-1} on the date of expiry.

In the 1930s, Arthur Burke, Head of the Dairy Department of Alabama Polytechnic Institute, published a book ('Practical Manufacture of Cultured Milks and Kindred Products'; Burke, 1938) which, according to its sub-title, provided 'A complete and practical treatise on the manufacture of commercial cultured buttermilks of all types – lactic, Bulgarian, Acidophilus, Kefir, Koumiss and yoghurt'. Kefir was manufactured and marketed in Los Angeles in 1979 and in New Jersey in 1985 (Kroger *et al.*, 1989).

2.7 Meanwhile, in Europe

In 1916, Isaac Carasso, a member of a prominent Sephardic Jewish family located in Ottoman Salonica (now Greek Thessaloniki), concerned about the unrest in the Balkans and the threat of an oncoming Greek army of occupation, decided to transfer his family back to Barcelona, which his ancestors had been forced to leave in or about 1492. On arrival, he was struck by how many children were suffering with intestinal infections. Like everyone born in the Balkans, he knew that such children were treated with traditional yoghurt. Meanwhile, he had become aware of Metchnikoff's work on the use of lactic acid bacteria to prevent and treat intestinal infections. In 1919, he set up a small laboratory to manufacture traditional yoghurt into which he introduced Metchnikoff's cultures ('*Lb. bulgaricus*' – presumed to be *Lb. delbrueckii* subsp. *bulgaricus* – and

Streptococcus thermophilus), purchased from the Pasteur Institute. Unlike in the Balkans, yoghurt was relatively unknown in Western Europe; therefore, Carasso decided to market the product as a medicine, which he introduced to doctors and distributed through pharmacies. Every morning, 400 porcelain pots containing his yoghurt were delivered to pharmacies in Barcelona. News about the health benefits of the product was spread by word of mouth, and the small porcelain pots became very successful in just a few years. During this time, it became necessary to give the product a name, so Carasso chose to call it Danone, being the diminutive for his young son Daniel. In due course, Daniel was sent to the Pasteur Institute to take a course in intestinal bacteriology and prepare him for his major role in the future development of what is now a highly successful global company (Grimes, 2009).

In 1917 during WWI, the German officer and bacteriologist Alfred Nissle isolated a strain of *Escherichia coli* from the stool of a soldier who, unlike his comrades, had survived an attack of *Shigella* dysentery. Impressed by this, Nissle cultured the organism and went on to treat shigellosis and salmonellosis with significant success. The strain, now designated *E. coli* Nissle, still used today, is a good example of a non-lactic acid-producing probiotic. For obscure reasons, prospective, controlled clinical studies do not appear to have been undertaken with the strain until the late 1990s and early 2000s. Comparisons with mesalazine for the treatment of ulcerative colitis revealed the strain's efficacy to be similar to that of the standard pharmaceutical treatment (Jacobi & Malfertheiner, 2011).

2.8 The ultimate breakthrough comes from Japan?

Coincidental with Cheplin & Rettger's (1920) dismissal of Metchnikoff's claims for '*B. bulgaricus*' (presumed to be *Lb. delbrueckii* subsp. *bulgaricus*), a young medical student named Minoru Shirota (Figure 2.2), who had enrolled in Kyoto Imperial University (now Kyoto University), Japan, in 1921, was excited by Metchnikoff's thesis and became convinced that a positive balance of the good bacteria in the gut was the basis of a long and healthy life. He was also very concerned about the widespread loss of life among children in Japan in the early 1900s, due to poor sanitary conditions and accompanying infectious disease, and decided to direct his career towards preventative medicine and microbiology. In 1924, he decided to dedicate his research to finding a strain of *Lactobacillus* that would pass safely through the intestines to the colon, where it could contribute positively to the balance of the gut microbiota. While still a student in the Faculty of Medicine, he started his research in the Bacteriology Laboratory, initially with Metchnikoff's '*B. bulgaricus*' (presumed to be *Lb. delbrueckii* subsp. *bulgaricus*), hoping this would lead to a simple, inexpensive and easily administered method of countering the effects of disease-producing bacteria. His early results were disappointing, as he too found that the strain could not survive passage through the gut. Then followed a meticulous screening of some 300 lactobacilli strains, looking for one or more capable of maintaining viability through to the large intestine. From these, 18 promising candidates were selected, including one particularly robust strain of *Lactobacillus casei* (then classified as *Lb. acidophilus*). The strain was later renamed

Figure 2.2 Minoru Shirota (1899–1982) in front of an electron microscope. Reproduced with permission of Yakult Europe B.V., The Netherlands.

Lb. casei Shirota in his honour. According to Kyoto Professor Emeritus Nakaya, 'Shirota laboured hard to find a lactic acid producing bacteria which could survive passage through the intestines. It took several years and intense passion'. Shirota was eventually successful in the winter of 1930, the year in which he had received his doctor of medicine degree (Anonymous, 2009).

Together with colleagues, in 1935, Shirota developed a milk-based suspension of his strain, producing it in his clinic and distributing it from there. Writing in his news-sheet in 1937 and addressing would-be users of his product, he explained how his bacterium worked:

> Following oral ingestion it decomposes sugars converted from food starch within the bowel to yield lactic acid. If there are pathogenic bacteria in the bowel, they are immediately inhibited or destroyed. Result – diseases like children's dysentery, typhus and cholera can be cured. At normal intestinal temperature a single bacterium divides every 17 minutes and therefore increases to 160 million in 24 hours.

He continued, 'readers may find it strange that the same product is effective against both constipation and diarrhea' (Shirota, 1937). His explanation was based on a study (possibly unpublished) by Matsuo and Yoshikawa of Kyoto University who had discovered that 'lactic acid slows small bowel peristalsis but speeds large bowel peristalsis'.

Demand grew to a point that it became necessary and desirable to scale up manufacture and begin serious marketing and sales activities. Possessing a good business sense, he set up a company to produce and market his product under the name 'Yakult'. He chose this name because of its similarity to the Esperanto word for yoghurt. Esperanto was an artificial language created by Zamenhof, a Polish ophthalmologist, in 1887; it was based on a variety of different language roots and designed to provide a simple-to-learn method of facilitating communication between peoples irrespective of their national tongues. The name's international implications probably appealed to Shirota.

To confirm that viable organisms of *Lb. casei* Shirota were reaching the large intestine, Shirota had used microscopic examination and testing of stool samples; the gut survival of this strain has since been confirmed in several human intervention studies using more modern methods (Yuki *et al.*, 1999; Tuohy *et al.*, 2007). In addition, Shirota conceived the idea of distributing his product by hand, in order to establish a personal relationship with customers, and so he engaged the help of a handful of nurses. The outbreak of WWII caused a lengthy interruption to his plan, but retaining his original enthusiasm he soon re-started operations after the war, and by 1963 the scheme was sufficiently successful for it to be extended. As of 2016, some 40 000 Yakult Ladies provide a nationwide personal service in which the sale and distribution of the product are accompanied by the provision of general health and lifestyle advice, especially to young mothers. For this service, the Yakult Ladies are provided with a continuous training programme (Ben & Soble, 2013).

Interest in the gut microbiota and probiotic research revived post-war, mainly due to the discovery that antibiotic administration could stimulate the growth of animals and the development of improved methods of rearing germ-free animals – both of which underlined the need for better understanding of the composition of the gut microbiota (Fuller, 1992). Pasteur had been the first to suggest a possible role for germ-free animals in research, and Nuttal & Thierfelder (1895) had successfully reared guinea pigs completely free of microbes. The technique became available for research groups throughout the world by the 1950s, helping to confirm Metchnikoff's hypothesis that gut bacteria could have an adverse effect on the host animals.

Another important contributor to the post-war revival of probiotic research was the acceptance by relevant authorities of the granting of patent rights for novel probiotics that were clearly identifiable and for hitherto unknown strains with special, desirable features. This prompted commercial interest and substantial research investment into discovering suitable bacterial strains. The most prolific patent assignees from 1950 to 2011 were Nestlé, Danisco, DSM, Unilever and Yakult, which were helped considerably by the realisations during the 1950s to 1960s by scientists and nutritionists that the gut played an important role in protecting the host (whether animal or human) against disease, that not all bacteria were bad and that many gut species were actually beneficial (Anonymous, 2011). Not surprisingly, therefore, probiotic research carried out between 1950 and 1980 concentrated on screening for potential probiotics from strain collections isolated from humans, animals or other natural sources. At the same time, scientists began understanding how the gut bacteria interact with their hosts and inhibit pathogens.

Important work was conducted in the 1950s by Bohnhoff *et al.* (1954) and Freter (1955, 1956), who showed that the administration of antibiotics to animals killed potentially beneficial commensal gut bacteria and rendered the animals much more susceptible to infection by *Salmonella typhimurium* and *Shigella flexneri*. In 1954, Vergin (1954) proposed that antibiotic-induced dysbiosis could be reversed by a probiotic-rich diet;

this suggestion is often cited as the first example of a probiotic application, as defined in the 2000s. Further insights came from the discoveries that growth-promoting antibiotics given to chickens increased their susceptibility to salmonella colonisation of the gut and, in human medicine, when it was found that antibiotic therapy could induce diarrhoea including that caused by *Clostridium difficile*. If more evidence was needed, this was provided by an animal study that showed a germ-free guinea pig could be killed by ten cells of a *Salmonella* spp., but that it required 10^9 cells to kill a guinea pig with a normal gut microbiota (Collins & Carter, 1978).

Animals obtain their protective microbiota from the mother, but modern perinatal care practices used for farm and domesticated animals tend to restrict maternal contact and may provide an unnatural pre-prepared diet and environment. This may affect the development of the gut microbiota. In addition, concerns about the use of antibiotics in animals has led to the restriction on antibiotics as growth promoters to only those not used clinically. In some countries (e.g. Sweden), restrictions have gone further with a ban on any antibiotics being used as growth promoters. All of this has prompted interest in probiotic use in animals, with research in the 1960s showing that *Lactobacillus* supplementation could stimulate growth in pigs (Kohler & Boehm, 1964). At that time, however, antibiotics seemed the more effective way of promoting rapid weight gain, coupled with lower feeding costs and earlier dispatch to the market. Probiotics failed to make an impression until efficacy was shown and the implications of antibiotic usage in animals emerged. Widespread evidence of resistance to antibiotic treatments in both animals and humans began to appear, followed by legislation designed to restrict the liberal and often inappropriate use of antibiotics. The problem was well illustrated in Canada, where the pig industry was the country's second largest agricultural export, contributing 42 000 jobs to the Ontario economy alone. Around 10–12% of pigs were dying before weaning, 50% through intestinal infections (Reid & Friendship, 2002). Antibiotics were clearly not solving the problem and could well have been exacerbating it by disrupting the intestinal microbiota, increasing infection susceptibility and permitting the emergence of antibiotic-resistant pathogens. Thus, agricultural scientists and farmers began turning to probiotics, with some marked success. In a comprehensive study of 296 strains of lactic acid bacteria from the gut of 50 chicks, 77 were found to inhibit the growth of *Salmonella enteritidis* and enteropathogenic *E. coli* (Carriga *et al.*, 1998).

The carriage of *E. coli* 0157:H7 is particularly alarming in cattle. In a challenge study screening 18 probiotic strains against this pathogen, the ability of *E. coli* 0157:H7 to colonise the gut was diminished, emphasising the importance of early gut colonisation by probiotics (Zhao *et al.*, 1998). For probiotics to be universally accepted in the farming industry, however, probiotic strains need to be carefully selected and scientifically validated.

2.9 Conclusions

The emergence of a highly specific group of bacteria with important common health-promoting properties owes its origin to Metchnikoff and his concept that installing harmless lactic acid-producing bacteria into the intestines would suppress or prevent pathogenic invasion. Although his recommended strain for the purpose proved unable to colonise the colon, the concept was carried forward with partial success, primarily by Leo Rettger and colleagues at Yale University in the USA from 1912 to 1934, and then with complete success by Minoru Shirota in Japan in 1930. Consolidation of this progress began in Europe in the 1970s and 1980s, and during this period, Parker (1974) gave an identity to that small, remarkable, highly specific range of bacteria with important common properties, by naming them 'probiotics'. Currently, bacterial strains considered to meet the probiotic definition belong to several different species of lactobacilli and bifidobacteria, as well as strains of *E. coli*, enterococci, streptococci, lactococci and one yeast (*Saccharomyces cerevisiae* var. *boulardii*).

One of the most striking facts to emerge is the characteristic change in the intestinal microbiota that occurs whenever milk constituted a large part of the diet. The beneficial effects of milk in the treatment of various intestinal diseases has been known for some time, but only comparatively recently has it been shown that the reduction in putrefactive organisms and increase of beneficial Gram-positive bacterial species are partly due to the lactose present in the milk.

Acknowledgements

I am grateful to New Century Health Publishers LLC for permission to use material from my earlier paper on the history of probiotics in the *International Journal of Probiotics and Prebiotics* (Levin, 2011).

References

Anonymous (2009) *Yakult Company Profile 2008/9*. Yakult Honsha Co. Ltd, Tokyo, Japan.

Anonymous (2011) *Probiotics*. Technology Insight Report, Patent iNSIGHT Pro., Gridlogics Technologies Pvt. Ltd, Pune, India. www.patentinsightpro.com

Ben, S. & Soble, J. (2013) On the trail of the Yakult Ladies. *Financial Times*, 27 November. http://www.ft.com/cms/s/0/e8e517c8-5387-11e3-9250-00144feabdc0.html#axzz4GvgxMBPH

Bertrand, G. & Weisweiler, G. (1906) Action du ferment bulgare sûre le lait. *Annals de l'Institut Pasteur*, **12**, 977–990.

Bienstock, E. (1902) Untersuchungen über die Aetiologie der Eiweissfaulnis. *Archiv für Hygiene und Bakteriologie*, **39**, 390–427.

Billroth, C.T. (1874) *Untersuchungen über die Vegetationsformen von* 'Coccobacteria septica' *und den Antheil welchen sie an der Entstehung und Verbreitung der accidentellen Wunderkrankheiten haben, Versuch einer Wissenschaftlichen Kritik der verschiedenen Methode antiseptischer Wundbehandlung*. Reimer, Berlin.

Bohnhoff, N., Drake, B.L. & Muller, C.P. (1954) Effect of streptomycin on susceptibility of the intestinal tract to experimental salmonella infection. *Proceedings of the Society of Experimental Biology and Medicine*, **86**, 132–137.

Bouchard, C.J. (1884) Recherches experimentales sur la toxicité des urines normales. *Comptes Rendus des Séances de la Société de Biologie*, **1**, 665–668.

Bouchard, C.J. (1887) *Leçons sur les auto-intoxications dans les maladies*. Librairie F. Savy, Paris.

Burke, A.D. (1938) *Practical Manufacture of Cultured Milks and Kindred Products*. The Olsen Publishing Company, Milwaukee.

Cannon, P.R. (1921) The effects of diet on the intestinal flora. *Journal of Infectious Diseases*, **29**, 369–375.

Carriga, M., Pascual, M., Montfort, J.M. & Hugas, M. (1998) Selection of lactobacilli for chicken probiotic adjuncts. *Journal of Applied Microbiology*, **84**, 125–132.

Cheplin, H.A. & Rettger, L.F. (1920) Studies on the transformation of the intestinal flora, with special reference to the implantation of *Bacillus acidophilus*: II. Feeding experiments on man. *Proceedings of the National Academy of Sciences of the United States of America*, **6**, 704–705.

Collins, F.M. & Carter, P.B. (1978) Growth of salmonellae in orally infected germfree mice. *Infection and Immunity*, **21**, 41–47.

Cohendy, M. (1906a) De la désinfection intestinale obtenue, sans régime special, par l'acclimatation d'un ferment lactique dans le gros intestine. *Comptes Rendus des Séances de la Société de Biologie*, **60**, 602–604.

Cohendy, M. (1906b) Essais d'acclimation microbienne persistante dans la cavite intestinale. Delimitation du siege probably de cette acclimatation. *Comptes Rendus des Séances de la Société de Biologie*, **60**, 364–366.

Distaso, A. & Schiller, J. (1914) Sur l'acclimatation de microbes étranger à flore intestinal. *Comptes Rendus des Séances de la Société de Biologie*, **76**, 243–244.

Döderlein, A. (1892) *Des Scheiden Sekret und Seine Bedeutung für das Puerperalfieber*. Verlag von Eduard Besold, Leipzig.

Douglas, L.M. (1911) *The Bacillus of Long Life: A Manual of the Preparation and Souring of Milk for Dietary Purposes, together with an Historical Account of the Use of Fermented Milks from the Earliest Times to the Present Day, and their Wonderful Effect in the Prolonging of Human Existence*. G.P. Putnam's Sons, New York.

Eberle, R. (1896) Zählung der bakterien im normalen zäuligskoth. *Zentralblatt für Bakteriologie*, **19**, 2–5.

Escherich, T. (1886) Die darmbakterien des neugeborenen und saunglings. *Fortschritt der Medizin*, **3**, 515–522.

Freter, R. (1955) The fatal enteric cholera infection in the guinea pig, achieved by inhibition of normal enteric flora. *Journal of Infectious Diseases*, **97**, 57–65.

Freter, R. (1956) Experimental enteric *Shigella* and *Vibrio* infections in mice and guinea pigs. *Journal of Experimental Medicine*, **104**, 411–418.

Frost, W.D., Butterworth, T.H. & Farr, S.M. (1931) Present status of acidophilus milk. *American Journal of Public Health*, **21**, 862–866.

Fuller, R. (ed.) (1992) *Probiotics: The Scientific Basis*. Chapman and Hall, London.

Griffith, I. & Matt, M.C. (1932) Intestinal implantation of the *Bacillus acidophilus* through the feeding of buttermilk containing acidophilus. *American Journal of Pharmacy*, **104**, 296–309.

Grigoroff, S. (1905) Étude sur un lait fermente comestible – Le 'Kisselomleko' de Bulgarie. *Revue Médicale de la Suisse Romande*, **35**, 714–720.

Grimes, W. (2009) Daniel Carasso, a pioneer of yogurt, dies at 103. *The New York Times*, 20 May.

Hamilton-Miller, J.M.T. (2008) Elie Metchnikoff and his probiotic legacy. *Medical Science History*, **24**, 1–5.

Hehewerth, F.H. (1900) Die mikroskopische zählungsmethode der bakterien von Alex Klein und einige anwendungen derselben. *Archiv für Hygiene*, **39**, 321–389.

Hirschler, A. (1886) Ueber den einfluss der kohlehydrate und eineger anderen körper der fettsäurereuhe und die eiweeissfäullniss. *Zeitschrift für Physioliche Chemie*, **10**, 806–817.

Hull, T.J. & Rettger, L.F. (1917) Influence of milk and carbohydrate feeding on the character of the intestinal flora. *Journal of Bacteriology*, **2**, 47–71.

Jacobi, C.A. & Malfertheiner, P. (2011) *Escherichia coli* Nissle 1917 (Mutaflor): new insights into an old probiotic bacterium. *Digestive Diseases*, **29**, 600–607.
James, L.H. (1927) Commercial *Bacillus acidophilus* and *Bacillus bulgaricus* cultures and preparations. *JAMA*, **89**, 89–92.
Kern, E. (1881) Ueber ein neues milchferment aus dem Kaukasus. *Bulletin de la Société Impériales des Naturalistes de Moscou*, **3**, 141–177.
Klein, A. (1900) Eine neue mikroskopische zählungsmethode der bakterien. *Archiv für Hygiene*, **27**, 834–835.
Kohler, E.M. & Boehm, E.M. (1964) Prophylaxis of diarrhea in newborn pigs. *Journal of the American Veterinary Medical Association*, **144**, 1294–1297.
Kopeloff, N. (1923) Is the action of *Bacillus acidophilus* a strictly bacteriologic phenomenon? *JAMA*, **80**, 602–604.
Kopeloff, N. & Beerman, P. (1924) Studies on the nature of *Bacillus acidophilus* therapy. *Archives of Internal Medicine*, **33**, 55–57.
Kopeloff, N. & Beerman, P. (1925) *L. acidophilus* versus *L. bulgaricus* milk feeding. *Proceedings of the Society for Experimental Biology & Medicine*, **22**, 318.
Kroger, N., Kurmann, J.A. & Rasic, J.L. (1989) Fermented milk – past and present. *Food Technology*, **43**(1), 92–99.
Kulp, W.L. & Rettger, L.F. (1924) Comparative study of *Lactobacillus acidophilus* and *Lactobacillus bulgaricus*. *Journal of Bacteriology*, **9**, 357–395.
Levin, R. (2011) Probiotics: the road map. *International Journal of Probiotics and Prebiotics*, **3**/4, 133–140.
Luerssen, A. & Kuhn, M. (1908) Yogurt, die Bulgarische sauermilch. *Zentralblatt für Bakteriologie*, **20**, 234–238.
Metchnikoff, E. (1907) *Prolongation of Life*. English trans. Heinemann, London.
Moro, E. (1900) Ueber den *Bacillus acidophilus*. *Jahrbuch für Kinderheilkunde und Physische Erzihung*, **52**, 38–55.
Nuttal, G.H.F. & Thierfelder, H. (1895) Theirisches leben ohne bakterien im verdauungskanal. *Zeitschrift für Physiolische Chemie*, **21**,109–121.
Ortweiller, L. (1886) Ueberie physiologische bedeutung des harnindicans. *Mittheilungen auss der Medicinischen Klinik zu Wüerzburg*, **2**, 153–188.
Parker, R.B. (1974) Probiotics: the other half of the antibiotic story. *Animal Nutrition and Health*, **29**, 4–8.
Pasteur, L. (1878) Lecture before the French Academy of Sciences, 29 April. *Comptes Rendus de l'Acadamie des Sciences*, **86**, 1037–1043.
Poehl, A.W. (1887) Bestimmung der darmfäulniss durch untersuchung des hams. *Jahres-Bericht über die Fortschitte der Thier-Chemie*, **17**, 277.
Reid, G. & Friendship, R. (2002) Alternatives to antibiotic use: probiotics for the gut. *Animal Biotechnology*, **13**(1), 97–112.
Rettger, L.F. & Cheplin, H.A. (1921a) *A Treatise on the Transformation of the Intestinal Flora with Special Reference to B.* acidophilus. Yale University Press, New Haven.
Rettger, L.F. & Cheplin, H.A. (1921b) Therapeutic applications of *Bacillus acidophilus*. *Proceedings of the Society of Experimental Biology and Medicine*, **19**, 72–76.
Rettger, L.F., Levy, M.N., Weinstein, I. & Weiss, J.E. (1935) *Lactobacillus acidophilus and its Therapeutic Applications*. Yale University Press, New Haven.
Rotch, T.M. & Kendall, A.J. (1911) A preparatory study of the *Bacillus acidophilus* in regard to its possible therapeutic use. *American Journal of Diseases of Children*, **2**, 30–38.
Rovighi, A. (1892) Die aetherschwefelsäuren im harn und die darmdes-infection. *Zeitschrift für Physiologische Chemie*, **16**, 20–46.
Senator, H. (1868) Ueber einen fall von hydrothionämie und über slebstinfectin durch abnorme verdauungsvorgänge. *Klinische Wochenschrift (Berlin)*, **5**, 254–256.

Shirota, M. (1937) Live bacterial therapy. *Shakai Eisei*, **2**, 1–5.

Solukha, I.P. (1896) *Beitrag zur Frage des Einflusses des milchzuckers auf den Eiwesstoffwechsel und die Darmfaulniss bei Gesunden Menschen*. Dissertation, St. Petersburg University.

Spiegel, L. (1911) Natur und Bedeutung des Yogurt. *Reichs-Medizinal-Anzeiger (Leipzig)*, **36**, 663.

Strasburger, J. (1902) Untersuchungen uber die bakterienmenge im menschlichen fäces. *Zeitschrift für Klinische Medizin*, **16**, 311–324.

Tauber, A.I. (1992) The birth of immunology – III. The fate of the phagocytosis theory. *Cell Immunology*, **139**, 505–530.

Tissier, H. (1900) *Recherches sur la Flore intestinale Normale et Pathologique du Nourison (état Normal et Pathologique)*. Thesis, G. Carré et C. Naud Publishers, Paris.

Tissier, H. & Gasching, P. (1903) Recherches sur la fermentation du lait. *Annals de l'Institut Pasteur*, **17**, 540–563.

Tissier, H. & Martelly, A. (1902) Recherches sur la putrefaction de la viande de boucherie. *Annals de l'Institut Pasteur*, **16**, 865–903.

Torrey, T.C. (1915) The fecal flora of typhoid fever and its reaction to various diets. *Journal of Infectious Diseases*, **16**, 72–108.

Tuohy, K.M., Pinart-Gilberga, M., Jones, M., Hoyles, L., McCartney, A.L. & Gibson, G.R. (2007) Survivability of a probiotic *Lactobacillus casei* in the gastrointestinal tract of healthy human volunteers and its impact on the faecal microflora. *Journal of Applied Microbiology*, **102**, 1026–1032.

Vergin, F. (1954) Antibiotics and probiotics (in German). *Hippokrates*, **25**, 115–119.

Wassilieff, N.P. (1882) Ueber die wirkung des calomel auf gährungsprozesse und das leben von mikroorganismen. *Zeitschrift für Physiologische Chemie*, **6**, 112–134.

Weiss, H. (1904) Zur kentniss der darmlflora. *Zentralblatt für Bakteriologie*, **36**, 13–18.

Winterberg, H. (1898) Zur methodik der bakterienzählung. *Zeitschrift für Hygiene*, **29**, 75–93.

Yuki, N., Watanabe, K., Mike, A., Tagami, Y., Tanaka, R., Ohwaki, M. & Morotomi, M. (1999) Survival of a probiotic, *Lactobacillus casei* strain Shirota in the gastrointestinal tract: selective isolation from faeces and identification using monoclonal antibodies. *International Journal of Food Microbiology*, **48**, 51–57.

Zhao, T., Doyle, M.P., Harmon, B.G., Brown, C.A., Mueller, P.O. & Parks, A.H. (1998) Reduction of carriage of enterohemorragic *Escherichia coli* O157:H7 in cattle by inoculation with probiotic bacteria. *Journal of Clinical Microbiology*, **36**, 641–647.

3 Genomic Characterisation of Starter Cultures and Probiotic Bacteria

G.E. Felis, S. Torriani, A.B. Flórez and B. Mayo

3.1 Introduction

The first publicly available genome sequence of a lactic acid bacterium strain commonly used as a starter in dairy fermentations is that of *Lactococcus lactis* subsp. *lactis* IL1403 (Bolotin *et al.*, 2001). Since then, a huge number of genome sequences of starter and probiotic lactic acid bacteria (LAB) has been determined. Indeed, genome sequencing has become the starting point of most microbial studies, since it can provide, at a reasonable cost, the framework of the biology of any strain of interest. Comparative genome analyses can provide insights into the genetic make-up, and thus the metabolic potential, of the strains, as well as into many aspects of their genome evolution and divergence. Furthermore, the sequencing of genomes can reveal horizontal gene transfer (HGT) events between species and strains of starter and probiotic cultures.

Knowledge of the complete genome sequence is useful not only in discovering novel genes and new properties, but also in determining the absence at the genetic level of undesirable traits. In addition, it is of much help in characterising the variants of modified or improved strains for both cultures with a technological role (i.e. necessary/useful for food fermentation processes) and a functional role (i.e. able to confer health benefits to the consumer/host) (FAO/WHO, 2001; Hill *et al.*, 2014). Genome sequencing can further allow precise identification down to the strain level, while providing strategies for its specific detection and quantification: important aspects for commercially relevant bacteria.

As an example, comparative sequence analysis led, at the beginning of this century, to the recognition of clustered regularly interspaced short palindromic repeats (CRISPR) and their CRISPR-associated (*cas*) genes (Jansen *et al.*, 2002). In slightly longer than a decade, these rudimentary immune systems of bacteria have also been identified in several industrial LAB strains (Horvath *et al.*, 2009). CRISPR-*cas* systems seem to be related to the prevention of lateral HGT processes which are involved in, for instance, phage resistance (Barrangou & Horvath, 2012), and thus they are considered as desirable traits for strain components of starter cultures. Moreover, CRISPR-*cas* loci could be used as a strain-level identification tool (Barrangou & Horvath, 2012), and have become a method for genome editing in a variety of genetic engineering applications (Barrangou, 2014).

Probiotic Dairy Products, Second Edition. Edited by Adnan Y. Tamime and Linda V. Thomas.

In general, comparative genomics (i.e. the comparison of genome sequence data for many strains, which can be performed with dedicated software platforms and using specific databases) is a very important strategy to reveal the peculiarities of specific strains or groups of strains. Therefore, genome sequencing efforts continuing to date on one hand might provide redundant information but, on the other hand, are necessary to continuously improve our knowledge of microbial biodiversity.

Since the first edition of this book (Dellaglio *et al.*, 2005), several important reports have been published that have provided new data on genome analyses of strains of interest for the food industry. First, in 2006, the comparative genomics of a bunch of LAB strains belonging to several genera and species were published (Makarova *et al.*, 2006) and, more relevant for probiotics, the term 'probiogenomics' was introduced by Ventura *et al.* (2009) to indicate that genome sequence analyses could provide insights into the genetic background related to the beneficial properties of probiotic organisms (as discussed in this chapter); this has been further investigated by Lukjancenko *et al.* (2011). Genome-wide analyses can indeed provide scientists with clues about the genetic background of the ability of strains to sense and adapt to their ecological niche where they exert specific functions (Siezen & Wilson, 2010), or it can reveal the genetic background relating to physiological properties of specific interest for industrial applications, as in the case of the proto-cooperation between *Streptococcus thermophilus* and *Lactobacillus delbrueckii* subsp. *bulgaricus* for yoghurt production (Hao *et al.*, 2011). This approach could be applied to all starter cultures (Garrigues *et al.*, 2013), making it possible to match the phenotypic properties and genetic make-up of strains, opening the way for targeted strain improvements as well as being the basis for in-depth safety assessments. In fact, the term 'pangenomics', namely the analysis of the pan-genome of a species that is devised from the genome sequences of different strains of the same species (Medini *et al.*, 2005), implies the importance of comparative analyses. In this respect, the GOLD database (i.e. the Genomes OnLine Database: http://www.genomesonline.org), currently at its fifth version (Reddy *et al.*, 2015), is like a gold mine, since it is a comprehensive online and regularly updated resource, cataloguing and monitoring ongoing genome studies worldwide. Information on genomes (and metagenomes) present in the database is classified in a four-level system: (a) *Studies*: which group one or more related organism, (b) *Biosamples*: individual samples of genetic material, from which the organism DNA has been isolated for downstream sequencing, and which are analogous to the BioProjects in the National Center for Biotechnology Information (NCBI), (c) *Sequencing projects*: the sequencing deliverables from the Biosamples, and (d) *Analysis projects*: methods of data processing which are applied to sequencing projects (Reddy *et al.*, 2015), all in compliance with the indications of the Genomic Standards Consortium about minimum information standards (Field *et al.*, 2011; Yilmaz *et al.*, 2011).

Performing a GOLD database search, when 'food industry' is selected among the filters for the relevance of the study, the website (in October 2016) showed that 2578 sequencing projects were ongoing worldwide for 2537 organisms, mainly related to the domain Bacteria.

Within bacteria, the most important species for dairy applications are of course LAB strains belonging to the genera *Lactococcus* and *Lactobacillus*, together with the species *Str. thermophilus*, which all belong to the phylum Firmicutes. Several LAB species also include probiotic strains, together with many strains of *Bifidobacterium* spp., this latter genus belonging to the phylum Actinobacteria. This has led to the accumulation of sequence data, details of which are shown in Table 3.1. It has to be pointed out that suppliers of commercial and probiotic cultures around the world may have additional data on larger numbers of industrial proprietary strains of these genera and species.

In the last 2 years, two important published papers have appeared reporting the efforts to determine the genome sequences of the type strains of *Bifidobacterium* (Milani *et al.*, 2014) and *Lactobacillus* species (Sun *et al.*, 2015), as well as LAB strains of

Table 3.1 Summary of genome sequences available to date for dairy starter cultures and probiotic strains retrieved using species names as organisms' names; Studies, Organisms, Sequencing projects and Analysis projects correspond to the four-level classification system used in the GOLD database[1] (see text for details).

Microbial species and subspecies	Number of studies performed	Number of micro-organisms tested	Sequencing projects	Analysis projects
Bifidobacteria				
Bifidobacterium animalis subsp. *animalis* and *Bifidobacterium animalis* subsp. *lactis*	26	192	35	28
Bifidobacterium bifidum	18	198	32	25
Bifidobacterium breve	22	185	42	35
Bifidobacterium longum subsp. *longum* and *Bifidobacterium longum* subsp. *infantis*	40	439	107	75
Lactobacilli				
Lactobacillus acidophilus	15	388	26	19
Lactobacillus casei	26	389	44	35
Lactobacillus delbrueckii subsp. *delbrueckii*, *Lactobacillus delbrueckii* subsp. *bulgaricus* and *Lactobacillus delbrueckii* subsp. *lactis*	20	837	30	29
Lactobacillus fermentum	17	400	20	17
Lactobacillus gasseri	12	361	26	17
Lactobacillus helveticus	19	297	30	22
Lactobacillus johnsonii	10	61	12	11
Lactobacillus paracasei	15	581	52	49
Lactobacillus plantarum	70	1106	118	100
Lactobacillus reuteri	15	184	20	17
Lactobacillus salivarius	18	119	20	15
Miscellaneous				
Lactococcus lactis subsp. *lactis* and *Lactococcus lactis* subsp. *cremoris*	53	1632	95	89
Streptococcus thermophilus	34	330	34	31

[1] https://gold.jgi.doe.gov.
Note: Data compiled from Reddy *et al.* (2015).

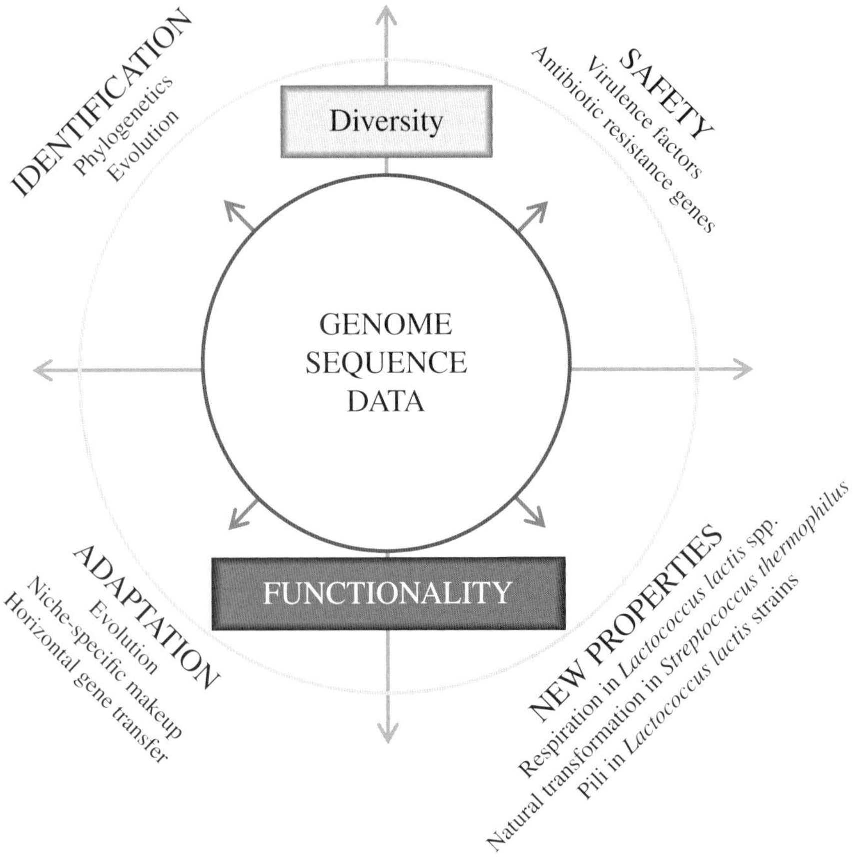

Figure 3.1 Compass rose of different levels and aspects of knowledge provided by genome sequence and data analysis.

related genera, including those relevant for dairy applications. This constitutes the framework for genome-based analyses at various taxonomic levels and provides a reference scheme to investigate the physiology, biochemistry, evolution and diversity of dairy-related starters and probiotics.

Since information on identification will be reported elsewhere in this book (see Chapter 6), here we report on the insights obtained from genome sequence data, on diversity and evolution at taxonomic ranks above the species level, as well as on safety aspects and on specific technologically and functionally important features (Figure 3.1).

3.2 Genome sequencing and comparative genomics: insights into evolution and adaptation to dairy environments

The most recent information gained from genome sequence analyses on the evolution and adaptation to dairy environments of the main microbial genera and species will be briefly reviewed. Due to the large amount of information available, an effort of synthesis and selection of information has been made. Information is structured following a taxonomic framework and focusing mainly on Firmicutes first, which includes *Lactococcus lactis* subsp. *lactis* and *Lactococcus lactis* subsp. *cremoris*, the most frequently employed starter cultures, and *Str. thermophilus*, as well as lactobacilli, which could be used as either starter, adjunct or probiotic cultures (Kelleher *et al.*, 2015). Moreover, some information on Actinobacteria, in particular bifidobacteria and dairy propionibacteria, will be provided.

It can be emphasised that adaptation to milk, which is a nutritionally rich medium containing carbohydrates, proteins, fats, vitamins and minerals at a neutral pH (Marshall, 1991), seems to have generally determined trends of loss or inactivation (mostly by mutation) of genes encoding non-essential properties for growth in milk. The loss of some phenotypic traits has further been complemented by the acquisition through horizontal transfer of key genes coding for critical functions in the milk environment, for example lactose and casein utilisation, phage resistance, exopolysaccharide (EPS) production and so on.

3.2.1 *Phylum Firmicutes*

Family Streptococcaceae: Lactococcus *and* Streptococcus

The taxonomic family of Streptococcaceae includes the three genera *Lactococcus*, *Lactovum* and *Streptococcus*, which share a high degree of sequence similarity (considering 16S rRNA gene sequences), and these can be found in a variety of environmental niches, including the dairy-related environments (Cavanagh *et al.*, 2015).

Lactococcus lactis spp. and *Str. thermophilus* constitute the most economically important dairy species. Consequently, they have been investigated in several studies from a genomic viewpoint; the impact of genome sequencing for the selection of dairy starters in these two taxa has also been recently reviewed (Kelleher *et al.*, 2015). A common element in industrial dairy Streptococcaceae genomes is the process of gene loss, along with a relatively high percentage of pseudogenes caused by nonsense mutations, deletions, truncations and/or frameshifts, which are thought to contribute to cell economy, and thus to milk adaptation (Cavanagh *et al.*, 2015).

At present, 11 species are recognised in the genus *Lactococcus* (http://www.bacterio.net/lactococcus.html). Of these, the most important is *Lac. lactis* spp., although taxonomically speaking, this species is divided into four subspecies: *Lac. lactis* subsp. *cremoris*, *Lac. lactis* subsp. *lactis*, *Lactococcus lactis* subsp. *hordniae* and *Lactococcus lactis* subsp. *tructae* (Pérez *et al.*, 2011), with the former two being the only subspecies of commercial relevance in the dairy industry. The evolution and adaptation of *Lac. lactis* strains to milk have been recently reviewed by Cavanagh *et al.* (2015), and will be briefly updated here. Besides *Lac. lactis* spp., there is a long list of newly described species, including *Lactococcus chungangensis*, *Lactococcus formosensis*, *Lactococcus fujiensis*, *Lactococcus garvieae*, *Lactococcus hircilactis*, *Lactococcus laudensis*, *Lactococcus piscium*, *Lactococcus plantarum*, *Lactococcus raffinolactis* and *Lactococcus taiwanensis*.

Since October 2016, 85 genomes for different strains of *Lac. lactis* subsp. *lactis* and *Lac. lactis* subsp. *cremoris* have been deposited in the GenBank database (https://www.ncbi.nlm.nih.gov/genome/?term=Lactococcus+lactis). Genomes of lactococci are usually relatively small, ranging from ~2.2 to 2.5 Mb (Kelleher *et al.*, 2015), with a protein-encoding gene content between ~2000 and ~2800 (Makarova *et al.*, 2006). At the genus level, the analysis of five *Lac. lactis* spp. genomes, 1 *Lac. raffinolactis* genome and 19 *Lac. garvieae* genomes has shown that 70% of the genes of the latter species were shared with other lactococci, constituting, most probably, the core genome of the genus (Ferrario *et al.*, 2013). In most strains, chromosomes are often complemented by a large array of 4–7 plasmids (Kelleher *et al.*, 2015). In fact, dairy lactococci of industrial use are characterised by the presence of large plasmids carrying genes for all pivotal properties of technological significance mentioned here (lactose utilisation and casein breakdown, bacteriophage resistance etc.) (Siezen *et al.*, 2005; Ainsworth *et al.*, 2014).

Dairy starters of *Lac. lactis* strains have been defined as 'domesticated' (Passerini *et al.*, 2010). Most probably, these domesticated strains derive from plant-associated micro-organisms (Kelly *et al.*, 2010) that became adapted to the dairy environment. In fact, adaptation has led to a reductive evolution of their genomes, mainly with respect to the biosynthesis of amino acids and the ability to ferment plant-derived carbohydrates (Kelly *et al.*, 2010; Cavanagh *et al.*, 2015). Comparison of a dairy strain and a sourdough strain, *Lac. lactis* subsp. *lactis* IL1403 versus *Lac. lactis* subsp. *lactis* A12, revealed that about a quarter of the *Lac. lactis* subsp. *lactis* A12 genes were strain-specific and mainly responsible for niche specialisation (Passerini *et al.*, 2013), indicating the importance of the 'dispensable genome' (i.e. of genes present only in a particular strain or group of strains). These genes are most commonly associated with phages, transposons/mobile elements and plasmids, and could be horizontally transferred, generating diversity within the species. Interestingly, lactose utilisation and citrate metabolism (the latter is a characteristic of *Lactococcus lactis* subsp. *lactis* biovar. *diacetylactis*) are plasmid-encoded traits in *Lac. lactis* strains (Van Hylckama Vlieg *et al.*, 2006; Kelleher *et al.*, 2015). This reductive evolution seems to be common not only to the other industrially important dairy bacterium of the same taxonomic family, *Str. thermophilus* (Hols *et al.*, 2005), but also to *Lb. delbrueckii* subsp. *bulgaricus* (van de Guchte *et al.*, 2006).

As already mentioned elsewhere, lactose utilisation in starter strains of *Lac. lactis* spp. is plasmid encoded and could be unstable (Ainsworth *et al.*, 2014). At least two alternatives for lactose metabolism have been described in this species, showing that phenotypic growth on lactose is not always exclusive of strains bearing a plasmid-encoded *lac* operon. These alternatives can easily be revealed by whole genome sequencing, demonstrating that this technique is more appropriate to completely characterise the strains compared to gene-specific detection and analysis (Kelleher *et al.*, 2015).

As for *Str. thermophilus*, to date 30 genome sequences are available in GenBank (https://www.ncbi.nlm.nih.gov/genome/?term=Streptococcus+thermophilus), and the species belongs to a genus which includes 85 species including several pathogens (http://www.bacterio.net/streptococcus.html). In more detail, *Str. thermophilus* belongs to the *salivarius* group of streptococci, which also includes *Streptococcus salivarius* and *Streptococcus vestibularis*, two commensals that may occasionally cause opportunistic infections in humans (Delorme *et al.*, 2015). Despite the genetic relatedness of these three species, they can be readily distinguished by multilocus sequence typing (MLST) and comparative genome analysis. These techniques confirm the status of separate species for *Str. thermophilus*, which was temporarily classified as a subspecies of *Str. salivarius* (Schleifer *et al.*, 1991). In fact, according to recent genome analyses, *Str. thermophilus* seems to have evolved relatively recently, from a common ancestor of *Str. vestibularis*, in parallel with the development of the agriculture (Delorme *et al.*, 2015).

Streptococcus thermophilus also possesses a relatively small genome (about 1.7–1.8 Mb), compared to strains of the other two species, and its genome is characterised by a relatively high number of pseudogenes, which is in line with its adaptation to milk (Bolotin *et al.*, 2004; Delorme *et al.*, 2015). Moreover, as observed for *Lac. lactis* strains, additional genes coding for thermotolerance are frequently carried on plasmids (Kelleher *et al.*, 2015). Regarding lactose metabolism, so important for milk adaptation, it differs from that of *Lac. lactis* starter culture strains, as lactose utilisation in *Str. thermophilus* is more similar to that of *Lactobacillus delbrueckii* subsp. *bulgaricus* (Kelleher *et al.*, 2015).

Interestingly, two other species of the genus *Streptococcus*, belonging to a distinct phylogenetic group, the *Streptococcus bovis–Streptococcus equinus* complex [i.e. *Streptococcus gallolyticus* subsp. *macedonicus* (Schlegel *et al.*, 2003; Whiley & Kilian, 2003) and *Streptococcus infantarius* subsp. *infantarius* (Jans *et al.*, 2013)], appear to be dominant in specific fermented (dairy) products (Papadimitriou *et al.*, 2014). In particular, *Str. gallolyticus* subsp. *macedonicus* seems to have all the characteristics indicative of reductive evolution: reduced genome size, increased number of pseudogenes as well as loss of genes and pathways related to plant carbohydrates catabolism. In addition, its genome make-up includes, among others, extra genes for lactose and galactose metabolism as well as casein hydrolysis, which supports once again its adaptation to milk (Papadimitriou *et al.*, 2014). However, data reported by Papadimitriou *et al.* (2014) suggest that neither *Str. gallolyticus* subsp. *macedonicus* nor *Str. infantarius* subsp. *infantarius* are specialised dairy bacteria like *Str. thermophilus*, and most probably represent intermediate evolutionary stages that could be compared to the ancestors of *Str. thermophilus*, and not to the highly dairy-specialised species that we know today.

Families Lactobacillaceae *and* Leuconostocaceae

Members of these two taxonomic families include strains that may be used as starter and adjunct cultures for dairy applications (Kelleher *et al.*, 2015).

With regards to the genus *Lactobacillus*, at present it includes about 180 validly described species (http://www.bacterio.net/lactobacillus.html), while 14 are described in the genus *Leuconostoc* (http://www.bacterio.net/leuconostoc.html). Recently, a genome-sequencing initiative has been completed focusing on 185 type strains of species of 11 genera related to the genus *Lactobacillus*, including *Oenococcus* and *Leuconostoc* genomes (Sun *et al.*, 2015). The report showed that, when analysing different parameters such as average nucleotide identity (ANI) and total nucleotide identity (TNI), the genus *Lactobacillus* appears to be more diverse than a well-defined 'standard' taxonomic family (Sun *et al.*, 2015). Moreover, the genus is paraphyletic (i.e. its evolutionary origin is not unique), and the genera *Pediococcus*, *Weissella*, *Leuconostoc*, *Oenococcus* and *Fructobacillus* appear to be subclades within the 'real' lactobacilli. For those reasons, therefore, the name *Lactobacillus* Genus Complex (LGC), which would include all these genera, has recently been proposed (Sun *et al.*, 2015). This implies that the separation of these bacteria into two families, namely *Lactobacillaceae* and *Leuconostocaceae*, should be revised. In contrast, the family *Streptococcaceae* appears clearly separated from the LGC.

More than 900 genome sequences are available to date (NCBI Genome search with '*Lactobacillus*[orgn]' as a query, October 2016). Of these, *Lactobacillus rhamnosus* and *Lactobacillus plantarum* strains account for 225 sequences (104 and 121, respectively). Due to the remarkable diversity of the genus, these sequences are very variable in size and differ in many genetic properties (the presence or absence of plasmids, rRNA operons, number of coding sequences etc.). Sun *et al.* (2015) reported that the smallest and the largest genomes for lactobacilli were 1.23 Mb (*Lactobacillus sanfranciscensis*) and 4.91 Mb (*Lactobacillus parakefiri*), respectively. Furthermore, the genome guanine+cytosine (GC) content appears to be very variable (between 31.93 and 57.02%) among 213 strains tested, and the core genome very small, estimated as 73 genes. Remarkably, the pangenome appears to be very large (44 668 gene families) and is continuously increasing as novel genomes are added (Sun *et al.*, 2015).

Among the dairy-relevant species, the smallest genome appeared to be that of *Lb. delbrueckii* spp. (about 1.8 Mbp), and the largest that of *Lb. plantarum* (about 3.2 Mbp) (Stefanovic *et al.*, 2017). Comparative genomics analyses have revealed that a mobile insertion element (IS), IS91, seems to be dairy specific (found in dairy *Lactobacillus casei* and *Lactobacillus paracasei* subsp. *tolerans*) (Sun *et al.*, 2015).

The most important dairy starter cultures among the lactobacilli belong to *Lactobacillus delbrueckii* spp. and *Lactobacillus helveticus*. These two species belong to the same phylogenetic group, *Lb. delbrueckii* strains which can also be considered the genus *Lactobacillus*, *sensu stricto*. Other useful adjunct cultures, like *Lb. casei*, *Lactobacillus paracasei* subsp. *paracasei*, *Lb. plantarum* and *Lb. rhamnosus*, are phylogenetically and metabolically distinct from the *Lb. delbrueckii* clade. Notably, *Lb. casei*, *Lb. paracasei* subsp. *paracasei* and *Lb. rhamnosus* form a group of closely related facultatively heterofermentative species, while *Lb. plantarum* belongs to another phylogenetic clade and is most similar to *Lactobacillus fabifermentans*, *Lactobacillus paraplantarum*, *Lactobacillus pentosus* and *Lactobacillus xiangfangensis* (Sun *et al.*, 2015). Probiotic strains most commonly associated with dairy products belong either to *Lactobacillus acidophilus*, a species within the *Lb. delbrueckii* clade, or to *Lb. rhamnosus* (Stefanovic *et al.*, 2017). However, other relevant strains belonging to *Lactobacillus fermentum*, *Lactobacillus gasseri*, *Lactobacillus johnsonii*, *Lb. plantarum*, *Lactobacillus reuteri* and *Lactobacillus salivarius* species are also used for the same purposes (Table 3.1).

Dairy-specialised strains have shown an abundance of genes encoding components of proteolytic systems, as well as sugar and amino acid transporters. These might result from duplications, which make it possible for organisms to uptake nutrients from milk, while substantial gene loss has been described for coding sequences not necessary in dairy ecosystems (Stefanovic *et al.*, 2017). Information on dairy-related genes is not always consistent among studies (O'Sullivan *et al.*, 2009; Kant *et al.*, 2010), but the studies by Broadbent *et al.* (2012) and Smokvina *et al.* (2013) on strains of *Lb. casei* group (*Lb. casei* and *Lb. paracasei* subsp. *paracasei*, respectively) have shown that dairy-specialised strains undergo reduction of the genome, which appears to be associated to a restricted capacity of carbohydrate degradation. In contrast, strains that are able to survive in dairy products but are not dairy specialists could contain genes useful for conferring flexibility as well as ability to proliferate in different niches, such as those of the gut or the environment (Senan *et al.*, 2015a).

In summary, dairy adaptation has had deep consequences at the genomic level and, although there has been a general trend towards genome reduction with respect to several functions, strains belonging to different species could show different sets of genes, either vertically or horizontally transmitted, that account for their performance in milk fermentation.

3.2.2 *Phylum Actinobacteria*

Genus Bifidobacterium *and related genera*

The genus *Bifidobacterium* includes bacterial strains highly relevant for their health-promoting properties. At present, the genus comprises 50 validly described species (http://www.bacterio.net/bifidobacterium.html), and the most important for the probiotic perspective are *Bifidobacterium animalis* subsp. *animalis*, *Bifidobacterium animalis* subsp. *lactis*, *Bifidobacterium bifidum*, *Bifidobacterium breve*, *Bifidobacterium longum* subsp. *longum* and *Bifidobacterium longum* subsp. *infantis* (Mattarelli *et al.*, 2008).

In the last 2 years, the sequencing initiative of the type strains for the to-date-described species has been completed (Milani *et al.*, 2014), and a comparative genome analysis on 62 genome sequences of Bifidobacteriales, the taxonomic order in which the genus *Bifidobacterium* is included, has been published by Zhang *et al.* (2016). The extensive and detailed comparative analysis performed by Milani *et al.* (2014) showed a general trend of horizontal gene gain during evolution, involving mainly determinants for complex carbohydrate transport and degradation, which are most probably related to their adaptation to the gut niche. The analysis at the order level (Zhang *et al.*, 2016) also revealed that, inside the family Bifidobacteriaceae, the genera *Bifidobacterium* and *Gardnerella* are not clearly separated; they appear more related to each other than to other genera (i.e. *Scardovia*, *Parascardovia* and *Alloscardovia*). Moreover, confirming previous findings, several phylogenetic groups can be devised within the genus *Bifidobacterium*, and species of human probiotic interest were spread among three groups: the *Bif. longum* group *Bif. breve* and '*Bif. longum*' – presumed to be *Bif. longum* subsp. *longum* and subsp. *infantis*), *Bif. bifidum* group (*Bif. bifidum*) and *Bifidobacterium pseudolongum* group (*Bif. animalis* subsp. *animalis* and subsp. *lactis*). Remarkably, the authors highlighted the presence of three *Bifidobacterium*-specific conserved proteins (conserved signature proteins, or CSPs) and two CSPs specific for the genera *Bifidobacterium* and *Gardnerella*, which all encode for unknown functions. This highlights the usefulness of genome sequencing and comparative analyses to reveal gene targets that are probably crucial for the biology of the micro-organism but, at present, completely unknown. Besides those evolutionary considerations and insights into the biology of organisms yet to be discovered, complete genome sequences have also been the basis for the definition of an accurate polymerase chain reaction (PCR)-based identification protocol (Ferrario *et al.*, 2015).

Genus Propionibacterium

Propionibacteria are Gram-positive, non-spore-forming, non-motile rods with high GC content (between 53 and 68%), and they include at present 14 species (http://www.bacterio.net/propionibacterium.html). Within the genus, several species, namely *Propionibacterium freudenreichii* (subsp. *freudenreichii* and subsp. *shermanii*), *Propionibacteriu acidipropionici*, *Propionibacterium jensenii* and *Propionibacterium thoenii*, represent the dairy propionibacteria (Poonam Pophaly *et al*., 2012); these can be used as starter and/or adjunct cultures. The first genome sequence of a *Propionibacterium* strain was only determined in 2010 (Falentin *et al*., 2010); it consisted of 2.7 Mb with 67% GC content. Six years later, 187 genome sequences of propionibacteria have been deposited in GenBank (https://www.ncbi.nlm.nih.gov/genome/?term=Propionibacterium+%5Borgn%5D). Of these, a large majority (120) focus on *Propionibacterium acnes* and therefore are not relevant for the topic of this chapter. The relatively low number of sequences from dairy propionibacteria suggests that their study is far behind that of other dairy-associated genera. However, since propionibacteria could produce antimicrobial compounds (bacteriocins and organic acids) as well as vitamins of the B group, conjugated linoleic acid, trehalose, propionic acid and bifidogenic factors (Poonam Pophaly *et al*., 2012), it can be expected that the study of this group of bacteria could progress in the coming years.

3.2.3 *Other micro-organisms*

Some other starters and, in particular, probiotic micro-organisms could further be considered, including strains belonging to other LAB genera and species, as well as strains from other bacterial species and yeasts. High numbers of *Enterococcus* spp. and *Weissella* spp. have been repeatedly reported in many fermented foods, particularly in cheese (Franz *et al*., 2011; Abriouel *et al*., 2015). However, strains of the genus *Enterococcus* have been proposed as starters (Giraffa, 2003), and some others (such as Symbioflor 1) have been used as probiotics since the 1950s (Domann *et al*., 2007). However, enterococci strains, in particular *Enterococcus faecalis* and *Enterococcus faecium*, have been largely considered to be opportunistic pathogens. Their dual-faced status might be based on a few set of genes whose absence needs to be demonstrated in species of this genus to be used in food and feed (EFSA, 2012). Strains of other species, such as *Escherichia coli* (Nissle 1917), *Bacillus subtilis* (Enterogermina), *Bacillus cereus* (Toyoi) and others, have been widely used as human and animal probiotics. Further micro-organisms could also include moulds, such as *Geotrichum candidum*, which has many different metabolic pathways that are of particular interest to the dairy industry (Boutrou & Guéguen, 2005), and only one yeast (*Saccharomyces cerevisiae* var. *boulardii*) that has been approved as a probiotic for human consumption (Hatoum *et al*., 2012). Representative strains of all these species have already been genome sequenced; however, for the sake of clarity and brevity, in this section we have focused our attention on the typical LAB genera and species.

3.3 Application of genome analysis to LAB and bifidobacteria

The availability of genome sequences has significantly increased our ability to unravel the fermentation pathways and biochemical routes of LAB and bifidobacteria species involved in industrial and probiotic applications (Klaenhammer *et al.*, 2005; Stanton *et al.*, 2005). Comparative genomics is now being used to find core genes, niche-specific genes and genes linked to specific probiotic traits, but can also be used to find or exclude the presence of virulence or antibiotic resistance genes (see Figure 3.1) or to find indications of chromosomal integration of horizontally acquired DNA, which could indicate the potential of HGT harbouring such undesirable traits. The knowledge gathered through genome analysis will ultimately allow full exploitation of the biotechnological potential of LAB and bifidobacteria, facilitating, at the same time, their genetic manipulation. Engineering of LAB and bifidobacteria is pivotal in the discovery of gene functionality of key phenotypic traits, as well as in the use of model strains as a cell factory for the expression of heterologous proteins (Hanniffy *et al.*, 2004), the synthesis of food-grade additives and nutraceuticals (Hugenholtz *et al.*, 2002) and the use of these bacteria as vaccine and therapeutic delivery systems (Wells & Mercenier, 2003). Genome sequence availability has recently allowed the replacement of former conventional genetic engineering techniques (Gasson & de Vos, 2004) by state-of-the-art genome editing methods, such as the single-stranded DNA recombination-mediated genetic engineering (ssDNA recombineering) (van Pijkeren & Britton, 2014), and the CRISPR-*cas* mentioned in this chapter (Oh & van Pijkeren, 2014).

3.3.1 In silico *safety assessment of LAB and bifidobacteria*

In silico safety assessment of starter and probiotics strains encompasses several aspects, such as the absence of transmissible antibiotic resistance (AR), and the genetic make-up for virulence factors (VFs) and other deleterious characteristics, required to be met by regulatory agencies, such as those of the European Food Safety Authority (EFSA, 2013).

Determination of the presence or absence of these genetic traits has become more rapid and cost-effective thanks to the growing availability of whole genome sequences and the development of bioinformatics tools (Alkema *et al.*, 2015). Thus, such determination could be considered as an indispensable criterion for (pre-)selection of strains with industrial and probiotic properties. Indeed, a deep and complete safety evaluation is required not only for each novel strain but also for strains that have been in use over a long period of time. As a consequence, whole genome analyses could serve as the standard in the safety evaluation process, and influence regulatory decisions regarding the commercial acceptability of the strain (Salvetti *et al.*, 2016).

In the last few years, a limited number of papers have been published regarding the safety assessments of LAB and *Bifidobacterium* strains based on the complete genome sequences. A summary of such studies, which have considered mainly strains used or proposed to be used as probiotics, is reported in Table 3.2.

Bennedsen *et al.* (2011) were the first to perform a screening for antimicrobial resistance (AR) genes and VFs via genome sequencing. Twenty-eight strains of LAB and

Table 3.2 Summary of studies on genome-based assessment of safety for lactic acid bacteria and *Bifidobacterium* strains.

Species	Strain	Origin	Genes of concern	References
Bifidobacterium animalis subsp. *lactis*	ATCC 27536	Chicken faeces	*tet*(W)	Bennedsen *et al.* (2011)
	CHCC13471	Food/natural source	*tet*(W)	
	IPLAIC4	Fermented milk	*tet*(W)	
Bifidobacterium longum subsp. *longum* and *Bifidobacterium longum* subsp. *infantis*	JDM301	Chinese commercial probiotic product	162 non-specific VF; 36 AR including *tet*(V), *tet*(W), *tet*(PB), *tet*(Q); 5 AM	Wei *et al.* (2012)
	CECT 7347	Infant faeces	No VF or AR	Chenoll *et al.* (2013)
Lactococcus lactis subsp. *lactis*	CHCC6005	Food/natural source	*tet*(S)	Bennedsen *et al.* (2011)
Leuconostoc mesenteroides subsp. *dextranicum*	LbE15	Italian cheese	*erm*(B)	Flórez *et al.* (2016)
	LbE16	Italian cheese	*tet*(S), *aadE*, *aphA-3*, *sat4*, *vat*(E)	
Lactobacillus plantarum	JDM1	Chinese commercial probiotic product	126 non-specific VF, 51 AR, 23 AM	Zhang *et al.* (2012)
Lactobacillus helveticus	MTCC 5463	Indian – vaginal tract	44 non-specific VF, 32 AR, 8 AM	Senan *et al.* (2015b)
Streptococcus salivarius	NU10 and YU10	Malaysian subjects	No VF	Barbour & Philip (2014)
Enterococcus faecium	NRRL B-2354	Dairy utensils	*scm*, *sagA*, *efaA*, *pilA*	Kopit *et al.* (2014)
Weissella cibaria	KACC 11862	Kimchi – Korean fermented product	2 VF, 4 AR	Abriouel *et al.* (2015)
Weissella confusa	LBAE C39-2	French wheat sourdough	4 VF, 2 AR, *vanZ*	

AM = adverse metabolites; AR = antibiotic resistance; VF = virulence factors.

bifidobacteria were tested for the presence of >250 AR genes and >400 toxin and VF genes. It was found that some strains contained AR genes, while no VF genes were detected. In particular, *Lac. lactis* subsp. *lactis* CHCC6005, showing high-level resistance to tetracycline (MIC >256 μg mL^{-1}), carried the gene *tet*(S) on a medium-copy-number plasmid and, therefore, the authors indicated that this strain should be plasmid-cured before use. The gene *tet*(W), encoding resistance to tetracycline, has been detected in all three *Bif. animalis* subsp. *lactis* strains analysed. This determinant is widespread in *Bif. animalis* subsp. *lactis*; however, since transfer of *tet*(W) from *Bif. animalis* subsp. *lactis* to other bacteria has never been demonstrated, this gene can be considered non-transmissible (Gueimonde *et al.*, 2010; Bennedsen *et al.*, 2011).

A comprehensive safety assessment of the commercial probiotic strain *Lb. plantarum* JDM1 based on the whole genome sequence was performed by Zhang *et al.* (2012). In total, 51 nonspecific AR-associated genes (as classified by the RAST annotating service;

http://rast.nmpdr.org/), 126 no-offensive virulence-associated genes and 23 adverse metabolism-associated genes were found; however, there were no toxin or hemolysin encoding genes, and safety-associated genes were rarely transferable, thus the generally regarded as safe (GRAS) status for *Lb. plantarum* JDM1 was confirmed. An aspect to be underlined is related to some discrepancies found between genotype and phenotype. For example, *Lb. plantarun* JDM1 was sensitive to the antibiotic chloramphenicol, but the genome contained a *cat* gene coding for a chloramphenicol O-acetyltransferase enzyme that is responsible for chloramphenicol resistance in bacteria. These data suggest that when the analysis of whole bacterial genome sequences reveals putative unfavourable genes, the biosafety of strains must be assessed more carefully and comprehensively through phenotypic analyses.

Whole genome sequencing was used by Wei *et al.* (2012) to assess the safety of *Bif. longum* subsp. *longum* JDM301, a commercial strain used widely in China with several probiotic functions. This strain is safe based on phenotype; however, bioinformatics analysis of its genome revealed several potential risk factors, that is, 36 genes associated with AR, including a tetracycline resistance gene with certain risk of transfer, and five putative genes associated with production of harmful metabolites. In addition, *Bif. longum* subsp. *longum* JDM301 contains 162 nonspecific VFs, mainly associated with transcriptional regulation, adhesion, and sugar and amino acid transport (Wei *et al.*, 2012). The discrepancies between phenotype and genotype underline the necessity to further experimentally evaluate the potential risk factors found in the genome of *Bif. longum* subsp. *longum* JDM301 (e.g. through *in vivo* assessment using animal models and/or clinical trials).

Chenoll *et al.* (2013) have demonstrated *in silico*, *in vitro* and *in vivo* the safety of *Bif. longum* subsp. *longum* CECT 7347, a probiotic strain reported to ameliorate gluten-related damage in celiac disease. Robust arguments can be found in the genome analysis that support the *in vitro* and *in vivo* results, confirming the safety status of *Bif. longum* subsp. *longum* CECT 7347. Indeed, its genome contains neither relevant VFs nor potential AR genes.

In silico safety assessment of the *Str. salivarius* strains NU10 and YU10 was performed by Barbour and Philip (2014). The two bacteriocin-producer strains were isolated from healthy Malaysian subjects and have a potential application as probiotics in the oral cavity. The *in vitro* tests and genome sequencing established the absence of virulence determinants known to be present in streptococcal pathogens. This finding indicates these strains to be potential candidates for probiotic development, as they pass the initial safety assessment described previously for *Str. salivarius* K12 (Burton *et al.*, 2006).

The safety of strain *Enterococcus faecium* NRRL B-2354 (ATCC 8459) based on its genomic and functional characteristics has been recently investigated (Kopit *et al.*, 2014). This strain has a long history of use as a surrogate of pathogens in food products and thermal process validation, and the findings of this study support its continued use. Indeed, the inspection of the genome sequences revealed that *Ent. faecium* NRRL B-2354 lacks AR genes, as well as enterococcal VFs, including *acm*, *cyl*, the *ebp* operon, *esp*, *gelE*, *hyl* and IS*16*. Accordingly, the strain is sensitive to clinically relevant antibiotics and does not present phenotypes associated with expression of VFs. It contains complete copies of *scm*, *sagA*, *efaA* and the *pilA* operon, but the roles of these genes in enterococcal virulence are not yet well understood. The technical guidance of the EFSA for establishing the safety of *Ent. faecium* strains intended as additives in animal nutrition recommends the investigation of the presence of *esp*, *hyl* and IS*16* as well as sensitivity to ampicillin as exclusion criteria (EFSA, 2012). *Enterococcus faecium* NRRL B-2354 does not contain these marker genes typical of hospital-associated isolates responsible for clinical infections, and it is susceptible to clinically relevant antibiotics, including ampicillin; therefore, this strain meets the requirements for safety by the EFSA guidelines.

Data mining of the whole genome sequences of the established probiotic *Lb. helveticus* MTCC 5463, a strain of Indian origin, was carried out by Senan *et al.* (2015b) in order to obtain assurance of its safety. Genome sequences were screened for genetic determinants associated with AR, production of harmful metabolites and VFs. *Lactobacillus helveticus* MTCC 5463 carried AR genes associated with β-lactam and fluoroquinolone resistance. However, there is no threat of transferability of such resistance as the genome lacks the presence of transmissible elements, such as plasmids, transposons and complete prophages. A haemolysin gene was discovered in the genome, revealing a theoretical risk of virulence. Overall, the results of *in silico* analyses have complemented the *in vitro* studies and human clinical trials and have provided supporting evidence for the safety of *Lb. helveticus* MTCC 5463 in the qualified presumption of safety (QPS) list of micro-organisms (EFSA, 2013), which strengthens the potential use of this strain as a probiotic.

In the frame of a survey on the genus *Weissella*, Abriouel *et al.* (2015) examined the safety aspects of *Weissella cibaria* KACC 11862 and *Weissella confusa* LBAE C39-2 based on *in silico* analyses of their whole genome sequences. Strains belonging to these species are widespread in fermented foods and have been proposed as a starter culture, and also as probiotics (Fusco *et al.*, 2015). In this survey, *W. cibaria* KACC11862 was found to harbour some VFs (hemolysins) and AR genes (coding for a multidrug transporter involved in fosfomycin resistance, MDT-FosB; three methicillin resistance proteins; and *vanZ*, encoding a glycopeptide resistance protein). *Weissella confusa* LBAEC39-2 showed the presence of VF genes (encoding collagen adhesion, hemolysin and mucus-binding proteins) and AR genes (MDT-FosB, MRP and *vanZ*). Further studies should elucidate the functionality of the AR genes of weissellaes, and the transferability of these genes to other bacteria. Therefore, the safety of *Weissella* strains intended for industrial use should be investigated in detail on a strain by strain basis, carefully selecting strains lacking pathogenic potential and which do not possess transferable AR genes.

Recently, genome analysis of three *Leuconostoc mesenteroides* strains isolated from Italian soft cheese samples provided a better understanding of the genetic bases of AR in such species and its transference capability among foodborne bacteria (Flórez *et al.*, 2016). Indeed, in the genome of the multi-resistant strain *Leu. mesenteroides* subsp. *mesenteroides* LbE16, genes that might be involved in tetracycline [*tet*(S)], aminoglycoside (*aadE*, *aphA-3* and *sat4*) and virginiamycin [*vat*(E)] resistance were found. *Leuconostoc mesenteroides* subsp. *dextranicum* LbE15, an erythromycin and clindamycin-resistant strain, harbours an *erm*(B) gene associated to a plasmid of ≈ 35 kbp. However, no known tetracycline resistance genes were detected in *Leu. mesenteroides* subsp. *cremoris* LbT16, displaying atypical resistance to this antibiotic [minimum inhibitory concentration (MIC) of 32 µg mL^{-1}] (EFSA, 2012), thus suggesting a new mechanism of resistance that can be due either to acquired genes or to a mutation of indigenous genes. Interestingly, analysis of the AR genes and their flanking regions revealed the potential that some determinants were horizontally transferred. Indeed, the erythromycin resistance was transferred by conjugation between *Leu. mesenteroides* subsp. *dextranicum* and *Enterococcus faecalis* both *in vitro* and in cheese (Flórez *et al.*, 2016), supplying novel proof that LAB can act as a reservoir of acquired AR genes, for which reason their safety should be carefully monitored.

Globally, these publications document that whole genome sequences of LAB and bifidobacteria provide a wealth of *in silico* information related to strain safety, which cannot be obtained by any other approach. Genome-wide screening could be an effective and time-saving tool for identifying prognostic biomarkers of biosafety, as this would provide relevant information to answer key safety questions required for marketing authorisation and approval of health claims (Miquel *et al.*, 2015; Salvetti *et al.*, 2016). Although many phenotypic tests can be replaced by whole genome analyses, the overall physiology of a strain should also be taken in consideration. Since there is still a lack of homogeneity regarding the genetic and phenotypic traits to be assayed and the proper use of the available bioinformatics tools, harmonisation of scientific procedures is needed to obtain an accurate characterisation of each strain and solid demonstration of its safety. This obvious need is underlined by recent publications, in which frameworks for appropriate evaluation schemes to determine the safety and efficacy of micro-organisms intentionally added to foods are being proposed (Miquel *et al.*, 2015; Pariza *et al.*, 2015; Salvetti *et al.*, 2016). The application of a comprehensive workflow, starting from the genome sequence of a probiotic or starter strain, is expected to increase the consistency of future safety assessment, ensuring stakeholders involved in this area (scientists, manufacturers, legislative bodies and consumers) have the ability to obtain complete and easily comparable information to meet regulatory requirements.

3.3.2 *Unravelling LAB and bifidobacteria properties*

Only a few genetic traits appear to be universally conserved among the different LAB genomes, including enzymes involved in glycolysis in 'true' LAB or those of the fructose 6-phosphate or bifidus shunt in bifidobacteria (Klijn *et al.*, 2005; Makarova *et al.*, 2006). Different genetic events (i.e. mutation, gene duplication, HGT, gene decay, gene loss, genome rearrangements etc.) are considered to contribute to the present genome shape and structure of LAB and bifidobacteria species. As already mentioned, adaptation to nutritionally rich environments (e.g. milk, plant material, and human and animal gastrointestinal tracts) has promoted progressive gene decay (van de Guchte *et al.*, 2006; Callanan *et al.*, 2008), but also acquisition through HGT events of genes involved in key properties for niche colonisation or rapid growth in different environments (Schell *et al.*, 2002; O'Sullivan *et al.*, 2009). In addition to the genetic characterisation of typical LAB and bifidobacteria features, sequencing and analysis of genomes have uncovered the genetic make-up of previously unnoticed technological- and probiotic-relevant phenotypic properties. As explained in this chapter, characterisation of these 'new' traits has occasionally driven the development of novel industrial processes, enhancing applicability and/or benefits of the use of LAB and bifidobacteria in food systems (see Figure 3.1).

Colonisation and niche-competition strategies

In addition to its contribution to metabolic and bioprocessing potential, the availability of LAB and bifidobacteria genome sequences has also expanded our knowledge of the molecular basis of the mechanisms by which species colonise and persist in the ecological niches they occupy. As such, genome analyses of well-known probiotic strains have given some clues as to the mechanism by which probiotic organisms colonise the gut and how they positively affect host health (Klaenhammer *et al.*, 2005). The properties must then be experimentally tested under laboratory conditions and in clinical trials. In this way, transcriptome analysis of the *Bif. breve* UCC2003 genome in a murine colonisation model has revealed differential expression of a type IVb tight adherence (*tad*) pilus-encoding gene cluster, which proved to be essential for efficient *in vivo* murine (and probably human) gut colonisation (O'Connell Motherway *et al.*, 2011). The *tad* pilus-encoding locus was shown to be conserved in different *Bifidobacterium* species, supporting the evidence of the general involvement of pili in gut colonisation and persistence (Turroni *et al.*, 2011).

In a similar approach, pili-like structures were reported for the first time in lactobacilli (Kankainen *et al.*, 2009). Pili are proteinaceous appendages (1–10 nm in diameter protruding 2–3 μm) localised at the cell surface that have been well characterised in Gram-positive and Gram-negative pathogens (Proft & Baker, 2009). Unlike Gram-negative pili, each Gram-positive pilus is an assembly of multiple pilin subunits (proteins having LPXTG motives) coupled to each other by covalent bonds by the transpeptidase activity of a pilin-specific sortase. Typically, one of the proteins is called the pilin backbone, as 100–200 of these subunits are assembled head to tail to form the pilus shaft, while one or two other pilins, called ancillary pilins, may exist decorating the base, the cap or all along the pilus shaft. Analysis of the genome of the human-derived *Lb. rhamnosus* GG strain identified two separate pilus clusters in its genome (*spaCBA-srtC1* and *spaFED*). The former operon is encoded in the *Lb. rhamnosus* GG-specific genomic island, while the latter operon was also shown to be present in the genome of *Lb. rhamnosus* Lc705, a dairy strain that does not produce pili. The purified *spaC* and *spaB* components of the first cluster and *spaF* of the second have been shown to bind mucus (Kankainen *et al.*, 2009; von Ossowski *et al.*, 2010). Functional analysis using knockout mutants further proved that the SpaCBA pili, the only pili produced under *in vivo* conditions (Reunanen *et al.*, 2012), were involved in biofilm formation and efficient adherence of *Lb. rhamnosus* GG to Caco-2 cells (Lebeer *et al.*, 2012; Rasinkangas *et al.*, 2014). The SrtC1 is the pilin-dedicated sortase enzyme recognising and polymerising the SpaA and SpaC pilin subunits. The sortase has also been found to be essential for pili formation (Rasinkangas *et al.*, 2014). In addition, reduced expression of the pro-inflammatory interleukin-8 (IL8) mRNA was induced when Caco-2 cells were incubated with the wild-type strain as compared with the mutant. This suggests that, while providing mucus-binding ability that may explain the colonisation and persistence of *Lb. rhamnosus* GG in the intestine, pili might also modulate IL8 expression through their interaction with surface molecules of the host cells (Lebeer *et al.*, 2012). Therefore, in addition to a role in colonisation, this system represents a previously undescribed mechanism for the interaction of probiotics with host immune tissue.

Genetic evidence for a putative pilus locus consisting in a sortase C gene flanked by three LPXTG protein-encoding genes (*yhgD*, *yhgE* and *yhhB*) organised in an operon-like structure has been recently reported in the laboratory, plasmid-free derivative *Lac. lactis* subsp. *lactis* IL 1403 strain (Dieye *et al.*, 2010). Although *Lac. lactis* subsp. *lactis* IL 1403 does not produce pili under standard growth conditions, overexpression of the pilus operon has been shown to result in the display of pili on the surface of the cells (Oxaran *et al.*, 2012). It is worth mentioning that the piliated strain of *Lac. lactis* subsp. *lactis* IL 1403 exhibited an auto-aggregating phenotype in liquid cultures and formed a thicker biofilm compared to the wild-type, non-piliated strain. Functional analysis of the biogenesis machinery indicated that the pilus shaft was formed by oligomers of the YhgE pilin, the pilus cap was formed by YhgD, and YhhB was the basal pilin that enabled the tethering of the pilus fibres to the cell wall (Oxaran *et al.*, 2012). Analysis of wild *Lac. lactis* strains isolated from plant and clinical environments showed that the majority of isolates could produce pili under normal culturing conditions, which suggests that these structures are spread among lactococci and might be pivotal for enabling *Lac. lactis* spp. to thrive in natural ecosystems.

The genus *Lactobacillus* is usually defined as not motile, although more than half of the species belonging to the *Lb. salivarius* clade (based on 16 rRNA gene phylogeny) have officially been recognised as motile by virtue of a sophisticated molecular structure, the flagellum (Cousin *et al.*, 2015). The motility of *Lactobacillus ruminis* ATTC 27782 has been particularly well studied at both the genomic and phenotypic levels (Forde *et al.*, 2011; Neville *et al.*, 2012). In this strain, all 45 genes required to produce a functional flagellum have been shown to be organised in a single operon (Forde *et al.*, 2011). The same structure has recently been identified in a single strain of *Lactobacillus curvatus*, a species of the '*Lactobacillus sakei*' (presumed to be *Lactobacillus sakei* subsp. *sakei*) clade (Cousin *et al.*, 2015). Motility might confer competitive advantages for niche colonisation, such as a superior acquisition of nutrients or superior biofilm formation capability, but it could also have an impact on the ecology in terms of host signalling and colonisation (Neville *et al.*, 2012).

Respiration in Lac. lactis *strains and other LAB species*

Sugar fermentation was long considered to be the sole means of energy metabolism available to LAB, with the production of organic acids (mainly lactic acid) as final end products. While this is generally still true, some LAB species exhibit a respiratory capability in the presence of oxygen and an exogenous haem supply (Pedersen *et al.*, 2012). LAB respiration results in the production of greatly reduced amounts of lactic acid, higher biomass yield and improved fitness of starters.

Strictly speaking, respiration is the coupling of a membrane potential to the reduction of oxygen. Early evidence for the respiratory capability of LAB was largely overlooked (Sijpesteijn, 1970), and initial reports on the respiration in *Lac. lactis* strains (Duwat *et al.*,

2001) were essentially based on the analysis of the complete genome sequence of *Lac. lactis* subsp. *lactis* IL 1403, which revealed the presence of genes encoding enzymes related to the aerobic (pyruvate oxidase, NADH-dependent oxidase and NADH-dependent peroxydase) and respiratory (cytochrome oxidase and ubiquinone/menaquinone biosynthesis C-methylase) metabolism (Bolotin *et al.*, 2001). Subsequent research confirmed that this species did indeed have the ability to respire in the presence of oxygen (Duwat *et al.*, 2001; Gaudu *et al.*, 2002), provided the growth medium contained haem because this bacterium lacks a functional biosynthetic pathway for this compound. Transcriptomic analysis of *Lac. lactis* subsp. *cremoris* MG 1363 showed that the pyruvate dehydrogenase complex (encoded by the *pdhABCD* operon) was upregulated fourfold under respiratory conditions (Vido *et al.*, 2004). Acetolactate syntase (*als*) and α-acetolactate decarboxylase (*aldC*) genes were also upregulated, facilitating the synthesis of both diacetyl and acetoin. The most highly upregulated gene under respiratory conditions was *ygfC*, which encodes a putative regulatory protein that increased almost 100-fold. In contrast, the expression of the pyruvate formate lyase (*pfl*) and the alcohol dehydrogenase (*adhE*) genes was reduced 2.5- and 50-fold, respectively (Vido *et al.*, 2004). Altogether, these gene expression changes under respiratory conditions cause a profound rerouting of the *Lac. lactis* spp. metabolism, producing acetate, acetoin and diacetyl from pyruvate at the expense of lactic acid (Duwat *et al.*, 2001).

These results are of industrial significance and have allowed the development of a patented process for the production of LAB starters (Duwat *et al.*, 1998). The patent was licensed to Chr. Hansen in 1999, and the initial results obtained with the well-characterised laboratory strain *Lac. lactis* subsp. *lactis* IL 1403 were optimised for different industrial *Lac. lactis* strains and also for some *Leuconostoc* species (Pedersen *et al.*, 2005). Industrial *Lac. lactis* strains were also assayed in aeration in the absence of added haem to distinguish simple aeration from true respiration. Numerous genes were differentially expressed under these two different conditions. Approximately half of these genes have an unknown function, indicating that more research is needed to fully understand the physiology of respiration. Starter cultures obtained by the respiration technology have been assayed in pilot-scale tests for Cheddar cheese production. Manufacture parameters were all within the normal range, and statistically significant differences between analytical parameters of the cheeses (moisture content, total soluble nitrogen and pH) made from respiration-grown cells or fermentation-grown cells were not observed. Indeed, sensory differences of ripened cheeses were not perceived by two trained sensory panels. Industrial-scale trials of Cheddar, Feta and Cottage cheeses have already been performed and, as before, no significant differences were revealed with regard to manufacturing parameters, cheese microbiology, chemistry, texture or flavour development (Pedersen *et al.*, 2005). More recently, other phenotypic properties of technological relevance, such as the production of the antimicrobial bacteriocin nisin, have been reported to be enhanced under respiratory conditions (Kördikanlıoğlu *et al.*, 2015). This might also have industrial significance for the development of nisin-based protective cultures.

Respiration in *Lb. plantarum* needs the exogenous addition of both haem and vitamin K_2, which act as a source of menaquinone (Brooijmans *et al.*, 2009). As this bacterium lacks superoxide dismutase, high levels of manganese are also needed for its aerobic growth. Under respiratory conditions, growth of *Lb. plantarum* results in higher biomass yields. It also affects the robustness of the cells, as strains from aerobic cultures have been shown to be more resistant to industrially relevant stress conditions (Watanabe *et al.*, 2012). A similar respiration process has been reported for some *Lb. casei* strains, also requiring both hemin and menaquinone (Zotta *et al.*, 2014). Respiration in this latter species resulted in the expression of phenotypes with enhanced technological properties, such as increased survival to stress and higher antioxidant capability (Ianniello *et al.*, 2015), which can have application in the use of these bacteria as starter or probiotic cultures. Remarkably, the genome of *Oenococcus oeni* also contains all the genes for aerobic respiration (Borneman *et al.*, 2012), but its functionality has yet to be tested.

Increased biomass after aerated incubation in the presence of haem has never been attained for *Str. thermophilus*, *Lb. delbrueckii* subsp. *bulgaricus* and *Lb. helveticus*. In agreement, analysis of the complete genome sequences of strains of these species (Bolotin *et al.*, 2004; van de Guchte *et al.*, 2006; Callanan *et al.*, 2008, respectively) revealed the presence of genes neither for cytochrome oxidase nor for the biosynthesis of quinones (Pedersen *et al.*, 2012), essential features for respiration.

Natural transformation of Str. thermophilus

Streptococcus thermophilus is considered the second most important industrial dairy starter species after *Lac. lactis* spp. due to its extensive use in the manufacture of dairy products such as yoghurt, hard-cooked cheeses of the Italian and Swiss types, soft cheeses and so on (Leroy & de Vuyst, 2004). As stated here, this bacterium is closely related to *Str. salivarius* and pathogenic species, such as *Streptococcus pyogenes* and *Streptococcus pneumoniae* (Bolotin *et al.*, 2004). It is worth mentioning that the latter species was the micro-organism in which natural competence was first discovered. Since then, this process has been described in many Gram-positive and Gram-negative bacteria, and – although most rarely – also in archaea (Johnston *et al.*, 2014). Competence is defined as a transient physiological state that enables cells to take up exogenous, naked DNA and to stably integrate this into the genome by homologous recombination (Chen & Dubnau, 2004). Currently, the process is considered as a stress response process that may increase bacterial adaptability and fitness under adverse conditions (Charpentier *et al.*, 2012). In response to specific environmental stresses, streptococci species synthesise and secrete peptide competence pheromones that, at a critical extracellular concentration, activate the master competence regulator (ComX). This protein is a sigma factor that associates with the RNA polymerase to redirect transcription towards genes required for DNA transport, processing and integration (Fontaine *et al.*, 2015).

Genome analysis of the first *Str. thermophilus* sequenced strains proved that strains of this species still share a substantial part of their overall genetic make-up with other streptococci, including all genes necessary for competence (Bolotin *et al.*, 2004). *Streptococcus thermophilus* contains a *comX*-like gene coding for a typical peptide pheromone-dependent two-component system that is similar to the competence control loci of *Streptococcus mutans* and *Str. pyogenes*. The regulatory pathway controlling expression of key components of competence in *Str. thermophilus* has recently been reported (Haustenne *et al.*, 2015). The pheromone (called XIP peptide) is synthesised as a large peptide that is exported through an unknown transport system and matured by specific proteases. The resulting extracellular XIP is reimported by the oligopeptide transporter Opp/Ami, an essential component of the proteolytic system for growth in milk that is also necessary for competence (Gardan *et al.*, 2009). In the cytoplasm, the intracellular mature XIP binds to the transcriptional regulator ComR, and the XIP–ComR complex activates the transcription of most early competence genes (including *comS* and *comX*). Thereafter, ComX induces transcription of all late competence genes necessary for DNA transformation (Haustenne *et al.*, 2015).

The mechanistic characterisation of natural competence in *Str. thermophilus* has allowed the development of genetic tools for the introduction of heterologous genes in this bacterium by natural transformation without the need of antibiotic resistance or other controversial markers for the selection of transformants (Blomqvist *et al.*, 2010; Fontaine *et al.*, 2010). Subsequently, insertion of the DNA into the chromosome takes place by double cross-over, homologous recombination. The ability of *Str. thermophilus* to be naturally transformable has allowed recombinant strains to be easily obtained (Lecomte *et al.*, 2016). In a proof-of-concept study, the gene encoding the cell-envelope-located proteinase PrtS, which is only present in certain fast milk-acidifying *Str. thermophilus* strains (Delorme *et al.*, 2010), has been experimentally transferred to a series of slow-acidifying starter strains (Dandoy *et al.*, 2011).

3.4 Concluding remarks

This review has described different aspects of genome-wide studies on strains relevant for dairy and probiotic products. These have shown that an incredibly high number of genome sequences have accumulated, which have been or are still to be analysed scientifically. Such studies are extremely important in understanding the evolution and defining the diversity of strains at different taxonomic levels, but, most importantly, this research could greatly enhance our knowledge on metabolic capabilities (e.g. lactose and citrate metabolism, proteolysis, lipolysis and bacteriocin production) that can eventually be linked to their genetic basis. This is the starting point of possible combinations with other ‘omics’ data (i.e. proteomics, transcriptomics, metabolomics etc.) that could be employed to devise genome-scale metabolic models. This, coupled with traditional culture-based characterisations and small-scale fermentations, could enable more precise strain selection for efficient fermentation and flavour development at an industrial-scale and probiotic ability (Stefanovic *et al.*, 2017).

Remarkably, the need for large sequence data manipulation has led to the development of a novel area of competence for microbiology (i.e. bioinformatics), with many microbiologists working in front of computers instead of at the laboratory bench. In line with this observation, it can be noted that an EU project recently concluded in 2015, funded in the FP7-SME framework (grant agreement no. 6054853 – FP7/2007–2013), aimed to develop a genomics toolbox to enhance business for small and medium-sized enterprises (SMEs) in the market of starter cultures and probiotics (http://www.genobox.eu/ns/). Therefore, genome sequencing is a useful tool not just for scientific purposes; it could also be pivotal for patent opportunities, reliable communication to authorities and approval of health claims (Danielsen & Johansen, 2009). Genome sequencing continues to provide an incredible amount of genetic information useful for the characterisation of starter cultures and probiotic bacteria, which will need to be deciphered in the coming decades.

References

Abriouel, H., Lavilla Lerma, L., Casado Muñoz, M.C., Pérez Montoro, B., Kabisch, J., Pichner, R., Cho, G.-S., Neve, H., Fusco, V., Franz, C.M.A.P., Gálvez, A. & Benomar, N. (2015) The controversial nature of the *Weissella* genus: technological and functional aspects versus whole genome analysis-based pathogenic potential for their application in food and health. *Frontiers in Microbiology*, **6**(1197), 1–14.

Ainsworth, S., Stockdale, S., Bottacini, F., Mahony, J. & Sinderen, D. (2014) The *Lactococcus lactis* plasmidome: much learnt, yet still lots to discover. *FEMS Microbiology Reviews*, **38**, 1066–1088.

Alkema, W., Boekhorst, J., Wels, M. & van Hijum, S.A.F.T. (2015) Microbial bioinformatics for food safety and production. *Briefings in Bioinformatics*, **17**, 283–292.

Barbour, A. & Philip, K. (2014) Variable characteristics of bacteriocin-producing *Streptococcus salivarius* strains isolated from Malaysian subjects. *PLoS One*, **9** (e100541).

Barrangou, R. (2014) Cas9 targeting and the CRISPR revolution. *Science*, **344**, 707–708.

Barrangou, R. & Horvath, P. (2012) CRISPR: new horizons in phage resistance and strain identification. *Annual Review of Food Science and Technology*, **3**, 143–162.

Bennedsen, M., Stuer-Lauridsen, B., Danielsen, M. & Johansen, E. (2011) Screening for antimicrobial resistance genes and virulence factors via genome sequencing. *Applied and Environmental Microbiology*, **77**, 2785–2787.

Blomqvist, T., Steinmoen, H. & Håvarstein, L.S. (2010) A food-grade site-directed mutagenesis system for *Streptococcus thermophilus* LMG 18311. *Letters in Applied Microbiology*, **50**, 314–319.

Bolotin, A., Wincker, P., Mauger, S., Jaillon, O., Malarme, K., Weissenbach, J., Ehrlich, S.D. & Sorokin, A. (2001) The complete genome sequence of the lactic acid bacterium *Lactococcus lactis* ssp. *lactis* IL1403. *Genome Research*, **11**, 731–753.

Bolotin, A., Quinquis, B., Renault, P., Sorokin, A., Ehrlich, S.D., Kulakauskas, S., Lapidus, A., Goltsman, E., Mazur, M., Pusch, G.D., Fonstein, M., Overbeek, R., Kyprides, N., Purnelle, B., Prozzi, D., Ngui, K., Masuy, D., Hancy, F., Burteau, S., Boutry, M., Delcour, J., Goffeau, A. & Hols, P. (2004) Complete sequence and comparative genome analysis of the dairy bacterium *Streptococcus thermophilus*. *Nature Biotechnology*, **22**, 1554–1558.

Borneman, A.R., McCarthy, J.M., Chambers, P.J. & Bartowsky, E.J. (2012) Comparative analysis of the *Oenococcus oeni* pan genome reveals genetic diversity in industrially-relevant pathways. *BMC Genomics*, **13**(373), 1–13.

Boutrou, R. & Guéguen, M. (2005) Interests in *Geotrichum candidum* for cheese technology. *International Journal of Food Microbiology*, **102**, 1–20.

Broadbent, J.R., Neeno-Eckwall, E.C., Stahl, B., Tandee, K., Cai, H., Morovic, W., Horvath, P., Heidenreich, J., Perna, N.T., Barrangou, R. & Steele, J.L. (2012) Analysis of the *Lactobacillus casei* supragenome and its influence in species evolution and lifestyle adaptation. *BMC Genomics*, **13**(533), 1–18.

Brooijmans, R.J., de Vos, W.M. & Hugenholtz, J. (2009) *Lactobacillus plantarum* WCFS1 electron transport chains. *Applied and Environmental Microbiology*, **75**, 3580–3585.

Burton, J.P., Wescombe, P.A., Moore, C.J., Chilcott, C.N. & Tagg, J.R. (2006) Safety assessment of the oral cavity probiotic *Streptococcus salivarius* K12. *Applied and Environmental Microbiology*, **72**, 3050–3053.

Callanan, M., Kaleta, P., O'Callaghan, J., O'Sullivan, O., Jordan, K., McAuliffe, O., Sangrador-Vegas, A., Slattery, L., Fitzgerald, G.F., Beresford, T. & Ross, R.P. (2008) Genome sequence of *Lactobacillus helveticus*, an organism distinguished by selective gene loss and insertion sequence element expansion. *Journal of Bacteriology*, **190**, 727–735.

Cavanagh, D., Fitzgerald, G.F. & McAuliffe, O. (2015) From field to fermentation: the origins of *Lactococcus lactis* and its domestication to the dairy environment. *Food Microbiology*, **47**, 45–61.

Charpentier, X., Polard, P. & Claverys, J.P. (2012) Induction of competence for genetic transformation by antibiotics: convergent evolution of stress responses in distant bacterial species lacking SOS? *Current Opinion in Microbiology*, **15**, 570–576.

Chen, I. & Dubnau, D. (2004) DNA uptake during bacterial transformation. *Nature Reviews Microbiology*, **2**, 241–249.

Chenoll, E., Codoñer, F.M., Silva, A., Ibáñez, A., Martinez-Blanch, J.F., Bollati-Fogolín, M., Crispo, M., Ramírez, S., Sanz, Y., Ramón, D. & Genovés, S. (2013) Genomic sequence and pre-clinical safety assessment of *Bifidobacterium longum* CECT 7347, a probiotic able to reduce the toxicity and inflammatory potential of gliadin-derived peptides. *Journal of Probiotics & Health*, **1**, 1–6.

Cousin, F.J., Lynch, S.N., Harris, H.M.B., McCann, A., Lynch, D.B., Neville, B.A., Irisawa, T., Okada, S., Endo, A. & O'Toole, P.W. (2015) Detection and genomic characterization of motility in *Lactobacillus curvatus*: confirmation of motility in a species outside of the *Lactobacillus salivarius* clade. *Applied and Environmental Microbiology*, **81**, 1297–1308.

Dandoy, D., Fremaux, C., de Frahan, M.H., Horvath, P., Boyabal, P., Hols, P. & Fontaine, L. (2011) The fast milk acidifying phenotype of *Streptococcus thermophilus* can be acquired by natural transformation of the genomic island encoding the cell-envelope proteinase PrtS. *Microbial Cell Factories*, **10**(Suppl. 1, S21), 1–9.

Danielsen, M. & Johansen, E. (2009) Functional genomics of dairy micro-organisms and probiotics in the era of low-cost DNA sequencing. *Australian Journal of Dairy Technology*, **64**, 102–105.

Dellaglio, F., Felis, G.E., Torriani, S., Sørensen, K. & Johansen, E. (2005) Genomic characterisation of starter cultures. In *Probiotic Dairy Products* (ed. A. Tamime), 16–38. Blackwell Publishing, Oxford.

Delorme, C., Abraham, A.L., Renault, P. & Guédon, E. (2015) Genomics of *Streptococcus salivarius*, a major human commensal. *Infection, Genetics and Evolution*, **33**, 381–392.

Delorme, C., Bartholini, C., Bolotine, A., Ehrlich, S.D. & Renault, P. (2010) Emergence of a cell wall protease in the *Streptococcus thermophilus* population. *Applied and Environmental Microbiology*, **76**, 451–460.

Dieye, Y., Oxaran, V., Ledue-Clier, F., Alkhalaf, W. Buist, G., Juillard, V., Lee, C.W. & Piard, J.C. (2010) Functionality of sortase A in *Lactococcus lactis*. *Applied and Environmental Microbiology*, **76**, 7332–7337.

Domann, E., Hain, T., Ghai, R., Billion, A., Kuenne, C., Zimmermann, K. & Chakraborty, T. (2007) Comparative genomic analysis for the presence of potential enterococcal virulence factors in the probiotic *Enterococcus faecalis* strain Symbioflor 1. *International Journal of Medical Microbiology*, **297**, 533–539.

Duwat, P., Sourice, S. & Gruss, A. (1998) Process for preparing starter cultures of lactic acid bacteria. French Patent Application, FR9809463.

Duwat, P., Sourice, S., Cesselin, B., Lamberet, G., Vido, K., Gaudu, P., Le Loir, Y., Violet, F., Loubière, P. & Gruss, A. (2001) Respiration capacity of the fermenting bacterium *Lactococcus lactis* and its positive effects on growth and survival. *Journal of Bacteriology*, **183**, 4509–4516.

EFSA (2007) Opinion of the Scientific Committee on a request from EFSA on the introduction of a Qualified Presumption of Safety (QPS) approach for assessment of selected microorganisms referred to EFSA. *EFSA Journal*, **587**, 1–16.

EFSA (2012) Guidance on the safety assessment of *Enterococcus faecium* in animal nutrition. *EFSA Journal*, **10**, 2682–2692.

EFSA (2013) Scientific opinion on the maintenance of the list of QPS biological agents intentionally added to food and feed (2013 update). *EFSA Journal*, **11**, 34–49.

Falentin, H., Deutsch, S.M., Jan, G., Loux, V., Thierry, A., Parayre, S., Maillard, M.B., Dherbecourt, J., Cousin, F.J. & Jardin, J. (2010) The complete genome of *Propionibacterium freudenreichii* CIRMBIA1^T, a hardy *Actinobacterium* with food and probiotic applications. *PLoS One*, **5** (e11748).

FAO/WHO (2001) Health and nutritional properties of probiotics in food including powder milk with live lactic acid bacteria. World Health Organization. http://www.who.int/foodsafety/publications/fs_management/en/probiotics.pdf

Ferrario, C., Ricci, G., Milani, C., Lugli, G.A., Ventura, M., Eraclio, G., Borgo, F. & Fortina, M.G. (2013) *Lactococcus garvieae*: where is it from? A first approach to explore the evolutionary history of this emerging pathogen. *PLoS One*, **8** (e84796).

Ferrario, C., Milani, C., Mancabelli, L., Lugli, G.A., Turroni, F., Duranti, S., Mangifesta, M., Viappiani, A., Sinderen, D. & Ventura, M. (2015) A genome-based identification approach for members of the genus *Bifidobacterium*. *FEMS Microbiology Ecology*, **91** (fiv009).

Field, D., Amaral-Zettler, L., Cochrane, G., Cole, J.R., Dawyndt, P., Garrity, G.M., Gilbert, J., Glöckner, F.O., Hirschman, L., Karsch-Mizrachi, I., Klenk, H.P., Knight, R., Kottmann, R., Kyrpides, N., Meyer, F., San Gil, I., Sansone, S.A., Schriml, L.M., Sterk, P., Tatusova, T., Ussery, D.W., White, O. & Wooley, J. (2011) The Genomic Standards Consortium. *PLoS Biology*, **9** (e1001088).

Flórez, A.B., Campedelli, I., Delgado, S., Alegría, Á., Salvetti, E., Felis, G.E., Mayo, B. & Torriani, S. (2016) Antibiotic susceptibility profiles of dairy leuconostoc, analysis of the genetic basis of atypical resistances and transfer of genes *in vitro* and in a food matrix. *PLoS One*, **11** (e0145203).

Fontaine, L., Dandoy, D., Boutry, C., Delplace, B., de Frahan, M.H., Fremaux, C., Horvath, P., Boyaval, P. & Hols, P. (2010) Development of a versatile procedure based on natural transformation for marker-free targeted genetic modification in *Streptococcus thermophilus*. *Applied and Environmental Microbiology*, **76**, 7870–7877.

Fontaine, L., Wahl, A., Fléchard, M., Mignolet, J. & Hols, P. (2015) Regulation of competence for natural transformation in streptococci. *Infection, Genetics and Evolution*, **33**, 343–360.

Forde, B.M., Neville, B.A., O'Donnell, M.M., Riboulet-Bisson, E., Claesson, M.J., Coghlan, A., Ros, R.P. & O'Toole, P.W. (2011) Genome sequences and comparative genomics of two *Lactobacillus ruminis* strains from the bovine and human intestinal tract. *Microbial Cell Factories*, **10**(Suppl. 1, S13), 1–15.

Franz, C.M., Huch, M., Abriouel, H., Holzapfel, W. & Gálvez, A. (2011) Enterococci as probiotics and their implications in food safety. *International Journal of Food Microbiology*, **151**, 125–140.

Fusco, V., Quero, G.M., Cho, G., Kabisch, J., Meske, D., Neve, H., Bockelmann, W. & Franz, C.M.A.P. (2015) The genus *Weissella*: taxonomy, ecology and biotechnological potential. *Frontiers in Microbiology*, **6** (155), 1–22.

Gardan, R., Besset, C., Guillot, A., Gitton, C. & Monnet, V. (2009) The oligopeptide transport system is essential for the development of natural competence in *Streptococcus thermophilus*. *Journal of Bacteriology*, **191**, 4647–4655.

Garrigues, C., Johansen, E. & Crittenden, R. (2013) Pangenomics – an avenue to improved industrial starter cultures and probiotics. *Current Opinion in Biotechnology*, **24**, 187–191.

Gasson, M.J. & de Vos, W.M. (eds.) (2004) *Genetics and Biotechnology of Lactic Acid Bacteria*. Springer, Berlin.

Gaudu, P., Vido, K., Cesselin, B., Kulakauskas, S., Tremblay, J., Rezaïki, L., Lamberret, G., Sourice, S., Duwat, P. & Gruss, A. (2002) Respiration capacity and consequences in *Lactococcus lactis*. *Antonie van Leeuwenhoek*, **82**, 263–269.

Giraffa, G. (2003) Functionality of enterococci in dairy products. *International Journal of Food Microbiology*, **88**, 215–222.

Gueimonde, M., Florez, A.B., van Hoek, A.H., Stuer-Lauridsen, B., Stroman, P., de los Reyes-Gavilan, C.G. & Margolles, A. (2010) Genetic basis of tetracycline resistance in *Bifidobacterium animalis* subsp. *lactis*. *Applied and Environmental Microbiology*, **76**, 3364–3369.

Hanniffy, S., Wiedermann, U., Repa, A., Mercenier, A., Daniel, C., Fioramonti, J., Tlaskolova, H., Kozakova, H., Israelsen, H., Madsen, S., Vrang, A., Hols, P., Delcour, J., Bron, P., Kleerebezem, M. & Wells, J. (2004) Potential and opportunities for use of recombinant lactic acid bacteria in human health. *Advances in Applied Microbiology*, **56**, 1–64.

Hao, P., Zheng, H., Yu, Y., Ding, G., Gu, W., Chen, S., Yu, Z., Ren, S., Oda, M., Konno, T., Wang, S., Li, X., Ji, Z.S. & Zhao, G. (2011) Complete sequencing and pan-genomic analysis of *Lactobacillus delbrueckii* subsp. *bulgaricus* reveal its genetic basis for industrial yogurt production. *PLoS One*, **6** (e15964).

Hatoum, R., Labrie, S. & Fliss, I. (2012) Antimicrobial and probiotic properties of yeasts: from fundamental to novel applications. *Frontiers in Microbiology*, **3**(421), 1–12.

Haustenne, L., Bastin, G., Hols, P. & Fontaine, L. (2015) Modeling of the ComRS signaling pathway reveals the limiting factors controlling competence in *Streptococcus thermophilus*. *Frontiers in Microbiology*, **6**(1413), 1–20.

Hill, C., Guarner, F., Reid, G., Gibson, G.R., Merenstein, D.J., Pot, B., Morelli, L., Canani, R.B., Flint, H.J., Salminen, S., Calder, P.C. & Sanders, M.E. (2014) Expert consensus document. The International Scientific Association for Probiotics and Prebiotics consensus statement on the scope and appropriate use of the term probiotic. *Nature Reviews Gastroenterology & Hepatology*, **11**, 506–514.

Hols, P., Hancy, F., Fontaine, L., Grossiord, B., Prozzi, D., Leblond-Bourget, N., Decaris, B., Bolotin, A., Delorme, C. & Dusko, E.S. (2005) New insights in the molecular biology and physiology of *Streptococcus thermophilus* revealed by comparative genomics *FEMS Microbiology Reviews*, **29**, 435–463.

Horvath, P., Couté-Monvoisin, A.C., Romero, D.A., Boyaval, P., Fremaux, C. & Barrangou, R. (2009) Comparative analysis of CRISPR loci in lactic acid bacteria genomes. *International Journal of Food Microbiology*, **131**, 62–70.

Hugenholtz, J., Sybesma, W., Groot, M.N., Wisselink, W., Ladero, V., Burgess, K., van Sinderen, D., Piard, J.C., Eggink, G., Smid, E.J., Savoy, G., Sesma, F., Jansen, T., Hols, P. & Kleerebezem, M. (2002) Metabolic engineering of lactic acid bacteria for the production of nutraceuticals. *Antonie van Leeuwenhoek*, **82**, 217–235.

Ianniello, R.G., Ricciardi, A., Parente, E., Tramutola, A., Reale, A. & Zotta, T. (2015) Aeration and supplementation with heme and menaquinone affect survival to stresses and antioxidant capability of *Lactobacillus casei* strains. *LWT – Food Science and Technology*, **60**, 817–824.

Jans, C., Follador, R., Hochstrasser, M., Lacroix, C., Meile, L. & Stevens, M.J.A. (2013) Comparative genome analysis of *Streptococcus infantarius* subsp. *infantarius* CJ18, an African fermented camel milk isolate with adaptations to dairy environment. *BMC Genomics*, **14**(200), 1–16.

Jansen, R., Embden, J.D., Gaastra, W. & Schouls, L.M. (2002) Identification of genes that are associated with DNA repeats in prokaryotes. *Molecular Microbiology*, **43**, 1565–1575.

Johnston, C., Martin, B., Fichant, G., Polard, P. & Claverys, J.P. (2014) Bacterial transformation: distribution, shared mechanisms and divergent control. *Nature Reviews Microbiology*, **12**, 181–196.

Kankainen, M., Paulin, L., Tynkkynen, S., von Ossowski, I., Reunanen, J., Partanen, P., Satokari, R., Vesterlund, S., Hendrickx, A.P., Lebeer, S., De Keersmaecker, S.C., Vanderleyden, J., Hämäläinen, T., Laukkanen, S., Salovuori, N., Ritari, J., Alatalo, E., Korpela, R., Mattila-Sandholm, T., Lassig, A., Hatakka, K., Kinnunen, K.T., Karjalainen, H., Saxelin, M., Laakso, K., Surakka, A., Palva, A., Salusjärvi, T., Auvinen, P. & de Vos, W.M. (2009) Comparative genomic analysis of *Lactobacillus rhamnosus* GG reveals pili containing a human-mucus binding protein. *Proceedings of the National Academy of Sciences of the United States of America*, **106**, 17193–17198.

Kant, R., Blom, J., Palva, A., Siezen, R.J. & de Vos, W.M. (2010) Comparative genomics of *Lactobacillus*. *Proceedings of the National Academy of Sciences of the United States of America*, **4**, 323–332.

Kelleher, P., Murphy, J., Mahony, J. & van Sinderen, D. (2015) Next-generation sequencing as an approach to dairy starter selection. *Dairy Science & Technology*, **95**, 545–568.

Kelly, W.J., Ward, L.J.H. & Leahy, S.C. (2010) Chromosomal diversity in *Lactococcus lactis* and the origin of dairy starter cultures. *Genome Biology and Evolution*, **2**, 729–744.

Klaenhammer, T.R., Barrangou, R., Buck, B.L., Azcaráte-Peril, M.A. & Altermann, E. (2005) Genomic features of lactic acid bacteria effecting bioprocessing and health. *FEMS Microbiology Reviews*, **29**, 393–409.

Klijn, A., Mercenier, A. & Arigoni, F. (2005) Lessons from the genomes of bifidobacteria. *FEMS Microbiology Reviews*, **29**, 491–509.

Kopit, L.M., Kim, E.B., Siezen, R.J., Harris, L.J. & Marco, M.L. (2014) Safety of the surrogate microorganism *Enterococcus faecium* NRRL B-2354 for use in thermal process validation. *Applied and Environmental Microbiology*, **80**, 1899–1909.

Kördikanlıoğlu, B., Şimşek, Ö. & Saris, P.E. (2015) Nisin production of *Lactococcus lactis* N8 with hemin-stimulated cell respiration in fed-batch fermentation system. *Biotechnology Progress*, **31**, 678–685.

Lebeer, S., Claes, I., Tytgat, H.L., Verhoeven, T.L., Marien, E., von Ossowski, I., Reunanen, J., Palva, A., de Vos, W.M., Keersmaecker, S.C. & Vanderleyden, J. (2012) Functional analysis of *Lactobacillus rhamnosus* GG pili in relation to adhesion and immunomodulatory interactions with intestinal epithelial cells. *Applied and Environmental Microbiology*, **78**, 185–193.

Lecomte, X., Gagnaire, V., Lortal, S., Dary, A. & Genay, M. (2016) *Streptococcus thermophilus*, an emerging and promising tool for heterologous expression: advantages and future trends. *Food Microbiology*, **53**, 2–9.

Leroy, F. & de Vuyst, L. (2004) Lactic acid bacteria as functional starter cultures for the food fermentation industry. *Trends in Food Science & Technology*, **15**, 67–78.

Lukjancenko, O., Ussery, D.W. & Wassenaar, T.M. (2012) Comparative genomics of *Bifidobacterium*, *Lactobacillus* and related probiotic genera. *Microbial Ecology*, **63**, 651–673.

Makarova, K., Slesarev, A., Wolf, Y., Sorokin, A., Mirkin, B., Koonin, E., Pavlov, A., Pavlova, N., Karamychev, V., Polouchine, N., Shakhova, V., Grigoriev, I., Lou, Y., Rohksar, D., Lucas, S., Huang, K., Goodstein, D.M., Hawkins, T., Plengvidhya, V., Welker, D., Hughes, J., Goh, Y., Benson, A., Baldwin, K., Lee, J.H., Díaz-Muñiz, I., Dosti, B., Smeianov, V., Wechter, W., Barabote, R., Lorca, G., Altermann, E., Barrangou, R., Ganesan, B., Xie, Y., Rawsthorne, H., Tamir, D., Parker, C., Breidt, F., Broadbent, J., Hutkins, R., O'Sullivan, D., Steele, J., Unlu, G., Saier, M., Klaenhammer, T., Richardson, P., Kozyavkin, S., Weimer, B. & Mills, D. (2006) Comparative genomics of the lactic acid bacteria. *Proceedings of the National Academy of Sciences of the United States of America*, **103**, 15611–15616.

Marshall, V.M. (1991) Inoculated ecosystem in a milk environment. *Journal of Applied Bacteriology*, **73**, 127S–135S.

Mattarelli, P., Bonaparte, C., Pot, B. & Biavati, B. (2008) Proposal to reclassify the three biotypes of *Bifidobacterium longum* as three subspecies: *Bifidobacterium longum* subsp. *longum* subsp. nov., *Bifidobacterium longum* subsp. *infantis* comb. nov. and *Bifidobacterium longum* subsp. *suis* comb. nov. *International Journal of Systematic and Evolutionary Microbiology*, **58**, 767–772.

Medini, D., Donati, C., Tettelin, H., Masignani, V. & Rappuoli, R. (2005) The microbial pan-genome. *Current Opinion in Genetics & Development*, **15**, 589–594.

Milani, C., Lugli, G.A., Duranti, S., Turroni, F., Bottacini, F., Mangifesta, M., Sanchez, B., Viappiani, A., Mancabelli, L., Taminiau, B., Delcenserie, V., Barrangou, R., Margolles, A., van Sinderen, D. & Ventura, M. (2014) Genomic encyclopedia of type strains of the genus *Bifidobacterium*. *Applied and Environmental Microbiology*, **80**, 6290–6302.

Miquel, S., Beaumont, M., Martín, R., Langella, P., Braesco, V. & Thomas, M. (2015) A proposed framework for an appropriate evaluation scheme for microorganisms as novel foods with a health claim in Europe. *Microbial Cell Factories*, **14**(48), 2–11.

Neville, B.A., Forde, B.M., Claesson, M.J., Darby, T., Coghlan, A., Nally, K., Ros, R.P. & O'Toole, P.W. (2012) Characterization of pro-inflammatory flagellin proteins produced by *Lactobacillus ruminis* and related motile strains. *PLoS One*, **7**(e40592).

O'Connell Motherway, M., Zomer, A., Leahy, S.C., Reunanen, J., Bottacini, F., Claesson, M.J., O'Brien, F., Flynn, K., Casey, P.G., Munoz, J.A., Kearney, B., Houston, A.M., O'Mahony, C., Higgins, D.G., Shanahan, F., Palva, A., de Vos, W.M., Fitzgerald, G.F., Ventura, M., O'Toole, P.W. & van Sinderen, D. (2011) Functional genome analysis of *Bifidobacterium breve* UCC2003 reveals type IVb tight adherence (Tad) pili as an essential and conserved host-colonization factor. *Proceedings of the National Academy of Sciences of the United States of America*, **108**, 11217–11222.

Oh, J.H. & van Pijkeren, J.P. (2014) CRISPR-Cas9-assisted recombineering in *Lactobacillus reuteri*. *Nucleic Acids Research*, **42**(e131), 1–11.

O'Sullivan, O., O'Callaghan, J., Sangrador-Vegas, A., McAuliffe, O., Slattery, L., Kaleta, P., Callanan, M., Fitzgerald, G.F., Ross, R.P. & Beresford, T. (2009) Comparative genomics of lactic acid bacteria reveals a niche-specific gene set. *BMC Microbiology*, **9**(50), 1–9.

Oxaran, V., Ledue-Clier, F., Dieye, Y., Herry, J.M., Péchoux, C., Meylheuc, T., Briandet, R., Juillard, V. & Piard, J.C. (2012) Pilus biogenesis in *Lactococcus lactis*: molecular characterization and role in aggregation and biofilm formation. *PLoS One*, **7**(e50989).

Papadimitriou, K., Anastasiou, R., Mavrogonatou, E., Blom, J., Papandreou, N.C., Hamodrakas, S.J., Ferreira, S., Renault, P., Supply, P., Pot, B. & Tsakalidou, E. (2014) Comparative genomics of the dairy isolate *Streptococcus macedonicus* ACA-DC 198 against related members of the *Streptococcus bovis/Streptococcus equinus* complex. *BMC Genomics*, **15**(272), 1–21.

Pariza, M.W., Gillies, K.O., Kraak-Ripple, S.F., Leyer, G. & Smith, A.B. (2015) Determining the safety of microbial cultures for consumption by humans and animals. *Regulatory Toxicology and Pharmacology*, **73**, 164–171.

Passerini, D., Beltramo, C., Coddeville, M., Quentin, Y., Ritzenthaler, P., Daveran-Mingot, M.L. & Le Bourgeois, P. (2010) Genes but not genomes reveal bacterial domestication of *Lactococcus lactis*. *PLoS One*, **5**(e15306).

Passerini, D., Coddeville, M., Le Bourgeois, P., Loubière, P., Ritzenthaler, P., Fontagné-Faucher, C., Daveran-Mingot, M.L. & Cocaign-Bousquet, M. (2013) The carbohydrate metabolism signature of *Lactococcus lactis* strain A12 reveals its sourdough ecosystem origin. *Applied and Environmental Microbiology*, **79**, 5844–5852.

Pedersen, M.B., Gaudu, P., Lechardeur, D., Petit, M.A. & Gruss, A. (2012) Aerobic respiration metabolism in lactic acid bacteria and uses in biotechnology. *Annual Review of Food Science and Technology*, **3**, 37–58.

Pedersen, M.B., Iversen, S.L., Sorensen, K.I. & Johansen, E. (2005) The long and winding road from the research laboratory to industrial applications of lactic acid bacteria. *FEMS Microbiology Reviews*, **29**, 611–624.

Pérez, T., Balcazar, J.L., Peix, A., Valverde, A., Velazquez, E., De Bleas, I. & Ruiz-Zarzuela, I. (2011) *Lactococcus lactis* subsp. *tructae* subsp. nov. isolated from the intestinal mucus of brown trout (*Salmo trutta*) and rainbow trout (*Oncorhynchus mykiss*). *International Journal of Systematic and Evolutionary Microbiology*, **61**, 1894–1898.

Poonam Pophaly, S.D., Tomar, S.K., De, S. & Singh, R. (2012) Multifaceted attributes of dairy propionibacteria: a review. *World Journal of Microbial Biotechnology*, **28**, 3081–3095.

Proft, T. & Baker, E.N. (2009) Pili in Gram-negative and Gram-positive bacteria: structure, assembly and their role in disease. *Trends in Microbiology*, **16**, 33–40.

Rasinkangas, P., Reunanen, J., Douillard, F.P., Ritari, J., Uotinen, V., Palva, A. & de Vos, W.M. (2014) Genomic characterization of non-mucus-adherent derivatives of *Lactobacillus rhamnosus* GG reveals genes affecting pilus biogenesis. *Applied and Environmental Microbiology*, **80**, 7001–7009.

Reddy, T.B.K., Thomas, A., Stamatis, D., Bertsch, J., Isbandi, M., Jansson, J., Mallajosyula, J., Pagani, I., Lobos, E. & Kyrpides, N. (2015) The Genomes OnLine Database (GOLD) v.5: a metadata management system based on a four level (meta) genome project classification. *Nucleic Acids Research*, **43**, D1099–D1106.

Reunanen, J., von Ossowski, I., Hendrickx, A.P., Palva, A. & de Vos, W.M. (2012) Characterization of the SpaCBA pilus fibers in the probiotic *Lactobacillus rhamnosus* GG. *Applied and Environmental Microbiology*, **78**, 2337–2344.

Salvetti, E., Orrù, L., Capozzi, V., Martina, A., Lamontanara, A., Keller, D., Cash, H., Felis, G.E., Cattivelli, L., Torriani, S. & Spano, G. (2016) Integrate genome-based assessment of safety for probiotic strains: *Bacillus coagulans* GBI-30, 6086 as a case study. *Applied Microbiology and Biotechnology*, **100**, 4595–4605.

Schell, M.A., Karmirantzou, M., Snel, B., Vilanova, D., Berger, B., Pessi, G., Zwahlen, M.C., Desiere, F., Bork, P., Delley, M., Pridmore, R.D. & Arigoni, F. (2002) The genome sequence of *Bifidobacterium longum* reflects its adaptation to the human gastrointestinal tract. *Proceedings of the National Academy of Sciences of the United States of America*, **99**, 14422–14427.

Schlegel, L., Grimont, F., Ageron, E., Grimont, P.A. & Bouvet, A. (2003) Reappraisal of the taxonomy of the *Streptococcus bovis/Streptococcus equinus* complex and related species: description of *Streptococcus gallolyticus* subsp. *gallolyticus* subsp. nov., *S. gallolyticus* subsp. *macedonicus* subsp. nov. and *S. gallolyticus* subsp. *pasteurianus* subsp. nov. *International Journal of Systematic and Evolutionary Microbiology*, **53**, 631–645.

Schleifer, K.H., Ehrmann, M., Krusch, U. & Neve, H. (1991) Revival of the species *Streptococcus thermophilus* (ex Orla-Jensen, 1919) nom. rev. *Systematic and Applied Microbiology*, **14**, 386–388.

Senan, S., Prajapati, J.B. & Joshi, C.G. (2015a) Whole-genome based validation of the adaptive properties of Indian origin probiotic *Lactobacillus helveticus* MTCC 5463. *Journal of the Science of Food and Agriculture*, **95**, 321–328.

Senan, S., Prajapati, J.B. & Joshi, C.G. (2015b) Feasibility of genome-wide screening for biosafety assessment of probiotics: a case study of *Lactobacillus helveticus* MTCC 5463. *Probiotics and Antimicrobial Proteins*, **7**, 249–258.

Siezen, R.J. & Wilson, G. (2010) Probiotics genomics. *Microbial Biotechnology*, **3**, 1–9.

Siezen, R.J., Renckens, B., Van Swam, I., Peters, S., van Kranenburg, R., Kleerebezem, M. & de Vos, W.M. (2005) Complete sequences of four plasmids of *Lactococcus lactis* subsp. *cremoris* SK11 reveal extensive adaptation to the dairy environment. *Applied and Environmental Microbiology*, **71**, 8371–8382.

Sijpesteijn, A.K. (1970) Induction of cytochrome formation and stimulation of oxidative dissimilation by hemin in *Streptococcus lactis* and *Leuconostoc mesenteroides*. *Antonie van Leeuwenhoek*, **36**, 335–348.

Smokvina, T., Wels, M., Polka, J., Chervaux, C., Brisse, S., Boekhorst, J., van Hylckama Vlieg, J.E. & Siezen, R.J. (2013) *Lactobacillus paracasei* comparative genomics: towards species pan-genome definition and exploitation of diversity. *PLoS One*, **8**(e68731).

Stanton, C., Ross, R.P., Fitzgerald, G.F. & van Sinderen, D. (2005) Fermented functional foods. *Current Opinion in Biotechnology*, **16**, 198–203.

Stefanovic, E., Fitzgerald, G. & McAuliffe, O. (2017) Advances in the genomics and metabolomics of dairy lactobacilli: a review. *Food Microbiology*, **61**, 33–49.

Sun, Z., Harris, H.M., McCann, A., Guo, C., Argimón, S., Zhang, W., Yang, X., Jeffery, I.B., Cooney, J.C., Kagawa, T.F., Liu, W., Song, Y., Salvetti, E., Wrobel, A., Rasinkangas, P.,

Parkhill, J., Rea, M.C., O'Sullivan, O., Ritari, J., Douillard, F.P., Ross, R.P., Yang, R., Briner, A.E., Felis, G.E., de Vos, W.M., Barrangou, R., Klaenhammer, T.R., Caufield, P.W., Cui, Y., Zhang, H. & O'Toole, P.W. (2015) Expanding the biotechnology potential of lactobacilli through comparative genomics of 213 strains and associated genera. *Nature Communications*, **6**(8322), 1–13.

Turroni, F., Milani, C., van Sinderen, D. & Ventura, M. (2011) Genetic strategies for mucin metabolism in *Bifidobacterium bifidum* PRL2010: an example of possible human-microbe co-evolution. *Gut Microbes*, **2**, 183–189.

van de Guchte, M., Penaud, S., Grimaldi, C., Barbe, V., Bryson, K., Nicolas, P., Robert, C., Oztas, S., Mangenot, S., Couloux, A., Loux, V., Dervyn, R., Bossy, R., Bolotin, A., Batto, J.M., Walunas, T., Gibrat, J.F., Bessières, P., Weissenbach, J., Ehrlich, S.D. & Maguin, E. (2006) The complete genome sequence of *Lactobacillus bulgaricus* reveals extensive and ongoing reductive evolution. *Proceedings of the National Academy of Sciences of the United States of America*, **103**, 9274–9279.

van Hylckama Vlieg, J.E., Rademaker, J.L., Bachmann, H., Molenaar, D., Kelly, W.J. & Siezen, R.J. (2006) Natural diversity and adaptive responses of *Lactococcus lactis*. *Current Opinion in Biotechnology*, **17**, 183–190.

van Pijkeren, J.P. & Britton, R.A. (2014) Precision genome engineering in lactic acid bacteria. *Microbial Cell Factories*, **13**(Suppl. 1, S10), 1–11.

Ventura, M., O'Flaherty, S., Claesson, M.J., Turroni, F., Klaenhammer, T.R., van Sinderen, D. & O'Toole, P.W. (2009) Genome-scale analyses of health-promoting bacteria: probiogenomics. *Nature Reviews Microbiology*, **7**, 61–71.

Vido, K., Le Bars, D., Mistou, M.Y., Anglade, P., Gruss, A. & Gaudu, P. (2004) Proteome analyses of heme-dependent respiration in *Lactococcus lactis*: involvement of the proteolytic system. *Journal of Bacteriology*, **186**, 1648–1657.

von Ossowski, I., Reunanen, J., Satokari, R., Vesterlund, S., Kankainen, M., Huhtinen, H., Tynkkynen, S., Salminen, S., de Vos, W.M. & Palva, A. (2010) Mucosal adhesion properties of the probiotic *Lactobacillus rhamnosus* GG SpaCBA and SpaFED pilin subunits. *Applied and Environmental Microbiology*, **76**, 2049–2057.

Watanabe, M., van der Veen, S., Nakajima, H. & Abee, T. (2012) Effect of respiration and manganese on oxidative stress resistance of *Lactobacillus plantarum* WCFS1. *Microbiology*, **158**, 293–300.

Wei, Y-X., Zhang, Z-Y., Liu, C., Malakar, P.K. & Guo, X-K. (2012) Safety assessment of *Bifidobacterium longum* JDM301 based on complete genome sequences. *World Journal of Gastroenterology*, **18**, 479–488.

Wells, J.M. & Mercenier, A. (2003) Lactic acid bacteria as mucosal delivery system. In *Genetics of Lactic Acid Bacteria*. (ed. B.J.B. Wood & P.J. Warner), 261–290. Kluwer Academic/Plenum Publishers, New York.

Whiley, R.A. & Kilian, M. (2003) International committee on systematics of prokaryotes subcommittee on the taxonomy of staphylococci and streptococci: minutes of the closed meeting, 31 July 2002, Paris France. *International Journal of Systematic and Evolutionary Microbiology*, **53**, 915–917.

Yilmaz, P., Kottmann, R., Field, D., Knight, R., Cole, J.R., Amaral-Zettler, L., Gilbert, J.A., Karsch-Mizrachi, I., Johnston, A., Cochrane, G., Vaughan, R., Hunter, C., Park, J., Morrison, N., Rocca-Serra, P., Sterk, P., Arumugam, M., Bailey, M., Baumgartner, L., Birren, B.W., Blaser, M.J., Bonazzi, V., Booth, T., Bork, P., Bushman, F.D., Buttigieg, P.L., Chain, P.S., Charlson, E., Costello, E.K., Huot-Creasy, H., Dawyndt, P., DeSantis, T., Fierer, N., Fuhrman, J.A., Gallery, R.E., Gevers, D., Gibbs, R.A., San Gil, I., Gonzalez, A., Gordon, J.I., Guralnick, R., Hankeln, W., Highlander, S., Hugenholtz, P., Jansson, J., Kau, A.L., Kelley, S.T., Kennedy, J., Knights, D., Koren, O., Kuczynski, J., Kyrpides, N., Larsen, R., Lauber, C.L., Legg, T., Ley, R.E., Lozupone, C.A., Ludwig, W., Lyons, D., Maguire, E., Methé, B.A., Meyer, F., Muegge,

B., Nakielny, S., Nelson, K.E., Nemergut, D., Neufeld, J.D., Newbold, L.K., Oliver, A.E., Pace, N.R., Palanisamy, G., Peplies, J., Petrosino, J., Proctor, L., Pruesse, E., Quast, C., Raes, J., Ratnasingham, S., Ravel, J., Relman, D.A., Assunta-Sansone, S., Schloss, P.D., Schriml, L., Sinha, R., Smith, M.I., Sodergren, E., Spo, A., Stombaugh, J., Tiedje, J.M., Ward, D.V., Weinstock, G.M., Wendel, D., White, O., Whiteley, A., Wilke, A., Wortman, J.R., Yatsunenko, T. & Glöckner, F.O. (2011) Minimum information about a marker gene sequence (MIMARKS) and minimum information about any (x) sequence (MIxS) specifications. *Nature Biotechnology*, **29**, 415–420.

Zhang, Z-Y., Liu, C., Zhu, Y-Z., Wei, Y-X., Tian, F., Zhao, G-P. & Guo, X-K. (2012) Safety assessment of *Lactobacillus plantarum* JDM1 based on the complete genome. *International Journal of Food Microbiology*, **153**, 166–170.

Zhang, G., Gao, B., Adeolu, M., Khadka, B. & Gupta, R.S. (2016) Phylogenomic analyses and comparative studies on genomes of the *Bifidobacteriales*: identification of molecular signatures specific for the order *Bifidobacteriales* and its different subclades. *Frontiers in Microbiology*, **7**(978), 1–17.

Zotta, T., Ricciardi, A., Ianniello, R.G., Parente, E., Reale, A., Rossi, F., Lucilla, I., Comi, G. & Coppola, R. (2014) Assessment of aerobic and respiratory growth in the *Lactobacillus casei* group. *PLoS One*, **9**(e99189).

4 Production and Maintaining Viability of Probiotic Micro-organisms in Dairy Products

A.Y. Tamime, M. Saarela, M. Wszolek, H. Ghoddousi, D.M. Linares and N.P. Shah

4.1 Introduction

Definitions of probiotic foods including dairy products have been reported by many researchers (FAO/WHO, 2001, 2002; Gardiner *et al.*, 2002b; Moeller & de Vrese, 2004; Malcata *et al.*, 2005; Sharma & Ghosh, 2006; Shah, 2007; Vasiljevic & Shah, 2008; Baker *et al.*, 2009; Ershidat & Mazahreh, 2009; Soccol *et al.*, 2010; Hati *et al.*, 2013; Shiby & Mishra, 2013; Hill *et al.*, 2014; Sharma & Devi, 2014; Santiago-Lopez *et al.*, 2015; Tunick & van Hekken, 2015), such as 'foods containing live micro-organisms believed to actively enhance health by improving the balance of microflora in the gut', and/or 'microbial cells preparations or components of microbial cells that have a beneficial effect on health and well-being of the host'. Several preparations, most containing strains of *Lactobacillus* spp. and/or *Bifidobacterium* spp., are well established in the market, and foods containing probiotic bacteria have been marketed in Japan since the 1930s. Since the publication of the first edition of this book in 2005, there has been a tremendous increase in the number of microbial species and strains included in probiotic dairy products (e.g. pasteurised milk, ice cream, cheeses and infant formula); fermented dairy products, however, remain the most common vehicle for probiotic organisms (Tamime *et al.*, 1995; Belem, 1999; Lourens-Hattingh & Viljoen, 2001a; Hawrelak, 2002; Salama, 2002; Stanton *et al.*, 2002; Shah, 2004; Roy, 2005; Sanchez *et al.*, 2009; Granato *et al.*, 2010; Homayouni *et al.*, 2012b; Khan, 2014; Kumar *et al.*, 2015; Reid, 2015; Kandylis *et al.*, 2016). Of these, yoghurt is by far the most common vehicle for probiotic organisms. A number of health benefits have been linked to the consumption of foods containing probiotic bacteria (for further details, see Chapter 8). More than 100 probiotic fermented milk products are available in markets worldwide. To confer health benefits, it is advisable that products sold with any health claims contain a minimum concentration of probiotic bacteria of 10^6 colony forming units (cfu) mL^{-1} or g^{-1} at the expiry date. The minimum therapeutic dose per day is suggested to be 10^8–10^9 cfu mL^{-1} or g^{-1}; or, alternatively, the health effects relate to dosage where the concentration is not important, but how many cells are delivered per portion (e.g. total cfu per container consumed) (Lee & Salminen, 1996; Shah, 2000). However, studies have demonstrated variations in the viable counts of probiotic organisms in fermented

Probiotic Dairy Products, Second Edition. Edited by Adnan Y. Tamime and Linda V. Thomas.

milks (e.g. yoghurt and other dairy products), where the probiotic effects are considered strain specific; thus, probiotic products should state the strain, not just species or genera. Other important points, which have to be considered, are not just variations in counts, but whether the counts are below 10^6 cfu mL^{-1} or g^{-1}, whether the counts match those stated on the label and/or whether the organisms in the product are those stated on the label. Furthermore, it is important that the label shows the full strain names of the probiotic(s) and their minimum live count, guaranteed at end of shelf-life (Anonymous, 1992; Iwana *et al.*, 1993; Shah *et al.*, 1995; Collins *et al.*, 1998; El-Rahman, 2000; Shah & Ravula, 2000a; Viderola *et al.*, 2000a; Collins, 2001; Lourens-Hattingh & Viljoen, 2002; La Torre *et al.*, 2003; Varga *et al.*, 2003; Masco *et al.*, 2005; Champagne *et al.*, 2011; Karimi *et al.*, 2011, 2012; Plessas *et al.*, 2012; Mani-Lopez *et al.*, 2014; Unno *et al.*, 2015). Several factors may affect the viability of probiotic cultures in fermented milks, including the final acidity of the product, availability of nutrients, dissolved oxygen and oxygen permeation through the package. The stability of probiotic cultures has been seen as an issue for dairy manufacturers and consumers and, in this chapter, the technical and scientific aspects of probiotic dairy products will be reviewed.

4.2 Probiotic micro-organisms

'Traditional' lactic acid bacteria (LAB) that are normally used during the manufacture of fermented milks and cheeses belong to the genera *Lactococcus*, *Leuconostoc*, *Pediococcus*, *Streptococcus*, *Lactobacillus* and *Propionibacterium*; the former two genera are mesophilic, whilst the latter types are thermophilic. In some applications, blue and white moulds are used in cheesemaking, and *Geotrichum candidum* is used in Viili production to produce a velvety surface on top of the product. In addition, a wide range of yeasts have been identified in the kefir grains, but the properties of these products will not be covered in this publication, as they were detailed by Tamime (2006a). However, *Saccharomyces cerevisiae* var. *boulardii* is the only yeast that has been identified as being probiotic, and limited data are available on the use of this species in dairy products (Lourens-Hattingh & Viljoen, 2001b; Surawicz, 2003; Stroehlein, 2004; Olivares & Xaus, 2007; Meile *et al.*, 2008; Rajkowska & Kunicka-Styczynska, 2009; Urkek *et al.*, 2014; Gil-Rodriguez *et al.*, 2015). The genus *Bifidobacterium* is widely used in mixed 'traditional' LAB starter cultures during the manufacture of probiotic dairy products.

4.2.1 *General characteristics*

Probiotic micro-organisms, which have been used in fermented and unfermented milk products including cheese, are shown in Table 4.1, together with their main metabolic products. This provides some information on their possible role in flavour production, but it should be noted that the traditional LAB (i.e. starter cultures) are mainly responsible for much of the flavour and aroma (Tamime *et al.*, 2006a).

The strains of pediococci, lactobacilli, enterococci and bifidobacteria that are used as probiotic micro-organisms in dairy products do not use the tricarboxylic acid cycle

Table 4.1 Some selected characteristics of probiotic micro-organisms used in dairy foods and their principle metabolic products.

Starter organism	Metabolic product	Lactose fermentation
I. Lactic acid bacteria		
Pediococcus acidilactici	DL lactate	Homofermentative
Lactobacillus acidophilus, gasseri, helveticus and *johnsonii*	DL lactate	Homofermentative
Lactobacillus casei, reuteri, plantarum, rhamnosus and *fermentum*	DL lactate	Heterofermentative
Bifidobacterium adolescentis, animalis subsp. *animalis, bifidum, breve, infantis*[1], *animalis* subsp. *lactis*, and *longum*[2]	L(+) lactate, acetate	Heterofermentative
Enterococcus faecium and *faecalis*	L(+) lactate	Homofermentative
II. Yeasts		
Saccharomyces cerevisiae var. *boulardii*	? Ethanol, CO_2	

[1] Presumed to be *Bifidobacterium longum* subsp. *infantis*.
[2] Presumed to be *Bifidobacterium longum* subsp. *longum*.
Data adapted from Masco *et al.* (2004) and Tamime *et al.* (2006).

when fermenting the milk, although some of its enzymes may be present. These strains also do not possess any cytochrome system for harnessing energy from electrons of the reduced form of nicotinamide adenine dinucleotide (NADH). Energy is largely obtained via substrate-level phosphorylation and the enzyme adenylpyrophosphatase (ATPase) of the cytoplasmic membrane. Carbohydrate is metabolised through either homofermentative or heterofermentative metabolic pathways; bifidobacteria metabolise the lactose in milk via the heterolactic fermentation (Marshall & Tamime, 1997). Details of all these metabolic pathways have been reviewed by Tamime *et al.* (2006a) and Lahtinen *et al.* (2012).

4.2.2 *Examples of commercial starter culture blends*

Yoghurt is normally manufactured using *Streptococcus thermophilus* and *Lactobacillus delbrueckii* subsp. *bulgaricus* as starter cultures. Probiotic micro-organisms grow slowly in milk; hence, the yoghurt starter culture (either the cocci or lactobacilli) is added to enhance the fermentation process, and probiotics, such as *Lactobacillus acidophilus*, *Bifidobacterium* spp. and *Lactobacillus casei*, are incorporated as dietary adjuncts (Leroy & de Vuyst, 2004; Minelli *et al.*, 2004; Saito, 2004). Products made with only *Lb. acidophilus* are known as 'Acidophilus milk' (e.g. sweet or fermented), or alternatively products with *Lb. acidophilus* LA-5 and *Bifidobacterium animalis* subsp. *lactis* BB-12 (known as AB cultures); *Lb. acidophilus* LA-5, *Bif. animalis* subsp. *lactis* BB-12 and *Lb. casei* 431 (known as ABC cultures) (Maiocchi, 2001; Anonymous, 2008, 2013); or *Lb. acidophilus* LA-5, *Bif. animalis* subsp. *lactis* BB-12. and *Str. thermophilus* (known as ABT cultures) (Martin-Diana *et al.*, 2003) could be manufactured. However, if milk is fermented with only AB, ABS or some blends of ABT cultures (e.g. ABT-1 to 7, 10 and 12), this increases the incubation period and affects product quality (i.e. milder flavour) (Anonymous, 2008). Other blends of probiotic starter cultures consisting of *Str. thermophilus*, *Lb. delbrueckii* subsp. *bulgaricus*, *Bif. animalis* subsp. *lactis* BB-12 and *Lactococcus lactis* subsp. *lactis* are known as BMY-1 and 2, but blend BY-700 does not contain any lactococci (Anonymous, 2008). Thus, the normal practice is to make yoghurt with *Str. thermophilus*, *Lb. delbrueckii* subsp. *bulgaricus* ('Y' culture) and probiotic micro-organisms, such as AB, ABT, ABC, BMY or BY cultures.

Although normally the choice of any probiotic strain to be used as a starter culture or together with a starter culture is based on the health benefits of the strain (Gardiner *et al.*, 2002b), the following technological aspects have to be considered from a starter culture manufacturer's point of view: (a) the ability of the probiotic micro-organisms to grow in a medium in which its cell counts increase, (b) the robustness of the organism to freezing and drying stages of preservation, and (c) the tolerance of the probiotic to gastric acidity and bile salts found in the gastrointestinal (GI) tract. Nevertheless, the blending of strains to make starter cultures for the manufacture of fermented milk products, including those that are probiotic, is a critical procedure. The measures needed to ensure the desired attributes of the final product have been detailed by Tamime *et al.* (2006a). The choice and the ratio of the strains in the starter culture are key factors in this respect, which may include considerations of the fermentation time, texture, mildness, sugar tolerance and post-acidification profiles (see Figure 4.1). However, an important feature for live cells of the probiotic products is the number and stability of the probiotic strains, bearing in mind the current trend towards products with longer shelf-life, which may be up to 52 d. The blends of probiotic bacteria and LAB starter cultures are typically tested by manufacturers in quality control laboratories to evaluate the stability of probiotic micro-organisms during 28 d of shelf-life at 8 °C. Usually, the primary aim is to have a high count of 1×10^9 cfu g^{-1} of probiotic strains at the end of the product's storage period.

In addition, interactions between the probiotic strains and the traditional starter cultures must be considered in order to guarantee the required probiotic count at the end of the product's shelf-life. For example, some strains of the Y culture may inhibit some strains of probiotic bacteria during the fermentation and storage of the product. Probiotic micro-organisms may be particularly influenced by other bacteria during long fermentations. In contrast, however, the growth of most probiotic bacteria hardly gets started during a short fermentation time, and the strain variability does not seem to be affected under these conditions. A positive interaction between some probiotic strains is also known, for example between *Bifidobacterium* spp. and *Lb. acidophilus* (Vinderola *et al.*, 2002).

Some probiotic micro-organisms may influence the flavour of the fermented product. For example, *Bifidobacterium* spp. will, when present in high numbers, produce a noticeable amount of acetic acid during a long fermentation time (Mahdi *et al.*, 1990; La Torre *et al.*, 2003); whilst *Lb. acidophilus* will produce acetaldehyde and lactic acid, which contributes to the characteristic of 'bio' yoghurt flavour. Highly proteolytic probiotic strains may produce peptides, which confer a cheesy flavour/taste to the fermented milk product (Rasic & Kurmann, 1983).

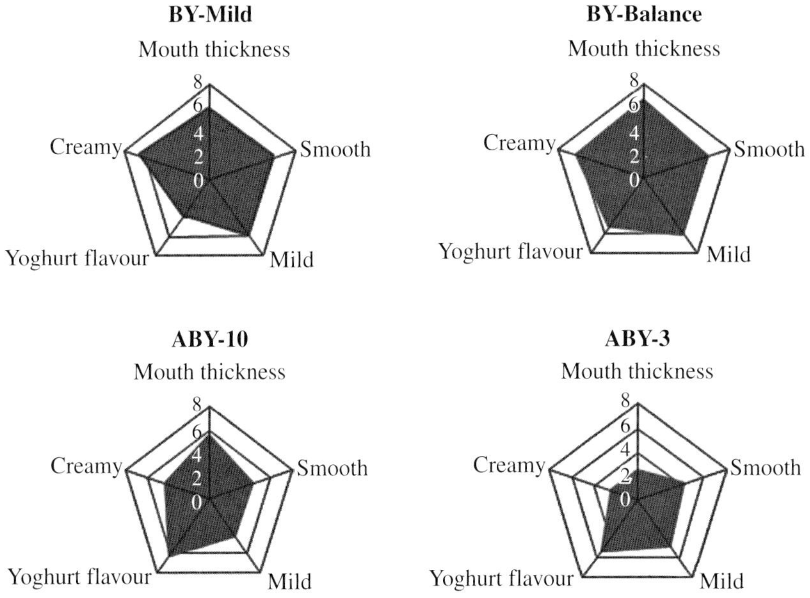

Figure 4.1 Sensory profiling of four different blends of nu-trish® probiotic starter cultures for drinking yoghurt.
Note: It is evident that starter cultures BY-Mild and BY-Balance have the highest mouth thickness attribute, whilst ABY-10 and ABY-3 had the highest yoghurt flavour attribute.
BY-Mild and BY-Balance: B = *Bifidobacterium animalis* subsp. *lactis* BB-12; Y = yoghurt cultures or *Streptococcus thermophilus* and *Lactobacillus delbrueckii* subsp. *bulgaricus*. ABY-10 and ABY-3: A = *Lactobacillus acidophilus* LA-5; B = Bif. *animalis subsp. lactis* BB-12; Y = yoghurt cultures or *Str. thermophilus* and *Lb. delbrueckii* subsp. *bulgaricus*.Data by permission of Chr. Hansen, Hørsholm, Denmark.

Probiotic bacteria currently used in commercial products mainly belong to the genera *Lactobacillus* and *Bifidobacterium*. Other species that have been identified as probiotic include *Pediococcus acidilatici*, *Enterococcus* spp. and the yeast *Sac. cerevisiae* var. *boulardii*, and the use of some of these species in dairy products has been reported by many researchers (Tamime *et al.*, 1995; Schillinger, 1999; Krishnakumar & Gordon, 2001; Lourens-Hattingh & Viljoen, 2001b; Tamime, 2002; Holm, 2003; Playne *et al.*, 2003; Masco *et al.*, 2004; FAO/WHO, 2006; Georgieva *et al.*, 2009; Franz *et al.*, 2011; Kongo & Malcata, 2016). However, *Sac. cerevisiae* var. *boulardii* is not used in foods; instead, it is sold as supplements. *Lactobacillus casei* Shirota is used as a single strain for the manufacture of Yakult® (a Japanese fermented milk beverage) (Tamime & Marshall, 1997). In addition, the properties of *Enterococcus faecium* as a probiotic micro-organism in a fermented milk product have been reviewed by Bertolami & Farnworth (2008).

The overall pattern of consumption of all types of fermented milks is steadily increasing in the majority of countries in the world, and this may be attributed to the nutritional and health aspects associated with these products (IDF, 2015). A detailed review of probiotic dairy products follows, and updates of newer probiotic strains are given in subsequent sections.

4.3 Economic value

For probiotic products, micro-organisms are selected for their various health benefits. Several preparations containing probiotic strains of *Lb. acidophilus* and bifidobacteria have become well established in the market. In France, products containing probiotic *Lb. acidophilus* and *Bifidobacterium* spp. have increased by approximately 300% to capture 4% of total fresh milk sales (Hughes & Hoover, 1991). In the mid-2000s, 11% of all the yoghurt sold in France contained *Bifidobacterium* spp. In Europe, probiotic applications are mainly restricted to fermented milk products and juices, and the economic values ($million) of such products in 1998 in some selected countries were 60 in Germany, 30 in The Netherlands, 28 in France, 24 in Spain and 18 in the United Kingdom (UK) (Shortt, 1999). However, the economic values of probiotic yoghurt and probiotic drinking yoghurt sold in the UK in 2002 were £320.2 million and £68 million, respectively (Anonymous, 2003). Currently, Kongo & Malcata (2016) reported that the fermented milk market in Europe, Canada, the United States of America (USA) and Asian countries was €billion 63.2, which accounted for 77% of the global market. The emergence of probiotic fermented milk products, however, has contributed to the growth of the dairy sector (i.e. consumption and economic value). In addition, probiotic micro-organisms are used as co-cultures with 'traditional' starter cultures for the manufacture of other dairy products, such as sweet milk, whey drinks, cheeses and infant formula; no worldwide data are available regarding the economic value of these probiotic products, but reports have been published (http://www.marketresearch.com/browse.asp?categoryid=510&SID=1498-1449-302165162-317076963 and http://www.ipaeurope.org/images/image/pdf/Euromonitor-Market-Data.pdf). It is safe to assume that the market value of these products has increased dramatically over the last decade, and Table 4.2 illustrates some examples of commercially available probiotic dairy products in some selected countries.

4.4 Unfermented probiotic milk

Limited data are available for probiotic liquid milk products (i.e. unfermented), and one such product is known as 'sweet Acidophilus milk' (Tamime & Marshall, 1997; Shortt, 1999). The manufacture of this product does not entail fermentation of the milk, because a concentrated *Lb. acidophilus* preparation is added to cold pasteurised milk before packaging; the anticipated viable count is ~5×10^6 cfu mL^{-1} (Salji, 1992). Young & Nelson (1978) reported that the level of *Lb. acidophilus* declined by a factor of 10 during a two-week storage period, which was attributed to the level of inoculum and the strain used.

Unfermented milk containing '*Bifidobacterium longum* ATCC 15708' (presumed to be *Bifidobacterium longum* subsp. *longum*) (Mattarelli *et al.*, 2008; Underwood *et al.*, 2015) at a dose of 5×10^8 cfu mL^{-1} improved lactose digestion *in vivo*, and the milk was better tolerated by lactose maldigesters (Jiang *et al.*, 1996). Another product similar to Acidophilus milk was known as BRA sweet milk, because it contained '*Bifidobacterium infantis*' (presumed to be *Bifidobacterium longum* subsp. *infantis*) (Mattarelli *et al.*,

Table 4.2 Examples of probiotic dairy products in some selected markets[1].

Trade name	Producer – country of origin	Probiotic micro-organisms present in the products as stated by the manufacturer[2]
Non-fermented milk		
Gefilus	Valio – Finland (FI)	*Lactobacillus rhamnosus* GG
Yoghurt/viscous products		
AB-Jogurtti	Juustoportti – FI	*Lactobacillus acidophilus*, *Bifidobacterium* spp.
Activia	Danone – France (FR)	*Lactobacillus casei* Immunitas
Benecol	Benecol – FI	*Bifidobacterium* spp.
Biola	Tine – Norway (NO)	*Lactobacillus rhamnosus* GG, *Lb. acidophilus*, *Bifidobacterium* spp.
Ekologisk	Änglamark – Sweden (SE)	*Lb. acidophilus*, *Bifidobacterium animalis* subsp. *lactis*
Fjällyoghurt	Milko – SE	*Lb. acidophilus*, *Bifidobacterium bifidum*
Gefilus	Valio – FI	*Lb. rhamnosus* GG
Lc1	Nestle – Switzerland (CH)	*Lactobacillus johnsonii*
L. Casei Piimä	Satamaito – FI	*Lb. casei*
Drinkable or beverage fermented milk/low-viscous products		
Actimel	Danone – FR	*Lb. casei* Immunitas
A-Fil	Arla – SE Skånemejerier – SE Norrmejerier – SE Milko – SE	*Lb. acidophilus*
AB-Piimä	Arla – FI	*Lb. acidophilus*
Aktifit Plus	Emmi – CH	*Lb. rhamnosus* GG
Benecol	Benecol – FI	*Bifidobacterium* spp.
Bifidus	Emmi – CH	*Lb. acidophilus*, *Bifidobacterium* spp.
Bifidus[3] or Activia	Danone – FR	*Lb. acidophilus*, Bifidus[4]
Biogarde	Strohmann – Germany (DE) Almhof – The Netherlands (NL) Albert Heijn – NL	*Lb. acidophilus*, *Bifidobacterium* spp.
Cultura Balance	Arla – Denmark (DK)	*Lb. casei* F19, *Lb. acidophilus* LA-5, *Bif. animalis* subsp. *lactis* BB-12
Gefilus	Valio – FI	*Lb. rhamnosus* GG
Öresundsfil	Skånemejerier – SE	*Lb. acidophilus*, *Bifidobacterium* spp.
Verum	NorrMejerier – SE	*Lb. rhamnosus* LB21
Vifit	Campina – NL	*Lb. rhamnosus* GG
Yakult	Yakult – NL	*Lb. casei* Shirota
Yosa[5]	Bioferme – FI	*Bif. animalis* subsp. *lactis* BB-12, *Lb. acidophilus* LA-5
Gefilus	Valio – FI	*Lb. rhamnosus* GG
Kefir	Bakoma – Poland (PL)	*Lb. acidophilus*, *Bifidobacterium lactis*[6]

(*Continued*)

Table 4.2 (Continued)

Trade name	Producer – country of origin	Probiotic micro-organisms present in the products as stated by the manufacturer[2]
Kefir	Bieluch – PL	*Lb. acidophilus* LA-5, *Bifidiobacterium* BB-12[6]
Concentrated fermented milk products		
Total	Fage – Greece (GR)	*Lb. acidophilus*, Bifidus[4], *Lb. casei*
Greek-style yoghurt	Morrison – United Kingdom (UK)	*Bif. lactis*[7], *Lb. acidophillus*
Authentic strained yoghurt	Chobani – UK	*Lb. acidophilus*, Bifidus[4], *Lb. casei*
Greek-style yoghurt	Fontera – New Zealand (NZ)	*Lb. acidophilus*, Bifidus[4]
Greek-style yoghurt	Roaming Cow – Australia (AU)	*Lb. acidophilus*, Bifidus[4], *Lb. casei*
Greek yoghurt	Jalna – AU	*Lb. acidophilus*, *Bifidobacterium* spp., *Lb. casei*
Okios	Stoneyfield – United States of America (USA)	*Lb. acidophilus*, Bifidus[4], *Lb. casei*
Greek yoghurt	Kroger – USA	*Lb. acidophilus*, *Bif. bifidum*
Greek-style yoghurt	Greek Gods – Canada (CA)	*Lb. acidophilus*, *Bifidobacterium* spp., *Lb. casei*
Greek yogourt	Liberté – CA	*Lb. acidophilus*, *Bif. lactis*[7], *Lb. casei*
Skyr	Siggis – Iceland (IS)	*Lb. acidophilus*, *Bif. lactis*[7]
Ymer	Arla – DK	*Lb. acidophilus*
Cheese		
Fitness Quark	Onken – DE	*Lb. acidophilus*, *Bifidobacterium* spp.

Data compiled from product labels and/or manufacturer websites.
[1] Data compiled from the websites of the commercial dairy companies.
[2] Strains and species of lactic acid starter cultures are not listed; in some cases, the strain identification of the probiotic organism was not stated.
[3] Data compiled from www.scienceforhealth.info.
[4] Presumed to be *Bif. animalis* subsp. *lactis* Bifidus ActiRegularis DN-173 010.
[5] This is an oat drink similar to a yoghurt-like product.
[6] Presumed to be *Bif. animalis* subsp. *lactis* BB-12.
[7] Presumed to be *Bif. animalis* subsp. *lactis*.

2008; Underwood *et al.*, 2015), *Lactobacillus reuteri* and *Lb. acidophilus* (Rothschild, 1995). Recently, strains of *Lactobacillus rhamnosus* GR-1 and *Lb. reuteri* RC-14 were shown to survive in milk containing inulin and yeast extract (Hekmat & Reid, 2007); although the milk was fermented overnight at 37 °C, the counts decreased by only 1 $\log_{10}$ cycle after 28 d of cold storage. In probiotic milk, Awaisheh *et al.* (2012) isolated strains (two of each) of *Lb. acidophilus*, *Lactobacillus gasseri* and *Lb. reuteri* from newly born Jordanian infants (i.e. breastfed). The isolates were inoculated into cold pasteurised milk (mono or mixed cultures) and, after 15 d storage at 4 °C, the viable counts of all the isolates ranged between 8.6 and 9.0 $\log_{10}$ cfu mL^{-1}. In the presence of isoflavones and phytosterols in the milk, and after incubation anaerobically at 37 °C for 18 h, the viability of single strains in the milk was enhanced, but not when they were mixed together (Awaisheh *et al.*, 2012). Pasteurised lactose hydrolysed and microfiltered skimmed milk inoculated aseptically with *Bif. animalis* subsp. *lactis* BB-12 and stored for 43 d at 5 °C had a viable count of >8 $\log_{10}$ cfu mL^{-1} after 28 d of cold storage (Alves *et al.*, 2016). In a carrot-flavoured pasteurised milk inoculated with different probiotic bacterial strains (*Lb. acidophilus* LA-5, *Lactobacillus plantarum*, *Lb. rhamnosus* GG or *Bif. animalis* subsp. *lactis* BB-12), *Lb. acidophilus* LA-5 seemed to be more stable than the other probiotic bacteria, with >98% viability during storage (20 d at 4 °C) compared to 88–92% viability for the other strains. Slight changes in the acidity of the product occurred, but this was strain related (Daneshi *et al.*, 2012). In a study by Zafari *et al.* (2013), chicory plant (*Dorema aucheri* – amount added ranged between 0.03 and 0.09 g 100 mL^{-1}) enhanced the growth of *Bif. bifidum* and *Lb. acidophilus* in milk, and the viable counts (cfu g^{-1}) of the probiotic bacteria after 21 d were 73.3×10^9 and 77.7×10^9, respectively.

Other unfermented probiotic dairy products are available in different markets, such as ice cream, butter and baby formula, and they will be reviewed separately in subsequent sections.

4.5 Probiotic fermented milks and beverages

A wide range of fermented milk products is made in many different countries. The classical example is yoghurt, which is manufactured as set, stirred and/or drinking types, and these products can be flavoured by adding fruit preparations or fruit essences plus colouring matter. Different mammalian milks have been used for the manufacture of fermented milks, including beverages (Hussein *et al.*, 2013). The fat content can be standardised, and the solids-not-fat (SNF) level can be fortified (e.g. with skimmed milk powder - SMP) to enhance rheological properties. In addition, yoghurt-related products, such as concentration of the fermentate (Labneh, Greek yoghurt, Ymer and Skyr), are made using a co-culture ('traditional' LAB as starter cultures, e.g. yoghurt and cheese) with probiotic bacteria. The technical and scientific aspects regarding the production of these products have been detailed by Kurmann *et al.* (1992), Tamime *et al.* (1995), Tamime & Robinson (1999, 2007), Bottazzi (2002), Tamime (2006a), Chandan *et al.* (2008), Yildiz (2010), Saad *et al.* (2013), Ozer & Tamime (2013) and Yerlikaya (2014). The manufacturing stages of probiotic yoghurt are very similar to those of 'classical' yoghurt, but natural/plain probiotic yoghurt is slightly sweeter in taste and the fermentation time is usually slightly longer when compared with the 'classical' product (Van de Water, 2003). It is interesting to note that there have been considerable development work and scientific publications in this field over the past decade; thus, these probiotic dairy products will be reviewed based on the proposed classification of fermented milks by Robinson *et al.* (2002), such as lactic acid fermentations, yeast–lactic acid fermentations and mould–lactic acid fermentations. Some examples follow.

4.5.1 *Lactic acid fermentations*

The group of products that fall within this category are mainly classified as:

- Mesophilic lactic acid fermentations (optimal growth at ~30 °C) with starter cultures belonging to the genera *Lactococcus*, *Leuconostoc* and *Pediococcus*. Some examples include 'general' types of fermented milks, Nordic sour milk products, fermented buttermilk (natural or cultured – the former type is a by-product of cultured buttermaking) and Dahi (an Indian fermented milk); the latter product is sometimes made using yoghurt starter cultures.
- Thermophilic lactic acid fermentations (optimal growth at ~42–45 °C) where the starter cultures are mainly the yoghurt organisms and *Lactobacillus helveticus*.
- Probiotic lactic acid fermentations (optimal growth at ~37 °C) where the starter cultures are mainly the yoghurt organisms and a wide range of probiotic bacteria.

Mesophilic probiotic fermented milks

Nordic cultured buttermilk (Piimä, Filmjölk, i.e. drinking type) is made by microbial fermentation of pasteurised whole milk or skimmed milk. Typical starter cultures in fermented buttermilk are mesophilic LAB, such as *Lactococcus lactis* subsp. *cremoris*, *Lac. lactis* subsp. *lactis*, *Lactococcus lactis* subsp. *lactis* biovar. *diacetylactis* and *Leuconostoc mesenteroides* subsp. *cremoris*. The processed milk base is fermented at 20 °C for ~20 h (final pH: 4.5–4.6), followed by stirring, cooling, flavouring (optional) and packaging (Mantere-Alhonen & Forsen, 1990; Tamime & Marshall, 1997; Leporanta, 2001; Mayo *et al.*, 2010).

Originally, cultured buttermilk was consumed as a drink during meals and did not contain any flavouring. This is still the case, but there are now fruit-flavoured varieties also on the market, for example in Finland and Sweden. These flavoured products are consumed during breakfast (often with cereals), or as a snack in a manner similar to yoghurt, whereas unflavoured buttermilks (especially the Finnish products that are less viscous) are typically consumed as drinks, together with foods.

In addition to lactic starter cultures, probiotic-cultured buttermilk may contain different micro-organisms, such as *Lb. rhamnosus* GG [e.g. Gefilus-Piimä in Finland and Gefilus Hapupiim in Estonia (http://www.valio.fi/tuotteet/haku/?haku=piim%C3%A4), and Syrnet Melk in Norway (http://www.tine.no/merkevarer/biola/produkter/biola-syrnet-melk-med-bl%C3%A5b%C3%A6r)], *Lb. casei* [Piimä in Finland (http://www.satamaito.fi/tuotteet)] and *Lb. acidophilus* alone or together with *Bifidobacterium* spp. (Fil in Sweden, Piimä in Finland and Syrnet Melk in Norway) (see Table 4.2).

Thermophilic probiotic fermented milks

Yoghurt and other fermented milks containing probiotic bacteria have become popular worldwide. In order to find new probiotic strains, researchers have tried screening traditional fermented milk products made in different countries, such as Nigeria and other African countries (Banwo *et al.*, 2013; Franz *et al.*, 2014), Portugal (Barbosa *et al.*,

2014), Iran (Heidarpour *et al.*, 2013; Emami *et al.*, 2014; Iranmanesh *et al.*, 2014; Sharafi *et al.*, 2015), Mongolia (Takeda *et al.*, 2011, 2015; Kimoto-Nira *et al.*, 2015; Kuda *et al.*, 2016), Kazakhstan (Kushugulova *et al.*, 2013), Italy (Pisano *et al.*, 2011) and Sardinia (Ortu *et al.*, 2007). Other potential sources of probiotic strains have been screened, such as dairy products, human faeces (Liu *et al.*, 2013; Archer & Halami, 2015) and human breast milk (Zacarias *et al.*, 2011). Other aspects of probiotic dairy products that have been reported in the scientific literature include general reviews on probiotic bacteria and the safety of lactobacilli (Chen *et al.*, 2006; Bernardeau *et al.*, 2008; Prasanna *et al.*, 2014; Chen *et al.*, 2015; van den Nieuwboer, 2016), enterococci (Ramakrishnan *et al.*, 2014), phage infection of probiotic strains (Capra *et al.*, 2009; Mercanti *et al.*, 2015), patents on probiotic dairy products (Nose *et al.*, 2007; te Biescbeke & de Vries, 2009; Peneva & Aleksandrov, 2013, 2015; Penhasi, 2013a, 2013b), the antibacterial effects of probiotics (Dabiza *et al.*, 2006), the use of whey and buttermilk to grow and freeze probiotic lactobacilli (Burns *et al.*, 2008) and the antibiotic resistance of commercially available probiotic starter cultures (Sharma *et al.*, 2014).

Nevertheless, '*Bifidobacterium bifidum* Bb-12' (presumed to be *Bif. animalis* subsp. *lactis* BB-12) (Anonymous, 2013, 2016) grew faster in goat's milk compared to cow's milk fortified with whey protein concentrate (WPC) powder and inulin, and had viable counts of 2.3×10^8 cfu mL^{-1} after 9 d at 5 °C (Bozanic & Tratnic, 2001), whilst the same milk did not enhance the growth of *Lb. acidophilus* LA-5 (Bozanic *et al.*, 2004). The growth rates of *Lb. acidophilus* JCM 11047 (human origin) co-cultured with *Str. thermophilus* 510 in goat's milk were higher than in cow's milk, and the viable count was 10^7 cfu mL^{-1} after 7 d at 5 °C (Masuda *et al.*, 2005). Similar counts were reported for *Lb. acidophilus* LA-5 and *Bif. animalis* subsp. *lactis* BB-12 in goat's milk fortified with inulin and transglutaminase (Tg-ase) (Mituniewicz-Malek *et al.*, 2014). The AB culture (*Lb. acidophilus* and *Bif. bifidum*) and BC culture (*Bif. bifidum* and *Lb. casei*) were grown separately as co-culture with a yoghurt starter (Y culture) (*Str. thermophilus* and *Lactobacillus bulgaricus* – presumed to be *Lb. delbrueckii* subsp. *bulgaricus*) for the production of a probiotic goat's yoghurt. The AB product had viable counts of lactobacilli and bifidobacteria of 3.5×10^7 and 3.4×10^7 cfu mL^{-1}, respectively; whilst the BC product had viable counts of bifidobacteria and lactobacilli of 7.6×10^7 and 5.6×10^7 cfu mL^{-1}, respectively. The optimum proportion of both probiotic bacteria and Y culture was 2:1:1 (Shu *et al.*, 2015).

The cell counts of '*Bif. longum* BB-536' (presumed to be *Bif. longum* subsp. *longum* BB-536) co-cultured with a Y culture was lower in goat's milk yoghurt fortified with inulin and SMP than in cow's milk, and a slight increase in counts of 1.7 $\log_{10}$ cycle mL^{-1} was observed in the latter product (Simunek & Evacic, 2009). Abe *et al.* (2009b) reported counts of $>1.0 \times 10^7$ cfu mL^{-1} after storage for 35 d at 5 °C by '*Bif. longum* BB-536' (presumed to be *Bif. longum* subsp. *longum* BB-536) (an isolate from an infant) co-cultured with a Y culture, but improved survival rate of bifidobacteria was obtained by reducing the incubation temperature to 37 °C and co-culturing with a *Lactococcus lactis* spp. The cell counts of *Lb. rhamnosus* GG in goat's milk yoghurt made and co-cultured with Y culture ranged between 10^8 and 10^9 cfu mL^{-1}; the addition of sugar (7 g 100 g^{-1}) resulted in reduced levels of short- and medium-chain fatty acids (FAs), and a less 'goaty' flavour, but the firmness of the coagulum was weak (Jia *et al.*, 2016).

Potentially enhanced therapeutic values of fermented milks (cow, goat and camel) with *Pediococcus pentosaceus* were achieved due to their increased antioxidant activity. This was highest in goat's milk (93%) > camel's milk (86%) > cow's milk (79 %); the FA profiles were also higher (Balakrishnan & Agrawal, 2014). Probiotic Dahi and yoghurts made from cow's and buffalo's milks with or without AB culture (*Lb. acidophilus* and '*Bif. bifidum*' – presumed to be *Bif. animalis* subsp. *lactis* BB-12) had different viable counts of probiotic bacteria, and lower count was observed in buffalo's milk ($P < 0.05$), but not the yoghurt organisms (Vijayendra & Gupta, 2014). In addition, comparative studies on the survival of probiotic bacteria in yoghurt made from camel's, cow's, goat's and sheep's milk or goat's and camel's milk were reported by Varga *et al.* (2014b) and Hussein *et al.* (2013), respectively,

Standardised buffalo's milk (4 fat and 10 g 100 g^{-1} SNF), fortified with *Aloe vera* (AV) juice (16 g 100 g^{-1}) and fermented with *Lac. lactis* subsp. *lactis* biovar *diacetylactis* NCDO 60 and *Lactobacillus paracasei* subsp. *paracasei* NCDO 627, resulted in enhanced probiotic counts in Dahi (Hussain *et al.*, 2016); whilst Farooq *et al.* (2013) isolated *Lb. acidophilus* strains from Dahi, which were potential probiotic bacteria. In a study by Kristo *et al.* (2003), *Lb. paracasei* subsp. *paracasei* B117 grew well in co-culture with a Y culture, but favoured a lower incubation temperature at 36–38 °C. A mixture of cow's and buffalo's milks (1:1) was used for fermentation by isolates from infant faeces (*Ent. faecium* NM 113 and NM 213, and *Lb. casei* NM 512) in co-cultures (1:1) with a Y culture; these isolates affected the rheological properties of the fermentate, but the viable counts of the probiotic bacteria were >10^7 cfu g^{-1} at the end of the storage period (Abdou *et al.*, 2015).

When *Lb. acidophilus* LA-5 and *Lb. rhamnosus* LR-35 were grown as single cultures or as co-culture with Y culture in milk containing two milk protein concentrates and two casein hydrolysates; *Lb. acidophilus* LA-5 grew well in the milk, but showed poor stability during storage. In contrast, *Lb. rhamnosus* LR-35 grew weakly in the milk, but was remarkably stable during storage. However, the growth of the same probiotic bacteria in the milk base containing casein hydrolysate required 11 h of incubation period, and the counts (cfu mL^{1}) were >10^6 for *Lb. acidophilus* LA-5 and 10^7 for *Lb. rhamnosus* LR-35 after 5 weeks' storage at 5 °C (Sodini *et al.*, 2002). Similarly, *Lb. acidophilus* LAC-4 and *Bif. animalis* subsp. *lactis* BL were blended with a Y culture (either as single or mixed) to ferment the milk at 42 °C until pH values reached 4.5. The counts of bifidobacteria and Y culture were high after 28 d storage, but *Lb. acidophilus* LAC-4 count decreased by 14 d and the final count was <10^6 cfu mL^{-1} after 28 d (Damin *et al.*, 2006). Replacing SMP with WPC and sodium caseinate (Na-Cn) affected the acidification rate of *Bif. animalis* subsp. *lactis* LAFTI® B94 and Y culture, and the counts of bifidobacteria were ~6.88 $\log_{10}$ cfu mL^{-1} after 28 d at 4 °C; the structure of the gel was more compact (Marafon *et al.*, 2011a, 2011b; Akalin *et al.*, 2012) probably due to the fusion of the casein micelles (Tamime *et al.*, 1984). Furthermore, supplementation of the milk base with whey protein isolate (WPI) and resistant starch improved the firmness of the probiotic yoghurt, and enhanced the rate of gelation, which was 98 min compared to 135 min for the control product (Skrzypczak & Gustaw, 2012). Supplementation of the milk base with ω-3-FA, isoflavones and phytosterols had no effects on monocultures of *Lb. gasseri* and '*Bif. infantis*' (presumed to be *Bif. longum* subsp. *infantis*) grown with a Y culture, resulting in high counts of each of the probiotic strains (i.e. 41.0 × 10^8 cfu mL^{-1}) after 15 d of storage (Awaisheh *et al.*, 2005).

The viability of *Lb. acidophilus* LA-5, *Lb. rhamnosus* LB-A and *Bif. animalis* subsp. *lactis* BL-04 (monocultures) in milk fermented with Y culture was found to be good, and the counts for both strains were similar (6.8 $\log_{10}$ cfu mL^{-1}) after 21 d at 4 °C (Saccaro *et al.*, 2009). The metabolic activity of *Bif. animalis* subsp. *lactis* HN 019 with Y culture in organic cow's milk yoghurt resulted in higher amounts of FAs, including conjugated linoleic acid (CLA) (Florence *et al.*, 2012, 2013). Similar work was reported by Bisig *et al.* (2007), Oliveira *et al.* (2009, 2011a), Rodriguez-Alcala *et al.* (2011) and do Espirito Santo *et al.* (2012a).

Using different strains of *Lb. plantarum* ACA-DC 146 and *Lb. paracasei* subsp. *tolerans* ACA-DC 4037 in probiotic yoghurt resulted in a low milk acidification activity. Viable counts were >7.0 $\log_{10}$ cfu g^{-1} after 14 d of storage, but increasing the microbial load further using concentrated and encapsulated inoculant (10–11 $\log_{10}$ cfu g^{-1}) resulted in yoghurts with long fermentation times and poor sensory properties (Maragkoudakis *et al.*, 2006a, 2006b; see also Elshaghabee, 2016). Other potentially mono probiotic strains (*Lb. plantarum* 14 or *Lactobacillus fermentum* 4a) acidified milk in 72 h at 37 °C, and maintained counts of each strain of 10^8 cfu mL^{-1} during 21 d of cold storage (Modzelewska-Kapitula *et al.*, 2008). Mirlohi *et al.* (2014) reported better survival of *Lb. plantarum* in yoghurt with the use of a slow-acid-producing strain of Y culture.

The quality of probiotic Bulgarian yoghurt, which is mainly made with a monoculture of *Lb. delbrueckii* subsp. *bulgaricus*, was influenced by the particular probiotic strain used (*Lb. plantarum*, *Lactobacillus salivarius* or *Lactobacillus brevis*), and only *Lb. plantarum* survived without affecting the sensory properties in the product (Tropcheva *et al.*, 2014). In a separate study, Makino *et al.* (2016) reported that exopolysaccharide (EPS) produced by a strain of *Lb. delbrueckii* subsp. *bulgaricus* OLL1073R-1 in Bulgarian yoghurt induced the production of interferon-γ (IFNγ) *in vitro*, which could exert immunostimulatory effects.

The fermentation and sensory properties of a probiotic isolate (*Lb. casei* Zhang; mono- or co-culture with *Str. thermophilus*) was reported by Wang *et al.* (2010, 2013). The sensory properties were influenced by storage time (28 d at 4 °C), and the highest scores were observed after 14 d of refrigerated storage. Wang *et al.* (2009) studied the transit tolerance of the same strain in soymilk and cow's milk during storage, whilst the lipid-lowering capability of five strains of *Lactobacillus* spp. was reported by Wang *et al.* (2015a). Co-culturing *Lb. casei* Zhang with a Y culture inhibited the growth of yeast and moulds in the product, and the *Lb. casei* Zhang count was 3×10^6 cfu g^{-1} (Li *et al.*, 2013). A South African fermented milk product (Yoba Mutandabota) made with *Lb. rhamnosus* Yoba was also shown to inhibit the growth of *Listeria monocytogenes*, *Campylobacter jejuni*, *Escherichia coli* O157:H7 and *Bacillus cereus*; *Salmonella* was the only pathogenic species to grow in the product (Mpofu *et al.*, 2014, 2016). The antimicrobial and antihypertensive activities of a probiotic *Lb. plantarum* Tensia DSM 21380 in fermented milk has been patented to be used in fermented milk; the patent also claimed that this product could suppress the growth of pathogens and non-starter lactobacilli, thus extending the shelf-life of food products at the end of the storage period (Songisepp *et al.*, 2009, 2014; Hutt *et al.*, 2015). It is of interest to note that *Lb. reuteri* RC-14 and *Lb. rhamnosus* GR-1 have also been used successfully in yoghurt making. These strains have been shown to colonise the intestine and vagina, and there is evidence that this combination can reduce recurrences of bacterial vaginosis, yeast vaginitis and urinary tract infections. Both probiotic strains have shown resistance to bile and have survived passage through the human GI tract without inducing systemic immune or inflammatory responses (Hekmat *et al.*, 2009).

Strain selection is important in probiotic yoghurt. For example, there was good stability of *Lb. acidophilus* strains PIM703 and SBT2062 ($\sim 6\times 10^7$ cfu g^{-1}) when either strain was co-cultured with a Y culture. Conversely, a study by Ng *et al.* (2011) showed that *Lb. acidophilus* ATCC 700396 and NCFM were slightly inhibited when co-cultured with *Lb. delbrueckii* subsp. *bulgaricus*. Other lactobacilli strains [*Lb. plantarum* DK 211 and DK 303, *Lb. paracasei* subsp. *paracasei* DK207 and DK215 and '*Lactobacillus sakei*' (presumed *Lactobacillus sakei* subsp. *sakei*); Woo *et al.*, 2010] have been isolated from Kimchi (a Korean fermented product). The isolate (*Lb. plantarum* DK 211) was used in yoghurt making, and the product was highly rated by the sensory panellists. The rest of the isolates from Kimchi (including *Lb. plantarum* DK 211) were acid and bile salt tolerant, inhibited the growth of certain pathogens except *E. coli* and could have potential as probiotic bacteria to be used during the manufacture of yoghurt (Baick & Kim, 2015). Further information on isolates from kimchi has been reported by Cho *et al.* (2013), Choi *et al.* (2015) and Khan & Kang (2016); however, Lee *et al.* (2015) have also characterised a probiotic lactococci isolated from kimchi.

In a different study, five commercial yoghurt-related fermented milk products containing probiotic bacteria – *Bifidobacterium* spp., *Lb. acidophilus*, '*Bif. lactis* Bb-12' (presumed to be *Bif. animalis* subsp. *lactis* BB-12) and *Lb. casei*, *Lb. casei* Shirota, *Lb. casei*, and *Lb. casei* and *Lb. rhamnosus* HN001 – were spiked with two different strains of *Mycobacterium avium* subsp. *paratuberculosis* (MAP). The MAP numbers decreased (averaged 2.3 $\log_{10}$ mL^{-1}) in the products after 6 weeks' storage (van Brandt *et al.*, 2011). Another probiotic isolate (*Bacillus indicus* HU 36, a carotenoid-producing organism) was used in yoghurt production; viable counts were ~5 and 3.5 $\log_{10}$ cfu mL^{-1} after 14 and 21 d at 4 °C, respectively. Although this strain increased the yellow colour of the product, it did not affect the sensory profiling and/or rheological properties of the yoghurt (Ersan *et al.*, 2016).

The use of different flavouring ingredients in probiotic yoghurts can affect the survival rate of these organisms. Examples include the following: (a) black locust honey (*Robinia pseudoacacia* L.) (5 g 100 mL^{-1}) improved the viability of the bifidobacteria during storage for 35 d at 4 °C in a product made with ABT starter culture (for details, refer to Section 4.2.2) (Varga *et al.*, 2014a), (b) chicory plant (*D. aucheri* – amount added ranged between 0.03 and 0.09 g 100 mL^{-1}) enhanced the growth of *Bif. bifidum* and *Lb. acidophilus* strains in yoghurt, and the viable counts (cfu g^{-1}) of the probiotic bacteria after 21 d were 88.3×10^9 and 71×10^9, respectively (Zafari *et al.*, 2013), and (c) strawberry juice, tomato paste or orange/carrot in probiotic yoghurt made with mono- or co-culture of '*Bif. bifidum* Bb-12' (presumed to be *Bif. animalis* subsp. *lactis* BB-12), *Lb. acidophilus* LA-5 and *Lb. casei* 01 increased the survival of the latter two strains, which had an average viable count of 10^7 cfu g^{-1} by the end of the storage period (Vinderola *et al.*, 2002; Taha *et al.*, 2007).

Other additives, which have been studied in relation to the survival rate and quality of probiotic yoghurts and fermented milks, are shown in Table 4.3. Some processing

Table 4.3 The effect of some selected additives on the quality and viable counts of probiotic yoghurt and fermented milks (viscous-type).

Probiotic bacteria	Additives	Comments/bacterial count colony forming units (cfu) mL^{-1} or g^{-1}	References
Lactobacillus acidophilus, *Bifidobacterium longum*[2]	Lao-chao[1]	The properties of the fermentate were different from yoghurt, and the counts of both organisms were 10^7–10^8 after 14 d at 4 °C.	Su *et al*. (2005)
Bifidobacterium lactis BB-12[3], *Lb. acidophilus* LA-5	Raffinose	The prebiotic helped to maintain viability of probiotic bacteria after 21 d at 4 °C.	Martinez-Villaluenga *et al*. (2006)
Bif. lactis BB-12[3], *Lb. acidophilus* LA-5 and *Lactobacillus paracasei* subsp. *paracasei*	L-cysteine	The additive enhanced the survival of all probiotic bacteria (>10^7) after 14 d at 4 °C, especially when the milk was fermented at 37 °C.	Guler-Akin & Akin (2007)
Bif. longum BL 05[2], *Lb. acidophilus* LA 14 and co-culture with *Lactobacillus delbrueckii* subsp. *bulgaricus* LB 3440 and *Streptococcus thermophilus* TA 040	Glucose oxidase (≤500 mg kg^{-1})	No effect on quality of product; all the probiotic counts were ~8 log_{10} after 30 d at 5 °C; however, Batista *et al*. (2015) achieved counts of >6 log_{10} cfu g^{-1} in glucose oxidase probiotic yoghurt.	Cruz *et al*. (2010a, 2010b, 2010c, 2011, 2012)
ABY-3 and ABT-5 (for details, refer to Section 4.2.2)	Rosehip extract	The viable counts of probiotic bacteria were 2.5×10^8 at the end of the storage period.	Mocanu *et al*. (2011)
Lactobacillus rhamnosus and *Str. thermophilus*	Inulin	The additive stimulated both biomass growth and levels of all end of the product's shelf life.	Oliveira *et al*. (2011b, 2011c, 2012)
Lb. acidophilus L 10 and NCFM, *Bifidobacterium animalis* subsp. *lactis* B 104 and HN 019 and co-cultured with yoghurt starter (CY 340)	Passion fruit peel powder	Fermentation time was reduced; rheological properties: all the skimmed milk yoghurts except the batches fermented with *Lb. acidophilus* L 10 and NCFM were improved, and the counts of bifidobacteria were about 1 log_{10} higher in full-fat yoghurt compared with the control after 28 d.	do Espirito Santo (2012b)
Lb. acidophilus and *Bif. animalis* subsp. *lactis*	Red or green lentils	The lentils exhibited strong antioxidant potential, and enhanced the viable counts during 28 d storage period at 4 °C.	Zare *et al*. (2012) and Agil *et al*. (2013)
Lb. acidophilus, *Bifidobacterium bifidum* (grown as monocultures)	Oregano, garlic, tea extract or coffee extract	The viable counts[5] averaged $\times 10^{10}$ in the 4 flavoured products for lactobacilli – 3.4, 7.3, 21.3 and 32.5, respectively – and for bifidobacteria were 3.25, 3.7, 31.8 and 23.3, respectively, after 21 d at 2 °C.	Marhamatizadeh *et al*. (2012b, 2012c, 2013, 2014)

(Continued)

Table 4.3 (Continued)

Probiotic bacteria	Additives	Comments/bacterial count colony forming units (cfu) mL^{-1} or g^{-1}	References
Bifidobacterium pseudocatenulatum G4, *Bif. longum* BB 536[2] and each was co-cultured with a yoghurt starter	*Mangifera pajang* fibrous polysaccharides (0.75 g 100 g^{-1}), inulin	The additives stimulated growth and improved activities (production of short-chain fatty acids and proteolysis) of both strains of bifidobacteria.	Al-Sheraji *et al.* (2012)
Many different strains of probiotic bacteria were used	Algae[4]	This is a review article: slight increase in probiotic counts was evident, and the added algae affected the sensory attributes of the probiotic yoghurts.	Beheshtipour *et al.* (2013)
Lb. acidophilus LA-5 and *Bif. animalis* subsp. *lactis* BB-12	Date syrup (up to 20 g 100 g^{-1})	Both organisms had viable counts of 10^6 in all date syrup levels used in the milk base except for bifidobacteria in fermented milk containing date syrup at 20 g 100 g^{-1}.	Al-Otaibi *et al.* (2013)
Lb. acidophilus CCDM 151 and *Enterococcus durans* CCDM 922; each was co-cultured with a yoghurt or mesophilic starter culture	Malt extract (5 g 100 g^{-1})	The additive did not affect the yoghurt or mesophilic starter culture, but inhibited growth of probiotic lactobacilli and not the enterococci; however, the fermentate was highly acceptable.	Nemeckova *et al.* (2013)
Lb. paracasei subsp. *paracasei* Lpc-37 and yoghurt starter culture	Inulin (2 g 100 mL^{-1})	The probiotic bacteria were mixed with banana purée (i.e. added on top of the fermentate containing inulin), and the viable count was 8.86 $\log_{10}$ cfu g^{-1} after 21 d, but the physical properties declined after 14 d.	Srisuvor *et al.* (2013); see also Yuksel & Bakirci (2014, 2015) and Kavaz & Bakirci (2014)
Lactobacillus helveticus 05-29 and *Lactobacillus casei* 05-211	Mungbean milk, soymilk and sugar	The additives were mixed with milk and cultured with probiotic bacteria isolated from dairy products, and the viable counts of both probiotic bacteria were ~10^8.	Zhihua *et al.* (2013)
Bif. animalis subsp. *lactis* BB-12 or *Lb. acidophilus* LA-5	Quinoa flour	Had no effect on fermentation time or the viable counts of the probiotic organisms after 28 d at 4 °C, but it did not have a positive effect on the adhesion of probiotic bacteria to Caco-2 cells *in vitro*.	Casarotti *et al.* (2014a)
Lb. acidophilus LA-5	Capuassu fruit and inulin	The additive (i.e. acidic and fibre-rich) improved the texture of the product, and the viable count was 7 after 28 d at <5 °C.	Costa *et al.* (2014, 2015)
Bifidobacterium breve ATCC 15701 or *Lactobacillus reuteri* ATCC DSM 20016; each was co-cultured with a yoghurt starter	Shiitake (mushroom) extract	The extract did not affect the growth of probiotic bacteria; the viable counts were 8.0 and 7.9 $\log_{10}$, respectively, after 35 d at 4 °C; and the shiitake enhanced the α- and β- galactosidase activities during storage.	Hassan *et al.* (2014)

Lb. acidophilus LA-5 or *Bifidobacterium animalis*[3]*; each was* co-cultured with a yoghurt starter	Oleoresins	Eight different yoghurts were made containing cardamom, cinnamon or nutmeg; the products had good sensory properties and acceptability, and the probiotic counts during refrigerated storage for 4 weeks were not affected.	Illupapalayam *et al.* (2014)
Lb. acidophilus NCDC-291 or *Bif. bifidum* NCDC 232; each was co-cultured with a yoghurt starter	Fructooligosaccharides (FOS) or inulin	The added supplements improved the growth and survival of the lactobacilli and bifidobacteria which averaged 7 and 7.3 $\log_{10}$, respectively, after 14 d at 4 °C.	Celestin *et al.* (2015)
Lb. acidophilus – ATCC 4356, *Lb. casei* – ATCC 393, *Lb. paracasei* subsp. *paracasei* – ATCC BAA52; each was co-cultured with a yoghurt starter	Pineapple peel powder (PPP) or inulin	The PPP additive improved the rheological properties and nutritional quality of probiotic yoghurts; the viable counts of probiotic bacteria ranged between 7.7 and 8.0 $\log_{10}$ after 28 d at 4 °C (i.e. 1 $\log_{10}$ cycle higher than the control).	Sah *et al.* (2015, 2016)
Lb. acidophilus LA-5, *Bif. lactis* BB-12[3]; each was co-cultured with a yoghurt starter	Phytosterols (18 g L^{-1})	The viable counts of probiotic lactobacilli and bifidobacteria were >6.55 and <8.90 $\log_{10}$, respectively, after 14 d at 5 °C.	Parsa *et al.* (2015)
Lb. casei ATCC 393	*Pistacia terebinthus* resin	Cells were encapsulated in the resin, which sustained their viability (7 $\log_{10}$) in yoghurt stored for 60 days at 4 °C; the resin also inhibited the growth of yeasts and moulds.	Schoina *et al.* (2015)
Lb. acidophilus NCFM	Onion juice (≥60 g kg^{-1})	Stimulated growth of the lactobacilli, enhanced antioxidant activity to <0.5 μmol TE g^{-1}, and viable count was ~9 $\log_{10}$ cfu g^{-1} after 14 d at <5 °C.	Li *et al.* (2016)
Lb. acidophilus, Bif. lactis[3]; each was co-cultured with a yoghurt starter	Sea buckthorn (strong antioxidant activity)	The viable counts of probiotic lactobacilli and bifidobacteria were 9.3 and 9.2 $\log_{10}$, respectively, after 21 d at 4 °C.	Gunenc *et al.* (2016)

[1] A Chinese fermented rice product using *Rhizopus javanicus* and *Saccharomyces cerevisiae*; the filtrate has milk-clotting activity.
[2] Presumed to be *Bifidobacterium longum* subsp. *longum*.
[3] Presumed to be *Bifidobacterium animalis* subsp. *lactis* BB-12.
[4] Strains used were *Spirulina platensis* and *Chlorella vulgaris*.
[5] The high counts obtained were due to the fact that the freeze-dried culture was grown in milk, and subsequently used as an active bulk starter culture.

factors, such as homogenisation pressures and/or type of milk, can have similar effects. The structure of probiotic buffalo's and cow's probiotic yoghurt containing *Lb. acidophilus* can have larger void spaces, which was shown to affect their physical properties, and resulted in a lower viable count of *Lb. acidophilus* (5.17 $\log_{10}$ cfu g^{-1}) in buffalo's yoghurt at the end of the storage period (Nguyen *et al.*, 2014). Similarly, the structures of probiotic yoghurts made using different commercial probiotic starter cultures (YO MIX 236 and DPL ABY – refer to Section 4.2.2 for details) were influenced by the milk base processing conditions, such as the addition of SMP and high hydrostatic pressure (HHP) at 676 MPa for 5 min (Penna *et al.*, 2007) – the normal processing pressure is 17.5 MPa, as reported by Tamime & Robinson (1999). In comparison, autoclaved skimmed milk was mixed with a Y culture and *Lb. acidophilus* LA-K, homogenised for five continuous passes at different pressures (0, 3.45, 6.90, 10.34 and 13.80 MPa) and then fermented (Muramalla & Aryana, 2011). Homogenisation pressures of 13.80 and 6.90 MPa improved acid and bile tolerances, respectively, of *Lb. acidophilus* LA-K. Another homogenisation pressure (60 MPa) was studied by Patrignani *et al.* (2016) during production of probiotic yoghurt (Y culture and *Lb. rhamnosus* BFE 5264). The homogenised batches developed acid more quickly, and there was better production of volatile compounds, as well as improved rheological properties and structure of the gel. After 60 d at 4 °C, the viable counts of *Lb. rhamnosus* BFE 5264 were 7.55 and 6.9 $\log_{10}$ cfu mL^{-1} in homogenised and non-homogenised yoghurts, respectively (Patrignani *et al.*, 2016; see also Massoud *et al.*, 2015). The effect of ultrasound treatment of the milk base on the survival rate of bifidobacteria in fermented milk has also been reported by Ljubic *et al.* (2015). Oxidoreduction of the milk using N_2 and N_2H_2 has also been shown to affect the survival rate of *Bif. bifidum* mixed with a Y culture; the bifidobacteria counts significantly increased during the storage period (Ebel *et al.*, 2011).

Drinking probiotic fermented milks

This category of products is different from the drinking yoghurt or Nordic fermented milks known in Europe and North America. In general, this type of drinking yoghurt is categorised as stirred yoghurt of low viscosity, and they are widely consumed in the Middle East as refreshing drinks. Some examples include Ayran in Turkey, Dough or Doogh in Iran and Yakult in Japan, a closely related product. Basically, the fermentate (i.e. the SNF level of the milk base is not fortified) is diluted with water, salted (optional), homogenised and packaged. Other related products, which are sometimes known as beverages (including those that are carbonated), are made from whey and/or a blend of whey and milk. The functional properties of probiotic dairy beverages have been reviewed by Ozer & Kirmaci (2010; see also Mercenier *et al.*, 2012a, 2013, ch. 8).

Four probiotic low-fat samples of Ayran containing Dairy-Lo® (fat replacer), inulin and SMP were fermented with a Y culture, *Lb. acidophilus* LA-5 and '*Bif. bifidum* BB-12' (presumed to be *Bif. animalis* subsp. *lactis* BB-12) (Kok-Tas & Guzel-Seydim, 2010; see also Uysal-Pala *et al.*, 2006). The viable counts ($\log_{10}$ cfu mL^{-1}) of *Lb. acidophilus* LA-5 and *Bif. animalis* subsp. *lactis* BB-12 in ayran ranged between 6.5–7.0 and 5.5–6.0, respectively, after 1 d at 4 °C. Similar counts for *Lb. acidophilus* in ayran were reported by Ayar & Burucu (2013), despite the fact that the milk base was fortified with different whey products.

Two different methods were evaluated during the production of probiotic Doogh (*Lb. acidophilus* LA-5, *Bif. animalis* subsp. *lactis* BB-12 and a Y culture): either diluting the milk with water followed by fermentation, or production of yoghurt from milk followed by dilution of the fermentate with water. In the latter production method, the viability of the probiotic bacteria was higher, a faster rate of acidification of the milk was observed and lower levels of acetic acid were produced, although the product was slightly more acidic (Mortazavian *et al.*, 2010). The use of microencapsulated probiotic bacteria (*Lb. acidophilus* LA-5 and *Bif. animalis* subsp. *lactis* BB-12) did not affect the quality of Doogh (Mortazavian *et al.*, 2008a, 2008b; Khosrokhavar & Mortazavian, 2010). Salt (NaCl) may be used in Doogh production, and substitution of part of the salt with a mixture of 0.5 g 100 g^{-1} NaCl/potassium chloride (KCl) before fermentation with *Lb. acidophilus* LA-5 and *Bif. animalis* subsp. *lactis* BB-12 and co-culturing with Y starter produced the best product (Arab *et al.*, 2016). The effect of KCl substitution on the viability of the selected probiotics, and the effect of variation of inoculum on the quality of Doogh, have been reported by Gandhi *et al.* (2014) and Ahmadi *et al.* (2012).

Fermented milk beverages using different combinations of probiotic bacteria (*Bif. bifidum*, *Str. thermophilus* and *Lb. casei*) were shown to be acceptable: high counts of *Lb. casei* and *Bif. bifidum* were maintained for 19 d at 4 °C (Real *et al.*, 2005), whilst buffalo's milk fermented with '*Bif. lactis*' (presumed to be *Bif. animalis* subsp. *lactis*) and flavoured with different ingredients had counts of ~10^7 cfu g^{-1} after 10 d at 5 °C (Salem *et al.*, 2006). Similar counts (~7 $\log_{10}$ cfu mL^{-1}) were found in fermented milk drinks for both probiotic strains (*Lb. acidophilus*, *Bif. animalis* subsp. *lactis* or *Lb. casei*) after 30 d (Yerlikaya *et al.*, 2013). However, fermented milks (goat and cow) made with '*Bif. longum* BB46' (presumed to be *Bif. longum* subsp. *longum* BB46) and spiked with *Serratia marcescens* or *Cam. jejuni* exhibited inhibitory effects on the growth of these pathogenic micro-organisms in the product (Pavlovic *et al.*, 2006).

Drinking yoghurts made with skimmed milk and different prebiotics ingredients (polydextrose, soluble corn fibre and inulin) were fermented with *Lb. acidophilus* LA-5, *Bif. animalis* subsp. *lactis* BB-12 and a Y culture; the fermentates were used to identify descriptive terms for sensory attributes. The viable counts of both probiotic bacteria decreased by 2 to 3 $\log_{10}$ cycle cfu mL^{-1}, respectively, after 30 d (Allgeyer *et al.*, 2010a, 2010b). The efficacy of using strains of *Weissella confusa* UI 006 and UI 007 and *Lb. paracasei* subsp. *paracasei* UI 014 and UI 022 in fermented milk was reported by Ayeni *et al.* (2011). All the strains increased their cell counts by 2 $\log_{10}$ cycles after 24 h of incubation at 37 °C, and the viable counts were ~10^7 cfu mL^{-1} after 4 weeks at 4 °C. The survival rate of these strains was good; thus, it was concluded they could be used as starter culture (lactobacilli) or as an adjunct culture (*Wei. confusa*) (Ayeni *et al.*, 2011). An acceptable strawberry-flavoured fermented milk beverage was made using *Lb. acidophilus*, *Bifidobacterium* spp. and *Str. thermophilus*, which included sucrose, the natural colouring cochineal carmine and fructooligosaccharide syrup; the probiotic counts of the lactobacilli and bifidobacteria were 1.7×10^7 and 4.8×10^6 cfu mL^{-1}, respectively (de Medeiros Burkert *et al.*, 2012).

In probiotic milk beverage made with wheat extract and cow's milk (plus added sugar) and fermented with a mixed starter culture [*Lb. acidophilus* ATCC 20552; '*Bif. lactis* Bb-12' (presumed to be *Bif. animalis* subsp. *lactis* BB-12) and *Str. thermophilus*], the counts of both probiotic bacteria were >4 log_{10} cfu mL^{-1} after 21 d at 4 °C (El-Zainy *et al.*, 2102). A fermented beverage (a mixture of buffalo's milk and whey with added starch) had a viable count of 9.28 log_{10} cfu mL^{-1} for '*Bif. longum*' (presumed to be *Bif. longum* subsp. *longum*) when fresh, but these counts decreased slightly to 9.26 log_{10} cfu mL^{-1} after 7 d of refrigerated storage (Abd-Elhamid, 2010). In addition, buffalo's milk, which was fortified with zinc sulphate or zinc acetate, was fermented at 37 °C for 5 h under continuous stirring using mixed starter culture (*Lb. acidophilus* LA-5, *Bif. bifidum* and *Lac. lactis* subsp. *lactis* biovar *diacetylactis* MD 099). The presence of the zinc in the milk enhanced bacterial growth, and the viable counts of both probiotic strains ranged between 10^6 and 10^7 cfu g^{-1} after 10 d at 5 °C (Seleet *et al.*, 2011). The gut survival of the probiotic bacteria in a goat's milk fermented with *Lb. rhamnosus* and '*Bif. longum*' (presumed to be *Bif. longum* subsp. *longum*) was investigated in a dynamic model of the human digestive system. There was a significant drop in viable numbers of the probiotic bacteria, but administration of a 175 mL serving of the product delivered >10^7 cfu g^{-1} viable cells to the colon section of the model (Kheadr *et al.*, 2011). Salva *et al.* (2011) also reported counts of 10^6 cfu g^{-1} for a fermented goat's milk beverage made using *Lb. rhamnosus* CRL 1505, and a study on Yakult containing *Lb. casei* Shirota prevented aflatoxin (AF M_1 and B_1) absorption in the human gut (Mohd Redzwan *et al.*, 2016).

Other types of probiotic fermented milk drinks are known as whey-based beverages, which are developed primarily for the utilisation of whey (sweet cheese whey, milk permeate, hydrolysed lactose whey and reconstituted demineralised whey powder); some examples are shown in Table 4.4. Shahabbaspour *et al.* (2013) used a mixture of cow's milk and soy milk (50:50), which was fermented with a Y culture and *Lb. acidophilus* LA-5 or *Lb. casei* L-01. The fermentate was flavoured with different fruit juices; the viability of *Lb. casei* L-01 was best in the apricot juice product (8.69 log_{10} cfu mL^{-1}), whilst the count for *Lb. acidophilus* LA-5 was 8.56 log_{10} cfu mL^{-1} after 21 d at 5 °C.

Different combinations of LAB (mesophilic and a Y culture) were co-cultured with *Lb. acidophilus* CCDM 151 or *Enterococcus durans* CCDM 922 to produce fermented milk-based beverages with added malt extract or saccharified malt grains. The malt extracts inhibited the growth of the probiotic strains and caused flavour deterioration during storage; however, the product containing saccharified malt fermented with *Ent. durans* was deemed highly acceptable by the sensory panellists (Kunova *et al.*, 2013; Nemeckova *et al.*, 2013). The combined use of *Lb. acidophilus* and *Propionibacterium freudenreichii* subsp. *shermanii* in different ratios was evaluated for potential development of a new fermented dairy beverage (Farhadi *et al.*, 2013; see also Foligne *et al.*, 2016). The viable rates of these organisms varied according to the culture ratio used and temperature of incubation. The maximum count of *Lb. acidophilus* occurred at a strain ratio of 1:8 and incubation temperature at 35 °C, but there was considerable decrease in cell counts of *Pro. freudenreichii* subsp. *shermanii* and *Lb. acidophilus* during the first and last weeks of storage, respectively.

Lastly, carbonated fermented milk beverages have been developed as a new type of probiotic product to increase consumer appeal. Carbonation (i.e. using carbon dioxide - CO_2) of these milk beverages has not yet become industrialised, but some literature includes the following examples. Jardim *et al.* (2012) developed a strawberry-flavoured

Table 4.4 Examples of some fermented probiotic drinking beverages and additives used in the formulation.

Probiotic bacteria	Ingredients used	Comments/viable counts (cfu mL^{-1})	References
Lactococcus spp., *Leuconostoc mesenteroides* subsp. *cremoris*, *Bifidobacterium animalis* subsp. *lactis* BB-12	SM, sucrose, fruit flavours	Bifidobacteria averaged 8.08 log_{10} after 28 d at 5 °C; the product is similar to buttermilk.	Antunes *et al.* (2007, 2009); see also El-Shafie (2003)
Lactobacillus reuteri, *Bifidobacterium bifidum*	WP, sucrose, pectin	Both probiotic organisms were at $>10^6$ after 30 d at 4 °C.	Hernandez-Mendoza *et al.* (2008)
Lactobacillus acidophilus LA-5, *Bif. animalis* subsp. *lactis* BB-12	WP	In fresh products, probiotic counts were 8.5 and 8.7 log_{10}, respectively.	Matijevic *et al.* (2008, 2011)
	HW	Lactobacilli counts were 9.45 log_{10}, and bifidobacteria growth was not enhanced in the hydrolysed whey.	
Lactobacillus rhamnosus NCDO 243, *Bif. bifidum* NCDO 2715, *Propionibacterium freudenreichii* subsp. *shermanii* MTCC 1371	Whey	All probiotic organisms were $>10^8$ after 10 d at 4 °C.	Maity *et al.* (2008)
Lb. acidophilus, *Bif. animalis* subsp. *lactis*, *Lb. rhamnosus*; each co-cultured with yoghurt starter culture[1]	Whey	Counts of bifidobacteria were 8.4 log_{10}, *Lb. acidophilus* were 6.7 log_{10}, and *Lb. rhamnosus* were 5.6 log_{10}.	Almeida *et al.* (2008, 2009)
Bifidobacterium spp., *Lb. acidophilus*	Milk, whey, soy extract, peach	Counts of both probiotic organisms were 1×10^6 for 22 d.	Pinto Kempka *et al.* (2009)
Lb. acidophilus M 92, *Lactobacillus plantarum* L4, *Enterococcus faecium* L3	MP, WR	In the fresh product, all organisms were ~10^8, decreasing to ~10^7 after 28 d at 4 °C.	Lebos Pavunc *et al.* (2009)
ABT-4 (for details, refer to Section 4.2.2)	Milk or whey, sucrose, oligofructose	Both probiotic organisms were ~10^6; oligofructose enhanced overall acceptability.	de Castro *et al.* (2009a, 2009b)
Lb. acidophilus LA-5 or *Bif. bifidum* BB-12 (presumed to be *Bif. animalis* subsp. *lactis* BB-12)	MP, skimmed milk	Both probiotic organisms: $>10^8$ after 21 d at 5 °C.	Marhamatizadeh *et al.* (2012a)
Lb. acidophilus	SM, whey, strawberry	All beverages were 8 log_{10}; whey >65 mL mL^{-1}, and had lower consumer acceptability.	Castro *et al.* (2013a, 2013b)

(*Continued*)

Table 4.4 (Continued)

Probiotic bacteria	Ingredients used	Comments/viable counts (cfu mL^{-1})	References
Bifidobacterium spp., *Lb. acidophilus*	Milk, yacon juice	The product had smooth texture, sweet and sour taste, and was rich in yoghurt flavour and taste of yacon juice.	Wang (2014)
Str. thermophilus TA-40, *Bif. animalis* subsp. *lactis* BB-12, *Lb. rhamnosus* Lr-32	Whey, goat's milk, guava or soursop pulp, PHGM	Probiotic organisms in PHGM beverage were >7 $\log_{10}$ after 21 d at refrigerated temperature.	Buriti *et al.* (2014)
Bif. animalis subsp. *lactis*	Goat's cheese whey, inulin, oligofructose, chocolate	Bifidobacteria counts were >7 $\log_{10}$ in milk base containing 6 g 100 mL^{-1} prebiotics and 45 mL 100 mL^{-1} whey; effects of other prebiotics in probiotic beverages were reported by de Dias *et al.* (2013) and Yi *et al.* (2014).	Fornelli *et al.* (2014) and da Silveira *et al.* (2015)
Lb. acidophilus, *Bif. animalis* subsp. *lactis*, *Str. thermophilus*	WP, low-fat milk	Milk-based beverage(s) was highly preferred to the whey product.	Akpinar *et al.* (2015)
Bifidobacterium longum DSM 20088, co-cultured with yoghurt starter culture (EPS producer)	MP, papaya and guava pulps	In the fresh product, bifidobacteria were ~7.5 $\log_{10}$ and dropped ~1 $\log_{10}$ cycle after 30 d at 4 °C.	Atallah (2015a)
'*Bif. longum*' (presumed to be *Bif. longum* subsp. *longum*), *Lb. delbrueckii* subsp. *bulgaricus* and *Str. thermophilus* (at a ratio of 1:1:1)	MP, sucrose, carrot, mango pulp	Bifidobacteria ranged between 10^6 and 10^7 after 30 d at 4 °C.	Atallah (2015b)
Lb. acidophilus LA-5, *Lactobacillus casei* LBC-81	DWP, soy isoflavones or phytosterols, κ-carrageenan, xanthan gum, sucrose	In the fresh product, *Lb. acidophilus* LA-5 counts were 28.4×10^8 and *Lb. casei* LBC-81 were 16.7×10^8; after 28 d at 4 °C, *Lb. acidophilus* LA-5 was 1.2×10^8 and *Lb. casei* LBC-81 was 7.8×10^7 (average counts).	Seyhan *et al.* (2016)

[1] *Streptococcus thermophilus* or *Lactobacillus delbrueckii* subsp. *bulgaricus*.
cfu = Colony forming units; WP = whey powder; HW = hydrolysed whey; MP = milk permeate; WR = whey retentate; DWP = demineralised whey powder; SM = skimmed milk; PHGM = hydrolysed galactomannan from *Caesalpinia pulcherrima* seeds; EPS = exopolysaccharides.

beverage made with *Lb. acidophilus* LA-5, *Bif. animalis* subsp. *lactis* BB-12 and *Str. thermophilus*, but the carbonation process did affect the viability of probiotic bacteria after 28 d at 4 °C. Another example is yoghurt beverages formulated containing pomegranate or vanilla, inulin and probiotic bacteria (*Lb. acidophilus* and *Bifidobacterium* spp.). The beverages were stabilised with high-methoxyl pectin and WPC and compared with carbonated samples; the viable counts of both probiotic bacteria were $>10^6$ cfu g^{-1} after 9 weeks at 4 °C in all the beverages regardless of carbonation (Walsh *et al.*, 2014).

Concentrated/strained and very viscous probiotic fermented milks

The manufacturing methods of concentrated yoghurt consist of the following methods: (a) traditional cloth bag, (b) nozzle separator, (c) membrane filtration (mainly ultrafiltration, or UF) and (d) product formulation. These types of products, which will be reviewed in this section, are concentrated yoghurt and Ymer.

Concentrated yoghurt is known under many names, such as Labneh, Greek yoghurt, Greek-style yoghurt, Matsou, Suzme and Zimme (Tamime & Robinson, 1999; Ozer & Tamime, 2013). Although several commercial probiotic products are available in different markets (Table 4.2) (Tamime *et al.*, 2014), limited data have been published on these types of products. In a UF Labneh, the viable cell count of '*Bif. bifidum* Bb-12' (presumed to be *Bif. animalis* subsp. *lactis* BB-12) ranged between 2×10^5 and 4×10^7 cfu g^{-1} depending on the type of milk used (Mahdi *et al.*, 1990). The effects of levels of milk solids, fats, fat substitutes and vegetable oils in the milk base on the quality of concentrated yoghurt made with ABT culture (for details, refer to Section 4.2.2) or enterococci species have been reported by Amer *et al.* (1997), Taha *et al.* (1997) and El-Samragy (1997). Recently, goat's milk Labneh was made using ABT-5 culture. The cold fermentate was concentrated using the cloth bag method, and the product was mixed with salt, fortified with different levels (3 to 15 g 100 g^{-1}) of textured soy protein (TSP) and stored for 21 d at 5 °C. The experimental batches of Labneh had different chemical compositions when compared with the control; the sensory profiling and rheological properties of the product resulted in a recommendation of TSP level of fortification to be between 3 and 6 g 100 g^{-1}. The viable counts of the lactobacilli and bifidobacteria at the recommended level of TSP in Labneh were $30–35 \times 10^5$ and $21–25 \times 10^5$ cfu g^{-1}, respectively (Basiony *et al.*, 2015). A low-fat probiotic goat's milk Labneh has been made using an EPS-producing starter culture (*Bif. bifidum*, *Lb. acidophilus* and a Y culture). This product was highly rated, and the viable counts for both probiotic strains were $\sim 22 \times 10^6$ cfu g^{-1} after 21 d at 6–8 °C (Ayyad *et al.*, 2015).

Ymer is a concentrated fermented milk product developed in Denmark in the 1930s, which is now produced from heat-treated milk that has been homogenised and ultrafiltrated (Mogensen, 1980). When using UF milk, the product is more concentrated (i.e. the protein content is 6 g 100 g^{-1} and the SNF level is 11 g 100 g^{-1}). The UF milk is reheated, treated, homogenised and fermented with *Lac. lactis* subsp. *cremoris* and *Lac. lactis* subsp. *lactis* biovar. *diacetylactis* at 18–20 °C for 18–20 h (final pH: 4.4–4.6). After fermentation, the product is stirred, cooled, left to stand at 5 °C for 1 d and stirred again prior to packaging (Delaney, 1977; Ulrich, 1980; Kurmann *et al.*, 1992; Oberman & Libudzisk, 1998). In Denmark, probiotic Ymer is produced by Arla Foods using *Lb. acidophilus* to ferment the milk base (Table 4.2) (see http://www.arla.dk/produkter/arla-a38-ymer-1000g-3681/).

Långfil is a variation of fermented milk product made with the same mesophilic LAB species as cultured buttermilk, but these strains produce large amounts of EPS, which makes the product much more viscous than cultured buttermilk. During the production of Långfil, the milk is heated to a high temperature, cooled to 18–20 °C, mixed with starter cultures in the tank, packaged and fermented for 18–20 h (Mantere-Alhonen & Forsen, 1990; Oberman & Libudzisk 1998; Leporanta, 2003). Långfil is mainly consumed in northern parts of Sweden and, similar to Filmjölk and Fil, it is eaten during breakfast (usually with berries or cereal) or as a snack. To our knowledge, there are no probiotic Långfil products on the market.

4.5.2 *Yeast–lactic acid fermentations*

In the past, Skyr was classified as a traditional and concentrated fermented milk product that originates from Iceland and Norway. The current Icelandic Regulation has reclassified Skyr, and now it is listed in Section F Dairy Products together with Quarg (Skyr and Kvarg) (Tamime *et al.*, 2014). In this review, however, the term Skyr will be retained as originally known: 'concentrated fermented milk product'. Nevertheless, 'traditional' Skyr was produced from skimmed milk by using two-stage fermentation with a Y culture and lactose-fermenting yeasts (*Saccharomyces* spp.). The milk was heated to 90–100 °C, then cooled to 40 °C and starter (Skyr from an earlier production batch) was added with a small amount of a cheese coagulant (i.e. chymosin – optional). The fermentation period was ~5 h or until pH dropped to 4.7. The fermentate was cooled to 18–20 °C and incubated for a further 18 h for the yeast to grow or until pH reached 4.2. The whey was then removed using a cloth bag and, after de-wheying for ~24 h, the pH of the concentrate was 3.8–4.0 (Gudmundsson, 1987; Wolpert, 1988; Kurmann *et al.*, 1992; Gudmundsson & Kristbergsson, 2016). The current industrial process for Skyr production differs from the traditional one in several aspects: the cheese coagulant and yeast are not used anymore, and the concentration of the fermentate is done using a nozzle separator followed by concentrating the whey by UF to the same solids content as Skyr. The whey retentate is then mixed with Skyr and, as a consequence, the yield of the end product is increased due to higher retention of protein from the milk base (Ozer & Tamime, 2013). Thus, today's Skyr is similar to concentrated yoghurt where the fermentate is concentrated using the UF method, which results also in higher yield (Gudmundsson & Kristbergsson, 2016). Skyr is consumed during breakfast or as a snack, and it is often flavoured with berries and fruits. During the past few years, Skyr products (natural or fruit flavoured) have become popular in Central and Northern Europe (http://blog.euromonitor.com/2015/11/skyr-trademark-or-noun-a-real-game-of-monopoly.html).

Recently, there has been a dispute as to whether Skyr is a brand name or a generic name. Skyr is a registered MS trademark in Norway and Finland, and thus, at present, only MS Iceland Dairies can sell a product called Skyr in these countries. In addition to Nordic countries, Skyr can be found in the markets in Switzerland, Belgium, Denmark, Germany, The Netherlands, the USA and the UK (http://icelandmonitor.mbl.is/news/news/2015/10/07/finland_bans_swedish_skyr/; http://icelandmag.visir.is/article/swedish-arla-banned-selling-skyr-finland; http://www.just-food.com/interview/ms-iceland-dairies-adds-uk-to-skyr-yoghurt-export-push-interview_id132516.aspx). To our knowledge, there are no probiotic-containing Skyr products in the European

Union (EU) market. In the USA, Siggi's company is producing Icelandic Skyr-style yoghurt with *Lb. acidophilus* and '*Bif. lactis*' (presumed to be *Bif. animalis* subsp. *lactis*) (http://www.wegmans.com/webapp/wcs/stores/servlet/ProductDisplay?productId=802554&storeId=10052&langId=-1).

Kefir is a carbonated fermented milk product made using a complex mixture of micro-organisms known as kefir grains. Typically, these include strains of *Lac. lactis* subsp. *cremoris*, *Lac. lactis* subsp. *lactis*, *Lactobacillus kefir*, *Lactobacillus kefiranofaciens*, *Lb. brevis*, *Lb. acidophilus*, *Leuconostoc* spp., *Acetobacter* spp., lactose-fermenting yeasts (*Kluyveromyces* spp.) and non-lactose-fermenting yeasts (*Saccharomyces* spp. and *Candida* spp.). However, kefir grains can contain a wide variety of microbial species in addition to those mentioned above (Marshall, 1987; Tamime & Marshall, 1997; Garrotte *et al.*, 2001; Leporanta, 2001; Simova *et al.*, 2002; Santos *et al.*, 2003; Witthuhn *et al.*, 2004; Wszolek *et al.*, 2006; Marsh *et al.*, 2013; Diosma *et al.*, 2014; Nalbantoglu *et al.*, 2014; Nielsen *et al.*, 2014; John & Deeseenthum, 2015; Anton *et al.*, 2016; Bourrie *et al.*, 2016). The mixture of bacteria and yeasts in kefir grains appears as clusters of microbes held together with a matrix consisting mainly of the polysaccharide 'kefiran' produced by *Lb. kefiranofaciens* and protein. The activity of the yeasts results in a product with a typical yeasty flavour, formation of carbon dioxide and some ethanol ($<2\,mL\ 100\,mL^{-1}$). Nowadays, there are two main methods to produce Kefir: (a) traditional, which uses kefir grains, and (b) modern, which uses direct-to-vat inoculation (DVI) starter cultures. In the traditional method, the processed milk is cooled to 18–25 °C, inoculated with kefir grains ($2–10\,g\ 100\,mL^{-1}$) and incubated for 18–24 h followed by stirring and cooling. After incubation, the grains are separated, washed (optional) and reused (Kurmann *et al.*, 1992; Wszolek *et al.*, 2001; Schoevers & Britz, 2003). When DVI cultures are used, the milk is heat-treated, cooled to inoculation temperature (32–35 °C), inoculated with DVI cultures and incubated for 10–15 h (final pH: 4.4–4.5); the fermented product is then stirred, cooled and packaged (K. Leporanta, Valio Ltd, personal communication). Industrial-scale Kefir production is typically done with DVI cultures because these make the process and product easier to control (Bourrie *et al.*, 2016). It could be argued, however, that despite the fact that commercial Kefir starter cultures are available in different blends of thermophilic and mesophilic LAB, yeast cultures and possibly probiotic bacteria, the fermentate lacks the 'traditional characteristics' (i.e. fizziness, alcohol production and taste) of Kefir made using kefir grains.

In countries where Kefir has not been traditionally consumed (e.g. in the USA), it is often advertised as a healthy product and also used in cooking; for example, see the websites Lifeway (http://lifewaykefir.com/what-is-kefir/), Wallaby (http://wallabyyogurt.com/our-products/kefir) and Green Valley (http://greenvalleylactosefree.com/products/kefir.php) for the products, and see https://authoritynutrition.com/9-health-benefits-of-kefir/ for Kefir's assumed health benefits. Kefir is often sold as probiotic without any information being provided on whether any probiotic strains have actually been added. However, Kefir products with added probiotic bacteria are on the market (Table 4.2), for example Valio's unflavoured and flavoured Kefir with *Lb. rhamnosus* GG in Finland (http://www.valio.fi/yritys/media/uutiset/valio-toi-markkinoille-vahempisokerisen-juotavan-valipalan-valio-kefir-lgg/) (see also Muir *et al.*, 1999; Farnworth & Mainville, 2008).

Nevertheless, in the past, commercial starter culture companies, such as DuPont Nutrition and Health (Danisco), have marketed probiotic Kefir cultures (HOWARU™ Kefir 1 & 2 Bifido that contain different probiotic strains that are unrelated bacteriophages), but currently these kefir cultures have been withdrawn from the market, and the company provides special blends of probiotic kefir cultures as required by customers (P. Kolakowski, personal communication). Probiotic cultures (Danisco strains '*Bif. lactis* NH019' – presumed to be *Bif. animalis* subsp. *lactis* NH019; *Lb. acidophilus* NCFM; and *Lb. rhamnosus* NH001) were used singly (inoculation rates 10^6 or 10^7 cfu mL^{-1}) in co-cultures with DC kefir starter during the manufacture of Kefir. The live probiotic counts (cfu mL^{-1}) surviving in the product averaged 1.3×10^7, 1.3×10^7 and 4.3×10^7 for *Bif. animalis* subsp. *lactis* NH019, *Lb. acidophilus* NCFM and *Lb. rhamnosus* NH001, respectively, after 21 d at 4 °C. However, using an inoculation level of $\geq 10^7$ cfu mL^{-1} affected the flavour of Kefir due to acetic acid production by the bifidobacteria (Kolakowski & Pawlikowski, 2012). Encapsulation of the *Bif. animalis* subsp. *lactis* BB-12 strain improved its survival rate in Kefir: the reduction in the count was 2 $\log_{10}$ cycle after 28 d at 3 °C, whilst the loss was greater (6 $\log_{10}$ cycle) in the product made with free cells of bifidobacteria. In addition, after subjecting these different Kefirs in a simulated gastric juice model, it was recommended not to store the product more than 14 d in order to maintain the count $>10^6$ cfu mL^{-1} (Gonzalez-Sanchez *et al.*, 2010).

For some years, there has been a lot of interest and 'hype' about 'traditional' Kefir as it has been suggested that there are several health benefits associated with this product because the kefir grains contain a diverse range and complex microbiota. Beneficial effects have been demonstrated in *in vitro* and animal studies (Bourrie *et al.*, 2016) but, as is often the case, there has been poor translation from preclinical animal models to human clinical studies (van den Nieuwboer *et al.*, 2016). Other studies and reviews (Nielsen *et al.*, 2014; Prado *et al.*, 2015) on the micro-organisms of Kefir, which may be of interest to the reader, are as follows:

- An isolate (*Lac. lactis* subsp. *lactis* CIDCA 8221) from Kefir was found to secrete heat-sensitive products able to protect eukaryotic cells from the cytopathic effect of *Clostridium difficile* toxins *in vitro* (Bolla *et al.*, 2013).
- *In vitro* demonstration of various probiotic properties of mixed bacteria and yeast isolates (i.e. freeze-dried) was reported by Bolla *et al.* (2010).
- Cellular injury of spray-dried *Lactobacillus* spp. isolated from Kefir, which may have probiotic potential, was reported by Golowczyc *et al.* (2011).
- A *Lb. plantarum* CIDCA 83114 isolate from kefir grains was evaluated as a potential strain to be used in probiotic fermented milk (Kakisu *et al.*, 2010).
- Potential probiotic isolates from kefir grains have also been reported by many researchers in different countries. Some examples include: (a) *Lb. kefiranofaciens* M1 from Taiwanese grains (Chen *et al.*, 2012, 2013), (b) *Lb. kefiranofaciens* 8U and '*Lb. paracasei* MRS59' (presumed to be *Lb. paracasei* subsp. *paracasei* MRS59) in Brazilian grains (Leite *et al.*, 2015; Zanirati *et al.*, 2015), and (c) lactobacilli, lactococci and *Pediococcus* spp. strains in Turkish grains (Sabir *et al.*, 2010).

A closely related product to Kefir is Koumiss (Wszolek *et al.*, 2006) but, to our knowledge, no probiotic Koumiss product has been produced. As above, isolates with potential probiotic characteristics have been reported by many researchers. Some examples include: (a) *Lb. helveticus* CAUH18 (Yang *et al.*, 2016), (b) '*Lb. paracasei* CAUH35' (presumed to be *Lb. paracasei* subsp. *paracasei* CAUH35) (Wang *et al.*, 2015b), (c) '*Lb. paracasei* TXW' (presumed to be *Lb. paracasei* subsp. *paracasei* TXW) (Zhang *et al.*, 2011) or '*Lb. paracasei* H9' (presumed to be *Lb. paracasei* subsp. *paracasei* H9) (Xie *et al.*, 2012), (d) *Lb. casei* Zhang (Wu *et al.*, 2010), and (e) different *Lactobacillus* strains (Danova *et al.*, 2005; Guo *et al.*, 2015).

Sameel milk (a traditional Saudi Arabian fermented milk) is made by the nomadic herders from unpasteurised cow's, sheep's, goat's or camel's milks in leather bags. It is an on-going fermentation, similar to traditional Koumiss production (Tamime & Marshall, 1997; Wszolek *et al.*, 2006). Recently, Al-Otaibi (2012) screened, isolated and identified the micro-organisms in sameel milk from different regions in Saudi Arabia; the most frequently isolated species were *Lb. plantarum*, *Lb. paracasei* subsp. *paracasei*, *Candida lusitania*, *Lactobacillus pentosus*, *Cryptococcus laurentii* and *Saccharomyces cerevisiae* (*Saccharomyces cerevisiae* subsp. *cerevisiae*). Some of the lactobacilli isolates may have probiotic potential, but this needs to be confirmed by further research.

4.5.3 *Mould–lactic acid fermentations*

Viili is a viscous fermented milk product which is manufactured in Finland. The industrial production of Viili began in the late 1950s. It is produced by fermenting milk with mesophilic starter cultures (*Lac. lactis* subsp. *cremoris*, *Lac. lactis* subsp. *lactis* biovar. *diacetylactis* and *Leu. mesenteroides* subsp. *cremoris*) together with a mould (*Geotrichum candidum*) in the retail container. The fermentation time is ~20 h at 20 °C (final pH is ~ 4.3). Traditionally, Viili was made from non-homogenised milk, which resulted in the formation of a cream layer on the surface of the milk. *G. candidum* grew on this layer and formed a velvety growth similar to Camembert and Brie. There is also a non-mouldy Viili-type product, and the high-fat variety is used only in cooking (Mantere-Alhonen & Forsen, 1990; Kurmann *et al.*, 1992; Leporanta, 2003; Ruas-Madiedo *et al.*, 2006).

A wide variety of Viili-type products are available in Finland, including low-fat, low-lactose and berry-flavoured variants. Viili is consumed mainly at breakfast and as a snack. A product containing *Lb. rhamnosus* GG is currently the only probiotic Viili product available on the market.

4.5.4 *Quality appraisal of probiotic fermented milks*

Probiotic micro-organisms can be incorporated into fermented milk using different methods. The most popular way is to add the probiotic bacteria together with the starter cultures. Since the fermentation rarely occurs in conditions optimal for probiotic species, such organisms do not usually grow well during a mixed fermentation with the 'traditional' starter cultures. Alternatively, the probiotic bacteria may be grown initially for

~2 h to achieve a high viable count, and the fermentation is completed with the 'traditional' starter cultures, which results in slightly longer fermentation time. Another method involves use of probiotic micro-organism(s) as a starter culture, but this means the fermentation time may take up to several days. A typical example of the use of a probiotic culture alone is the manufacture of Yakult, which is fermented with *Lb. casei* Shirota (Heimbach, 2012). The health properties of Yakult have been reviewed by Miyazaki & Matsuzaki (2008) (see also Chapter 8), whilst the growth activity of *Lb. acidophilus* in different mammalian milks has been reported by Drakoularakou *et al.* (2003).

During the production of probiotic fermented milks, several aspects have to be considered, including the following: (a) many probiotic strains grow slowly in milk with no added growth factors, such as peptides (Casarotti *et al.*, 2014b), (b) the production conditions (especially the traditional fermentation temperatures) are often unsuitable for the probiotic's growth (Kearney *et al.*, 2008), (c) there are difficulties in the enumeration of probiotic strains in products because some cells may be in a viable but non-culturable state due to the stress of processing and formulation, or present in mixed cultures (Savoie *et al.*, 2007; Davis, 2014), and (d) some metabolites of probiotic strains may be undesirable due to the formation of off-flavours (e.g. bifidobacteria produce acetic acid, which gives a vinegar-like taste) (Gomes & Malcata, 1999; Saxelin *et al.*, 1999; Saarela *et al.*, 2000; Ostlie *et al.*, 2003; Chandan & O'Rell, 2008). However, if the food matrix supports the growth of probiotic micro-organisms and no off-flavour formation occurs, growth during the production of fermented milks can lower processing costs and enhance the ability of the probiotic to survive alive in the product during storage. When both probiotic and traditional starter micro-organisms are present during the fermentation stage, it is important to use compatible and suitable blends of probiotic/starter cultures (Saxelin *et al.*, 1999; Ouwehand *et al.*, 2000; Champagne, 2014). In extreme cases, starter cultures may produce inhibitory compound(s) (e.g. hydrogen peroxide or high amounts of lactic acid) that are harmful to the probiotic culture(s), causing a decrease in the probiotic's viable count in the product (Katla *et al.*, 2001; Vinderola *et al.*, 2002; Ng *et al.*, 2011). Nevertheless, certain starter cultures may enhance the growth and survival of probiotic micro-organisms by producing growth-promoting substrates or by reducing the oxygen content in the milk (Dave & Shah, 1997a, 1997b; Kailasapathy & Rybka, 1997; Saarela *et al.*, 2000; Vinderola *et al.*, 2002; Homayouni *et al.*, 2012a). Another important factor, which should not be overlooked, is the growth temperature of mixed fermentations. Some of the traditional products described in this chapter are fermented at 20 or 30 °C, which are sub-optimal temperatures for the growth of probiotic micro-organisms and, in particular, the strains that originate from the human GI tract (optimum growth temperature of 37 °C). Increasing the fermentation temperature to favour the growth of probiotic micro-organisms is not recommended, however, because it can lead to an unacceptable flavour profile in the products (Mantere-Alhonen & Forsen, 1990). Therefore, mixed fermentation with probiotic micro-organisms has the best chance to succeed when the probiotic strain is combined with a thermophilic starter (e.g. a blend of *Lb. acidophilus* and/or bifidobacteria and yoghurt starter cultures) (Gardini *et al.*, 1999; Saxelin *et al.*, 1999; Saarela *et al.*, 2000). Alternatively, the probiotic micro-organisms may be added at high numbers to a 'traditional' starter culture to produce a fermented milk product irrespective of sub-optimal growth temperature for the probiotic species (see Baron *et al.*, 2000).

4.6 Probiotic cheeses

The success of using probiotic bacteria in fermented liquid milk products has inspired the development of other dairy products with probiotics (Heller, 2001; Ross *et al.*, 2002; Fonden *et al.*, 2003; Heller *et al.*, 2003; Boylston *et al.*, 2004; Ibrahim *et al.*, 2010). The production of cheeses, especially the matured types, with probiotic bacteria presents unique challenges because of the need for co-survival of these bacteria with the 'traditional' LAB, mould or yeasts that are used for cheesemaking. The latter micro-organisms may be antagonistic or competitive, or may possess the characteristics of associative growth towards each other. Some key characteristics of cheeses and cheesemaking that are relevant to the inclusion of probiotics are as follows:

- Relatively low or reduced moisture content, depending on the cheese variety;
- Presence of salt [i.e. the salt-in-moisture (S/M) ratio];
- Mesophilic and/or thermophilic LAB used (e.g. acid production, flavour production during the maturation stage and competition for nutrients); and
- Extent of the maturation period over 3 months, which can influence the biochemical activities, alter the redox potential and change the matrix/structure of the cheese.

It could be argued, however, that certain cheese varieties have been considered as good carriers of probiotic bacteria because of lower acidity and the existence of a complex cheese matrix of protein and fat that could potentially protect the probiotic strains during their passage through the GI tract (Stanton *et al.*, 1998). Some early studies on cheeses, which were reviewed in this book's first edition and contained viable counts of probiotic bacteria, are as follows: (a) Turkish White brined, Feta-type and other related cheeses (Ghoddusi & Robinson, 1996; Psomas *et al.*, 2001; Yilmaztekin *et al.*, 2004; Awaisheh, 2011; Dimitrellou *et al.*, 2014), (b) Kareish – an Egyptian variety (Murad *et al.*, 1998; Abou-Dawood, 2002), (c) Cheddar (Gardiner *et al.*, 1998, 1999a, 1999b), (d) soft cheese (Barraquio *et al.*, 2001; Shehata *et al.*, 2001; Mehanna *et al.*, 2002; El-Kholy *et al.*, 2003; Kasimoglu *et al.*, 2004), (e) Ras – an Egyptian variety (Osman & Abbas, 2001; Abdou *et al.*, 2003; Shehata *et al.*, 2004a, 2004b, 2004c), (f) Edam and semi-hard (Antonsson *et al.*, 2002; Rogelj *et al.*, 2002; Tungjaroenchai *et al.*, 2004), (g) Emmental and Swiss-type (Weinrichter *et al.*, 2004a, 2004b), (h) cheese-based dips (Tharmaraj & Shah, 2004), (i) Quarg (Milanovic *et al.*, 2004), and (j) Queso Fresco (Viderola *et al.*, 2000b; Suarez-Solis *et al.*, 2002).

4.6.1 *Methods of introduction of probiotics in cheese*

The inclusion of probiotic bacteria in cheesemaking is challenging in terms of retaining the main characteristics of the product and the viability of these organisms. Taking into account the cheese variety and strain(s) of probiotic bacteria, some selected methods developed to be used in cheesemaking are as follows:

- Probiotic bacteria are used as adjunct culture when they are added with LAB starter culture (Stanton *et al.*, 1998; McBrearty *et al.*, 2001; Perko *et al.*, 2002); another approach is to add spray-dried probiotic milk powder (Gardiner *et al.*, 2002a).

- Fermentation of the cream with '*Bif. infantis*' (presumed to be *Bif. longum* subsp. *infantis*) is used as a dressing during the manufacture of Cottage cheese (Blanchette *et al.*, 1996; Daigle *et al.*, 1998), or such a fermentate is used in Cheddar-type cheesemaking to standardise the cheese milk (Daigle *et al.*, 1999). *Lb. rhamnosus* GG is used to ferment the cream for Cottage cheese (Tratnik *et al.*, 2000).
- Microencapsulation techniques have been used to protect the probiotic bacteria [*Bif. bifidum*, '*Bif. infantis*' (presumed to be *Bif. longum* subsp. *infantis*) and '*Bif. longum*' (presumed to be *Bif. longum* subsp. *longum*)] and improve their viability in Kareish, Cheddar and Crescenza cheeses (Gobetti *et al.*, 1998; Abou-Dawood, 2002; Kailasapathy, 2002; Godward & Kailasapathy, 2003a; Picot & Lacroix, 2003; see also the reviews by Boylston *et al.*, 2004; Hayes *et al.*, 2006).
- Addition of a dried culture of *Bif. bifidum* (i.e. immobilised by forming gelled beads in carrageenan and then freeze-dried) during salting of the curd during the manufacture of semi-hard and hard cheeses (Dinakar & Mistry, 1994); the viable count of bifidobacteria was $\sim 10^7$ cfu g^{-1} after a 24-week maturation period.
- Probiotic bacteria are grown in milk hydrolysate (i.e. to increase the biomass of the cells) before using them in cheesemaking (Gomes *et al.*, 1998).
- A spray-dried probiotic milk powder (Gardiner *et al.*, 2002a) containing '*Lb. paracasei* NFBC 338' (presumed to be *Lb. paracasei* subsp. *paracasei* NFBC 338) had a survival rate of 84.5% of the probiotic strains; and, when used in the manufacture of Cheddar cheese as an adjunct culture, the initial probiotic count was 2×10^7 cfu g^{-1}, increasing to 3.3×10^7 cfu g^{-1} after 3 months.

Although some previous work demonstrated a limited impact of added bifidobacteria on cheese quality (Dinakar & Mistry, 1994; Daigle *et al.*, 1999), other studies have suggested that bifidobacteria may affect sensory qualities due to the formation of acetic acid in cow's and goat's milk cheeses (Gomes *et al.*, 1995; Gomes & Malcata, 1998). More recently, with better probiotic strain selection in cheesemaking, the sensory properties of the product are no longer affected (see further in this chapter).

4.6.2 *Probiotic strain selection for cheesemaking*

The capacity of probiotic bacteria to perform beneficial health-related effects, as well as its capacity to survive the GI stress and technological conditions inherent to cheese elaboration, are strain specific (Gilliland, 2001; Papadimitriou *et al.*, 2016). This highlights the need to carry out a careful strain selection, depending on the purpose and cheese model to be elaborated (Gilliland, 2001; Castro *et al.*, 2015). The first main requirement for a probiotic strain to be used in cheesemaking is to show a well-established therapeutic effect on consumer health and the absence of pathogenicity (Abou-Dawood, 2002; Linares *et al.*, 2016). For cheese applications, the probiotic strains must also be compatible with the cheese starter cultures, and should not adversely alter sensory attributes of cheese, such as aroma, flavour, colour or texture. For example, in one study that used two probiotic strains for Cheddar cheesemaking, it was observed that, with *Bif. animalis* subsp. *lactis* BB-12 or *Bif. animalis* subsp. *animalis* (Masco *et al.*,

2004), the moisture content of the cheese was higher, and there was more proteolysis and better flavour development in the product than with '*Bif. longum* BB-536' (presumed to be *Bif. longum* subsp. *longum* BB-536) (McBrearty *et al.*, 2001). The major challenge associated with the application of probiotic cultures in the manufacture of foods is their survival during processing (thermal treatments, salt exposure etc.) and storage/maturation time (Castro *et al.*, 2015). It should be noted that some cheese processes, such as Pasta Filata (in which the cheese curd is heated to 55 °C and stretched at 70 °C in hot brine), might impose restrictive conditions for probiotic bacteria (Ortakci *et al.*, 2012). A prerequisite of probiotic cheese manufacture is that the cultures must survive the relatively long maturation period (months or even years, depending on the cheese type) (Castro *et al.*, 2015). As an example, probiotic strains *Lb. acidophilus*, *Lb. casei*, *Lb. paracasei* subsp. *paracasei* and *Bif. animalis* subsp. *lactis* added to full-fat, reduced-fat and low-fat Cheddar cheeses survived over 270 d of maturation (Ganesan *et al.*, 2014). In general, the dairy industry has established a minimum number of probiotic bacteria ($\geq 10^6$ cfu g^{-1}), which is required at the moment of ingestion, in order to ensure a favourable impact on consumer health (Bezerra *et al.*, 2016). This means that probiotic bacteria should be cultivable to high cell density for inoculation into the cheese vat or be able to proliferate during the manufacturing process or the maturation period (Karimi *et al.*, 2011).

Other technological hurdles that might affect the viability of probiotic bacteria in commercial cheese include the intrinsic product conditions, such as pH, acidity, molecular oxygen (especially for bifidobacteria), salt and sugars, food additives, moisture content, availability of nutrient(s), growth promoters and inhibitors, and interaction with other strains (Roy *et al.*, 1997; see also Champagne *et al.*, 2005). In addition, cultures used for probiotic cheeses should be selected in a way that minimises the antagonistic relationship among the non-probiotic and probiotic starters. This inhibitory activity among strains could be caused by several factors, such as production of lactic acid and/or other organic acids, hydrogen peroxide, bacteriocins, antibiotics or nutrient competition and depletion. Thus, in order to select strains able to maintain their viability under these conditions, in particular cheese, preliminary small-scale cheesemaking trials are recommended. In general, each strain has adapted to its particular environment, and thereby the selection of autochthonous probiotic bacteria may represent a valuable approach to overcome many of these technological hurdles (Ferrari *et al.*, 2016). The inoculation level is one of the parameters the manufacturer can tune to optimise the numbers of viable bacteria in probiotic cheeses. Two-stage fermentation for cultured dairy products has been shown to be effective in increasing the viability of probiotic bacteria by allowing these strains to become dominant prior to the addition of the starter cultures (Karimi *et al.*, 2011).

Apart from the viability of probiotics in cheeses until the time of consumption, their survival after exposure to GI tract conditions is also crucial to guarantee the therapeutic effect. Therefore, the tolerance of probiotic strains after exposure to GI tract conditions should be investigated and considered a critical criterion for strain selection. Food matrices possess significant effects in successful delivery of probiotics into the intestine (Mattila-Sandholm *et al.*, 2002). The denser matrix of the cheese texture may protect bacteria more efficiently than a fluid environment during its transit through the human

GI tract (Karimi *et al.*, 2011; Oh *et al.*, 2016). The dense matrix, high buffering capacity and high fat content of cheeses, such as Cheddar, may offer added protection to probiotic bacteria against bile salts and low pH conditions in the stomach (Dinakar & Mistry 1994; Gardiner *et al.*, 1999a, 1999b). Microencapsulation seems to be a promising technique for bacterial protection against GI tract and technological stress (Kailasapathy 2002; Rodrigues *et al.*, 2012). For example, this technique increased survivability of the probiotic bacteria (*Lb. paracasei* subsp. *paracasei* LBC-1e) through the Pasta Filata process during manufacture of Mozzarella cheese and simulated gastric digestion (Ortakci *et al.*, 2012).

In practice, probiotic *Lactobacillus* spp. and *Bifidobacterium* spp. are the most common micro-organisms included in cheeses. Because of their physiology, they are very well suited to this matrix. Different cheeses have been used to deliver a variety of probiotic bacteria (Corbo *et al.*, 2001; Vinderola *et al.*, 2009; Karimi *et al.*, 2011). Some cheese examples include the use of *Bif. bifidum* Bb02 and '*Bif. longum* Bb46' (presumed to be *Bif. longum* subsp. *longum* BB46) in Canestrato Pugliese (Corbo *et al.*, 2001); '*Bif. infantis*' (presumed to be *Bif. longum* subsp. *infantis*) in Cheddar (Daigle *et al.*, 1999); *Lb. casei* I90, *Lb. plantarum* I91 or *Lb. rhamnosus* I73 and I75 in Cremoso – an Argentinian soft cheese (Milesi *et al.*, 2009); and *Bif. bifidum* in Edam (Sabikhi & Mathur, 2000, 2002; Karimi *et al.*, 2011; Sabikhi *et al.*, 2014).

Within the lactobacilli group, probiotic cheeses with human-derived *Lb. paracasei* subsp. *paracasei* have been manufactured with no impact on cheese composition (Gardiner *et al.*, 1998; Stanton *et al.*, 1998; Caggia *et al.*, 2015). In one study, *Lb. paracasei* subsp. *paracasei* NFBC 338 and NFBC 364 grew to 2.9×10^8 cfu g^{-1} in matured cheese in 3 months, and maintained the numbers up to 200 d (Stanton *et al.*, 1998). Bacteria other than lactobacilli and bifidobacteria may play essential roles in probiotic cheeses. Propionibacteria have also been suggested to have probiotic properties (Mantere-Alhonene, 1995; Jan *et al.*, 2002, de Freitas *et al.*, 2015); some strains are used in the manufacture of Swiss cheese and provide an added value to the product. *Propionobacterium freudenreichii* subsp. *shermanii* SI41 is able to survive acid and bile salts and function as a probiotic (Jan *et al.*, 2002). This species also has a bifidogenic effect, which promotes the growth of bifidobacteria strains (Kaneko, 1999; de Freitas *et al.*, 2015). Other species, such as *Ent. faecium* PR88, which is believed to be useful in the treatment of irritable bowel syndrome, have been used in cheesemaking (Gardiner *et al.*, 1999b; Dos Santos *et al.*, 2015). Yeast from infant faeces has been isolated and has potential for use in industrial fermentation (Psomas *et al.*, 2001).

It is recognised that probiotic bacteria have certain health-promoting properties, but they may also serve a useful function by producing compounds of 'health' value (Gobetti *et al.*, 2010; Joshi *et al.*, 2015). For example, in Finland, *Lb. acidophilus* and *Bifidobacterium* spp. were used to produce a low-fat cheese (Ryhanen *et al.*, 2001). These organisms apparently produced bioactive peptides with anti-hypertensive properties during the maturation of the cheese, thus adding to the nutritional and probiotic values of the product. The inclusion of probiotic cultures, such as *Lb. rhamnosus* GG, in cheese may also have an impact on dental caries (Abou-Dawood, 2002). Some strains of bifidobacteria produce anti-microbial compounds, which reduce the levels of *Pseudomonas* in Cottage cheese (O'Riordan & Fitzgerald, 1998). There also are commercial probiotic cultures of *Lb. rhamnosus* and *Pro. freudenreichii* subsp. *shermanii* with anti-clostridial effects and activity against contaminating yeasts and moulds (Hansen, 1997; Ephraim *et al.*, 2013).

4.6.3 *Very hard and hard cheese varieties*

Cheeses in this category contain ≤38 g 100 g^{-1} moisture and require a long maturation period; the starter cultures are mainly blends of thermophilic and mesophilic LAB, including *Propionibacterium* spp. In subsequent sections of this chapter, it is concluded by many researchers that using encapsulated probiotic bacteria and cheese coating materials enhances their survival rates in cheeses and other dairy products. The reader is referred to various reviews for further discussion (Jung *et al.*, 2007; Su *et al.*, 2007; Bosnea *et al.*, 2009, 2014; de Vos *et al.*, 2010; Islam *et al.*, 2010; Burgain *et al.*, 2011; de Menezes *et al.*, 2013; Feucht & Kwak, 2013; Huq *et al.*, 2013; Riaz & Masud, 2013; Dhewa *et al.*, 2014; de Prisco *et al.*, 2015). Examples of probiotic cheeses follow.

Pecorino, *Pecorino Siciliano* and *Ragusano* are types of Italian grating cheeses, mainly made from raw or heat-treated sheep's milk and using traditional lamb rennet paste (RP; i.e. control). Santillo & Albenzio (2008) used different blends of probiotic bacteria [*Lb. acidophilus* LA-5 (RP-L), '*Bif. lactis* BB-12' (RP-B) (presumed to be *Bif. animalis* subsp. *lactis* BB-12) and '*Bif. longum*' (presumed to be *Bif. longum* subsp. *longum*) (RP-B)]. During the maturation period, the counts of lactobacilli and bifidobacteria were 8 $\log_{10}$ and 9 $\log_{10}$ cfu g^{-1}, respectively; cheeses made with *Bifidobacterium* spp. enhanced proteolytic activity in the product, resulting in the highest values of non-casein nitrogen (Cn N), water-soluble N and α_s-Cn at 60 d old cheese. The RP-L cheese displayed intermediate levels of N fractions, and the percentage of γ-Cn in RP and RP-L cheeses at 60 d was twofold higher than in the cheese curd of the same groups compared to 3-fold higher in RP-B cheese. Lower hardness in RP-B cheese was observed at the end of the maturation period, which could be attributed to greater proteolysis in the product, but there were no differences in the sensory profiles (i.e. smell and taste) of all the cheeses made. However, the effect of encapsulation of the same probiotic bacteria (i.e. in RP) on the quality of Pecorino cheese was recently reported by Santillo *et al.* (2012), where the viable counts of the lactobacilli and total bifidobacteria were 6.92 and 5.44 $\log_{10}$ cfu g^{-1}, respectively, after 120 d of storage.

In a study by Caggia *et al.* (2015), *Lactobacillus* strains (177) were isolated from Ragusano and Pecorino Siciliano cheeses, and were screened *in vitro* for probiotic traits in comparison to *Lb. rhamnosus* GG (a commercial strain). Of these isolates, only 13 lactobacilli strains were selected, and multiplex-PCR application revealed that nine strains belonged to *Lb. rhamnosus* and four to *Lb. paracasei* subsp. *paracasei*. These strains were investigated in simulated GI tract that showed some of the isolates (*Lb. rhamnosus* FS10 and *Lb. paracasei* subsp. *paracasei* PM8) had the potential for further investigation.

Sao Jorge and *Parmigiano Reggiano* are hard varieties of Portuguese and Italian cheeses, respectively, where the microbiota in the mature products has been screened for potentially new probiotic bacteria (Dias *et al.*, 2014). The main isolates were

Lb. rhamnosus, *Lb. paracasei* subsp. *paracasei*, *Lactobacillus buchneri*, *Lactobacillus curvatus* and *Bif. animalis* subsp. *lactis*, but further research is needed.

Canestrato Pugliese is an Italian sheep's milk cheese (Corbo *et al.*, 2001), and the probiotic types have been made using the strains of *Bif. bifidum* BB02 and '*Bif. longum* BB46' (presumed to be *Bif. longum* subsp. *longum* BB46) (either single or mixed; initial count was 7.0 $\log_{10}$ cfu mL^{-1}) as co-culture with *Str. thermophilus*. After 56 d maturation, the survival rate of the bifidobacteria strains was 6.0 and 5.0 $\log_{10}$ cfu g^{-1}, respectively. The probiotic cheeses had a higher concentration of acetic acid, enhanced α- and β-galactosidase activities, a slight increase in soluble nitrogen (N) levels and higher levels of amino acids and fatty acids; also, all the lactose content was catabolised. The sensory attributes were not significantly different from those of the control cheese.

Ras cheese is an Egyptian variety similar to the Greek Kefalotiri. In a study by Dabiza & El-Deib (2007), single strains of *Lb. reuteri*, *Lb. casei* and *Lb. gasseri* were used, or in combination with a yoghurt starter culture. All the probiotic cheeses produced more soluble N, essential and non-essential amino acids, α- and β-galactosidase, amino peptidase and di-peptidase enzymes than the cheese produced using the mixed starters. When compared to the control cheese using sodium dodecyl sulphate–polyacrylamide gel electrophoresis (SDS-PAGE), the probiotic cheeses exhibit more proteolytic activity than the yoghurt starter culture. In all the cheeses, the starter cultures inhibited the growth of coliforms, *Staphylococcus* spp., yeasts and other fungi. Abd El-Salam *et al.* (2011) also made probiotic Ras cheese using *Lb. acidophilus* or *Lb. casei* mixed with a yoghurt starter culture, and matured for 3 months at 12 °C. The probiotic strains were good producers of conjugated linoleic (CLA), and their counts averaged 8 $\log_{10}$ cfu g^{-1} during the storage period. Recently, a probiotic Ras cheese was made using '*Bif. bifidum* Bb_{12}' (presumed to be *Bif. bifidum*), '*Bif. lactis* Bb_{12}' (presumed to be *Bif. animalis* subsp. *lactis* BB-12) and *Lb. acidophilus* LA-5 with a co-culture of a yoghurt starter and '*Lactobacillus lactis*' (presumed to be *Lactobacillus delbrueckii* subsp. *lactis*), with animal studies showing evidence of potential benefit for lowering cholesterol and serum lipids when fed to rats (El-Zahar *et al.*, 2015).

Cheddar cheese batches were made containing different combinations of commercially available probiotic bacteria ['*Bif. lactis* LAFTI B94' and '*Bif. lactis* Bb12' (presumed to be *Bif. animalis* subsp. *lactis* LAFTI B94 and *Bif. animalis* subsp. *lactis* BB-12, respectively) and *Bifidobacterium* spp. HOWARU DR10, *Lb. acidophilus* LAFTI® L10 and LA-5, *Lb. casei* LC-1, '*Lb. paracasei* LAFTI L26' (presumed to be *Lb. paracasei* subsp. *paracasei* LAFTI® L26) and/or *Lb. rhamnosus* HOWARU DR20] (Phillips *et al.*, 2006). The viability (cfu g^{-1}) of these probiotic strains during the maturation of Cheddar cheese over 32 weeks was as follows: (a) all three bifidobacteria strains remained at high numbers (i.e. 4×10^7, 1.4×10^8 and 5×10^8, respectively), (b) *Lb. casei* at 2×10^7, (c) *Lb. paracasei* subsp. *paracasei* at 1.6×10^7, and (d) *Lb. rhamnosus* at 9×10^8. However, the two *Lb. acidophilus* strains performed poorly, with both decreasing to levels of 3.6×10^3 and 4.9×10^3 cfu g^{-1}. Phillips *et al.* (2006) concluded that Cheddar cheese is a good vehicle for a variety of commercial probiotic bacteria, but the survival of two *Lb. acidophilus* strains still needs to be improved. In a similar study, Ong & Shah (2009) produced Cheddar cheese made with '*Bif. longum* 1941' (presumed to be *Bif. longum* subsp. *longum* 1941), *Bif. animalis* subsp. *lactis* LAFTI B94, *Lb. casei* 279, *Lb. casei* LAFTI

L26, *Lactobacillus acidophilus* 4962 or *Lb. acidophilus* LAFTI L10 as adjunct cultures, and the products were matured for 24 weeks at 4 and 8 °C. The counts of the lactococci starter culture in cheeses produced with strains B94, L26 or 4962 and matured at 8 °C were significantly lower than those ripened at 4 °C ($P<0.05$) after 24 weeks. All the probiotic bacteria remained viable (>7.50 $\log_{10}$ cfu g^{-1}) and were not affected by the maturation temperatures. There were significant effects of the type of probiotic bacteria used, maturation time, maturation temperatures and their interactions on the concentration of lactic and acetic acids in the cheeses ($P<0.05$). The acetic acid concentrations in cheeses made with *Bifidobacterium* spp. or L26 were significantly higher than those of the control cheese ($P<0.05$). Citric, propionic and succinic acids contents of the cheeses were not significantly affected by the type of probiotic bacteria or maturation temperatures ($P>0.05$) (see also Ong *et al.*, 2007). The difference in the counts of *Lb. acidophilus* LAFTI L10 in Cheddar cheeses in the studies by Phillips *et al.* (2006) and Ong & Shah (2009) could have been due to the duration of the maturation period (i.e. long or short, respectively) or the enumeration methods used. Oberg *et al.* (2011) also studied the survival and enumeration of probiotic bacteria in cheeses.

In a different study, Ristagno *et al.* (2012) produced probiotic Cheddar cheese using *Lb. casei* DPC 2048^{CM} (chloramphenicol-resistant and bacteriocins-sensitive, i.e. an indicator strain) and *Lb. paracasei* subsp. *paracasei* DPC 4715 (bacteriocins-producing strain) as co-cultures with mesophilic cheese starter culture. During the maturation period (6 months) at 8 °C, no inhibition of the indicator strain was observed and no bacteriocins production was detected. However, *Lb. paracasei* subsp. *paracasei* 4715 was sensitive to chymosin and cathepsin D, and it may have been hydrolysed by the coagulant used in cheesemaking or by indigenous milk proteases. When *Lb. rhamnosus* 6134 and iso-malto-oligosaccharide were used in Cheddar cheesemaking, the viable count of lactobacilli was >8 $\log_{10}$ cfu g^{-1} after 2 months; this is a very young product as normally matured Cheddar is ~12 months old (Liu *et al.*, 2015a). Desfosses-Foucault *et al.* (2012) reported the use of lactococci starter culture and three probiotic strains (*Bif. animalis* subsp. *lactis* BB-12, *Lb. rhamnosus* RO011, or *Lactobacillus helveticus* RO052 and/or mixed combinations of these organisms) in Cheddar cheesemaking. They reported the following observations in 6-month-old matured product: (a) the addition of probiotic strains seemed to accelerate the loss of lactococci viability in comparison to the control cheese samples, especially when *Lb. helveticus* RO052 was used, (b) the viability of all three probiotic bacteria was also significantly reduced in the product when using mixed starter culture ($P<0.0001$), with *Bif. animalis* subsp. *lactis* BB-12 being the most sensitive to the presence of other strains, and (c) all probiotic bacteria did retain their viability (9 $\log_{10}$ cfu $^{-1}$g) throughout the maturation period. Lastly, Zhang *et al.* (2013) reported the benefit(s) of using probiotic *Lb. plantarum* K25 in Cheddar cheesemaking by reducing the cholesterol level in mice.

In low-fat probiotic Cheddar (Rayan *et al.*, 2015), *Lac. lactis* subsp. *lactis* R604 and *Lactobacillus mucosae* DPC 6426 (previously shown to have promising hypocholesterolaemic activity in a murine model, and an EPS producer) were used as an adjunct culture at a level of ~10^6 cfu mL^{-1} in the cheese milk (subsequently present in the cheese curd at >10^7 cfu g^{-1}). After 6 months of maturation, the viable count of lactobacilli remained >5×10^7 cfu g^{-1}. The use of such adjunct culture had no significant effect on the sensory attributes of the Cheddar cheese, including its chemical composition, proteolysis, pH or instrumentally quantified textural characteristics.

Swiss-type cheese was made using the same adjunct culture (*Lb. mucosae* DPC 6426) as reported by Rayan *et al.* (2015) for Cheddar cheesemaking, *Str. thermophilus* Th3, *Lb. helveticus* LHBO2 and '*Propionibacterium freudenreichii* DPC 6451' (presumed to be *Pro. freudenreichii* subsp. *shermanii* DPC 6451). The viable counts of the adjunct culture were similar to those reported here for Cheddar cheesemaking; however, the presence of the adjunct culture imparted a more appealing appearance to the product. In a study by O'Sullivan *et al.* (2016), the presence in the cheese milk of the strain *Lb. casei* DPC-6987 (which was isolated from a cheese plant environment), in the presence of propionic acid bacteria and in the absence of *Lb. helveticus* to mimic starter failure, led to excessive eye formation during the maturation period. The availability of excess amounts of lactose, galactose and citrate during the initial maturation stages probably provided the *Lb. casei* DPC-6987 strain with sufficient substrates for gas formation. These results demonstrated the commercial importance of both the viability of starter populations and control of the specific nonstarter LAB in ensuring appropriate eye formation in Swiss-type cheese.

An Emmental cheese environment was shown to enhance the probiotic-type characteristic(s) of *Pro. freudenreichii* subsp. *shermanii* CIRM BIA 1 with regard to its stress tolerance (Gagnaire *et al.*, 2015); the surface protein of the same strain has been associated with anti-inflammatory properties (Le Marechal *et al.*, 2015).

Silter is a traditional Italian hard cheese; it was investigated in a study where a total of 426 lactic acid strains were isolated, of which 274 strains were found to produce bacteriocins against a wide range of pathogens tested. This led to the conclusion that this cheese variety could be an important source of novel probiotic strains (Losio *et al.*, 2015). Pisano *et al.* (2011) also reported the diversity and functional/probiotic properties of *Lb. plantarum* strains isolated from traditional Italian cheeses.

4.6.4 *Semi-hard varieties*

The moisture content of cheeses in this category averages ~40 g 100 g^{-1}, and some probiotic varieties include the following.

Pategras is an Argentinian semi-hard variety of cheese, for which a probiotic variety was made using *Str. thermophilus*, *Lb. acidophilus* LA-5, *Lactobacillus paracasei* subsp. *paracasei* DSM and *Bif. animalis* subsp. *lactis* DSM. The product was matured for 60 d. The free fatty acids (FFAs) level in the cheese increased in a manner similar to that of Edam and Port Salut, whilst the probiotic counts ranged between 7.5 and 9.1 $\log_{10}$ cfu g^{-1} without affecting the sensory attributes of the product. Multivariate analysis, however, showed clear differences between the probiotic and control cheeses (Perotti *et al.*, 2009). In another study, Bergamini *et al.* (2010) produced Pategras cheese using a single-strain starter culture (*Str. thermophilus*) to ferment the milk to make cheese. Three different groups of probiotic strains were used as adjunct cultures: (a) the casei group – *Lb. casei*, *Lb. paracasei* subsp. *paracasei* and *Lb. rhamnosus*, (b) the acidophilus group – *Lb. acidophilus*, and (c) the bifido group – *Bif. animalis* subsp. *lactis*. These were added to the milk as freeze-dried or were pre-cultured in a substrate. The counts of these organisms in the cheeses after a 60 d maturation period ranged between 10^7 and 10^9 cfu g^{-1}. More details regarding the proteolytic behaviour of these probiotic bacteria in Pategras cheese were reported by Bergamini *et al.* (2005, 2006, 2009).

St. Paulin cheese (30 d old) containing *Lb. paracasei* subsp. *paracasei* was investigated in a dynamic model of the human digestive system to assess its survival ability. There was a significant reduction in the numbers of probiotic bacteria surviving alive through the model, but administration of a 175 g serving of cheese delivered >10^7 cfu g^{-1} viable cells to the colon section of the model (Kheadr *et al.*, 2011).

Coalho is a Brazilian semi-hard goat's milk cheese; the survival of probiotic bacteria (*Lb. acidophilus* LA-5, *Lb. paracasei* subsp. *paracasei*-01 and *Bif. animalis* subsp. *lactis* BB-12) in this product was studied in an *in vitro* model simulating the conditions of digestion, where the viable counts of all the strains decreased from 7–8 $\log_{10}$ to 5.5–6.0 $\log_{10}$ cfu g^{-1}. The probiotic strains inhibited *L. monocytogenes* and *Staphylococcus aureus* in the product during the maturation period (de Oliveira *et al.*, 2014).

Dutch- and Gouda-type cheeses were studied in order to determine the influence of probiotic bacteria (*Lb. paracasei* subsp. *paracasei* LPC-37, *Lb. acidophilus* NCFM and *Lb. rhamnosus* HN-001) on the counts of lactic starter culture and pathogens; a decrease in counts of the pathogens was observed during the maturation period (Aljewicz & Cichosz, 2015a). In separate studies, (a) Aljewicz & Cichosz (2015b) observed the effect(s) of probiotic *Lb. rhamnosus* HN001 on the *in vitro* availability of minerals in the cheese, and (b) Aljewicz *et al.* (2014, 2016) observed the following aspects of probiotic Gouda-type cheeses including substitution of milk fat with palm oil:

- The counts of conventional starter culture (*Lactococcus* spp.) were lower in cheeses containing the probiotic strain *Lb. acidophilus* NCFM than in the cheese made with *Lb. paracasei* subsp. *paracasei* LPC-37.
- The viability of *Lb. acidophilus* NCFM significantly correlated with cheese type and storage time. Counts were higher in matured cheeses and, overall, greater viability was observed with *Lb. paracasei* subsp. *paracasei* LPC-37 compared to *Lb. acidophilus* NCFM.

In a study of yoghurt and Prato cheese (an Argentinian variety similar to Gouda and Edam), counting *Lb. acidophilus* NCFM (Howaru Dophilus – as an adjunct culture) was successful by using selective media (MRS agar containing sorbitol or bile salt) and incubating the agar plates at 45 °C and 37 °C for 72 h, respectively, under anaerobic conditions (Gebara *et al.*, 2015).

Caciotta is an Italian farmhouse semi-hard cheese made from goat's or cow's milk. A selective medium was developed to detect members of the *Lb. casei* group isolated from probiotic milk and cheeses, and to monitor the probiotic strain *Lb. paracasei* subsp. *paracasei* CRL 431 in Caciotta cheese; this would help with the correct labelling of bacterial species in the product (di Lena *et al.*, 2015).

4.6.5 *Brined cheeses*

The technical and scientific aspects of the manufacture of brined cheese, including mechanisation systems used, have been detailed by Tamime (2006b) and, in general, the moisture content of these cheeses is 50–55 g 100 g^{-1}. Many varieties are produced in the Middle Eastern region, and the reported probiotic types are as follows.

Beyaz is a Turkish Feta-like cheese, normally made from cow's, goat's or sheep's milks, and produced as soft, semi-soft or hard cheese. Probiotic cheeses were made using blends of probiotic strains of human origin (*Lactobacillus fermentum* AB5-18 and AK4-120, and *Lb. plantarum* AB16-65 and AC18-82) in co-culture with *Lac. lactis* subsp. *cremoris* and *Lac. lactis* subsp. *lactis*. The experimental cheeses were stored for 120 d at 4 °C, and their quality was similar to that of the control; at the end of the storage period, the total probiotic count was 7.42×10^7 cfu g^{-1} (Kilic *et al.*, 2009). In a separate study, Yerlikaya & Ozer (2014) produced Beyaz cheese using different strains of lactobacilli (*Lb. plantarum*, *Lb. rhamnosus*, *Lb. acidophilus* and *Lb. casei*) in co-culture with *Str. thermophilus* and, after 28 d at 4 °C, the viable counts of the probiotic bacteria ranged between 7.4 and 9.1 $\log_{10}$ cfu g^{-1}. Sensory profiling of the cheeses was highest for the product made with *Lb. casei* and *Str. thermophilus*, whereas the remaining three probiotic cheeses had slightly lower scores for taste and appearance. A closely related product known as Turkish white brined was investigated by Gursoy *et al.* (2014); '*Bif. longum*' (presumed to be *Bif. longum* subsp. *longum*) was added with the lactic starter culture, and after 90 d of maturation, the count of the probiotic bacteria was $>10^7$ cfu g^{-1}.

Iranian White brined is similar to the product made in Turkey, and it is made using different mammalian milks. Pasteurised cow's milk spiked with *E. coli* O157:H7 (10^4 cfu mL^{-1}) was made into cheese using a mixed lactic acid starter culture and two strains, *Lb. acidophilus* 4962 and LA-5. During cheesemaking, the number of coliforms increased to 10^7 cfu g^{-1}, but decreased significantly ($P<0.05$) during the maturation period (Darehabi & Nikmaram, 2011). In a study, Zomorodi *et al.* (2011) used the ultrafiltration (UF) method to produce a cheese containing *Lb. casei* ATCC 39392, *Lb. plantarum* ATCC or *Bif. bifidum* ATCC 29521 either in a free form or microencapsulated. In the latter form, the survival of all the probiotic bacteria was high (10^6–10^7 cfu g^{-1}). There was no significant difference in rheological properties between the probiotic and control cheeses, but the sensory profiling of the encapsulated probiotic cheeses scored higher for flavour compared to products containing un-encapsulated strains. A summary of observations on probiotic Iranian White brined cheese is as follows:

- The use of *Lb. paracasei* subsp. *paracasei* and *Pediococcus inopinas* resulted in cheese that was highly rated by the sensory panel. The viable counts in the product, however, were not reported; this study focused on the physicochemical changes in the probiotic cheeses (Barouel *et al.*, 2011).
- Probiotic bacteria ('*Bif. animalis* ATCC 25527' – presumed to be *Bif. animalis* subsp. *animalis* ATCC 25527 – and *Lb. rhamnosus* ATCC 7469) were mixed separately with lactic starter culture for the production of cheese. The viability of bifidobacteria ($\log_{10}$ cfu g^{-1}) in the cheese after being at 6–8 °C after 45 d was between 6 and 7 and increased to >8 after 60 d, whilst the lactobacilli remained at a constant 6–7 during the storage period (Mahmoudi *et al.*, 2012).
- Cow's milk (spiked with either *Staph. aureus* or *L. monocytogenes*) was fortified with *Mentha longifolia* L. essential oil (EO) and fermented with lactic starter and *Lb. casei* (10^8–10^9 cfu mL^{-1}) to produce Iranian White brined cheese. The presence of EO, even at low concentration, inhibited the growth of the pathogenic organisms in the product (Ehsani & Mahmoudi, 2013; Mahmoudi *et al.*, 2013).

Akkawi is a brined cheese popular in Lebanon, Syria and Jordan. Partial replacement of NaCl with KCl (at ratios of 3:1, 1:1 and 1:3, respectively) in probiotic products (with added *Lb. casei* and *Lb. acidophilus*) significantly affected the level of water- and phosphotungsic-soluble nitrogen, calcium contents and growth of *Lb. delbrueckii* subsp. *bulgaricus* in the experimental cheeses during a 30 d storage period. The growth of *Str. thermophilus* and the probiotic bacteria was significantly affected in the experimental cheeses but, after 30 d, counts of *Lb. casei* and *Lb. acidophilus* averaged 7.4 and 7.3 $\log_{10}$ cfu g^{-1}, respectively. No significant differences were observed regarding the physical properties and sensory attributes over the same storage period (Ayyash *et al.*, 2012). A related study by Gandhi *et al.* (2014) reported the effect of KCl substitution on survival of *E. coli* ATCC 25922 in the presence of selected probiotic bacteria.

A closely related product is Nabulsi cheese (Tamime, 2006b). Yamani *et al.* (1998) isolated the following potentially probiotic strains from cow's and sheep's milks: '*Lb. paracasei*' (presumed to be *Lb. paracasei* subsp. *paracasei*), *Lb. rhamnosus*, *Ent. faecalis*, *Ent. faecium* and *Ent. durans*. Nabulsi cheeses made from these isolates, including mesophilic LAB, were acceptable.

Feta-type cheese was made using encapsulated (using either an extrusion or emulsion technique) and 'free' forms of '*Bif. bifidum* BB-12' (presumed to be *Bif. animalis* subsp. *lactis* BB-12) and *Lb. acidophilus* LA-5. Both encapsulation techniques were effective in protecting the probiotic bacteria, and counts were $>10^7$ cfu g^{-1} at the end of the storage period. Counts of probiotic bacteria in 'free' form (i.e. un-encapsulated) decreased by ~3 $\log_{10}$ cycle compared to the encapsulated probiotic where the cells decreased by only 1 $\log_{10}$ cycle. The levels of medium- and long-chain FFAs and carbonyl compounds in the cheeses with immobilised probiotics were much higher than in the control cheese, but the sensory properties of the control and experimental cheeses were similar (Ozer *et al.*, 2009). In a study by Xanthopoulos *et al.* (2000), 32 *Lb. plantarum* strains isolated from Feta cheese throughout its maturation period were able to grow at low pH and in the presence of bile; thus, these strains (if health benefits could be demonstrated) might have probiotic potential. The results also suggest that the presence of specific *Lb. plantarum* strains in Feta cheese may have interesting biotechnological properties.

Domiati cheese (an Egyptian brined variety) was made using single probiotic strains of *Lb. acidophilus* LA-5 or '*Bif. longum* ATCC 15707' (presumed to be *Bif. longum* subsp. *longum* ATCC 15707) to study the inhibition of *Staph. aureus* and *E. coli* O157:H7 after the pathogens were spiked into the milk when making the product. Both probiotic bacteria inhibited the growth of the test pathogens, but the rate of inactivation was greater for *Staph. aureus* than *E. coli* O157:H7. The antibacterial activity of *Lb. acidophilus* LA-5 was significantly higher ($P<0.05$) than '*Bif. longum* ATCC 15707' (presumed to be *Bif. longum* subsp. *longum* ATCC 15707). The survival rate in the cheeses of both probiotic organisms was considered satisfactory (El-Kholy, 2014).

4.6.6 *Soft cheeses*

Fresh soft cheese was made from UF milk retentate, and fermented with ABT starter culture (see Section 4.2.2); the product had an acceptable flavour and smooth body, and it could be easily cut and handled. The viable counts of bifidobacteria and lactobacilli were ~1×10^7 cfu g^{-1} after 3 to 5 d (El-Shibiny *et al.*, 2005). In a study by Cocolin *et al.* (2010), fresh soft cheese was made with different probiotic strains (*Lb. acidophilus* LA-5, *Lb. rhamnosus* GG or *Lb. casei* Shirota) whose growth was evident during production and storage (8 d at 4 °C). The viable counts of all the probiotic strains reached 10^8 cfu g^{-1} after 15 d; however, due to the relatively low rate of acidification during the manufacture of the cheese, the product was susceptible to spoilage by yeasts and coliforms. The combined action of mesophilic lactic starter culture, '*Bif. longum*' (presumed to be *Bif. longum* subsp. *longum*) and potassium sorbate was found to be effective to reduce the counts of *Pseudomonas fluorescens* at the end of storage period by 99.5% in fresh cheese without affecting the sensory properties of the product (Deeb & Ahmed, 2010). In a study by Cardenas *et al.* (2014), fresh cheese was produced using two strains of *Lactobacillus salivarius* CECT 5713 and PS2 (isolated from human milk) and lactococci starter culture, and after 28 d at 4 °C, the viable counts were 6.7 and 6.6 $\log_{10}$ cfu g^{-1}, respectively, representing a ~1.3 $\log_{10}$ cycle reduction. The product was highly rated, but the body characteristic was slightly hard.

Tallaga (an Egyptian soft cheese variety) was made with a mixed starter culture of *Lb. rhamnosus* and *Lac. lactis* subsp. *lactis* biovar *diacetylactis*. The probiotic bacteria reduced the counts of *B. cereus*, and no toxin was detected in the product (Sadek *et al.*, 2006). In a goat's milk soft cheese, a co-culture of *Ent. faecalis* CECT 7121 (5.0×10^4 cfu mL^{-1}) and a starter culture inhibited the growth of *Staph. aureus*, but not the lactococci or lactobacilli. Thus, it was concluded that the probiotic culture might be helpful for producing safer products (Sparo *et al.*, 2012).

Panela (a Mexican soft cheese variety) was made using *Bif. breve* ATCC 15700 and *Lb. rhamnosus* GG ATCC 53103 as mono or mixed cultures, and the milk was fortified with fava bean starch, which was considered to be prebiotic (Escobar *et al.*, 2012). After 30 d of storage at 4 °C, the viability of single and mixed cultures in the cheeses was 7.1, 8.8 and 9.0 $\log_{10}$ cfu g^{-1}, respectively, which did not affect the taste of the product. The sensory panel did not favour the addition of fava bean starch in Panela cheese, however; the starch also affected the structure of the product, making an open amorphous matrix with some void spaces, and the starch granules was observed embedded in the protein matrix.

Cottage cheese was made with a mixed starter culture (YO-MIX TM 205), *Lb. casei* and *Lb. rhamnosus* GG. The viable counts of both probiotic bacteria were $>10^6$ cfu g^{-1} during 28 d of storage at 8 °C. The cheese showed an increased metabolic activity with higher levels of lactic and acetic acids, and higher numbers of potentially bioactive peptides that were associated with reduced counts of *L. monocytogenes* by ~1 $\log_{10}$ cycle after 20 d of storage (Garcia *et al.*, 2013).

Petit-Suisse was made with ABT culture (see Section 4.2.2), and the milk was fortified with different sweeteners (sucrose, aspartame, Neotame®, sucralose, stevia and refined sugar). The survival of lactobacilli and bifidobacteria were >7.5 and >7.0 $\log_{10}$ cfu g^{-1}, respectively. None of the sweeteners exerted any negative effect, although aspartame caused a slight reduction in bifidobacteria counts (Esmerino *et al.*, 2013, 2015). The same probiotic culture was used to compare the effect of soy-based (SP), milk-based (MP) and mixed ingredients (milk, soy and cream – MSP) on the quality of Petit-Suisse. The counts of *Bif. animalis* subsp. *lactis* BB-12 were >8 $\log_{10}$ cfu g^{-1}, but the viability of *Lb. acidophilus* in these products varied (MP: 7.6 $\log_{10}$ cfu g^{-1}; MSP: 6.5 $\log_{10}$ cfu g^{-1}; and SP: 6.8 $\log_{10}$ cfu g^{-1}). Hardness and gumminess characteristics were higher in the soy-based products compared with milk-based products, as well as the sensory attributes (Matias *et al.* (2014). In a study by Pereira *et al.* (2016), the effect of added antioxidant (ascorbic acid, glucose oxidase, cysteine and jabuticaba extract) on the quality of Petit-Suisse was investigated, whilst the effect of hydrocolloids on the quality of probiotic Petit-Suisse was reported by Maruyama *et al.* (2008). In the latter study, the viable counts of *Lb. acidophilus* and '*Bif. longum*' (presumed to be *Bif. longum* subsp. *longum*) were >6.40 and >7.30 $\log_{10}$ cfu g^{-1}, respectively, during the storage period at 4 °C.

Minas Frescal is a Brazilian soft cheese variety that originated in Minas Gerais state. Enzymatic and direct acidification methods were used to coagulate the milk in the presence of *Lb. casei* Zhang; the bacterium reduced the pH to 4.94 and hydrolysed the protein (high proteolysis indexes – 0.470 to 0.702 absorbance at 340 nm) during storage. The viable counts of the lactobacilli reached 8.1 and 9.0 $\log_{10}$ cfu g^{-1} in the direct acidification and enzymatic coagulation methods, respectively, after 21 d of refrigerated storage. All the cheeses showed more viscous-like behaviour, with the rigidity tending to decrease during storage; lower luminosity values were also observed. Consumer acceptability of the control cheese made by direct acidification was highly rated, but both probiotic cheeses had lower scores for all sensory attributes, in particular flavour and overall liking. The addition of *Lb. casei* Zhang led to changes in all parameters and had a negative effect on sensory acceptance, but this could be controlled by reducing the inoculation rate (Dantas *et al.*, 2016). A combined mixture of *Lac. lactis* subsp. *lactis* and *Lb. acidophilus* was used during the manufacture of Minas Frescal cheese; average counts of the probiotic bacteria were 9 $\log_{10}$ cfu g^{-1}. The addition of arginine and reduction of salt content did not affect the quality and acceptability of the product (Felicio *et al.*, 2016). Other aspects of this type of probiotic cheese have been reported as follows: (a) Buriti *et al.* (2005a) found the ranges of viable counts of *Lb. acidophilus* in cultured and direct-acidified Minas Frescal after 21 d of storage to be 6.0 to 6.9 and 5.4 to 6.5 $\log_{10}$ cfu g^{-1}, respectively; similar counts were found in Minas Frescal made with *Lb. paracasei* subsp. *paracasei* (Buriti *et al.*, 2005b). (b) A high inoculation rate of *Lb. acidophilus* during the manufacture of Minas Frescal cheese resulted in high viable counts (9.1 to 9.4 $\log_{10}$ cfu g^{-1}) in the product, which affected the sensory scores for appearance, aroma and texture compared with conventional cheeses (Gomes *et al.*, 2011). (c) Lollo *et al.* (2012) reported that probiotic cheese attenuates exercise-induced immune suppression in Wistar rats. (d) Costa *et al.* (2013) isolated LAB strains from an artisanal product, which had probiotic potential. And (e) Andrade *et al.* (2014) studied the *in vitro* probiotic properties of *Lactobacillus* spp. isolated from Minas artisanal cheese.

Kalari/Kradi is a Himalayan soft cheese. The shelf life (at 4 °C for 30 d) of the product was extended/improved by using different probiotic bacteria (*Lb. casei*, *Lactobacillus plantarum* and *Lactobacillus brevis*). The addition of probiotic strains did not affect the physico-chemical properties of the product except that the acidity was found to be significantly higher than in control cheese. The probiotic strains, however, exhibited greater antioxidant activity compared with the control cheese. The content of the flavour compounds increased significantly during the storage period, but their concentration was significantly lower in the probiotic cheese, where the counts of probiotic bacteria were ~6 $\log_{10}$ cfu g^{-1}. Psychrotrophic bacteria, yeasts and mould counts were found to increase significantly in the control cheese, whereas these counts decreased significantly in the probiotic cheeses (Mushtaq *et al.*, 2016).

4.6.7 *Pasta Filata cheeses*

Kasar cheese (a Turkish variety) was made using encapsulated and 'free' form *Lb. acidophilus* LA-5 and '*Bif. bifidum* BB-12' (presumed to be *Bif. animalis* subsp. *lactis* BB-12) as co-culture with lactococci during manufacture. As expected, scalding of the curd caused a drastic decline of the 'free' probiotic bacteria, but not the encapsulated cultures. After 90 d of storage at 10 °C, the counts of both probiotic bacteria were $>10^7$ cfu g^{-1} (Ozer *et al.*, 2008).

Mozzarella cheese has been made from semi-skimmed milk, with *Lb. paracasei* subsp. *paracasei* LBC-1 added to the milk either as an alginate-microencapsulated (LBC-1e) or in a 'free' form (LBC-1f) at a level of 10^8 and 10^7 cfu g^{-1}, respectively. Survival rates of the probiotic bacteria LBC-1f and LBC-1e during cheesemaking (i.e. in the curd, after stretching and storage for 42 d at 4 °C) and the counts (cfu g^{-1}) were 5.9×10^7 and 5.4×10^8, 2.1×10^7 and 3.2×10^8, and 3.2×10^7 and 2.5×10^8, respectively. The cheeses were subjected to a simulated gastric digestion, which affected the survival rate of the probiotic bacteria in the product (Ortakci *et al.*, 2012).

Fior di Latte is a high-moisture cow's milk Mozzarella cheese. Minervini *et al.* (2012) screened 18 probiotic strains belonging to the species *Lb. casei*, *Lb. delbrueckii* subsp. *bulgaricus*, *Lb. paracasei* subsp. *paracasei*, *Lb. plantarum*, *Lb. rhamnosus* and *Lb. reuteri*. Prior exposure of the lactobacilli cells to 42 °C for 10 min increased their heat resistance at 55 °C for 10 min; two strains (*Lb. delbrueckii* subsp. *bulgaricus* SP5 and *Lb. paracasei* subsp. *paracasei* BGP1) were thought suitable because of their higher survival rate when mimicking the stretching of the curd. The physicochemical properties and sensory attributes of the cheeses made with these two strains using either the biological or direct acidification methods were very good, and the viable count of each was ~8.0 $\log_{10}$ cfu g^{-1}.

Scamorza is a Mozzarella-type cheese made from sheep's milk in southern Italy. Albenzio *et al.* (2010, 2013a, 2013b, 2015) used probiotic organisms (either a mixture of *Bif. longum* and *Bif. animalis* subsp. *lactis* or *Lb. acidophilus*) to produce this type of cheese. At pH 4.6, both bifidobacteria strains showed greater proteolytic activity (i.e. testing for water-soluble nitrogen extract) and ability to generate peptides with potential bio-functionality.

4.6.8 *Miscellaneous cheeses*

A few other cheese varieties (e.g. whey, blue vein, dairy tofu, dips, slurry etc.) have been investigated and/or developed over the past decade by many researchers, and a summary of the viability of probiotic bacteria in these products is shown in Table 4.5.

Table 4.5 Examples of miscellaneous probiotic cheese types.

Name	Probiotic bacteria	Comments	References
French onion cheese-base dip	*Lactobacillus acidophilus*, *Lactobacillus paracasei* subsp. *paracasei*, *Lactobacillus rhamnosus*, *Bifidobacterium animalis* subsp. *lactis* and *Propionibacterium freudenreichii* subsp. *shermanii*	Each of 8 different combinations of probiotic bacteria were used in cheese dips, and were stored for 10 weeks at 4 °C. To obtain >6 $\log_{10}$ cfu g^{-1} of each strain at the end of the shelf life of the product, recommended inoculation levels ($\log_{10}$ cfu g^{-1}) were 8 for *Lb. acidophilus* and *Bif. animalis* subsp. *lactis*, and 7 for *Lb. paracasei* subsp. *paracasei* and *Lb. rhamnosus*.	Tharmaraj & Shah (2004)
Dairy tofu	*Lb. acidophilus*, *Lactobacillus casei*, *Bifidobacterium bifidum* and '*Bifidobacterium longum*' (presumed to be *Bifidobacterium longum* subsp. *longum*)	This cheese was made by coagulating milk mixed with prebiotics (iso-malto-oligosaccharides), probiotic bacteria, skimmed milk powder, peptides and glucono-δ-lactone (GDL) to form a smooth milk gel. The viable counts for the probiotic bacteria were >6 cfu g^{-1} throughout the storage period.	Chen *et al.* (2004)
Ras (an Egyptian variety) cheese slurry	*Bif. bifidum*, *Lb. acidophilus* and *Streptococcus thermophilus*	Ultrafiltered milk retentate was fermented using a mixed culture, and the viable counts of probiotic bacteria were ~1×10^7 cfu g^{-1} after 3 to 5 d.	El-Shibiny *et al.* (2005)
Whey cheese	*Bif. animalis* subsp. *lactis*, *Lb. acidophilus* and *Lb. casei*	Whey cheese was made with single strains of probiotic cultures. The water-soluble extracts were analysed, and the fraction (i.e. <3 kDa) that inhibited angiotensin-converting enzyme was mainly produced by *Bif. animalis* subsp. *lactis* and *Lb. casei*, rather than the product made with *Lb. acidophilus*. The presence of these organisms in the cheese improved its safety against certain pathogens.	Madureira *et al.* (2008, 2011a, 2011b, 2013)
Cream cheese	*Lb. casei*	The fermented product had probiotic counts of 2.2×10^7 cfu g^{-1}; the sensory properties of the cream cheese were highly rated.	de Oliveira Gaino *et al.* (2012)

(*Continued*)

Table 4.5 (Continued)

Name	Probiotic bacteria	Comments	References
	Str. thermophilus, *Lb. acidophilus* LA-5 and '*Bif. animalis* Bb-12' (presumed to be *Bif. animalis* subsp. *lactis* BB-12)	Fat standardised milk (8 g $100\,g^{-1}$) was fermented with *Str. thermophilus* to produce cream cheese and, after de-wheying, the curd mass was mixed with different ingredients including freeze-dried probiotic bacteria. The lactobacilli and bifidobacteria counts ranged between 3.1–5.4 and 6.12–6.93 $\log_{10}$ cfu g^{-1}, respectively, after 25 d at 4 °C.	Alves *et al.* (2013)
Toma Piemontese	*Lactobacillus plantarum* S11T3E and S2T10D and *Lactobacillus pentosus* S3T60C	These bacteria, which were isolated from olive fermentations, were previously characterised for probiotic properties, and were used as adjunct cultures to manufacture Toma Piemontese, a Protected Denomination of Origin (PDO) cheese. All lactobacilli isolated, during the maturation period and after digestion of the cheeses, were deemed to be putative probiotics. Although the organic acid composition of the cheeses made with the adjunct culture differed from the control, such differences did not negatively affect the organoleptic properties of the final product.	Botta *et al.* (2015)
Blue Vein cheese	*Lb. acidophilus* LA-5	The cheese was spiked with *Yersinia enterocolitica* (at a level of 10^3 cfu g^{-1}), and made using mesophilic lactic starter culture and probiotic bacteria. The counts of the pathogen in products stored between 6 and 12 °C were lower compared to the control cheese (i.e. no probiotic), but it did not guarantee the microbiological safety of the product.	Zadernowska *et al.* (2015)

4.7 Probiotic ice-cream, frozen desserts and frozen yoghurt

4.7.1 *Background*

Ice-cream and frozen desserts have the potential to be carriers of probiotic bacteria, but freeze stress must be considered with respect to their viability during manufacture and extended storage. Frozen yoghurt technology may be adopted for the inclusion of probiotic cultures into ice cream and frozen desserts (Tamime *et al.*, 2006). Various combinations of lactobacilli and bifidobacteria have been used in these products (Tamime *et al.*, 1995). Addition may be direct (i.e. blending of ice-cream mix and probiotic cells immediately prior to freezing); it may involve fermentation of the milk to increase the probiotic cell count prior to blending with the ice-cream mix (Christiansen *et al.*, 1996; Ravula & Shah, 1998; Haynes & Playne, 2002), fermentation of the ice-cream mix prior to freezing, or blending probiotic yoghurt with the ice-cream mix prior to freezing (Soukoulis *et al.*, 2010).

4.7.2 *Ice-cream*

The approach of fermenting the ice-cream mix for proliferation of the probiotic bacteria was evaluated by Hekmat & McMahon (1992), who reported that the viable counts of *Lb. acidophilus* and *Bifi. bifidum* were 4×10^6 and 1×10^7 cfu mL^{-1}, respectively, after 17 weeks of storage at −29 °C. For either probiotic strain, protection of cells against freeze damage is of key importance. Encapsulation and freeze drying, and co-encapsulation of different micro-organisms [e.g. *Lb. acidophilus* 2401, 2404, 2409 and 2415; '*Bif. infantis* 1912' (presumed to be *Bif. longum* subsp. *infantis* 1912); *Bif. animalis* subsp. *lactis* 1941, 920 and B12-12; and '*Bif. longum* 5581' (presumed to be *Bif. longum* subsp. *longum* 5581)], have been evaluated (Godward & Kailasapathy, 2000, 2003b; Kailasapathy & Sultana, 2003; Talwalker & Kailasapathy, 2003a; Masco *et al.*, 2004): free cells and freshly encapsulated cells without freeze drying have shown the best survival rates. Reviews of various methods of encapsulation of probiotic bacteria to be used, for example in ice-cream production to enhance the survival rate and viability, have been reported by Mohammadi *et al.* (2011) and Acu *et al.* (2014). Using *Lb. casei* LAFTI L26 ('free' and encapsulated in an alginate–whey protein capsule) and inulin in ice-cream production affected probiotic survivability: the viable counts of free cells and encapsulated cells ranged between 7.5–7.9 and 6.6–7.5 $\log_{10}$ cfu mL^{-1}, respectively, after 30 d at −18 °C (Naeemi *et al.*, 2013). Freezing at −15 °C and storage at −25 °C of *Lb. acidophilus* LMGP 21381 resulted in a significant decrease in the viability of the freeze-dried culture, but not the active culture of lactobacilli (Nousia *et al.*, 2011).

Properly selected strains widely used in commercial dairy applications, such as *Lactobacillus johnsonii* LA-1, are able to survive the relatively high sugar content of ice cream as well as the sublethal injuries caused by freezing. In the frozen product, counts of 10^7 cfu g^{-1} were maintained for 10 weeks (Kebary *et al.*, 1998, 2004; El Shazly *et al.*, 2004; El Tahra *et al.*, 2004a, 2004b; Hamed *et al.*, 2004; Rao & Prakash, 2004) or 8 months of storage (Alamprese *et al.*, 2002). Some strains, however, cannot survive the freezing and churning that occur during ice-cream manufacture (Hagen & Narvhus, 1999; Haynes & Playne, 2002), but others, such as '*Bif. longum*' (presumed to be *Bif. longum* subsp. *longum*) and '*Bif. infantis*' (presumed to be *Bif. longum* subsp. *infantis*), are able to survive these processes and storage for up to 11 weeks (Davidson *et al.*, 2000) and even 52 weeks (Haynes & Playne, 2002), and are not influenced by the fat content of the product. A study involving 13 strains of *Lb. acidophilus* and 11 strains of bifidobacteria demonstrated that these effects were strain dependent (Ravula & Shah, 1998). Because ice-cream is a non-fermented product, the impact of the probiotic bacteria on flavour is an important consideration. Some organisms, such as *Lb. reuteri* (Hagen & Narvhus, 1999) and *Bif. bifidum* (Ma, 1995), produce a slight acetic acid flavour due to fermentation (see also Anisimov *et al.*, 2013). Manufacturing conditions that limit fermentation may be adopted to minimise such flavours (Ordonez *et al.*, 2000a, 2000b).

Ice-cream mix fortified with inulin and fermented with probiotic bacteria (*Lb. acidophilus* LA-14 and '*Bacillus lactis* BL-01' – presumed to be *Bif. animalis* subsp. *lactis* BL-01) and a yoghurt starter culture at 37 °C had viable counts of both probiotic bacteria ranging between 10^6 and 10^7 cfu g^{-1} after 90 d at −18 °C; the addition of prebiotic (inulin) enhanced their growth in the mix (Akin, 2005). Similar counts (>10^6 cfu g^{-1}) were observed in ice-cream made with different combinations of '*Bif. Longum*' (presumed to be *Bif. longum* subsp. *longum*), '*Bif. lactis*' (presumed to be *Bif. animalis* subsp. *lactis*) and yoghurt starter culture (Favaro-Trindade *et al.*, 2006). In different ice-cream mixes containing blends of inulin and/or lactulose, *Bif. animalis* subsp. *lactis* was added to each mix before freezing and had viable counts that ranged between 6.1 and 7.0 $\log_{10}$ cfu g^{-1} after 90 d at −18 °C (Hashemi *et al.*, 2015). Partial replacement of cow's milk, however, with soy milk, coconut milk or combinations of both in ice-cream mix fermented with '*Bif. bifidum* BB-12' (presumed to be *Bif. animalis* subsp. *lactis* BB-12) and *Lb. acidophilus* LA-5 resulted in products that had different physical properties, but appreciable counts of probiotic bacteria (Aboulfazli & Baba, 2015; Aboulfazli *et al.*, 2015, 2016). Similarly, partial replacement of full-fat and skimmed milk mixes (i.e. five batches) with soy milk and Simplesse 100® (fat replacer), and fermentation with a monoculture of '*Bif. longum* ATCC 15707' (presumed to be *Bif. longum* subsp. *longum* ATCC 15707), resulted in frozen products that were highly rated by the sensory panel; viable counts averaged 7.8 $\log_{10}$ cfu g^{-1} after 30 d at −20 °C (Tawfek *et al.*, 2016).

4.8 Dried probiotic dairy products

4.8.1 *Introduction*

There is a significant market for dried probiotic pharmaceutical products, and a variety of products are available in the form of dietary supplements. This review will not focus on these products (Tamime & Marshall, 1997; Kaur *et al.*, 2002), but rather on other dried products that are subsequently used in the manufacture of dairy and other products.

Dried products are manufactured by either freeze drying or spray drying. While the cost of production is an important consideration, survival of the probiotic bacteria during the drying process and subsequent storage is equally important. Unless appropriate drying conditions are selected (Lian *et al.*, 2002), cell damage and loss of viability will occur during spray drying; viability during storage is inversely related to the storage temperatures (Gardiner *et al.*, 2000). Various factors should be considered with respect to the viability of probiotic micro-organisms in dried products (Prajapati *et al.*, 1986; Gilliland *et al.*, 2001), including:

- Drying method;
- Type and size of packaging;
- Temperature and humidity of storage;
- Powder quality;
- Rehydration procedure; and
- Handling of rehydrated product.

Protective compounds, such as gum acacia, were used to protect '*Lb. paracasei* NFBC 338' (presumed to be *Lb. paracasei* subsp. *paracasei* NFBC 338) during spray drying and storage (Desmond *et al.*, 2001, 2002). This method improved survival during drying and storage as well as their resistance to bile. On the other hand, the same organism was spray-dried without any protection with survival rates of 84.5% for use in Cheddar cheesemaking, as described earlier in this chapter (Gardiner *et al.*, 2002a), and the effect of drying on survival rate of probiotic bacteria was examined (Meng *et al.*, 2008; Kitamura *et al.*, 2009; Mercenier *et al.*, 2012b; Paez *et al.*, 2013). Compression coating of *Lb. acidophilus* containing powders in combination with sodium alginate and hydroxyl-propyl cellulose was used to increase storage stability by 10 times after 30 d of storage at 25 °C, compared to free cells (Chan & Zhang, 2002). Spray-dried whey containing microencapsulated *Bif. animalis* subsp. *lactis* BB-12 resulted in high viable counts after 12 weeks of storage at 4 °C. The dried bacteria were added to a dairy dessert, where the probiotic count remained at >7 $\log_{10}$ cfu g^{-1} after 6 weeks of storage (de Castro-Cislaghi *et al.*, 2012). An alternative method to produce biomass of *Lb. plantarum* is to use the microfiltration (MF) technique as described by Alfano *et al.* (2015).

4.8.2 *Infant formula*

Dried preparations of probiotics are of particular interest in the manufacture of infant formulae, where an important objective is to achieve products that are functionally similar to human milk. Processing technology is available for the manufacture of infant formulae with a gross composition similar to that of human milk, but methods continue to be refined for other factors (Lonnerdal, 2003). For example, the gut microbiota of infants fed human milk is different from that of formulae-fed infants (Edwards *et al.*, 2002; Guaraldi & Salvatori, 2012; O'Sullivan *et al.*, 2015). Various methods have been proposed to introduce probiotic organisms into the infant gut, including incorporating probiotics in dried preparations (see Saavedra *et al.*, 2004). In one example, blending freeze-dried preparations of *Lb. reuteri*, *Lb. acidophilus* and *Bif. bifidum* developed an infant formula intended for the prevention of diarrhoea. This type of formulation was shown to be effective in a study into the prevention of diarrhoea in infants, when the consumption of the three organisms was between 10^8 and 10^{10} cfu d^{-1} (Halpin-Dohnalek *et al.*, 1999). An example of a commercialised milk powder (i.e. Neslac®) containing *Bif. animalis* subsp. *animalis* BB-12 for older infants has been reported by Playne *et al.* (2003; see also Chouraqui *et al.*, 2004; Masco *et al.*, 2004). Another approach is to include dried prebiotics such as oligosaccharides (Goni-Cambrodon & Gudiel-Urbano, 2001; Kunz & Rudolff, 2002) and lactulose (Strohmaier, 1997) in infant formulae.

In a co-ordinated survey (290 samples) of *Cronobacter* spp. and related organisms in dried infant formula, follow-up formula and young children formula undertaken in seven countries, only three samples contained probiotic bacteria and the strains were not identified (Chap *et al.*, 2009). Stability tests of bifidobacteria ('*Bif. longum* BB-536' – presumed to be *Bif. longum* subsp. *longum* BB-536; *Bif. breve* M-16V; and '*Bif. infantis* M-63' – presumed to be *Bif. longum* subsp. *infantis* M-36) in infant and follow-up formula were studied by Abe *et al.* (2009a). In a survey in commercial products, formulae for toddlers containing bifidobacteria sold in the Indonesian market were analysed. When the inactivation rate constant of each stored sample, which was used as an index of the loss rate, was determined from the stability tests, '*Bif. longum*' (presumed to be *Bif. longum* subsp. *longum*) was the most stable strain. The mean inactivation rate constant of commercial products was significantly lower than those obtained in strain comparison, although the same strains ('*Bif. longum* BB-536' – presumed to be *Bif. longum* subsp. *longum* BB-536; and *Bif. breve* M-16V) were used. A possible reason was the lower water activity of commercial products compared to the follow-up formula. Also, higher storage temperature yielded lower stability in all strains or samples, which obeys Arrhenius theory.

Twenty infant formulae available in the Chinese market were checked for the viable count levels of probiotic bacteria (e.g. *Bif. animalis* subsp. *lactis* strains) when the products were opened and stored at 4, 25 and 37 °C for up to 28 d. The survival rate of the probiotic bacteria was negatively correlated with storage temperature and time (Lui *et al.*, 2015b). Nevertheless, two infant formulae were supplemented with '*Bif. lactis* BB-12' (presumed to be *Bif. animalis* subsp. *lactis* BB-12) or *Lactobacillus reuteri* ATCC 55730 and a probiotics-free formula to study the safety and tolerance of these formulae in infants <4 months old; it was concluded that the use in early infancy formulae supplemented with either probiotic bacteria strain in early infancy was safe and well tolerated, and did not adversely affect the infants' growth, bowel habits or behaviour (Weizman *et al.*, 2005; Weizman & Alsheikh, 2006).

4.8.3 *Dairy-based dried products*

Goat's milk yoghurt was fermented with *Bif. animalis* subsp. *lactis* BI-07 and a yoghurt starter culture. The fermentate was mixed with maltodextrin (10 g 100 g^{-1}), and then spray-dried. Using an inlet temperature of 130 °C in the drier, a good-quality powder was produced with high levels of viable bifidobacteria (de Medeiros *et al.*, 2014).

A traditional type of dried yoghurt product, such as kishk and trahana, is made using a low-fat fermentate, which is mixed with parboiled cracked wheat and then sun-dried. Although to our knowledge probiotic dried yoghurts and related products are not produced, the *in vitro* antimutagenic and anticancer effects of 25 strains of lactobacilli species isolated from trahana have been investigated. Only four isolates possessed such activities; these were identified as strains of the species *Lb. casei*, *Lb. plantarum* and *Lb. brevis* (Ahmadi *et al.*, 2014).

4.9 Miscellaneous probiotic dairy products

4.9.1 *Fat-based products*

A prototype reduced-fat ($60\,g\ 100\,g^{-1}$) edible table biospread was made from milk fat and soy oil containing mixed cultures of *Lb. casei* ACA-DC 212.3 and '*Bif. infantis* ATCC 25962' (presumed to be *Bif. longum* subsp. *infantis* ATCC 25962) (Charteris *et al.*, 2002). The process was modified to enhance the viability of the probiotic micro-organisms, so that both cultures showed 1 $\log_{10}$ cycle decline after processing. The rate of decrease in viability of the bifidobacteria during shelf life was greater than that of the lactobacilli. More development work is required and, to our knowledge, this type of probiotic product has not been commercially produced.

Whipped-cream batches of different formulations were prepared using cream, palm oil and cream, non-dairy ingredients, starter cultures (mesophilic and thermophilic types) and probiotic bacteria (e.g. *Lb. acidophilus* and *Bifidobacterium* spp.). The ingredients were blended together with addition of flavours (cocoa powder or vanilla) before whipping of the product and storage for 9 d at 5 °C. The whipping time, serum separation and over-run were influenced by the formulation used. *Lactobacillus acidophilus* and *Str. thermophilus* showed better survival (10^6 cfu mL^{-1}) compared to the bifidobacteria and *Lb. delbrueckii* subsp. *bulgaricus*. The vanilla-flavoured whipped cream was highly rated compared to the chocolate product, and the products supplemented with starter cultures scored higher than the other formulations (Hussein & Abo-El-Fetoh, 2010).

Cultured cream ($52\,g\ 100\,g^{1}$) and vegetable oils (sunflower, soybean or hazelnut) at a rate of $2\,g\ 100\,g^{1}$ blends were fermented with a yoghurt starter culture and a monoculture of probiotic bacteria (*Lb. acidophilus*, *Bif. Bifidum* and *Propionibacterium thoenii* P 126 and B 1264), or a yoghurt starter culture and mixed probiotic bacteria (*Lb. acidophilus* and *Bif. bifidum*). All the probiotic bacteria had viable counts >10^6 cfu g^{-1}, and the product made with *Bif. bifidum* demonstrated the highest concentration of CLA ($0.73\,mg\,g^{-1}$ fat) (Ekinci *et al.*, 2008). When single strains of probiotic bacteria (*Bif. animalis* subsp. *lactis*, *Lb. acidophilus* and *Lb. rhamnosus*) were used to ferment cream, different profiles of FAs were observed in the products during the storage period (15 d). All the probiotic bacteria increased the medium-chain and polyunsaturated FA contents in the products; however, *Bif. animalis* subsp. *lactis* increased levels of linoleic acid, α-linolenic acids and monounsaturated and polyunsaturated FAs, whilst the highest levels of saturated FAs were found in cream fermented with *Lb. acidophilus* (Yilmaz-Ersan, 2013).

4.9.2 *Long shelf-life fermented milk drinks or beverages*

As mentioned in Section 4.5.1, the fresh probiotic fermented drinks and beverages market is booming in the UK and worldwide, but long shelf-life dairy products do not contain either lactic starter cultures or the probiotic micro-organisms. This problem can be overcome, however, by using a specially designed straw (known as a LifeTop™ Straw, or alternatively referred to as a probiotic straw), which was developed and patented by BioGaia in Sweden (Anonymous, 2001; Thorball *et al.*, 2001; see also www.biogaia.com or www.reuteri.com). A freeze-dried *Lb. reuteri* culture (patented by BioGaia as Reuteri™) is suspended in oil droplets and attached to the inside of a two-jacketed straw. The straw is packed in an outer packaging container that is made of laminated aluminum foil (i.e. impermeable to moisture and oxygen). According to the manufacturer, the straw contains 10^8 cfu of the probiotic culture, has a shelf life of 12 months at 25 °C and, when the consumer drinks 100 mL of the beverage, 99% of the probiotic bacteria are released. It is possible that the same concept could be applied to other types of probiotic bacteria, and may be used, especially by children, when drinking pasteurised or flavoured milk drinks.

4.9.3 *Milk- and water-based cereal puddings*

Lactobacillus rhamnosus GG, *Lb. acidophilus* LA-5 and 1748, and *Bif. animalis* subsp. *animalis* BB-12 have been used successfully for the production of milk- and water-based puddings with and without prebiotics (e.g. polydextrose and Litesse™) (Helland *et al.*, 2004). All the strains showed good growth and survival in milk-based puddings (e.g. viable counts ranging between 8 and 9.1 $\log_{10}$ cfu g^{-1}), significantly ($p<0.05$) higher concentrations of lactic acid were produced after storage with pH levels reduced to <4.4 and the highest concentration of diacetyl (18 mg kg^{-1}) was detected in puddings inoculated with *Lb. rhamnosus* GG. In addition, puddings prepared with or without the addition of Litesse were not significantly different.

4.9.4 *Mousses, desserts and spreads*

Some prototype probiotic dairy-based products are discussed in this section.

Mousses

Chocolate mousses made with *Lb. paracasei* subsp. *paracasei* LBC, with or without inulin, had a firmer and more adhesive texture than the control product, and were stored for 28 d at 4 °C. Differences were noted in sensory attributes (flavour, aroma and texture) after 14 d of storage, and the colour effect was only observed in the mousse containing inulin. Probiotic counts were stable during the storage period, ranging between 7.3 to 7.4 $\log_{10}$ cfu g^{-1} (Cardarelli *et al.*, 2008).

ProPearls

This is a new idea that was developed using a fruit filling surrounded by a double gel layer. The inner layer consisted of calcium (Ca^{+2}) alginate, whilst the outer coloured layer contained probiotic bacteria (e.g. *Lactobacillus* spp.) and was encapsulated in milk protein. It was suggested that these ProPearls could be used as topping of Quark, yoghurt, and other dairy desserts. The viable count of the probiotic lactobacilli was ~10^{-8} cfu g^{-1}. In addition, the pearls contained vitamins, polyphenols, Ca^{+2} and antioxidants (Flockerzie *et al.*, 2014).

Milk-based dessert

Milk (3.2 g 100 g^{-1} fat) was mixed with SMP, sucrose, dextrose, starch, carrageenan and sodium diphosphate; heated between 80 and 90 °C; cooled to 65 °C; homogenised at 15 MPa; and mixed with cultures (*Lb. casei* Shirota and *Bif. animalis* subsp. *lactis*) and cranberry sauce. The product was stored for 21 d. The counts of the bifidobacteria and lactobacilli at the end of the storage period were 2.0×10^6 and 2.1×10^7 cfu g^{-1}, respectively (Magarinos *et al.*, 2008).

Oblea

This is a wafer-type dehydrated traditional Mexican dessert, made as follows: goat's sweet whey is fermented with '*Bif. infantis* ATCC 1793' (presumed to be *Bif. longum* subsp. *infantis* ATCC 1793) or *Lb. acidophilus* ATCC 521, mixed with prebiotic carbohydrates (inulin and resistant starch) and gelatine, and dried in a convection oven at different temperatures. Finally, it is dehydrated at a low relative humidity at room temperature (23 °C). Viable counts of both probiotic bacteria were >9 $\log_{10}$ 10 cfu g^{-1} when the Oblea was dried at 55 °C for 2.7 ± 0.2 h (Trujillo-de Santiago *et al.*, 2012).

Cheese-based tomato spread

Three starter culture formulations were used during the manufacture of cheese base (i.e. similar to Quarg) as follows: (a) *Str. thermophilus* ST-M6 (control), (b) same culture as control + *Lb. acidophilus* NCFM and *Bif. animalis* subsp. *lactis* BB-12, and (c) same as control + *Bif. animalis* subsp. *lactis* BB-12 and *Lb. sakei* subsp. *sakei* 2a. Different ingredients were used (pasteurised tomato pulp, dried tomato, commercial sterilised cream, dried basil, grated Parmesan cheese, salt, olive oil, sucrose, guar gum, xanthan gum, carrageenan gum, inulin, and whey from the cheese base); these were added to the cheese base and then mixed in a blender. The spread was stored for 28 d at 4 °C. The viability of all the strains was ≥7.9 $\log_{10}$ cfu g^{-1}, the pH decreased during the storage period and consumer acceptability of the product was high (Staliano *et al.*, 2015; see also Buriti & Saad, 2014).

4.10 Viability of probiotic micro-organisms

In order to obtain the desired health effects, probiotic bacteria must be able to grow in milk and milk products, and survive in sufficient numbers to the end of the shelf life. It has been suggested that probiotic organisms should be present in a food to a minimum concentration of 10^6 cfu g^{-1} or daily intake should be about 10^9 cfu g^{-1}. Such high numbers have been suggested to compensate for the possible losses in the numbers of the probiotic organisms during passage through the stomach and the intestine. In Japan, the

Fermented Milks and Lactic Acid Bacteria Beverages Association have developed a standard that requires a minimum of 10^7 viable cfu mL^{-1} to be present in dairy products. Studies have demonstrated that several probiotic micro-organisms grow poorly in milk, and the viability of these organisms is often low in yoghurt. A number of brands of commercial yoghurt have been analysed in Australia and Europe for the presence of *Lb. acidophilus* and *Bifidobacterium* spp. Most of the products contained very low numbers of these organisms, especially bifidobacteria (see Tamime, 2002). The viability and activity of the bacteria are important considerations, because these bacteria must survive in the food during shelf life, then transit through the acidic conditions of the stomach and resist degradation by hydrolytic enzymes and bile salts in the small intestine.

The viability of probiotic bacteria in yoghurt depends on the strains used, interaction between species present, production of hydrogen peroxide due to bacterial metabolism, and the final acidity of the product. Viability also depends on the availability of nutrients, growth promoters and inhibitors; the concentration of sugars; dissolved oxygen and oxygen permeation through the package (especially for *Bifidobacterium* spp.); inoculation level; and fermentation time (see also Oliveira & Damin, 2003). Bifidobacteria are anaerobic in nature; therefore, a high oxygen content may affect their growth and viability. *Lb. acidophilus* is reported to have high cytoplasmic buffering capacity (pH 3.72–7.74), which allows it to resist changes in cytoplasmic pH and remain stable under acidic conditions. *Lactobacillus acidophilus* is more tolerant of acidic conditions than *Bifidobacterium* spp., and the growth of the latter is significantly retarded below pH 5.0. The tolerance of *Bifidobacterium* spp. to acidic conditions is strain specific. '*Bifidobacterium longum*' (presumed to be *Bifidobacterium longum* subsp. *longum*) has shown better survival in acidic conditions and bile concentrations compared with '*Bif. infantis*' (presumed to be *Bif. longum* subsp. *infantis*), *Bifidobacterium adolescents* and *Bif. bifidum*. Furthermore, '*Bif. longum*' (presumed to be *Bif. longum* subsp. *longum*) is also easier to grow in milk, whilst *Bif. animalis* subsp. *animalis* survives well in fermented milks. It should be noted, however, that the latter species is not of human origin (Lankaputhra & Shah, 1996). However, Matto *et al.* (2004) confirmed that the tolerance of *Bifidobacterium* spp. to acidic conditions and bile is strain specific, but *Bif. animalis* subsp. *animalis* has shown better survival compared to other bifidobacterial species.

The presence of *Lb. delbrueckii* subsp. *bulgaricus* affects the survival of *Lb. acidophilus* and *Bifidobacterium* spp. due to the acid and hydrogen peroxide produced during the fermentation stage. Due to its proteolytic nature, *Lb. delbrueckii* subsp. *bulgaricus* grows fast and produces acid rapidly, and it appears to liberate essential amino acids like valine, glycine and histidine that are required to support the growth of bifidobacteria. *Streptococcus thermophilus* does not inhibit the growth of probiotic organisms, and in fact may stimulate their growth due to its consumption of oxygen.

4.10.1 *Composition of the fermentation medium*

Probiotic bacteria are used for the fermentation of milk to a limited extent because of their slow growth in milk. Although *Lb. acidophilus* and *Bifidobacterium* spp. show some level of β-galactosidase activity, the reason for poor growth is related to the low concentration of free amino acids and small peptides in milk, which are insufficient to support the growth of these organisms. Therefore, the addition of casein or whey protein hydrolysates, yeast extract, glucose and vitamins can enhance the growth of *Lb. acidophilus* and *Bifidobacterium* spp. in milk (see also Desai *et al.*, 2004; Lucas *et al.*, 2004). The addition of milk protein also increases the buffering capacity of fermented milks and allows better survival of probiotic organisms. Bifidobacteria are capable of utilising lactulose and oligosaccharides and, as other intestinal bacteria, are unable to utilise these complex carbohydrates; these compounds are considered to be prebiotics or 'bifidus factors'. Prebiotics are included in most probiotic products in order to promote the growth of bifidobacteria in the intestine (for further details, refer to Chapters 7 and 8).

In general, probiotic bacteria grow better in rich synthetic media, such as tryptose peptone yeast (TPY) and de Man, Rogosa and Sharpe (MRS) broths, than in milk (Shah, 2000). However, the culture media are complex, are costly for large-scale propagation of probiotic bacteria and may impart off-flavour(s) unless extensively washed before incorporation. To manufacture a quality product, in terms of both texture and viability of probiotic bacteria, a milk-based medium is usually required because of the presence of casein.

The slow growth of probiotic micro-organisms in milk increases the risk of overgrowth of undesirable micro-organisms, and strains that do not grow well tend to produce unpleasant flavours. Normally, it takes 4 h to complete the fermentation process with *Lb. delbrueckii* subsp. *bulgaricus* and *Str. thermophilus* as compared to 20 or 24 h with probiotic cultures alone. For this reason, fermented milk products containing *Lb. acidophilus* and bifidobacteria are often produced in conjunction with other cultures such as *Lb. delbrueckii* subsp. *bulgaricus* and *Str. thermophilus* in the case of yoghurt manufacture. Both groups of cultures are added together, or fermentation takes place in two steps (see Section 4.11.4).

4.10.2 *Viability as affected by oxygen*

The full strain name and minimum count of live cells of the probiotic bacteria should be stated on the product label to maintain confidence in probiotic products; it is important that manufacturers can demonstrate adequate survival of the bacteria in products throughout the shelf life of the product. Since bifidobacteria are anaerobic, oxygen toxicity is an important and critical problem. During yoghurt production, oxygen can easily invade and dissolve in the milk. To exclude oxygen during the production of probiotic milk products, special equipment is required to provide an anaerobic environment. Oxygen can also enter the product through packaging materials during storage. A satisfactory growth of a number of *Bifidobacterium* spp. in aerobic conditions has been reported in a whey-based medium containing L-cysteine (0.05 g 100 mL^{-1}) and yeast extract (0.3 g 100 mL^{-1}) (Dave & Shah, 1997d, 1998). L-cysteine is added to reduce redox potential in order to allow the growth of bifidobacteria.

Oxygen affects probiotic cultures in two ways. Firstly, it is directly toxic to the cells: certain probiotic cultures are sensitive to oxygen and they die in its presence. Secondly, in the presence of oxygen, certain cultures, particularly *Lb. delbrueckii* subsp. *bulgaricus*, produce peroxide. A synergistic inhibition of probiotic cultures due to acid and hydrogen peroxide has been reported (Lankaputhra & Shah, 1996); for this reason, the removal of *Lb. delbrueckii* subsp. *bulgaricus* from some starter cultures (i.e. ABT starter cultures; refer to Section 4.2.2 for details) has had some success in improving the survival of probiotic organisms. Several studies have focused on preventing the detrimental effects of oxygen on probiotic cultures, including the use of antioxidants or oxygen scavengers (Dave & Shah, 1997c; see also Talwalkar & Kailasapathy, 2003b; Talwalker *et al.*, 2004).

4.11 Approaches to improve the viability of the probiotic micro-organisms in the product

4.11.1 *Selection of bacterial strain(s)*

The parameters for screening micro-organisms for potentially valuable probiotic strains should include the fact that there is a necessity for the strain to be viable and metabolically active within the GI tract. In addition, it is important that viability of the strain and stability of their desirable characteristics be maintained during commercial production as well as the storage period of the final product (Godward *et al.*, 2000; see also Talwalker & Kailasapathy, 2004). High viable counts and survival rates during passage through the stomach are necessary to allow live probiotics from the fermented milk products to play a biological role in the human intestine. Surviving the acid conditions of the stomach and bile salts is, thus, of prime importance.

Tolerance to acid and bile is strain specific. Many strains of *Lb. acidophilus* and *Bifidobacterium* spp. intrinsically lack the ability to survive harsh conditions in the gut, and may not be suitable for use as dietary adjuncts in fermented milks. Some strains of *Lb. acidophilus* are reported to survive best under acidic conditions, and at modest bile concentrations (Clark *et al.*, 1993; Clark & Martin, 1994). '*Bifidobacterium longum*' (presumed to be *Bifidobacterium longum* subsp. *longum*) and *Bifidobacterium pseudolongum* subsp. *pseudolongum* have shown the best tolerance to acid and bile salts (Lankaputhra & Shah, 1995). Thus, selection of appropriate strains on the basis of acid and bile tolerance would help to improve the viability of these probiotic bacterial strains (Takahashi *et al.*, 2004).

4.11.2 *Type of packaging container*

Lactobacillus acidophilus is microaerophilic, and bifidobacteria are anaerobic. Since bifidobacteria are anaerobic, oxygen toxicity is an important consideration; oxygen can easily dissolve in milk. Dave & Shah (1997b) studied the survival of *Lb. delbrueckii* subsp. *bulgaricus*, *Str. thermophilus* and probiotic organisms in yoghurt stored in glass bottles and plastic containers. The increase in numbers and survival of *Lb. acidophilus* during storage was directly affected by the dissolved oxygen content, which was higher in yoghurts stored in plastic containers compared to glass. Counts remained higher for the products stored in glass bottles compared to plastic cups. Bifidobacteria also showed better growth in glass bottles compared to the plastic cups: initial counts and survival of bifidobacteria were substantially higher in yoghurt prepared in glass bottles than in plastic cups. Better survival and viability of bifidobacteria in de-aerated milk have also been observed (Klaver *et al.*, 1993). Thus, it may be important to store products in glass containers or to increase the thickness of the packaging materials (i.e. decrease the oxygen permeability rate) used for AB, ABC or ABT products (refer to Section 4.2.2 for details).

4.11.3 *Rate of inoculation*

Since probiotic organisms grow poorly in milk, a larger inoculum size (5–10 mL $100\,mL^{-1}$) is required compared to the 1 mL $100\,mL^{-1}$ used for the yoghurt starter cultures. Similarly, probiotic organisms do not grow well in the presence of certain other bacteria, including *Lb. delbrueckii* subsp. *bulgaricus* and *Str. thermophilus*. The size of the inoculum of the primary starter culture may also influence the eventual numbers of the probiotic bacteria: a small inoculum of the yoghurt organisms can result in over-acidification of the product, which can result in poor survival of the probiotic bacteria.

Dave & Shah (1997a) studied the effect of concentration of starter addition on the viability of *Lb. delbrueckii* subsp. *bulgaricus*, *Str. thermophilus* and probiotic bacteria in yoghurt made from four commercial starter cultures. Two starter cultures contained *Str. thermophilus* and *Lb. delbrueckii* subsp. *bulgaricus*, *Lb. acidophilus* and bifidobacteria, whereas the other two cultures were ABT types. The starter cultures were in the freeze-dried direct-to-vat set (DVS) form, and were used at a rate of 0.5, 1.0, 1.5 or 2.0 g $10\,L^{-1}$ in separate containers. The incubation conditions were as recommended by the starter cultures' manufacturers. *Streptococcus thermophilus* multiplied better with the lower level of inoculum; however, the final counts of this organism remained slightly higher with the higher level of inoculum, which was probably primarily due to the higher initial numbers.

The final pH at the end of fermentation is the most crucial factor for the survival of probiotic organisms. A pH below 4.4 at the end of fermentation results in a substantial decrease in the number of probiotic bacteria; hence, the inoculum level must be carefully adjusted and monitored.

4.11.4 *Two-stage fermentation*

Inhibitory substances, such as acid and hydrogen peroxide produced by yoghurt starter bacteria, are responsible for poor survival of probiotic cultures. Although the yoghurt starter cultures produce substances inhibitory to probiotic cultures, the former are essential in yoghurt manufacture to speed up the fermentation process, and to provide the typical yoghurt flavour. Generally, yoghurt starter bacteria grow faster than probiotic bacteria during the fermentation period and produce acids, which could reduce the viability of the probiotic bacteria.

One approach to improving the viability of probiotic organisms is to add them after the fermentation of the milk. This allows the strains of probiotic bacteria to be used that cannot grow in the presence of other organisms; however, growth or survival of probiotic organisms may be reduced if they are added post fermentation.

Another approach is to carry out the initial fermentation with the probiotic cultures followed by completion of fermentation with, for example, *Lb. delbrueckii* subsp. *bulgaricus* and *Str. thermophilus* (Lankaputhra & Shah, 1997). With this approach, the fermentation time can be slightly longer than the traditional fermentation process. Initial fermentation could be carried out with probiotic cultures for 2 h, followed by fermentation by the yoghurt starter cultures for 4 h. This allows the probiotic organisms to be in the final stage of their lag phase or the early stage of their log phase, and it results in higher probiotic counts at the end of 6 h of fermentation. Probiotic counts have been found to increase substantially in products made using a two-step fermentation process.

4.11.5 *Microencapsulation technique*

The numbers of probiotic bacteria in frozen fermented dairy desserts or frozen yoghurt are reduced significantly by acid, freeze injury, the sugar concentration of the product and oxygen toxicity (Tamime *et al.*, 1995). About 16 g sugar $100\,g^{-1}$ of product are added to frozen fermented dairy desserts, and the addition of sugar has been found to affect the growth of probiotic bacteria. Microencapsulation is a process whereby the cells are retained within the encapsulating membrane in order to reduce cell injury or cell loss (see Figure 4.2), and it may have applications in several products, such as cheese (Godward & Kailasapathy, 2003a), yoghurt (Adhikari *et al.*, 2000; ChienJung, 2000; Sultana *et al.*, 2000; WenRong & Griffiths, 2000; Hansen *et al.*, 2002; Godward & Kailasapathy, 2003c; Picot & Lacroix, 2004) and simulated gastric juice and bile solutions (WenChian *et al.*, 2003; see also Kailasapathy & Sultana, 2003; Krasaekoopt *et al.*, 2003, 2004; Sridar *et al.*, 2003; Anjani & Kailasapathy, 2004; Chandramouli *et al.*, 2004; Hsiao *et al.*, 2004; HungChi *et al.*, 2004).

Gelatin or vegetable gums have been used for the microencapsulation of bacteria, and have been reported to provide protection to acid-sensitive probiotic organisms. Entrapment of living microbial cells in calcium alginate is simple and low cost. Furthermore, alginate is non-toxic, so it may be safely used in foods. Alginate gels can be solubilised by sequestering calcium ions and, thus, releasing entrapped cells (Rao *et al.*, 1989; Sheu & Marshall 1993).

Encapsulated probiotic organisms incorporated into fermented frozen dairy desserts showed better viability ($>10^5$ cfu g^{-1}) in the products compared to encapsulated organisms ($<10^3$ cfu g^{-1}) (Shah & Ravula, 2000b, 2004).

4.11.6 *Supplementation of the milk with nutrients*

During yoghurt making, *Str. thermophilus* dominates the early stage of fermentation. As the redox potential of milk is reduced and the pH lowered from 6.5 to 5.5, growth of *Lb. delbrueckii* subsp. *bulgaricus* is stimulated. *Str. thermophilus* dominates the early stage of fermentation; then, when the pH falls below 5.0, *Lb. delbrueckii* subsp. *bulgaricus* dominates the fermentation, producing acetaldehyde and lactic acid, and giving the characteristic of green apple flavour. Continued acid production lowers the pH of

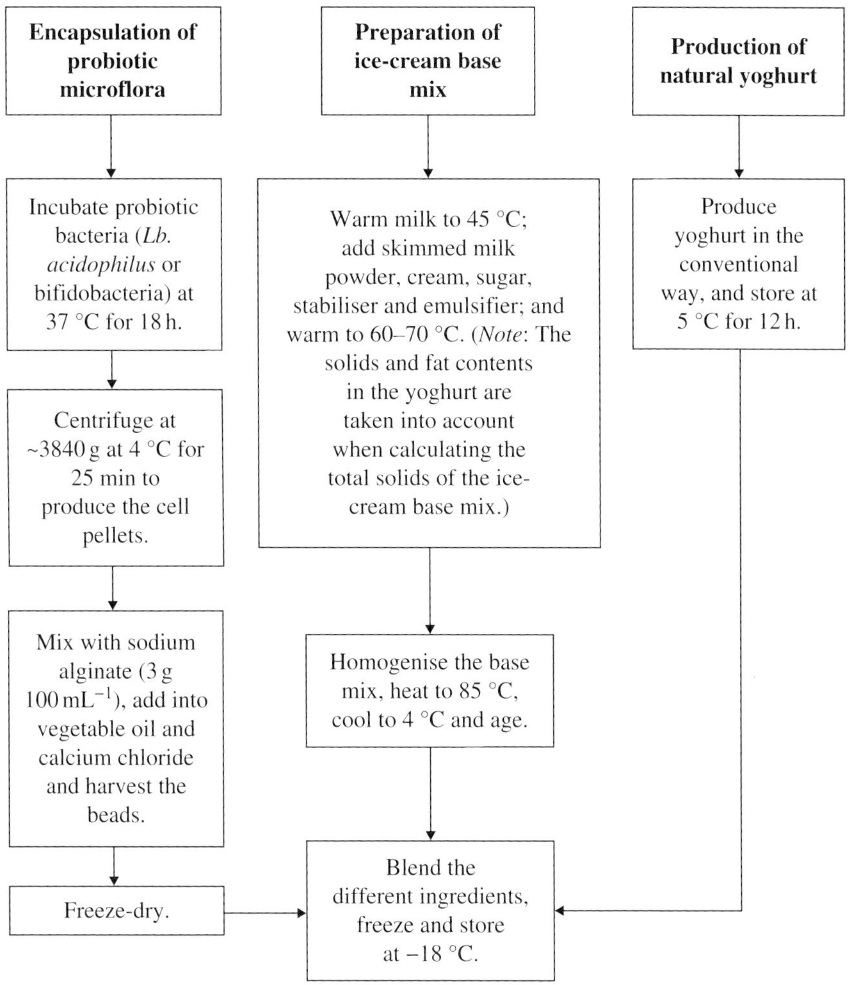

Figure 4.2 Preparation of probiotic fermented frozen dessert.
Note: Mix the yoghurt, ice-cream base and syrup (i.e. sucrose at 65% brix) at the following proportions: 45, 45 and 10, respectively.
Adapted from Shah & Ravula (2004).

yoghurt to values near 4.6, the iso-electric point of casein, which induces gelation. The fermentation is terminated at pH 4.5.

Due to its proteolytic nature, *Lb. delbrueckii* subsp. *bulgaricus* produces essential amino acids (Shihata & Shah, 2000, 2002), and its associative growth relationship with *Str. thermophilus* is well established. The streptococci also produce growth factors for the former organism. *Lactobacillus delbrueckii* subsp. *bulgaricus*, however, also produces lactic acid during refrigerated storage, which is a process known in the industry as 'post-acidification'. If this occurs during refrigerated storage, it may cause loss of viability of the probiotic bacteria.

To overcome any loss of viability of probiotic bacteria due to acid produced during fermentation and storage (post-acidification), the present trend is to use starter cultures devoid of *Lb. delbrueckii* subsp. *bulgaricus*, such as ABT. *Streptococcus thermophilus*, which is less proteolytic than *Lb. delbrueckii* subsp. *bulgaricus*, is the main organism responsible for the fermentation in some of the ABT cultures (e.g. ABT-1 and ABT-2; refer to Section 4.2.2 for further details). Such blends of starter cultures increase the fermentation time significantly (up to 10 h), as there is no associative growth without *Lb. delbrueckii* subsp. *bulgaricus*. ABT starter cultures require the incorporation of micronutrients (peptides and amino acids) through casein hydrolysates to reduce the fermentation time and improve the viability of probiotic organisms.

Dave & Shah (1998) studied the effects of some dairy and non-dairy ingredients, such as WP, WPC and acid casein hydrolysates (ACnH), on the viability of *Lb. acidophilus* and *Bifidobacterium* spp. in yoghurt made from four commercial starter cultures. The added ingredients considerably affected incubation time, because some of them provided peptides and amino acids for the bacterial growth. Addition of WP, WPC and ACnH improved the viability of *Bifidobacterium* spp., as did nitrogen sources in the form of peptides and amino acids. The addition of milk protein (casein and whey) hydrolysates, however, enhanced the acidification rate of *Str. thermophilus* and reduced the growth rate of the probiotic bacteria (*Lb. acidophilus* LA-5 and *Lb. rhamnosus* LR-35) in fermented milks during the manufacturing stages, but improved the survival of the latter bacteria after storage (Lucas *et al.*, 2004; see also Section 4.5.1).

4.11.7 *The use of oxygen scavengers*

Oxygen content and redox potential have been shown to be important factors for the viability of *Bifidobacterium* spp. during storage. Ascorbic acid (vitamin C) acts as an oxygen scavenger, and it is permitted in fruit juices and other products as a food additive. Furthermore, milk and milk products supply only 10–15% of the daily requirements of vitamin C (Rasic & Kurmann, 1978) and, as a result, fortification of yoghurt with ascorbic acid increases its nutritional value.

The oxygen content and redox potential were reported to gradually increase during the storage of yoghurt in plastic cups, but the redox potential remained lower with ascorbic acid (Dave & Shah, 1997c). *Streptococcus thermophilus* is aerobic, and its counts would be expected to be reduced in the presence of ascorbic acid. In contrast, the viability of *Lb. delbrueckii* subsp. *bulgaricus*, a micro-aerophilic to anaerobic species, would be expected to improve with increasing concentration of ascorbic acid. The addition of ascorbic acid also helps to improve the survival of *Lb. acidophilus*, but its oxygen-scavenging effect may not be sufficient to improve the viability of anaerobic *Bifidobacterium* spp. (Dave & Shah, 1997c).

4.11.8 *The addition of cysteine*

Media used for enumeration of bifidobacteria often contain L-cysteine (0.5–0.1 g 100 mL^{-1}) in order to improve recovery. Cysteine, a sulphur-containing amino acid, provides amino nitrogen as a growth factor while reducing the redox potential, both of which favour the growth of anaerobic bifidobacteria.

Cysteine at 250 mg L^{-1} appears to improve survival of *Lb. acidophilus* and *Bifidobacterium* spp. It should be noted that a low level of cysteine (50 mg L^{-1}) would promote the growth of *Str. thermophilus* and decrease incubation time, particularly in ABT starter culture (Dave & Shah, 1997d). A slight decrease in redox potential is beneficial for the survival of *Str. thermophilus* but, if the concentration of cysteine is increased above 50 mg L^{-1}, the reduced redox potential has a negative effect on growth. The growth of *Lb. delbrueckii* subsp. *bulgaricus* would be improved with a low concentration of cysteine, but would be suppressed at higher levels.

4.12 Future developments and overall conclusions

It is evident over the past three decades that there has been growing interest in the incorporation of probiotic micro-organisms into dairy products, which is mainly attributed to the health-associative benefits of probiotics for consumers (see Chapter 8). This has been reflected by the increased number of such products in different markets worldwide, and also the increase in consumption of probiotic dairy products, in particular fermented milks. Greater knowledge has been acquired regarding the growth and survival of probiotic strains, the discovery of new isolates of probiotic bacteria, and the development in technologies required for the manufacture of different dairy products.

Although the possible application of probiotic bacteria in dairy products other than fermented milks has been studied by many researchers, in fact the commercial realisation of these products has had limited applications; two examples are cheeses and ice cream. In the former product, it could be argued that in countries where hard and semi-hard cheeses are widely used in food preparations, using these products as a vehicle to introduce probiotic strains into the human gut may have limited potential for the following reasons: (a) consumers should be educated to consume 'natural' cheeses directly rather than using them in cooking preparations, and (b) the reluctance of the cheese industry to use probiotic bacteria that might affect flavour. Furthermore, the survival of probiotic bacteria in semi-hard and hard cheeses should be monitored in products more than 6–9 months old, rather than 1–2 months old as reported for an example study in Cheddar cheese.

According to Mattila-Sandholm *et al.* (2002) and Saarella (2011), future technological aspects that have to be considered or addressed in view of the 'functionality' of probiotic micro-organisms may include the following: (a) fermentation and drying technologies; (b) microencapsulation; (c) strain characterisation including daily dosage, stability, viability and non-viability; (d) food matrix formulation and (e) use of prebiotics together with probiotics. All these aspects have been studied extensively over the past decade, and the knowledge acquired will help commercial developments of probiotic dairy products in the future.

The dairy industry should work closely with regulatory agencies in different countries and with the medical profession and researchers to substantiate the health effects associated with probiotic bacteria in dairy products, the probiotic efficacy of new isolates, and how best to responsibly communicate any proven health benefits to consumers. These aspects will help to ensure the acceptability of probiotic dairy products by the consumer and legislative bodies, and thus safeguard the future of the industry.

Over the coming years, however, possible areas requiring further attention may include the following:

- In some commercial probiotic dairy products labels (Raeisi *et al.*, 2013; Tamime *et al.*, 2014), the nomenclature of the starter/probiotic cultures did not conform to recommendations of the International Union of Microbiological Societies (IUMS). In addition, many product labels do not provide the species name of the probiotic bacteria. For example, *Bifidobacterium* spp. is used instead of *Bif. animalis* subsp. *lactis*, probably to avoid the use of the word '*animalis*'.
- Although labelling requirements normally state that the bacterial strain name should appear on the package, the so-called name(s) given to certain bacterial strains, such as 'Bifidus' (presumed to be *Bif. animalis* subsp. *lactis* Bifidus ActiRegularis), may cause some confusion among scientists with the older nomenclature of *Lactobacillus bifidus*. In research papers, the latest accepted nomenclature of probiotic bacteria should be given in order to minimise the confusion in this field. To our knowledge, only a few labels give counts of any probiotic strains used, but it would be more informative if the minimum viable counts are given at the end of shelf-life.
- More properly designed clinical trials and studies are required to establish the proper health benefits to humans of newly isolated LAB of probiotic potential from traditional fermented milk products in different countries. Also, more *in vitro* and *in vivo* studies should be performed to substantiate and provide supporting evidence regarding the health benefits of certain dairy products, such as Kefir and related products.
- Lastly, the naming and classification of the product skyr (i.e. fermented milk or soft cheese) are somewhat puzzling – is it only confined to products made in Iceland or universally produced in all Scandinavian countries? According to Gudmundsson & Kristbergsson (2016), the current commercially produced skyr in Iceland resembles concentrated yoghurt where the 'traditional' production method with two-stage fermentation is not widely used anymore. Similarly, kefir production has changed considerably: instead of complex kefir grains, simple starter cultures are used. Thus, the trend seems to be that although traditional complex fermented milks are not commonly produced industrially, the original product name is still used even if the product is not the same anymore. The issue of (re-)classification of these kinds of products should be considered.

Acknowledgement

One of us (AYT) gratefully acknowledges Dr V.V. Mistry for partly reproducing his original contribution on probiotic cheeses that he reviewed in the first edition of *Probiotic Dairy Products*.

References

Abd-Elhamid, A.M. (2010) Using cheese whey in the preparation of fermented beverage. *Egyptian Journal of Dairy Science*, **38**, 201–210.

Abd El-Salam, M.H., Hippen, A.R., Assem, F.M., El-Shafei, K., Tawfik, N.F. & El-Aassar, M. (2011) Preparation and properties of probiotic cheese high in conjugated linoleic acid content. *International Journal of Dairy Technology*, **64**, 64–74.

Abdou, S.M., El-Alfy, M.B., Hefny, A.A., & Nasr, W.I.A. (2003) Probiotic Ras cheese made from concentrated milk by ultrafiltration or recombination techniques. *Egyptian Journal of Dairy Science*, **31**, 319–333.

Abdou, S.M., Shenana, M.E., Mansour, N.M. & Zakaria, M.K. (2015) Making bioyogurt using newly isolated lactic acid bacteria with probiotic features. *Egyptian Journal of Dairy Science & Technology*, **43**(Suppl.) of 12th Egyptian Conference of Dairy Science & Technology, 1-15.

Abe, F., Tomita, S., Yaeshima, T. & Iwatsuki, K. (2009a) Effect of production conditions on the stability of a human bifidobacterial species *Bifidobacterium longum* in yogurt. *Letters in Applied Microbiology*, **49**, 715–720.

Abe, F., Miyauchi, H., Uchijima, A., Yaeshima, T. & Iwatsuki, K. (2009b) Stability of bifidobacteria in powdered formula. *International Journal of Food Science and Technology*, **44**, 718–724.

Abou-Dawood, S.A. (2002) Survival of nonencapsulated and encapsulated *Bifidobacterium bifidum* in probiotic Kareish cheese. *Egyptian Journal of Dairy Science*, **30**, 43–52.

Aboulfazli, F. & Baba, A.S. (2015) Effect of vegetable milk on survival of probiotics in fermented ice cream under gastrointestinal conditions. *Food Science and Technology Research*, **21**, 391–397.

Aboulfazli, F., Baba, A.S. & Misran, M. (2015) The rheology and physical properties of fermented probiotic ice creams made with dairy alternatives. *International Journal of Food Engineering*, **11**,493–504.

Aboulfazli, F., Baba, A.S. & Misran, M. (2016) Replacement of bovine milk with vegetable milk: Effects on the survival of probiotics and rheological and physicochemical properties of frozen fermented dessert. *International Journal of Dairy Technology*, **69**, 71–80.

Acu, M., Yerlikaya, O. & Kinik, O. (2014) Microencapsulation in dairy technology. *Akademik Gida*, **12**(1), 97–107.

Adhikari, K., Mustapha, A., Grün, I.U. & Fernando, L. (2000) Viability of microencapsulated bifidobacteria in set yogurt during refrigerated storage. *Journal of Dairy Science*, **83**, 1946–1951.

Agil, R., Gaget, A., Gliwa, J., Avis, T.J., Willmore, W.G. & Hosseinian, F. (2013) Lentils enhance probiotic growth in yogurt and provide added benefit of antioxidant protection. *LTW - Food Science and Technology*, **50**, 45–49.

Ahmadi, E., Mortazavian, A.M., Fazeli, M.R., Ezzatpanah, H. & Mohammadi, R. (2012) The effects of inoculant variables on the physicochemical and organoleptic properties of doogh. *International Journal of Dairy Technology*, **65**, 274–281.

Ahmadi, M.A., Ebrahimi, M.T., Mehrabian, S., Tafvizi, F., Bahrami, H. & Dameshghian, M. (2014) Antimutagenic and anticancer effects of lactic acid bacteria isolated from trahana through ames test and phylogenetic analysis by 16S rDNA. *Nutrition and Cancer*, **66**, 1406–1413.

Akalin, A.S., Unal, G., Dinkci, N. & Hayaloglu, A.A. (2012) Microstructural, textural, and sensory characteristics of probiotic yogurts fortified with sodium calcium caseinate or whey protein concentrate. *Journal of Dairy Science*, **95**, 3617–3628.

Akin, M.S. (2005) Effects of inulin and different sugar levels on viability of probiotic bacteria and the physical and sensory characteristics of probiotic fermented ice-cream. *Milchwissenschaft*, **60**, 297–300.

Akpinar, S., Torunoglu, F.A., Yerlikaya, O., Kinik, O., Akbulut, N. & Uysal, H.R. (2015) Fermented probiotic beverages produced with reconstituted whey and cow milk. *Agro Food Industry Hi-Tech*, **26**(4), 24–27.

Alamprese, C., Foschino, R., Rossi, M., Pompei, C. & Savani, L. (2002) Survival of *Lactobacillus johnsonii* La1 and influence of its addition in retail-manufactured ice cream produced with different sugar and fat concentrations. *International Dairy Journal*, **12**, 201–208.

Albenzio, A., Santillo, A., Caroprese, M., Marino, R., Trani, A. & Faccia, M. (2010) Biochemical patterns in ovine cheese: Influence of probiotic strains. *Journal of Dairy Science*, **93**, 3487–3496.

Albenzio, A., Santillo, A., Caroprese, M., Ruggieri, D., Napolitano, F. & Sevi, A. (2013a) Physicochemical properties of Scamorza ewe milk cheese manufactured with different probiotic cultures. *Journal of Dairy Science*, **96**, 2781–2791.

Albenzio, A., Santillo, A., Caroprese, M., Braghieri, A., Sevi, A. & Napolitano, F. (2013b) Composition and sensory profiling of probiotic Scamorza ewe milk cheese. *Journal of Dairy Science*, **96**, 2792–2800.

Albenzio, M., Santillo, A., Marino, R., Malva, A.D., Caroprese, M. & Sevi, A. (2015) Identification of peptides in functional Scamorza ovine milk cheese. *Journal of Dairy Science*, **98**, 842–8432.

Alfano, A., Donnarumma, G., Cimini, D., Fusco, A., Marzaioli, I., de Rosa, M. & Schiraldi, C. (2015) *Lactobacillus plantarum*: Microfiltration experiments for the production of probiotic biomass to be used in food and nutraceutical preparations. *DOI 10.1002/btpr.2037* (wileyonlinelibrary.com).

Aljewicz, M. & Cichosz, G. (2015a) Protective effects of *Lactobacillus* cultures in Dutch-type cheese-like products. *LWT - Food Science and Technology*, **63**, 52–56.

Aljewicz, M. & Cichosz, G. (2015b) The effect of probiotic *Lactobacillus rhamnosus* HN001 on the *in vitro* availability of minerals from cheeses and cheese-like products. *LWT - Food Science and Technology*, **60**, 841–847.

Aljewicz, M., Siemianowska, W., Cichosz, G. & Tonska, E. (2014) The effect of probiotics (*Lactobacillus rhamnosus* HN001, *Lactobacillus paracasei* LPC-37, and *Lactobacillus acidophilus* NCFM) on the availability of minerals from Dutch-type cheese. *Journal of Dairy Science*, **97**, 4824–4831.

Aljewicz, M., Cichosz, G., Nalepa, B. & Bielecka, M. (2016) The effect of milk fat substitution with palm fat on lactic acid bacteria counts in cheese-like products. *LWT - Food Science and Technology*, **66**, 348–354.

Allgeyer, L.C., Miller, M.J. & Lee, S.-Y. (2010a) Sensory and microbiological quality of yogurt drinks with prebiotics and probiotics. *Journal of Dairy Science*, **93**, 4471–4479.

Allgeyer, L.C., Miller, M.J. & Lee, S.-Y. (2010b) Drivers of liking for yogurt drinks with prebiotics and probiotics. *Journal of Food Science*, **75**, S212–S219.

Almeida, K.E., Tamime A.Y. & Oliveira, M.N. (2008) Acidification rates of probiotic bacteria in *Minas frescal* cheese whey. *LTW - Food Science and Technology*, **41**, 311–316.

Almeida, K.E., Tamime A.Y. & Oliveira, M.N. (2009) Influence of total solids contents of milk whey on the acidifying profile and viability of various lactic acid bacteria. *LTW - Food Science and Technology*, **42**, 672–678.

Al-Otaibi, M.M. (2012) Isolation and identification of lactic acid bacteria and yeasts from Sameel milk; A Saudi traditional fermented milk. *International Journal of Dairy Science*, **7**(4), 73–83.

Al-Otaibi, M.M., Saleha, F.A. & Al-Obaib, S. (2013) Effect of date syrup (*dips*) and survival of probiotic bacteria in milk. *International Journal of Dairy Science*, **8**(1), 12–20.

Al-Sheraji, S.H., Ismail, A., Manap, M.Y., Mustafa, S. & Yusof, R.M. (2012) Viability and activity of bifidobacteria during refrigerated storage of yoghurt containing *Mangifera pajang* fibrous polysaccharides. *Journal of Food Science*, **77**, M624–M630.

Alves, L.L., Richards, N.S.P.S., Milani, L.I.G., Cruz, A.G. & Faria, J.A.F. (2013) Cream cheese as a symbiotic food carrier using *Bifidobacterium animalis* Bb-12 and *Lactobacillus acidophilus* La-5 and inulin. *International Journal of Dairy Technology*, **66**, 63–69.

Alves, A.T.S.E, Antunes, A.E.C., Trento, F.K.H.S., Zacarchenco, P.B., Ormenese, R.C.S.C. & Spadoti, L.M. (2016) Pasteurised, microfiltered and lactose hydrolysed skimmed milk with added probiotics: Development and storage stability. *International Journal of Dairy Technology*, **69**, 22–30.

Amer, E.S.N., Girgis, S.S., Taha, H. & Abed-El-Moeety, S.H. (1997) Effect of milk total solids and type of starter on the quality of Labneh. *Egyptian Journal of Dairy Science*, **25**, 179–192.

Andrade, C.R.G., Souza, M.R., Penna, C.F.A.M., Acurcio, L.B., Sant'Anna, F.M., Castro, R.D. & Oliveira, D.L.S. (2014) *In vitro* probiotic properties of *Lactobacillus* spp. isolated from Minas artisanal cheese from Serra da Canastra - MG. *Arquivo Brasileiro de Medicina Veterinaria e Zootecnia*, **66**, 1592–1600.

Anisimov, G.S., Anisimov, S.V., Ryabtseva, S.A., Evdokimov, I.A. & Ahmedova, V.R. (2013) Bio-fermented ice-cream with functional properties. *Molochnaya Promyshlennost*, **8**, 51–52.

Anjani, K. & Kailasapathy, K. (2004) Survival of co-encapsulated probiotics and prebiotics in yoghurt. *Milchwissenschaft*, **59**, 396–399.

Anonymous (1992) Yoghurt and probiotics. *Choice*, **11**, 32–35.

Anonymous (2001) Choose life top. *Dairy Industries International*, **66**(11), 45. Note to publisher - one page only

Anonymous (2003) *Short Life Dairy Products*, Müller Market Report 2003, Müller Dairy (UK) Ltd., Market Drayton.

Anonymous (2008) *Nu-trish Technical Information*, Revised Edition, Ch. Hansen A/S, Bøge Allé 10-12, DK 2970 Hørsholm, Denmark.

Anonymous (2013) The science behind *Bifidobacterium* BB-12®, Technical Bulletin, Ch. Hansen A/S, Bøge Allé 10-12, DK 2970 Hørsholm, Denmark.

Anonymous (2016) American Type Culture Collection (ATCC) product sheet – *Bifidobacterium animalis subsp. animalis* (ATCC 25527™). https:www.atcc.org/~/25527ashx. Accessed on 21st April 2016.

Anton, D., Raudsepp, P., Roasto, M., Meremae, K., Kuusik, S., Toomik, P., Elias, P., Laikoja, K., Kaart, T., Lepiku, M. & Tonu Pussa, T. (2016) Comparative study of microbiological, chemical and sensory properties of kefirs produced in Estonia, Latvia and Lithuania. *Journal of Dairy Research*, **83**, 89–95.

Antonsson, M., Ardö, Y., Nilsson, B.F. & Molin, G. (2002) Screening and selection of *Lactobacillus* strains for use as adjunct cultures in production of semi-hard cheese. *Journal of Dairy Research*, **69**, 457–472.

Antunes, A.E.C., Marasca, E.T.G., Moreno, I., Dourado, F.M., Rodrigues, L.G. & Lerayer, A.L.S. (2007) Development of a probiotic buttermilk. *Ciencia e Tecnologia de Alimentos*, **27**, 83–90.

Antunes, A.E.C., Silva, E.R.A., van Dender, A.G.F., Marasca, E.T.G., Moreno, I., Faria, E.V., Padula, M. & Lerayer, A.L.S. (2009) Probiotic buttermilk-like fermented milk product development in a semi-industrial scale: Physicochemical, microbiological and sensory acceptability. *International Journal of Dairy Technology*, **62**, 556–563.

Arab, M., Mortazavian, M. A., Azadnia, E., Komeyli, R., Alimiri, T. & Najafi, M. (2016) Effects of salt percentage (0.5 or 1%), combination (NaCl or NaCl/KCl) and stage of adding salt on biochemical, microbiological and sensory characteristics of probiotic doogh. *Iranian Journal of Food Science and Technology*, **13**(55), 13–22.

Archer, A. C. & Halami, P.M. (2015) Probiotic attributes of *Lactobacillus fermentum* isolated from human feces and dairy products. *Applied Microbiology and Biotechnology*, **99**, 8113–8123.

Atallah, A.A. (2015a) Development of new functional beverages from milk permeate using some probiotic bacteria and fruit pulp. *Egyptian Journal of Dairy Science*, **43**, 25–39.

Atallah, A.A. (2015b) Preparation and properties of functional beverages based on probiotic milk permeate with carrot or mango pulp. *Egyptian Journal of Dairy Science*, **43**, 147–158.

Awaisheh, S.S. (2011) Development of probiotic soft cheese manufactured using goat's milk with the addition of thyme. *Milchwissenschaft*, **66**, 51–54.

Awaisheh, S., Haddadin, M.S.Y. & Robinson, R.K. (2005) Incorporation of selected nutraceuticals and probiotic bacteria into a fermented milk. *International Dairy Journal*, **15**, 1184–1190.

Awaisheh, S., Al-Dmoor, H., Omar, S., Hawari, A. & Alroyli, M. (2012) Impact of selected nutraceuticals on viability of probiotic strains in milk during refrigerated storage at 4 °C for 15 days. *International Journal of Dairy Technology*, **65**, 268–273.

Ayar, A. & Burucu, H. (2013) Effect of whey fractions on microbial and physicochemical properties of probiotic ayran (drinkable yogurt). *International Food Research Journal*, **20**, 1409–1415.

Ayeni, F.A., Adeniyi, B.A., Ogunbanwo, S.T., Nader-Macias, M.E. & Ruas-Madiedo, P. (2011) Survival of *Weissella confusa* and *Lactobacillus paracasei* strains in fermented milks under cold storage and after freeze-drying. *Milchwissenschaft*, **66**, 61–64.

Ayyad, K.M., Addul Ghaffar, I., Salem,A.S. & Ismail, M.M. (2015) Production of probiotic low fat Labneh using exopolysaccharide-producing strains. *Egyptian Journal of Dairy Science*, **43**, 41–51.

Ayyash, M.M., Sherkat, F. & Shah, N.P. (2012) The effect of NaCl substitution with KCl on Akawi cheese: Chemical composition, proteolysis, angiotensin-converting enzyme-inhibitory activity, probiotic survival, texture profile, and sensory properties. *Journal of Dairy Science*, **95**, 4747–4759.

Baick, S.-C. & Kim, C.-H. (2015) Assessment of characteristics and functional properties of *Lactobacillus* species isolated from Kimchi for dairy use. *Korean Journal for Food Science of Animal Resources*, **35**,339–349.

Baker, H.C., Tran, D.N. & Thomas, L.V. (2009) Health benefits of probiotics for the elderly: A review. *Journal of Foodservice*, **20**, 250–262.

Balakrishnan, G. & Agrawal, R. (2014) Antioxidant activity and fatty acid profile of fermented milk prepared by *Pediococcus pentosaceus*. *Journal of Food Science and Technology*, **51**, 4138–4142.

Banwo, K., Sanni, A. & Tan, H. (2013) Functional properties of *Pediococcus* species isolated from traditional fermented cereal gruel and milk in Nigeria. *Food Biotechnology*, **27**(1), 14–38.

Barbosa, J. Borges, S. & Teixeira, P. (2014) Selection of potential probiotic *Enterococcus faecium* isolated from Portuguese fermented food. *International Journal of Food Microbiology*, **191**, 144–148.

Baron, M., Roy, D., & Vuillemard, J.-C. (2000) Biochemical characteristics of fermented milk produced by mixed-cultures of lactic starters and bifidobacteria. *Lait*, **80**, 465–478.

Barouel, J., Karbassi, A., Ghodddusi, H.B., Mortazavi, A., Ramezani, R. & Mousssavi, M. (2011) Impact of native *Lactobacillus paracasei* subsp. *paracasei* and *Pediococcus* spp. as adjunct cultures on sensory quality of Iranian White brined cheese. *International Journal of Dairy Technology*, **64**, 526–535.

Barraquio, V.L., Militate, S.T., Gonzaga, J.O. & Emata, O.C. (2001) Fermented milk drink as starter adjunct in the manufacture of probiotic white soft cheese. *The Phillippine Agricultural Scientist*, **84**, 382–387.

Basiony, M.M.M., Hamad, M.N.F. & Ismail, M.M. (2015) Effect of fortification with texturized soy protein on the chemical, microbial and sensorial attributes of Labneh made for goat's milk. http://www.researchgate.net/publication/277716568. Accessed on 22nd February 2016.

Batista, A.L.D., Silva, R., Cappato, L.P. Almada, C.N., Garcia, R.K.A., Silva, M.C., Raices, R.S.L., Arellano, D.B., Sant'Ana, A.S., Junior, C.A.C., Freitas, M.Q. & Cruz, A.G. (2015) Quality parameters of probiotic yogurt added to glucose oxidase compared to commercial products through microbiological, physical-chemical and metabolic activity analyses. *Food Research International*, **77**, 627–635.

Beheshtipour, H., Mortazavian, A.M., Mohammadi, R., Sohrabvandi, S. & Khosravi-Darani, K. (2013) Supplementation of *Spirulina platensis* and *Chlorella vulgaris* algae into probiotic fermented milks. *Comprehensive Reviews in Food Science and Food Safety*, **12**, 144–154.

Belem, M.A.F. (1999) Application of biotechnology in the product development of nutraceuticals in Canada. *Trends in Food Science & Technology*, **10**, 101–106.

Bergamini, C.V., Hynes, E.R., Quiberoni, A., Suarez, V. & Zalazar, C.A. (2005) Probiotic bacteria as adjunct starters: Influence of the addition methodology on their survival in a semi-hard Argentinean cheese. *Food Research International*, **38**, 597–604.

Bergamini, C.V., Hynes, E. & Zalazar, C.A. (2006) Influence of probiotic bacteria on the proteolysis profile of a semi-hard cheese. *International Dairy Journal*, **16**, 856–866.

Bergamini, C.V., Hynes, E.R., Candioti, M.C. & Zalazar, C.A. (2009) Multivariate analysis of proteolysis patterns differentiated the impact of six strains of probiotic bacteria on a semi-hard cheese. *Journal of Dairy Science*, **92**, 2455–2467.

Bergamini, C., Hynes, E., Meinardi, C., Suarez, V., Quiberoni, A. & Zalazar C. (2010) Pategras cheese as a suitable carrier for six probiotic cultures. *Journal of Dairy Research*, **77**, 265–272.

Bernardeau, M., Vernoux, J.P., Henri-Dubernet, S. & Gueguen, M. (2008) Safety assessment of dairy microorganisms: The *Lactobacillus* genus. *International Journal of Food Microbiology*, **126**, 278–285.

Bertolami, M.C. & Farnworth, E.R. (2008) The properties of *Enterococcus faecium* and the fermented milk product - Gaio®. *Handbook of Fermented Functional Foods*, (ed. E.R. Farnworth), 2nd Edition, pp. 71–88, CRC Press (Taylor & Francis Group), Boca Raton.

Bezerra, T.K., de Araujo, A.R., do Nascimento, E.S., de Matos Paz, J.E., Gadelha, C.A., Gadelha, T.S., Pacheco, M.T., do Egypto Queiroga Rde, C., de Oliveira, M.E. & Madruga, M.S. (2016) Proteolysis in goat "Coalho" cheese supplemented with probiotic lactic acid bacteria. *Food Chemistry*, **196**, 359–366.

te Biesebeke, R. & de Vries, Y.P. (2009) Long-life probiotic food product. *European Patent Application*, **EP 2 103 226 A1**.

Bisig, W., Eberhard, P., Collomb, M. & Rehberger, B. (2007) Influence of processing on the fatty acid composition and the content of conjugated linoleic acid in organic and conventional dairy products - A review. *Le Lait*, **87**, 1–19.

Blanchette, L., Roy, D., Belanger, G. & Gauthier, S.F. (1996) Production of Cottage cheese using dressing fermented by bifidobacteria. *Journal of Dairy Science*, **79**, 8–15.

Bolla P.A., de los Angeles Serradell, M., Urraza P.J. & de Antoni, G.L. (2010) Effect of freeze-drying on viability and *in vitro* probiotic properties of a mixture of lactic acid bacteria and yeasts isolated from kefir. *Journal of Dairy Research*, **78**, 15–22.

Bolla, P.A., Carasi, P., de los Angeles Serradell, M., & de Antoni, G.L. (2013) Kefir-isolated *Lactococcus lactis* subsp. *lactis* inhibits the cytotoxic effect of *Clostridium difficile in vitro*. *Journal of Dairy Research*, **80**, 96–102.

Bosnea, L.A., Kourkoutas, Y., Albantaki, N., Tzia, C., Koutinas, A.A. & Kanellaki, M. (2009) Functionality of freeze-dried *L. casei* cells immobilized on wheat grains. *LWT - Food Science and Technology*, **42**, 1696–1702.

Bosnea, L.A., Moschakis, T. & Biliaderis, C.G. (2014) Complex coacervation as a novel micro-encapsulation technique to improve viability of probiotics under different stresses. *Food and Bioprocess Technology*, **7**, 2767–2781.

Botta, C., Bertolino, M., Zeppa, G. & Cocolin, L. (2015) Evaluation of Toma Piemontese PDO cheese as a carrier of putative probiotics from table olive fermentations. *Journal of Functional Foods*, **18**(Part A), 106–116.

Bottazzi, V. (2002) Siamo e molto bravi in biochimica: Produciamo lo yoghurt (in Italian). *Industria del Latte*, **XXXVIII**(1-2), 5–64.

Bourrie, B.C.T., Willing, B.P. & Cotter, P.D. (2016) The microbiota and health promoting characteristics of the fermented beverage Kefir. *Frontiers in Microbiology*, **7**, Article 647.

Boylston, T.D., Vinderola, C.G., Ghoddusi, H.B. & Reinheimer, J.A. (2004) Incorporation of bifidobacteria into cheeses: challenges and rewards. *International Dairy Journal*, **14**, 375–387.

Bozanic, R. & Tratnic, L. (2001) Quality of cow's and goat's fermented bifido milk during storage. *Food Technology and Biotechnology*, **39**, 109–114.

Bozanic, R., Tratnik, L., Herceg, Z. & Maric, O. (2004) The influence of milk powder, whey protein concentrate and inulin on the quality of cow and goat acidophilus milk. *Acta Alimentaria*, **33**, 337–346.

Burgain, J., Gaiani, C., Linder, M. & Scher, J. (2011) Encapsulation of probiotic living cells: from laboratory scale to industrial applications. *Journal of Food Engineering*, **104**, 467–483.

Buriti, F.C.A. & Saad, S.M.I. (2014) Chilled milk-based desserts as emerging probiotic and prebiotic products. *Critical Reviews in Food Science and Nutrition*, **54**, 139–150.

Buriti, F.C.A., da Rocha, J.S. & Saad, S.M.I. (2005a) Incorporation of *Lactobacillus acidophilus* in Minas fresh cheese and its implications for textural and sensorial properties during storage. *International Dairy Journal*, **15**, 1279–1288.

Buriti, F.C.A., Rocha, J.S., Assis, E.G. & Saad, S.M.I. (2005b) Probiotic potential of Minas fresh cheese prepared with the addition of *Lactobacillus paracasei*. *LWT - Food Science and Technology*, **38**, 173–180.

Buriti, F.C.A., Freitas, S.C., Egito, A.S. & dos Santos, K.M.O. (2014) Effects of tropical fruit pulps and partially hydrolysed galactomannan from *Caesalpinia pulcherrima* seeds on the dietary fibre content, probiotic viability, texture and sensory features of goat dairy beverages. *LWT - Food Science and Technology*, **59**, 196–203.

Burns, P., Vinderola, G., Molinari, F. & Reinheimer, J. (2008) Suitability of whey and buttermilk for the growth and frozen storage of probiotic lactobacilli. *International Journal of Dairy Technology*, **61**, 156–164.

Caggia, C., De Angelis, M., Pitino, I., Pino, A.& Randazzo, C.L. (2015) Probiotic features of *Lactobacillus* strains isolated from Ragusano and Pecorino Siciliano cheeses. *Food Microbiology*, **50**, 109–117.

Capra, M.L., Binetti, A.G., Mercanti, D.J., Quiberoni, A. & Reinheimer, J.A. (2009) Diversity among *Lactobacillus paracasei* phages isolated from a probiotic dairy product plant. *Journal of Applied Microbiology*, **107**, 1350–1357.

Cardarelli, H.R., Aragon-Alegro, L.C., Alegro, J.H.A., de Castro, I.A. & Saad, S.M.I. (2008) Effect of inulin and *Lactobacillus paracasei* on sensory and instrumental texture properties of functional chocolate mousse. *Journal of the Science of Food and Agriculture*, **88**, 1318–1324.

Cardenas, N., Calzada, J., Peirotén, A., Jimenez, E., Escudero, R., Rodríguez, J.M., Medina, M. & Fernandez, L. (2014) Development of a potential probiotic fresh cheese using two *Lactobacillus salivarius* strains isolated from human milk. *BioMed Research International*, 10.1155/2014/801918, Accessed on 1st of July 2014.

Casarotti, S.N., Carneiro, B.M. & Penna, A.L.B. (2014a) Evaluation of the effect of supplementing fermented milk with quinoa flour on probiotic activity. *Journal of Dairy Science*, **97**, 6027–6035.

Casarotti, S.N., Monteiro, D.A., Moretti, M.M.S., & Penna, A.L.B. (2014b) Influence of the combination of probiotic cultures during fermentation and storage of fermented milk. *Food Research International*, **59**, 67–75.

Castro, W.F., Cruz, A.G., Bisinotto, M.S., Guerreiro, L.M.R., Faria, J.A.F., Bolini, H.M.A. Cunha, R.L. & Deliza, R. (2013a) Development of probiotic dairy beverages: rheological properties and application of mathematical models in sensory evaluation. *Journal of Dairy Science*, **96**, 16–25.

Castro, W.F., Cruz, A.G., Rodrigues, D., Ghiselli, G., Oliveira, C.A.F., Faria, J.A.F. & Godoy, H.T. (2013b) *Short communication:* Effects of different whey concentrations on physicochemical characteristics and viable counts of starter bacteria in dairy beverage supplemented with probiotics. *Journal of Dairy Science*, **96**, 96–100.

Castro, J.M., Tornadijo, M.E., Fresno, J.M. & Sandoval, H. (2015) Biocheese: A food probiotic carrier. *BioMed Research International*, **2015**, Article ID 723056, can be accessed at: https://www.hindawi.com/journals/bmri/2015/723056/

de Castro, F.P., Cunha, T.M., Barreto, P.L.M., Amboni, R.D.de M.C. & Prudencio, E.S. (2009a) Effect of oligofructose incorporation on the properties of fermented probiotic lactic beverages. *International Journal of Dairy Technology*, **62**, 68–74.

de Castro, F.P., Cunha, T.M., Ogliari, P.J., Teofilo, R.F., Ferreira, M.M.C. Prudencio, E.S. (2009b) Influence of different content of cheese whey and oligofructose on the properties of fermented lactic beverages: Study using response surface methodology. *LWT - Food Science and Technology*, **42**, 993–997.

de Castro-Cislaghi, F.P., dos Silva, C.R.E., Fritzen-Freire, C.B., Lorenz, G. & J Sant'Anna, E.S. & (2012) *Bifidobacterium* Bb-12 microencapsulated by spray drying with whey: Survival under simulated gastrointestinal conditions, tolerance to NaCl, and viability during storage. *Journal of Food Engineering*, **113**, 186–193.

Celestin, S., Thorat, S.S., Desale, R.J. & Chavan, U.D. (2015) Effect of milk Supplementation with fructooligosaccharides and inulin on viable counts of probiotic bacteria in goat and cow milk yoghurts. *Journal of Environmental Science, Toxicology and Food Technology*, **9**(7), 6–12.

Champagne, C.P. (2014) Development of fermented milk products containing probiotics. *Dairy Microbiology and Biochemistry: Recent Developments* (eds, B. Ozer and G. Akdemir-Evrendilek), pp. 214–244, CRC Press (Taylor & Francis Group), Boca Raton.

Champagne, C.P., Gardner, N.J. & Roy, D. (2005) Challenges in the Addition of Probiotic Cultures toFoods. *CriticalReviewsinFoodScienceandNutrition*,**45**(1)DOI:10.1080/10408690590900144

Champagne, C.P., Ross, R.P., Saarela, M., Hansen, K.F. & Charalampopoulos, D. (2011) Recommendations for the viability assessment of probiotics as concentrated cultures and in food matrices. *International Journal of Food Microbiology*, **149**, 185–193.

Chan, E.S. & Zhang, Z. (2002) Encapsulation of probiotic bacterial *Lactobacillus acidophilus* by direct compression. *Trans IChemE*, **80**(C), 78–82.

Chandan, R.C. & O'Rell, K.R. (2008) Principle of yoghurt processing. *Manufacturing Yogurt and Fermented Milks* (eds. R.C. Chandan and A. Kilara), 2nd edition, pp. 195–210, John Wiley & Sons, New York.

Chandan, R.C., Kilara, A & Shah.N.P. (2008) *Dairy Processing & Quality Assurance*. Wiley-Blackwell, Ames.

Chandramouli, V., Kailasapathy, K., Peiris, P. & Jones, M. (2004) An improved method of micro-encapsulation and its evaluation to protect *Lactobacillus* spp. in simulated gastric conditions. *Journal of Microbiological Methods*, **56**, 27–35.

Chap, J., Jackson, P., Siqueira, R., Gaspar, N., Quintas, C., Park, J., Osaili, T., Shaker, R., Jaradat, Z., Hartantyo, S.H.P., Abdullah Sani, N., Estuningsih, S. & Forsythe, S.J. (2009) International survey of *Cronobacter sakazakii* and other *Cronobacter* spp. in follow up formulas and infant foods. *International Journal of Food Microbiology*, **136**, 185–188.

Charteris, W.P., Kelly, P.M., Morelli, L. & Collins, J.K. (2002) Edible table (bio) spread containing potentially probiotic *Lactobacillus* and *Bifidobacterium* species. *International Journal of Dairy Technology*, **55**, 44–56.

Chen, M.-J., Chen, K.-N. & Lin, C.-W. (2004) Development of a new probiotic dairy tofu with glucono-delta-lactone. *Journal of Food Science*, **69**, E344–E350.

Chen, M.-J., Chen, K.-N. & Lin, C.-W. (2006) Development and verification of an optimum composition model for a synbiotic fermented milk using sequential quadratic programming techniques. *Asian-Australasian Journal of Animal Sciences*, **19**, 1490–1495.

Chen, Y.P., Hsiao, P.J., Hong, W.S., Dai, T.Y. & Chen, M.J. (2012) *Lactobacillus kefiranofaciens* M1 isolated from milk kefir grains ameliorates experimental colitis *in vitro* and *in vivo*. *Journal of Dairy Science*, **95**, 63–74.

Chen, Y.P., Lee, T.Y., Hong, W.S., Hsieh, H.H. & Chen, M.J. (2013) Effects *Lactobacillus kefiranofaciens* M1 isolated from milk kefir grains on enterohemorrhagic *Escherichia coli* infection using mouse and intestinal cells models. *Journal of Dairy Science*, **96**, 7467–7477.

Chen, C., Song, L.-J., Qi, P.-P., Wang, M., Feng, S. Zhang, J. & Zhao, J. (2015) Review on the probiotic properties of *Lactobacillus casei* in dairy products. *Food and Fermentation Technology*, **4**, 88–91.

ChienJung, H. (2000) (Studies on the manufacture of yoghurt by microentrapment of bifidobacteria – in Chinese). *Journal of the Chinese Society of Animal Science*, **29**, 125–131.

Cho, Y.-H., Hong, S.-M. & Kim, C.-H. (2013) Isolation and characterization of lactic acid bacteria from kimchi, Korean traditional fermented food to apply into fermented dairy products. *Korean Journal for Food Science of Animal Resources*, **33**, 75–82.

Choi, H.-J., Lee, N.-K. & Paik, H.-D. (2015) Health benefits of lactic acid bacteria isolated from kimchi, with respect to immunomodulatory effects. *Food Science and Biotechnology*, **24**, 783–789.

Chouraqui, J.-P., van Egroo, L.D. & Ficot, M.C. (2004) Acidified milk formula supplemented with *Bifidobacterium lactis*: impact on infant diarrhea in intestinal care settings. *Journal of Pediatric Gastroenterology and Nutrition*, **38**, 288–292.

Christiansen, P.S., Edelsten, D., Kristiansen, J.R. & Nielsen, E.W. (1996) Some properties of ice cream containing *Bifidobacterium bifidum* and *Lactobacillus acidophilus*. *Milchwissenschaft*, **51**, 502–504.

Clark, P.A. & Martin, J.H. (1994) Selection of bifidobacteria for use as dietary adjuncts in cultured dairy foods: III - Tolerance to simulated bile concentrations of human small intestines. *Cultured Dairy Products Journal*, **29**(3), 18–21.

Clark, P.A., Cotton, L.N. & Martin, J.H. (1993) Selection of bifidobacteria for use as dietary adjuncts in cultured dairy foods: II - tolerance to simulated pH of human stomachs. *Cultured Dairy Products Journal*, **28**(4), 11–14.

Cocolin, L., Valente, A., Alessandria, V. & Zeppa, G. (2010) Probiotic bacteria in fresh cheese: technological characterization and vitality. *Industrie Alimentari*, **49**(498), 11–15.

Collins, J.K. (2001) Demonstration of functional properties of probiotic lactic acid bacteria. *Industria del Latte*, **XXXVII**(1-2), 33–60.

Collins, J.K., Thornton, G. & Sullivan, G.O. (1998) Selection of probiotic strains for human applications. *International Dairy Journal*, **8**, 487–490.

Corbo, M.R., Albenzio, M., de-Angelis, M., Sevi, A., Gobbetti, M. & de-Angelis, M. (2001) Microbiological and biochemical properties of Canestrato Pugliese hard cheese supplemented with bifidobacteria. *Journal of Dairy Science*, **84**, 551–561.

Costa, H.H.S., Souza, M.R., Acurcio, L., Cunha, A.F., Resende, M.F.S. & Nunes, A.C. (2013) Probiotic potential of lactic acid bacteria isolated from Minas artisanal cheese from Serra da Canastra, MG. *Arquivo Brasileiro de Medicina Veterinaria e Zootecnia*, **65**, 1858–1866.

Costa, M.P., Balthazar, C.F., Franco, R.M., Marsico, E.T., Cruz, A.G. & Junior, C.A.C. (2014) Changes on expected taste perception of probiotic and conventional yogurts made from goat milk after rapidly repeated exposure. *Journal of Dairy Science*, **97**, 2610–2618.

Costa, M.P., Frasao, B.S., Silva, A.C.O., Freitas, M.Q., Franco, R.M. & Conte-Junior, C.A. (2015) Cupuassu (*Theobroma grandiflorum*) pulp, probiotic, and prebiotic: Influence on color, apparent viscosity, and texture of goat milk yogurts. *Journal of Dairy Science*, **98**, 5995–6003.

da Cruz, A.G., Buriti, F.C.A., de Souza, C.H.B., Faria, J.A.F & Saad, S.M.I. (2009) Review - Probiotic cheese: health benefits, technological and stability aspects. *Trends in Food Science & Technology*, **20**, 344–354.

Cruz, A.G., Cadena, R.S., Faria, J.A.F., Bolini, H.M.A., Dantas, C., Ferreira, H.M.C. & Deliza, R. (2010a) PARAFAC: adjustment for modelling consumer study covering probiotic and conventional yogurt. *Food Research International*, **45**, 211–215.

Cruz, A.G., Faria, J.A.F., Walter, E.H.M., Andrade, R.R., Cavalcanti, R.N., Oliveira, C.A.F. & Granato, D. (2010b) Optimization of the processing of probiotic yogurt added with glucose oxidase using response surface methodology. *Journal of Dairy Science*, **93**, 5059–5068.

Cruz, A.G., Walter, E.H.M., Cadena, R.S., Faria, J.A.F., Bolini, H.M.A., Pinherio, H.P. & Santa'Ana, A.S. (2010c) Survival analysis methodology to predict the shelf-life of probiotic flavoured yoghurt. *Food Research International*, **43**, 1444–1448.

Cruz, A.G., Cadena, R.S., Faria, J.A.F., Oliveira, C.A.F., Cavalcanti, R.N., Bona, E., Bolini, H.M.A. & da Silva, A.A.P. (2011) Consumer acceptability and purchase intent of probiotic yoghurt with added glucose oxidase using sensometrics, artificial neural networks and logistic regression. *International Journal of Dairy Technology*, **64**, 549–556.

Cruz, A.G., Castro, W.F., Faria, J.A.F., Lollo, P.C.B., Amaya-Farfan, J., Freitas, M.Q., Rodrigues, D., Oliveira, C.A.F. & Godoy, H.T. (2012) Probiotic yogurts manufactured with increased glucose oxidase levels: Postacidification, proteolytic patterns, survival of probiotic microorganisms, production of organic acid and aroma compounds. *Journal of Dairy Science*, **95**, 2261–2269.

Dabiza, N. & El-Deib, K. (2007) Biochemical evaluation and microbial quality of Ras cheese supplemented with probiotic strains. *Polish Journal of Food and Nutrition Sciences*, **57**, 295–300.

Dabiza, N.M.A., Effat, B.A. & Sharaf, O.M. (2006) Antibacterial effect of probiotic bacteria isolated from dairy products. *Deutsche Lebensmittel-Rundschau*, **102**(3), 114–121.

Daigle, A., Roy, D. & Vuillemard, J.C. (1998) Growth response and metabolic activity of *Bifidobacterium infantis* ATCC 27920G in creams enriched with native phosphocaseinate retentate. *Milchwissenschaft*, **53**, 625–629.

Daigle, A., Roy, D., Belanger, G. & Vuillemard, J.C. (1999) Production of probiotic cheese (Cheddar-like cheese) using enriched cream fermented by *Bifidobacterium infantis*. *Journal of Dairy Science*, **82**, 1081–1901.

Damin, M.R., Minowa, E., Alcantara, M.R., Oliveira, M.N. (2006) Chemical and viability changes during fermentation and cold storage of fermented milk manufactured using yogurt and probiotic bacteria. 10.1051/IUFoST:20060635. Accessed 15th March 2016.

Daneshi, M., Ehsani, M.E., Razavi, S.H., Labbafi, M. & Rezaee, M.S. (2012) Effect of cold storage on viability of probiotic bacteria in carrot fortified milk. *Journal of Nutrition & Food Sciences*, **2**(9). http://www.omicsonline.org/nutrition-food-sciences.php Accessed on 17th March 2016.

Danova, S., Petrov,K., Pavlov, P. & Petrova, P. (2005) Isolation and characterization of *Lactobacillus* strains involved in koumiss fermentation. *International Journal of Dairy Technology*, **58**, 100–105.

Dantas, A.B., Jesus, V,F., Silva, R., Almada, C.N., Esmerino, E.A., Cappato, L.P., Silva, M.C., Raices, R.S.L., Cavalcanti, R.N., Carvalho, C.C., Sant'Ana, A.S., Bolini, H.M.A., Freitas, M.Q. & Adriano G. Cruz, A.G. (2016) Manufacture of probiotic Minas Frescal cheese with *Lactobacillus casei* Zhang. *Journal of Dairy Science*, **99**, 18–30.

Darehabi, H.K. & Nikmaram, P. (2011) Assessment of the growth and survival of *Escherichia coli* O157:H7 during the manufacture and storage of Iranian White cheese and probiotic cheese. *Global Veterinaria*, **6**, 228–232.

Dave, R.I. & Shah, N.P. (1997a) Effect of level of starter culture on viability of yoghurt and probiotic bacteria in yoghurts. *Food Australia*, **49**, 164–168.

Dave, R.I. & Shah, N.P. (1997b) Viability of yoghurt and probiotic bacteria in yoghurts made from commercial starter cultures. *International Dairy Journal*, **7**, 31–41.

Dave, R.I. & Shah, N.P. (1997c) Effectiveness of ascorbic acid as an oxygen scavenger in improving viability of probiotic bacteria in yoghurts made with commercial starter cultures. *International Dairy Journal*, **7**, 435–443.

Dave, R.I. & Shah, N.P. (1997d) Effectiveness of cysteine as redox potential reducing agent in improving viability of probiotic bacteria in yoghurts made with commercial starter cultures. *International Dairy Journal*, **7**, 537–545.

Dave, R.I. & Shah, N.P. (1998) Ingredient supplementation effects on viability of probiotic bacteria in yoghurt. *Journal of Dairy Science*, **81**, 2804–2816.

Davidson, R.H., Duncan, S.E., Hackney, C.R., Eigel, W.N. & Boling, J.W. (2000) Probiotic cultures survival and implications in fermented frozen yogurt characteristics. *Journal of Dairy Science*, **83**, 666–673.

Davis, C. (2014) Enumeration of probiotic strains: Review of culture-dependent and alternative techniques to quantify viable bacteria. *Journal of Microbiological Methods*, **103**, 9–17.

Deeb, A.M.M. & Ahmed, H.F. (2010) Effect of potassium sorbate and/or probiotic bacteria on spoilage bacteria during cold storage of soft cheese. *Global Veterinaria*, **4**, 483–488.

Delaney, A.-G.O. (1977) Factory-scale experiments on the production of ymer by ultrafiltration. *Milchwissenschaft*, **32**, 651–653.

Desai, A.R., Powell, I.B. & Shah, N.P. (2004) Survival and activity of probiotic lactobacilli in skim milk containing prebiotics. *Journal of Food Science*, **69**, FMS57–FMS60.

Desfosses-Foucault, E., Dussault-Lepage, V., Le Boucher, C., Savard, P., LaPointe, G. & Roy, D. (2012) Assessment of probiotic viability during Cheddar cheese manufacture and ripening using propidium monoazide-PCR quantification. *Frontiers in Microbiology*, **3**, Article 350-1, doi: 10.3389/fmicb.2012.00350, can be obtain on-line (www.frontiersin.org). Accessed on 7th August 2016.

Desmond, C., Stanton, C., Fitzgerald, G.F., Collins, K. & Ross, R.P. (2001) Environmental adaptation of probiotic lactobacilli towards improvement of performance during spray drying. *International Dairy Journal*, **11**, 801–808.

Desmond, C., Ross, R.P., O'Callaghan, E., Fitzgerald, G. & Stanton, C. (2002) Improved survival of *Lactobacillus paracasei* NFBC 338 in spray-dried powders containing gum acacia. *Journal of Applied Microbiology*, **93**, 1003–1011.

Dhewa, T., Pant, S. & Mishra, V. (2014) Development of freeze dried synbiotic formulation using a probiotic strain of *Lactobacillus plantarum*. *Journal of Food Science and Technology*, **51**(1), 83–89.

de Dias, M.L.L.A., Salgado, S.M., Guerra, N.B., Livera, A.V.S., Andrade, S.A.C. & da Ximenes, G.N.C. (2013) Physicochemical, sensory, and microbiological evaluation and development of symbiotic fermented drink. *Ciencia e Tecnologia de Alimentos*, **33**, 805–811.

Dias, S., Oliveira, M., Semedo-Lemsaddek, T. & Fernando Bernardo, F. (2014) Probiotic potential of autochthone microbiota from *Sao Jorge* and *Parmigiano-Reggiano* cheeses. *Food and Nutrition Sciences*, **5**, 1793–1799.

Dimitrellou, D., Kandylis, P., Sidira, M., Koutinas, A.A. & Kourkoutas, Y. (2014) Free and immobilized *Lactobacillus casei* ATCC 393 on whey protein as starter cultures for probiotic Feta-type cheese production. *Journal of Dairy Science*, **97**, 4675–4685.

Dinakar, P. & Mistry, V.V. (1994) Growth and viability of *Bifidobacterium bifidum* in Cheddar cheese. *Journal of Dairy Science*, **77**, 2854–2864.

Diosma, G., Romanin, D.E, Rey-Burusco M. F., Londero, A. & Garrote, G.L. (2003) Yeasts from kefir grains: Isolation, identification, and probiotic characterization. *World Journal of Microbiology & Biotechnology*, **30**, 43–53.

Dos Santos, K.M., Vieira, A.D., Salles, H.O., Oliveira, J.S., Rocha, C.R., Borges Mde, F., Bruno, L.M., Franco, B.D. & Todorov, S.D. (2015) Safety, beneficial and technological properties of *Enterococcus faecium* isolated from Brazilian cheeses. *Brazilian Journal of Microbiology*, **46**, 237–249.

Drakoularakou, A.P., Kehagias, C., Karakanas, P.N., Koulouris, S. & Timbis, D. (2003) A study of the growth of *Lactobacillus acidophilus* in bovine, ovine and caprine milk. *International Journal of Dairy Technology*, **56**, 59–61.

Ebel, B., Martin, F., Le, L.D.T., Gervais, P. & Cachon, R. (2011) Use of gases to improve survival of *Bifidobacterium bifidum* by modifying redox potential in fermented milk. *Journal of Dairy Science*, **94**, 2185–2191.

Edwards, C.A., Parrett, A.M., Stephen, A.M., Henry, C. J. Marks, J. & Shortt, C. (2002) Intestinal flora during the first months of life: new perspectives. *British Journal of Nutrition*, **88**(Suppl. 1), S11–S18.

Ehsani, A. & Mahmoudi, R. (2013) Effects of *Mentha longifolia* L. essential oil and *Lactobacillus casei* on the organoleptic properties and on the growth of *Staphylococcus aureus* and *Listeria monocytogenes* during manufacturing, ripening and storage of Iranian white-brined cheese. *International Journal of Dairy Technology*, **66**, 70–76.

Ekinci, F.Y., Okur, O.D., Ertekin, B. & Guzel-Seydim, Z. (2008) Effects of probiotic bacteria and oils on fatty acid profiles of cultured cream. *European Journal of Lipid Science and Technology*, **110**, 216–224.

El-Kholy, W.I., El-Shafei, K. & Sadek, Z.I. (2003) The use of immobilized *Lactobacillus* metabolites as a preservative in soft cheese. *Arab Universities Journal of Agricultural Sciences*, **11**, 607–622.
El-Kholy, A.M., El-Shinawy, S.H., Meshref, A.M.S. & Korany, A.M. (2014) Microbiological quality of domiati cheese and the influence of probiotics on the behavior of *Staphylococcus aureus* and *Escherichia coli* 0157:H7 in Domiati cheese. *Journal of Food Safety*, **34**, 396–406.
El-Rahman, A.E.R.M.A. (2000) Survival of probiotic bacteria in yoghurt and their antimicrobial effect. *Alexandria Journal of Agricultural Research*, **45**, 63–80.
El-Samragy, Y.A. (1997) Labneh or yoghurt-cheese: a review. *Egyptian Journal of Dairy Science*, **25**, 165–178.
El-Shafei, K. (2003) Production of cultured buttermilk using mixed cultures containing polysaccharide producing *Leuconostoc mesenteroides*. *Egyptian Journal of Dairy Science*, **31**, 43–60.
Elshaghabee, F.M.M. (2016) Viability of *Bifidobacterium longum* encapsulated in different capsule materials under simulated gastrointestinal tract conditions. *Egyptian Journal of Dairy Science*, **44**, 37–43.
El-Shazly, A., El Tahra, M.A.A. & Abo-Sera, M.M. (2004) Effect of different methods for the manufacture of frozen yogurt on its properties. *9th Egyptian Conference for Dairy Science and Technology – Cairo 9–11 October 2004, Research Papers I*, 183–194.
El-Shibiny, S., Metwally, M.M., Abd-El-Gani, S., Abd-El-Fattah, A.M. & Okda, A.Y.M. (2005) Manufacture of some probiotic dairy products from ultrafiltered milk retentate. *Egyptian Journal of Dairy Science*, **33**, 215–227.
El Tahra, M.A.A., El Shazley, A. & Abo-Sera, M.M. (2004a) Effect of fat content on the properties of fresh and stored frozen yoghurt. *9th Egyptian Conference for Dairy Science and Technology – Cairo 9–11 October 2004, Research Papers I*, 195–207.
El Tahra, M.A.A., El Shazley, A. & Abo-Sera, M.M. (2004b) Effect of fat content and probiotic cultures on the properties of frozen yoghurt. *9th Egyptian Conference for Dairy Science and Technology – Cairo 9–11 October 2004, Research Papers I*, 209–225.
El-Zahar, K.M., Alam, R.T.M. & Shalaby, H.S. (2015) Effect of feeding probiotic Ras cheeses on some metabolic and histopathologic parameters of hypercholestrolemic rats. *Egyptian Journal of Dairy Science*, **43**, 77–89.
El-Zainy, A.R.M., El-Zamzamy F.M. & Mostafa, M.Y.A. (2012) Manufacture and evaluation of four novel wheat fermented milks beverages. *International Journal of Dairy Science*, **7**(4), 84–94.
Emami, H., Rabbani, M. & Majid Bouzari, M. (2014) Probiotic characteristics of lactobacilli isolated from various native yoghurts made by local and traditional dairy producers of Isfahan, Iran. *Journal of Pure and Applied Microbiology*, **8**, 797–806.
Ephraim, E., Schultz, R.D. & Safdar, N. (2013) *Lactobacillus rhamnosus* GG protects cells from *Clostridium difficile* toxins. *British Microbiology Research Journal*, **3**, 165–175.
Ersan, S.E., Gultekin-Ozguven, M., Berktas, I., Erdem, O., Tuna, H.E., Gunes, G. & Ozcelik, B. (2016) Use of *Bacillus indicus* BU36 as a probiotic culture in set-type, recombined nonfat yoghurt production and its effect on quality. *International Journal of Dairy Technology*, **69**, 81–88.
Ershidat, O.T.M. & Mazahreh, A.S. (2009) Probiotics bacteria in fermented dairy products. *Pakistan Journal of Nutrition*, **8**, 1107–1113.
Escobar, M.C., van Tassell, M.L., Martínez-Bustos, F., Singh, M., Castano-Tostado, E., Amaya-Llano, S.L. & Miller, M.J. (2012) Characterization of a Panela cheese with added probiotics and fava bean starch. *Journal of Dairy Science*, **95**, 2779–2787.
Esmerino, E.A., Cruz, A.G., Pereira, E.P.R., Rodrigues, J.B., Faria, J.A.F. & Bolini, H.M.A. (2013) The influence of sweeteners in probiotic Petit Suisse cheese in concentrations equivalent to that of sucrose. *Journal of Dairy Science*, **96**, 5512–5521.
Esmerino, E.A., Paixao, J.A., Cruz, A.G., Garitta, L., Hough, G. & H.M.A. Bolini, H.M.A. (2015) Survival analysis: A consumer-friendly method to estimate the optimum sucrose level in probiotic petit Suisse. *Journal of Dairy Science*, **98**, 7544–7551.

do Espirito Santo, A.P., Cartolano, N.S., Silva, T.F., Soares, F.A.S.M., Gioielli, L.A., Perego, P., Converti, A. & Oliveira, M.N. (2012a) Fibers from fruit by-products enhance probiotic viability and fatty acid profile and increase CLA content in yoghurts. *International Journal of Food Microbiology*, **154**(3), 135–144.

do Espirito Santo, A.P., Perego, P., Converti, A. & Oliveira, M.N. (2012b) Influence of milk type and addition of passion fruit peel powder on fermentation kinetics, texture profile and bacterial viability in probiotic yoghurts. *LWT - Food Science and Technology*, **47**, 393–399.

FAO/WHO (2001) *Evaluation of Health and Nutritional Properties of Probiotics in Food Including Powder Milk with Live Lactic Acid Bacteria*, Report of a Joint FAO/WHO Expert Consultation, Food and Agriculture Organization of the United Nations, Rome.

FAO/WHO (2002) *Guidelines for the Evaluation of Probiotics in Food*, Report of a Joint FAO/WHO Working Group on Drafting Guidelines for the Evaluation of Probiotics in Food, London Ontario in Canada, Food and Agriculture Organization of the United Nations, Rome.

FAO/WHO (2006) *Probiotics in Food: Health and Nutritional Properties and Guidelines for Evaluation*, Food and Nutrition Paper No. 85, Food and Agriculture Organization of the United Nations, Rome.

Farhadi, Sh., Darani K.K., Mashayekh, M. Mortazavian, A.M., Mohammadi, A. & Shahraz, F. (2013) Study of the microbiological characteristics of fermented dairy probiotic beverage containing propionic acid after fermentation and during cold storage. *Iranian Journal of Food Science and Technology*, **9**(37), 147–155.

Farnworth, E.R. & Mainville, I. (2008) Kefir: A fermented milk product. *Handbook of Fermented Functional Foods*, (ed. E.R. Farnworth), 2nd Edition, pp. 89–127, CRC Press (Taylor & Francis Group), Boca Raton.

Farooq, U., Liu. X., Masud, T. & Zhang, H. (2013) Probiotic potential of *Lactobacillus acidophilus* strains isolated from dahi, a traditional fermented milk product. *Journal of Pure and Applied Microbiology*, **7**, 927–931.

Favaro-Trindade, C.S., Bernardi, S., Bodini, R.B., de Carvalho Balieiro, J.C. & de Almeida, E. (2006) Sensory acceptability and stability of probiotic microorganisms and vitamin C in fermented acerola (*Malpighia emarginata* DC) ice cream. *Journal of Food Science*, **71**, S492–S495.

Felicio, T.L., Esmerino, E.A., Vidal, V.A.S., Cappato, L.P., Garcia, R.K.A., Cavalcanti, R.N., Freitas, M.Q., Conte Jr., C.A., Padilha, M.C., Silva, M.C., Raices, R.S.L., Arellano, D.B., Bollini, H.M.A., Pollonio, M.A.R. & Cruz, A.G. (2016) Physico-chemical changes during storage and sensory acceptance of low sodium probiotic Minas cheese added with arginine. *Food Chemistry*, **196**, 628–637.

Ferrari, I.S., de Souza, J.V., Ramos, C.L., da Costa, M.M., Schwan, R.F. & Dias, F.S. (2016) Selection of autochthonous lactic acid bacteria from goat dairies and their addition to evaluate the inhibition of *Salmonella typhi* in artisanal cheese. *Food Microbiology*, **60**, 29–38.

Feucht, A. & Kwak, H.-S. (2013) Microencapsulation of lactic acid bacteria (LAB). *Korean Journal for Food Science of Animal Resources*, **33**, 229–238.

Flockerzie, M., Fischbein, O., Jakobi, S., Blattert, I., Samtlebe, M. & Hinrichs, J. (2014) Propearls: fruit pearls enclosed in milk protein-encapsulated probiotics. *DMW - Die Milchwirtschaft*, **5**(8), 298–301.

Florence, A.C.R., Beal, C., Silva, R.C., Bogsan, C.S.B., Pilleggi, A.L.O.S., Gioielli, L.A. & Oliveira, M.N. (2012) Fatty acid profile, trans-octadecenoic, α-linolenic and conjugated linoleic acid contents differing in certified organic and conventional probiotic fermented milks. *Food Chemistry*, **135**, 2207–2214.

Florence, A.C.R.D., Beal, C., Roberta Claro da Silva, R.C. & Oliveira, M.N. (2013) Survival of three *Bifidobacterium animalis* subsp. *lactis* strains is related to trans-vaccenic and α-linolenic acids contents in organic fermented milks. *LWT - Food Science and Technology*, **46**, 1–6.

Foligne, B., Parayre, S., Cheddani, R., Famelart, M.-H., Madec, M.-N., Ple, C., Breton, J., Dewulf, J., Jan, G. & Deutsch, S.-M. (2016) Immunomodulation properties of multi-species fermented milks. *Food Microbiology*, **53**, 60–69.

Fonden, R., Saarela, M., Mätto, J. & Mattila-Sandholm, T. (2003) Lactic acid bacteria (LAB) in functional dairy products. *Functional Dairy Products*, (eds. T. Mattila-Sandholm & M. Saarela), pp. 244–262, Woodhead Publishing Ltd., Cambridge.

Fornelli, A.R., Bandiera, N.S., de Costa, M.R., de Souza, C.H.B., de Santana, E.H.W., Sivieri, K. & Aragon-Alegro, L.C. (2014) Effect of inulin and oligofructose on the physicochemical, microbiological and sensory characteristics of symbiotic dairy beverages. *Semina: Ciencias Agrarias*, **35**, 3099–3112.

Franz, C.M.A.P., Huch, M., Abriouel, H., Holzapfel, W. & Galvez, A. (2011) Enterococci as probiotics and their implications in food safety. *International Journal of Food Microbiology*, **151**, 125–140.

Franz, C.M.A.P., Huch, M., Mathara, J.M., Abriouel, H., Benomar, N., Reid, G., Galvez, A. & Holzapfel, W.H. (2014) African fermented foods and probiotics. *International Journal of Food Microbiology*, **190**, 84–96.

de Freitas, R., Chuat, V., Madec, M.N., Nero, L.A., Thierry, A., Valence, F. & de Carvalho, A.F. (2015) Biodiversity of dairy *Propionibacterium* isolated from dairy farms in Minas Gerais, Brazil. *International Journal of Food Microbiology*, **203**, 70–77.

Gagnaire, V., Jardin, J., Rabah, H., Briard-Bion, V. & Jan, G. (2015) Emmental cheese environment enhances *Propionibacterium freudenreichii* stress tolerance. *PLoS ONE*, **10**, 1932–6203.

Gandhi, A., Cui, Y., Zhou, M. & Shah, N.P. (2014) Effect of KCl substitution on bacterial viability of *Escherichia coli* (ATCC 25922) and selected probiotics. *Journal of Dairy Science*, **97**, 5939–5951.

Ganesan, B., Weimer, B.C., Pinzon, J., Dao Kong, N., Rompato, G., Brothersen, C. & McMahon, D.J. (2014) Probiotic bacteria survive in Cheddar cheese and modify populations of other lactic acid bacteria. *Journal of Applied Microbiology*, **116**, 1642–1656.

Garcia, L.A., Cardador,A. del Campo, S.T.M., Arvizu, S.M. Tostado, E.C. Gonzalez, C.R. Almendarez, B.G. & Llano, S.L.A. (2013) Influence of probiotic strains added to cottage cheese on generation of potentially antioxidant peptides, anti-listerial activity, and survival of probiotic microorganisms in simulated gastrointestinal conditions. *International Dairy Journal*, **33**, 191–197.

Gardiner, G.E., Ross, R.P., Collins, J.K., Fitzgerald, G. & Stanton, C. (1998) Development of a probiotic Cheddar cheese containing human-derived *Lactobacillus paracasei* strains. *Applied and Environmental Microbiology*, **64**, 2192–2199.

Gardiner, G.E., Stanton, C., Lynch, P.B., Collins, J.K., Fitzgerald, G. & Ross, R.P. (1999a) Evaluation of Cheddar cheese as a food carrier for delivery of a probiotic strain to the gastrointestinal tract. *Journal of Dairy Science*, **82**, 1379–1387.

Gardiner, G.E., Ross, R.P., Wallace, J.M., Scanlan, F.P., Jagers, P.P.J.M., Fitzgerald, G.F., Collins, J.K. & Stanton, C. (1999b) Influence of a probiotic adjunct culture of *Enterococcus faecium* on the quality of Cheddar cheese. *Journal Agriculture and Food Chemistry*, **47**, 4907–4916.

Gardiner, G.E., O'Sullivan, E., Kelly, J., Auty, M.A.E., Fitzgerald, G.F., Collins, J.K., Ross, R.P. & Stanton, C. (2000) Comparative survival rates of human-derived probiotic *Lactobacillus paracasei* and *Lactobacillus salivarius* strains during heat treatment and spray drying. *Applied and Environmental Microbiology*, **66**, 2605–2616.

Gardiner, G.E., Ross, R.P., Kelly, P.M. & Stanton, C. (2002a) Microbiology of therapeutic milks. *Dairy Microbiology Handbook*, (ed. R.K. Robinson), 3rd edn. pp. 431–478, John Wiley & Sons Inc., New York.

Gardiner, G.E., Bouchier, P., O'Sullivan, E., Kelly, J., Collins, J.K., Fitzgerald, G., Ross, R.P. & Stanton, C. (2002b) A spray-dried culture for probiotic cheese manufacture. *International Dairy Journal*, **12**, 749–756.

Gardini, F., Lanciotti, R.M., Guerzoni, M.E. & Torriani, S. (1999) Evaluation of aroma production and survival of *Streptococcus thermophilus, Lactobacillus delbrueckii* subsp. *bulgaricus* and *Lactobacillus acidophilus* in fermented milks. *International Dairy Journal*, **9**, 125–143.

Garrotte, G.L., Abraham, A.G. & De Antoni G.L. (2001) Chemical and microbiological characterisation of kefir grains. *Journal of Dairy Research*, **68**, 639–652.

Gebara, C., Ribeiro, M.C.E., Chaves, K.S., Gandara, A.L.N. & Gigante, M.L. (2015) Effectiveness of different methodologies for the selective enumeration of *Lactobacillus acidophilus* La-5 from yoghurt and Prato cheese. *LWT - Food Science and Technology*, **64**, 508–513.

Georgieva, R, Iliev, I., Haertle, T., Chobert, J.-M., Ivanova, I. & Danova, S. (2009) Technological properties of candidate probiotic *Lactobacillus plantarum* strains. *International Dairy Journal*, **19**, 696–702.

Ghoddusi, H.B. & Robinson, R.K. (1996) The test of time. *Dairy Industries International*, **63**(7), 25–28.

Gilliland, S.E. (2001) Probiotics and prebiotics. *Applied Dairy Microbiology* (eds. E.H. Marth & J.L. Steele), 2nd Edition, pp. 327–344, Marcel Dekker, New York.

Gilliland, S.E., Morelli, L. & Reid, G. (2001) Health and nutritional properties of probiotics in food including powder milk with live lactic acid bacteria. *Evaluation of Health and Nutritional Properties of Probiotics in Food Including Powder Milk with Live Lactic Acid Bacteria* - Report of joint FAO/WHO of expert consultation, pp. 1-34, Food and Agriculture Organization of the United Nations, Rome. http://www.fao.org/es/ESN/probio/report.pdf

Gil-Rodriguez, A.M., Carrascosa, A.V. & Requena, T. (2015) Yeasts in foods and beverages: *In vitro* characterisation of probiotic traits. *LWT - Food Science and Technology*, **64**, 1156–1162.

Giraffa, G. (2012) Selection and design of lactic acid bacteria probiotic cultures. *Engineering in Life Sciences*, **12**(4), 391–398.

Guaraldi, F. & Salvatori, G. (2012) Effect of breast and formula feeding on gut microbiota shaping in newborns. *Frontiers in Cellular and Infection Microbiology*, **2**, 94. (published online doi: 10.3389/fcimb.2012.00094). Accessed 25th August 2016.

Gobbetti, M., Corsetti, A., Smacchi, E., Zocchetti, A. & de Angelis, M. (1998) Production of Crescenza cheese by incorporation of bifidobacteria. *Journal of Dairy Science*, **81**, 37–47.

Gobbetti, M., Cagno, R.D. & De Angelis, M. (2010) Functional microorganisms for functional food quality. *Critical Reviews in Food Science and Nutrition*, **50**, 716–727.

Godward, G. & Kailasapathy, K. (2000) Viability and survival of free, encapsulated and co-encapsulated probiotic bacteria in ice cream. *Milchwissenschaft*, **58**, 161–164.

Godward, G. & Kailasapathy, K. (2003a) Viability and survival of free and encapsulated probiotic bacteria in Cheddar cheese. *Milchwissenscaft*, **58**, 624–627.

Godward, G. & Kailasapathy, K. (2003b) Viability and survival of free, encapsulated and co-encapsulated probiotic bacteria in ice cream. *Milchwissenscaft*, **58**, 161–164.

Godward, G. & Kailasapathy, K. (2003c) Viability and survival of free, encapsulated and co-encapsulated probiotic bacteria in yoghurt. *Milchwissenscaft*, **58**, 396–399.

Godward, G., Sultana, K., Kailasapathy, K., Peiris, P., Arumugaswamy, R. & Reynolds, N. (2000) The importance of strain selection on the viability and survival or probiotic bacteria in dairy foods. *Milchwissenschaft*, **55**, 441–445.

Golowczyc, M.A., Silva, J., Teixeira, P., de Antoni, G.L., Abraham, A.G. (2011) Cellular injuries of spray-dried *Lactobacillus* spp. isolated from kefir and their impact on probiotic properties. *International Journal of Food Microbiology*, **144**, 556–560.

Gomes, A.M.P. & Malcata, F.X. (1998) Development of probiotic cheese manufactured from goat milk: response surface analysis via technological manipulation. *Journal of Dairy Science*, **81**, 1492–1507.

Gomes, A.M.P. & Malcata, F.X. (1999) *Bifidobacterium* spp. and *Lactobacillus acidophilus*: biological, biochemical, technological and therapeutical properties relevant for use as probiotics. *Trends in Food Science & Technology*, **10**, 139–157.

Gomes, A.M.P., Malcata, F.X. & Klaver, F.A.M. (1995) Growth enhancement of *Bifidobacterium lactis* Bo and *Lactobacillus acidophilus* Ki by milk hydrolysate. *Journal of Dairy Science*, **81**, 2811–2825.

Gomes, A.M.P., Malcata, F.X., Klaver, F.A.M. & Grande, H.J. (1998) Incorporation and survival of *Bifidobacterium* spp. strain Bo and *Lactobacillus acidophilus* strain Ki in a cheese product. *Netherlands Milk and Dairy Journal*, **49**, 71–95.

Gomes, A.A., Braga, S.P., Cruz, A.G., Cadena, R.S., Lollo, P.C.B., Carvalho, C., Amaya-Farfan, J., Faria, J.A.F. & Bolini, H.M.A. (2011) Effect of the inoculation level of *Lactobacillus acidophilus* in probiotic cheese on the physicochemical features and sensory performance compared with commercial cheeses. *Journal of Dairy Science*, **94**, 4777–4786.

Goni-Cambrodon, I. & Gudiel-Urbano, M. (2001) Oligosaccharide supplementation in infant formula: natural prebiotics. *Alimentaria*, **38**, 77–81.

Gonzalez-Sanchez, F., Azaola, A., Gutierrez-Lopez, G.F. & Hernandez-Sanchez, H. (2010) Viability of microencapsulated *Bifidobacterium animalis* subsp. *lactis* BB-12 in kefir during refrigerated storage. *International Journal of Dairy Technology*, **63**,431–436.

Granato, D., Branco, G.F., Gomes-Cruz, A., de Fonseca Faria, J.A. Shah, N.P. (2010) Probiotic dairy products as functional foods. *Comprehensive Reviews in Food Science and Food Safety*, **9**, 455–470.

Gudmundsson, B. (1987) *Skyr. Scandinavian Dairy Industry*, **1**, 240–242.

Gudmundsson, G & Kristbergsson, K. (2016) Modernization of Skyr processing: Icelandic acid-curd soft cheese. *Modernization of Traditional Food Processes and Products* (eds. A. McElhatton and M.M. El Idrissi), pp. 45–53, Springer, New York.

Guler-Akin, M.B. & Akin, M.S. (2007) Effects of cysteine and different incubation temperatures on the microflora, chemical composition and sensory characteristics of bio-yogurt made from goat's milk. *Food Chemistry*, **100**, 788–793.

Gunenc, A., Khoury, C., Legault, C., Mirrashed, H., Rijke, J. & Hosseinian, F. (2016) Seabuckthorn as a novel prebiotic source improves probiotic viability in yogurt. *LWT - Food Science and Technology*, **66**, 490–495.

Guo, J.-F., Zhang, S., Yuan, Y.-H., Yue, T.-L. & Li, J.-Y. (2015) Comparison of lactobacilli isolated from Chinese suan-tsai and koumiss for their probiotic and functional properties. *Journal of Functional Foods*, **12**, 294–302.

Gursoy, O., Gokce, R., Con, A.H. & Kinik, O. (2014) Survival of *Bifidobacterium longum* and its effect on physicochemical properties and sensorial attributes of white brined cheese. *International Journal of Food Sciences and Nutrition*, **65**, 816–820.

Hagen, M. & Narvhus, J.A. (1999) Production of ice cream containing probiotic bacteria. *Milchwissenschaft*, **54**, 265–268.

Halpin-Dohnalek, M.I., Hilty, M.D. & Bynum, D.G. (1999) Method and formula for the prevention of diarrhea. *United States of America Patent Application* **5 902 578**.

Hamed, A.I., Zedan, M.A., Salem, O.M., Moussa, A.M. & Yousef, E.T.A. (2004) Impact of frozen yoghurt ingredients on its quality and survival of bifidobacteria: III Effect of milk solids not fat sources. *9th Egyptian Conference for Dairy Science and Technology – Cairo 9–11 October 2004, Research Papers I*, 227–241.

Hansen, S.L. (1997) Protective cultures for cheese. *Meieriposten*, **86**, 261–262.

Hansen, L.T., Allan-Wojtas, P.M., Jin, Y.-L. & Paulson, A.T. (2002) Survival of Ca-alginate microencapsulated *Bifidobacterium* spp. in milk and simulated gastrointestinal conditions. *Journal of Food Microbiology*, **19**, 35–45.

Harzallah, D. & Belhadj, H. (2013) Lactic acid bacteria as probiotics: Characteristics, selection criteria and role in immunomodulation of human GI muccosal barrier. *Lactic Acid Bacteria - R & D for Food, Health and Livestock Purposes* (ed. J. Marcelino Kongo (Ed.), InTech, DOI: 10.5772/50732. Available from: http://www.intechopen.com/books/lactic-acid-bacteria-r-d-for-food-health-and-livestock-purposes/lactic-acid-bacteria-as-probiotics-characteristics-selection-criteria-and-role-in-immunomodulation-o. Accessed on 5th August 2016.

Hashemi, M., Gheisar, H.R. & Shekarforoush, S. (2015) Preparation and evaluation of low-calorie functional ice cream containing inulin, lactulaose and *Bifidobacterium lactis*. *International Journal of Dairy Technology*, **68**, 183–189.

Hassan, O., Abughazaleh, A.A., Ibrahim, S.A., Isikhuemhen, O.S., Awaisheh, S.S. & Tahergrabi, R. (2014) Viability of α- and β-galactosidase activity of *Bifidobacterium breve* and *Lactobacillus reuteri* in yoghurt products supplemented with shiitake mushroom extract during refrigerated storage. *International Journal of Dairy Technology*, **67**, 570–576.

Hati, S., Mandal, S. & Prajapati, J.B. (2013) Novel starters for value added fermented dairy products. *Current Research in Nutrition and Food Science*, **1**, 83–91.

Hawrelak, J. (2002) Probiotics: Are supplements really better than yoghurt? *Journal of the Australian Traditional-Medicine*, **8**, 11–23.

Hayes, M., Coakley, M., O'Sullivan, L., Stanton, C., Hill, C., Fitzgerald, G.F., Murphy, J.J. & Ross, R.P. (2006) Cheese as a delivery vehicle for probiotics and biogenic substances. *The Australian Journal of Dairy Technology*, **61**, 132–141.

Haynes, I.N. & Playne, M.J. (2002) Survival of probiotic cultures in low-fat ice-cream. *Australian Journal of Dairy Technology*, **57**, 10–14.

Heidarpour, M., Mokhtari, F. & Heidarzadeh, M. (2013) Safety and properties of bifidobacteria isolated from traditional dairy products from Iran. *Journal of Pure and Applied Microbiology*, **7**, 2205–2211.

Heimbach, J.T. (2012) *Generally Recognized as Safe (GRAS) Determination for the Use of Lactobacillus casei Strain Shirota as a Food Ingredient*, GRAS Notice No. 000429. Yakult Honsha Co., Japan.

Hekmat, S. & McMahon, D.J. (1992) Survival of *Lactobacillus acidophilus* and *Bifidobacterium bifidum* in ice cream for use as a probiotic food. *Journal of Dairy Science*, **75**, 1415–1422.

Hekmat, S. & Reid, G. (2007) Survival of *Lactobacillus reuteri* RC-14 and *Lactobacillus rhamnosus* GR-1 in milk. *International Journal of Food Science and Technology*, **42**, 615–619.

Hekmat, S., Soltani, H. & Reid, G. (2009) Growth and survival of *Lactobacillus reuteri* RC-14 and *Lactobacillus rhamnosus* GR-1 in yogurt for use as a functional food. *Innovative Food Science and Emerging Technologies*, **10**, 293–296.

Helland, M.H., Wicklund, T. & Narvhus, J.A. (2004) Growth and metabolism of selected strains of probiotic bacteria in milk- and water-based cereal puddings. *International Dairy Journal*, **14**, 957–965.

Heller, K.J. (2001) Probiotic bacteria in fermented foods: product characteristics and starter organisms. *American Journal of Clinical Nutrition*, **73**(S supplement), 374S-379S.

Heller, K.J., Bockelmann, W., Schrezenmeir, J. & de Vrese, M. (2003) Cheese and its Potential as a Probiotic Food. *Handbook of Fermented Functional Foods*, (ed. E.R. Farnworth), pp. 203–225, CRC Press, Boca Roton.

Hernandez-Mendoza, A., Robles, V.J., Angulo, J.O., de la Cruz, J. & Garcia, H.S. (2008) Preparation of a whey-based probiotic product with *Lactobacillus reuteri* and *Bifidobacterium bifidum*. *Food Technology and Biotechnology*, **45** 27–31.

Hill, C., Guarner, F., Reid, G., Gibson, G.R., Merenstein, D.J., Pot, B., Morelli, L., Canani, R.B., Flint, H.J., Salminen, S., Calder, P.C. & Sanders, M.E. (2014) The International Scientific Association for Probiotics and Prebiotics consensus statement on the scope and appropriate use of the term probiotic. *Gastroenterology & Hepatology*, **11**, 506–514.

Holm, F. (2003) Gut health and diet: the benefits of probiotic and prebiotics on human health. *The World of Ingredients*, **2**, 52–55.

Homayouni, A., Payahoo, L. & Azizi, A. (2012a) Effects of probiotics on lipid profile: A review. *American Journal of Food Technology*, **7**, 251–265.

Homayouni, A., Alizadeh, M., Alikhah, H. & Zijah, V. (2012b) Functional dairy probiotic food development: Trends, concepts, and products. *Probiotics* (ed. E.C. Rigobelo), pp. 197–212, InTechOpen.

Hsiao, H.-C., Lian, W.-C. & Chou, C.-C. (2004) Effect of packaging conditions and temperature on viability of microencapsulated bifidobacteria during storage. *Journal of the Science of Food and Agriculture*, **84**, 134–139.

Hughes, D.B. & Hoover, D.G. (1991) Bifidobacteria: Their potential for use in American dairy products. *Food Technology*, **45**(4), 74–75, 78-80, 83.
HungChi, H., WenChian, L. & ChengChun, C. (2004) Effect of packaging conditions and temperature on viability of microencapsulated bifidobacteria during storage, *Journal of the Science of Food and Agriculture*, **84**, 134–139.
Huq, T., Khan, A., Khan, R.A., Riedl, B. & Lacroix, M. (2013) Encapsulation of probiotic bacteria in biopolymeric system. *Critical Reviews in Food Science and Nutrition*, **53**, 909–916.
Hussain, S.A., Patil, G.R., Yadav, V., Singh, R.R.B. & A.K. (2016) Ingredient formulation effects on physico-chemical, sensory, textural properties and probiotic count of *Aloe vera* probiotic *dahi*. *LWT - Food Science and Technology*, **65**, 371–380.
Hussein, G.A.M. & Abo-El-Fetoh, S.M. (2010) Production of functional dairy and non-dairy vanilla and chocolate whipped cream enriched with viable probiotic or traditional starter cultures. *Egyptian Journal of Dairy Science*, **38**, 253–265.
Hussein, G.A.M., Desouky, M.M. & Soryal, K.A. (2013) Preparation and properties of fermented goat's and camel's milk beverages. *Egyptian Journal of Dairy Science*, **41**, 45–57.
Hutt, P., Songisepp, E., Ratsep, M., Mahlapuu, R., Kilk, K. & Mikelsaar, M. (2015) Impact of probiotic Lactobacillus plantarum TENSIA in different dairy products on anthropometric and blood biochemical indices of healthy adults. *Beneficial Microbes*, **6**, 233–243.
Ibrahim, F., Ruvio, S., Granlund, L., Salminen, S., Viitanen, M. & Ouwehand, A.C. (2010) Probiotics and immunosenescence: cheese as a carrier. *FEMS Immunology & Medical Microbiology*, **59**, 53–59.
IDF (2015) *World Dairy Situation 2015*, Document No. 481, International Dairy Federation, Brussels.
Illupapalayam, V.V., Smith, S.C. & Gamlath, S. (2014) Consumer acceptability and antioxidant potential of probiotic-yogurt with spices. *LWT - Food Science and Technology*, **55**, 255–262.
Iranmanesh, M., Ezzatpanah, H. & Mojgani, N. (2014) Antibacterial activity and cholesterol assimilation of lactic acid bacteria isolated from traditional Iranian dairy products. *LWT - Food Science and Technology*, **58**, 355–359.
Islam, M.A. Yun, C.-H., Choi, Y.-J. & Cho, C.-S. (2010) Microencapsulation of live probiotic bacteria. *Journal of Microbiology and Biotechnology*, **20**, 1367–1377.
Iwana, H., Masuda, H., Fujisawa, T., Suzuki, H. & Mitsuoka, T. (1993) Isolation and identification of *Bifidobacterium* spp. in commercial yoghurts sold in Europe. *Bifidobacteria Microflora*, **12**, 39–45.
Jan, G., Leverrier, P., Proudy, I. & Roland, N. (2002) Survival and beneficial effects of propionibacteria in the human gut: *in vivo* and in vitro investigations. *Lait*, **82**, 131–144.
Jardim, F.B.B., Santos, E.N.F., Rossi, D.A., de Melo, R.T., Miguel, D.P., Rossi, E.A. & de Sylos, C.M. (2012) Development of potential probiotic carbonated dairy beverages: Physico-chemical, microbiological and sensory characteristics. *Alimentos e Nutricao*, **23**, 275–286.
Jia, R., H., Chen, H. & Ding, W. (2016) Effects of fermentation with *Lactobacillus rhamnosus* GG on product quality and fatty acids of goat milk yogurt. *Journal of Dairy Science*, **99**, 221–227.
Jiang, T., Mustapha, A. & Saviano, D.A. (1996) Improvement of lactose digestion in humans by ingestion of unfermented milk containing *Bifidobacterium longum*. *Journal of Dairy Science*, **79**, 750–757.
John, S.M. & Deeseenthum, S. (2015) Properties and benefits of kefir - A review. *Songklanakarin Journal of Science and Technology*, **37**, 275–282.
Joshi, V.K. (2015) Indigenous (ed. V.K. Joshi), pp. 1–67, CRC Press (Taylor & Francis Group) Boca Raton.
Jung, J.-K., Kil, J.-H., Kim, S.-K., Jeon, J.-T. & Park, K.-Y. (2007) Survival of double-microencapsulated *Bifidobacterium breve* in milk in simulated gastric and small intestinal conditions. *Journal of Food Science and Nutrition*, **12**, 58–63.

Kailasapathy, K. (2002) Microencapsulation of probiotic bacteria: technology and potential applications. *Current Issues in Intestinal Microbiology*, **3**(2), 39–48.
Kailasapathy, K. & Rybka, S. (1997) *L. acidophilus* and *Bifidobacterium* spp. – Their therapeutic potential and survival in yogurt. *Australian Journal of Dairy Technology*, **52**, 28–35.
Kailasapathy, K. & Sultana, K. (2003) Survival of β-D-galactosidase activity of encapsulated and free *Lactobacillus acidophilus* and *Bifidobacterium lactis* in ice-cream. *Australian Journal of Dairy Technology*, **58**, 223–227.
Kakisu, E., Irigoyen, A., Torre, P., de Antoni, G.L. & Abraham, A.G. (2010) Physicochemical, microbiological and sensory profiles of fermented milk containing probiotic strains isolated from kefir. *Journal of Dairy Research*, **78**, 456–463.
Kandylis, P., Pissaridi, K., Bekatorou, A., Kanellaki, M. & Koutinas, A.A. (2016) Dairy and non-dairy probiotic beverages. *Current Opinion in Food Science*, **7**, 58–63.
Kaneko, T. (1999) A novel bifidogenic growth stimulator produced by *Propionibacterium freudenreichii*. *Bioscience and Microflora*, **18**, 73–80.
Karimi, R., Mortazavian, A.M. & da Cruz, A.G. (2011) Viability of probiotic microorganisms in cheese during production and storage: A review. *Dairy Science & Technology*, **91**, 283–308.
Karimi, R., Sohrabvandi, S. & Mortazavian, A.M. (2012) Review article: Sensory characteristics of probiotic cheese. *Comprehensive Reviews in Food Science and Food Safety*, **11**, 437–452.
Kasimoglu, A., Goncuoglu, M. & Akgun, S. (2004) Probiotic white cheese with *Lactobacillus acidophilus*. *International dairy Journal*, **14**, 1067–1073.
Katla, A.K., Kruse, H. Johnsen, G. & Herikstad, H. (2001) Antimicrobial susceptibility of starter culture bacteria used in Norwegian dairy products. *International Journal of Food Microbiology*, **67**, 147–152.
Kaur, I.P., Chopra, K. & Saini, A. (2002) Probiotics: potential pharmaceutical applications, *European Journal of Pharmaceutical Sciences*, **15**(1), 1–9.
Kavaz, A. & Bakirci, I. (2014) Influence of inulin and demineralised whey powder addition on the organic acid profiles of probiotic yoghurts. *International Journal of Dairy Technology*, **67**, 577–583.
Kearney, N., Stanton, C., Desmond, C., Coakley, M., Collins, J.K., Fitzgerald, G. & Ross, R.P. (2008) Challenges associated with the development of probiotic-containing functional foods. *Handbook of Fermented Functional Foods* (ed. E.R. Farnworth), 2nd Edition, pp. 25–70, CRC Press (Taylor & Francis Group), Boca Raton.
Kebary, K.M.K., Hussein, S.A. & Badawi, R.M. (1998) Improving the viability of bifidobacteria and their effect on frozen ice milk. *Egyptian Journal of Dairy Science*, **26**, 319–337.
Kebary, K.M.K, Hamed, A.I., Salem, O.M. & Yousef, E.T.A. (2004) Impact of frozen yoghurt ingredients on its quality and survival of bifidobacteria:- I Effect of sweeteners. *9th Egyptian Conference for Dairy Science and Technology – Cairo 9–11 October 2004, Research Papers I*, 163–165.
Khan, S.U. (2014) Probiotics in dairy foods: A review. *Nutrition and Food Science*, **44**, 71–88.
Khan, I. & Kang, S.C. (2016) Probiotic potential of nutritionally improved *Lactobacillus plantarum* DGK-17 isolated from Kimchi - a traditional Korean fermented food. *Food Control*, **60**, 88–94.
Kheadr, E.E., Dabour, N., Petit, G. & Vuillemard, J.C. (2011) Probiotic-delivering capacity of dairy products: *in vitro* assessment using a gastro-intestinal dynamic model. *International Journal of Probiotics and Prebiotics*, **6**(2) 73–80.
Khosrokhavar, R. & Mortazavian, A.M. (2010) Effects probiotic-containing microcapsules on viscosity, phase separation and sensory attributes of drink based on fermented milk. *Milchwissenschaft*, **65**, 177–179.
Kilic, G.B., Kuleasan, H., Eralp, I. & Karahan, A.G. (2009) Manufacture of Turkish Beyaz cheese added with probiotic strains. *LWT - Food Science and Technology*, **42**, 1003–1008.

Kimoto-Nira, H., Yamasaki, S., Sasaki, K., Moriya, N., Takenaka, A. & Suzuki, C. (2015) New lactic acid bacterial strains from traditional Mongolian fermented milk products have altered adhesion to porcine gastric mucin depending on the carbon source. *Animal Science Journal*, **86**, 325–332.

Kitamura, Y., Itoh, H., Echizen, H. & Satake, T. (2009) Experimental vacuum spray drying of probiotic foods included with lactic acid bacteria. *Journal of Food Processing and Preservation*, **33**, 714–726.

Klaver, F.A.M., Kingma, F. & Weerkamp, A.H. (1993) Growth and survival of bifidobacteria in milk. *Netherlands Milk and Dairy Journal*, **47**, 151–164.

Kok-Tas, T. & Guzel-Seydim, Z. (2010) Determination of effects of using fat replacers and probiotic on Ayran quality. *Gida*, **35**, 105–111.

Kolakowski, P. & Pawlikowski, K. (2012) Kefir as a probiotic vehicle. *Milchwissenschaft*, **67**, 159–162.

Kongo, J.M. & Malcata, F.X. (2016) Acidophilus milk. *Encyclopedia of Food and Health* (eds. B. Caballero, P.M. Finglas and F. Toldra), Vol. A-Che, pp. 6–14, Elsevier, London.

Kosin, B. & Rakshit, S.K. (2006) Microbial and processing criteria for production of probiotics: A review. *Food Technology & Biotechnology*, **44**, 371–379.

Krasaekoopt, W., Bhandari, B. & Deeth, H. (2003) Evaluation of encapsulation techniques of probiotics for yoghurt. *International Dairy Journal*, **13**, 3–13.

Krasaekoopt, W., Bhandari, B. & Deeth, H. (2004) The influence of coating materials on some properties of alginate beads and survival of microencapsulated probiotic bacteria. *International Dairy Journal*, **14**, 737–743.

Krishnakumar, V. & Gordon, I.R. (2001) Probiotics: challenges and opportunities. *Dairy Industries International*, **66**(2), 38–40.

Kristo, E., Biliaderis, C.G & Tzanetakis, N. (2003) Modelling of rheological, microbiological and acidification properties of a fermented milk product containing a probiotic strain of *Lactobacillus paracasei*. *International Dairy Journal*, **13**, 517–528.

Kuda, T., Sarengaole, H., Takahashi, H. & Kimura, B. (2016) Alcohol-brewing properties of acid- and bile-tolerant yeasts co-cultured with lactic acid bacteria isolated from traditional handmade domestic dairy products from inner Mongolia. *LWT - Food Science and Technology*, **65**, 62–69.

Kumar, B.V., Vijayendra, S.V.N. & Reddy, O.V.S. (2015) Trends in dairy and non-dairy probiotic products - A review. *Journal of Food Science and Technology*, **52**, 6112–6124.

Kunova, G., Bohacenko, I., Pechacova, M., Pinkrova, J. & Peroutkova, J. (2013) Development of synbiotic milk-based fermented beverages and yogurts. *Mlekarske Listy*, **24**(137 - Suppl. Veda Vyzkum), V-X.

Kunz, C. & Rudolff, S. (2002) Health benefits of milk-derived carbohydrates. *Fresh Perspectives on Bioactive Dairy Products*, Document No. 375, pp. 72-79, International Dairy Federation, Brussels.

Kurmann, J.A., Rašić, J.Lj. & Kroger, M. (1992) Encyclopedia of Fermented Milk Products. Van Nostrand Reinhold, New York.

Kushugulova, A., Kozhakhmetov, S., Supiyev, A., Shakhabayeva, G., Sadukhasova, S., Sabitkyzy, S., Gulayev, A., Nurgozhin, T., Zhumadilov, Zh. & Sharman, A. (2013) Isolation and characterization of lactobacilli from traditional Kazakh dairy products. *International Journal of Probiotics and Prebiotics*, **8**(2-3), 95–100.

Lahtinen, S., Ouwenhand, A.C., Salminen, S. & von Wright, A. (eds.) (2012) *Lactic Acid Bacteria, Microbiological and Functional Aspects*, 4th Edition, CRC Press (Taylor & Francis Group), Boca Raton.

Lankaputhra, W.E.V. & Shah, N.P. (1995) Survival of *Lactobacillus acidophilus* and *Bifidobacterium* spp. in the presence of acid and bile salts. *Cultured Dairy Products Journal*, **30**(3), 2–7.

Lankaputhra, W.E.V. & Shah, N.P. (1996) A simple method for selective enumeration of *Lactobacillus acidophilus* in yogurt supplemented with *L. acidophilus* and *Bifidobacterium* spp. *Milchwissenschaft*, **51**, 446–451.

Lankaputhra, W.E.V. & Shah, N.P. (1997) Improving viability of *L. acidophilus* and bifidobacteria in yoghurt using two step fermentation and neutralized mix. *Food Australia*, **49**, 363–369.

La Torre, L., Tamime, A.Y. & Muir, D.D. (2003) Rheology and sensory profiling of set-type fermented milks made with different commercial probiotic and yoghurt starter cultures. *International Journal of Dairy Technology*, **56**, 163–170.

Lebos Pavunc, A., Turk, J., Beganovic, J., Frece, J. Mahnet, S., Kirin, S. & Suskovic, J. (2009) Production of fermented probiotic beverages from milk permeate enriched with whey retentate and identification of present lactic acid bacteria. *Mljekarstvo*, **59**(1), 11–19.

Lee, Y.L. & Salminen, S. (1996) The coming of age of probiotics. *Trends in Food Science and Technology*, **6**, 24–21.

Lee, N.-K., Han, K.-J., Son, S.-H., Eom, S.-J., Lee, S.-K. & Paik, H.-D. (2015) Multifunctional effect of probiotic *Lactococcus lactis* KC24 isolated from kimchi. *LWT - Food Science and Technology*, **64**, 1036–1041.

Leite, A.M.O., Miguel, M.A.L., Peixoto, R.S., Ruas-Madiedo, P., Paschoalin, V.M.F., Mayo, B. & Delgado, S. (2015) Probiotic potential of selected lactic acid bacteria strains isolated from Brazilian kefir grains. *Journal of Dairy Science*, **98**, 3622–3632.

Le Marechal, C., Peton, V., Ple, C., Vroland, C., Jardin, J., Briard-Bion, V., Durant, G., Chuat, V., Loux, V., Foligne, B., Deutsch, S.M., Falentin, H. & Jan, G. (2015) Surface proteins of *Propionibacterium freudenreichii* are involved in its anti-inflammatory properties. *Journal of Proteomics*, **113**, 447–461.

di Lena, M., Quero, G.M., Santovito, E. Verran, J., de Angelis, M. & Fusco, V. (2015) A selective medium for isolation and accurate enumeration of *Lactobacillus casei*-group members in probiotic milks and dairy products. *International Dairy Journal*, **47**, 27–36.

Leporanta, K. (2001) The world of fermented milks - 1: Drinking type products originating from Finland and Russia. *LGG Action*, **2**, 10–12.

Leporanta, K. (2003) The world of fermented milks - 4: Viili and Långfil – exotic fermented products from Scandinavia. *Valio Foods & Functionals*, **2**, 3–5.

Leroy, F. & de Vuyst, L. (2004) Lactic acid bacteria as functional starter cultures for the food fermentation industry. *Trends in Food Science & Technology*, **15**, 67–78.

Li, J., Wang, X., Gao, P.-F., Yao, G.-Q., Zhao, J., Wang, X.-W., Zhao, X.-Y., Menhebilige, W. & Zhang, H.-P. (2013) Probiotics in the yogurt productive process to control the pollution of yeast and mold. *China Dairy Industry*, **41**(9), 21–24.

Li, S., Ma, C., Gong, G., Liu, Z., Chang, C. & Xu, Z. (2016) The impact of onion juice on milk fermentation by *Lactobacillus acidophilus*. *LWT - Food Science and Technology*, **65**, 545–548.

Lian, W.-C., Hsiao, H.-C. & Chou, C.-C. (2002) Survival of bifidobacteria after spray-drying. *International Journal of Food Microbiology*, **74**, 79–86.

Linares, D.M., Ross, P. & Stanton, C. (2016) Beneficial microbes: The pharmacy in the gut. *Bioengineered*, **7**, 11–20.

Liu, W.J., Chen, Y.F., Kwok, L.Y., Li, M.H., Sun, T., Sun, C.L., Wang, X.N., Dan, T., Menghebilige, T., Zhang, H.P. & Sun, T.S. (2013) Preliminary selection for potential probiotic *Bifidobacterium* isolated from subjects of different Chinese ethnic groups and evaluation of their fermentation and storage characteristics in bovine milk. *Journal of Dairy Science*, **96**, 6807–6817.

Liu, L., Li, X., Bi, W., Zhang, L., Ma, L., Ren, H. & Li, M. (2015a) Isomaltooligosaccharide increases the *Lactobacillus rhamnosus* viable count in Cheddar cheese. *International Journal of Dairy Technology*, **68**, 389–398.

Liu, Y., Yu, Y., Duan, W., Qu, Q., Zhang, Q., Zhao, M. & Zhu, Q. (2015b) Home storage significantly impairs bifidobacteria survival in powered formula for infants and young children in the Chinese market. *International Journal of Dairy Technology*, **68**, 495–502.

Ljubic, A., Juric, A. & Jakopovic, K.L. (2015) Effect of high intensity ultrasound on the milk fermentation by bifidobacteria. *Mljekarstvo*, **65**, 71–80.
Lollo, P.C.B., Cruz, A.G., Morato, P.N., Moura, C.S., Carvalho-Silva, L.B., Oliveira, C.A.F., Faria, J.A.F. & Amaya-Farfan, J. (2012) Probiotic cheese attenuates exercise-induced immune suppression in Wistar rats. *Journal of Dairy Science*, **95**, 3549–3558.
Lonnerdal, B. (2003) Breast milk and breastfed infants: implications for improving infant formula. *American Journal of Clinical Nutrition*, **77**, 1537S–1543S.
Losio, M.N., Bozzo, G., Galuppini, E., Martella, V., Bertasi, B., Pavoni, E. & Finazzi, G. (2015) Silter cheese, a traditional Italian dairy product: A source of feasible probiotic strains. *International Journal of Food Properties*, **26**, 218–230.
Lourens-Hattingh, A. & Viljoen, B.C. (2001a) Yogurt as probiotic carrier food. *International Dairy Journal*, **11**, 1–17.
Lourens-Hattingh, A. & Viljoen, B.C. (2001b) Growth and survival of a probiotic yeast in dairy products. *Food Research International*, **34**, 791–796.
Lourens-Hattingh, A. & Viljoen, B.C. (2002) Survival of probiotic bacteria in South African commercial bio-yogurt. *South African Journal of Science*, **98**, 298–300.
Lucas, A., Sodini, I., Monnet, C., Jolivet, P. & Corrieu, G. (2004) Probiotic cell counts and acidification in fermented milks supplemented with milk protein hydrolysates. *International Dairy Journal*, **14**, 47–53.
Ma, G. (1995) An ice cream containing *Lactobacillus bifidus*. *Journal of Food Industry*, **1**, 15–16.
Madureira, A.R., Soares, J.C., Pintado, M.E., Gomes, A.M.P., Freitas, A.C. & Malcata, F.X. (2008) Sweet whey cheese matrices inoculated with the probiotic strain *Lactobacillus paracasei* LAFTI® L26. *Dairy Science and Technology*, **88**, 649–665.
Madureira, A.R., Pintado, M.E., Gomes, A.M.P. & Malcata, F.X. (2011a) Incorporation of probiotic bacteria in whey cheese: Decreasing the risk of microbial contamination. *Journal of Food Protection*, **74**, 1194–1199.
Madureira, A.R., Pintado, A.I., Gomes, A.M., Pintado, M.E. & Malcata, F.X. (2011b) Rheological, textural and microstructural features of probiotic whey cheese. *LWT - Food Science and Technology*, **44**, 75–81.
Madureira, A.R., Soares, J.C., Amorim, M., Tavares, T., Gomes, A.M., Pintado, M.M. & Malcata, F.X. (2013) Bioactivity of probiotic whey cheese: characterization of the content of peptides and organic acids. *Journal of the Science of Food and Agriculture*, **93**, 1458–1465.
Magarinos, H., Cartes, P., Fraser, B., Selaive, S., Costa, M., Figuerola, F. & Pizarro, O. (2008) Viability of probiotic micro-organisms (*Lactobacillus casei* Shirota and *Bifidobacterium animalis* subsp. *lactis*) in a milk-based dessert with cranberry sauce. *International Journal of Dairy Technology*, **61**, 96–101.
Mahdi, H.A., Tamime, A.Y. & Davies, G. (1990) Some aspects of the production of "Labneh" by ultrafiltration using cow's, sheep's and goat's milk. *Egyptian Journal of Dairy Science*, **18**, 345–367.
Mahmoudi, M., Asl, A.K. & Zomorodi, S. (2012) The influence or probiotic bacteria on the properties of Iranian White cheese. *International Journal of Dairy Technology*, **65**, 561–567.
Mahmoudi, R., Tajik, H., Ehsani, A., Farshid, A.A., Zare, P. & Hadian, M. (2013) Effects of *Mentha longifolia* L. essential oil on viability and cellular ultrastructure of *Lactobacillus casei* during ripening of probiotic Feta cheese. *International Journal of Dairy Technology*, **66**, 77–82.
Maiocchi, G. (2001) YOMO ABC: Functional food for consumers well-being and satisfaction. *Industria del Latte*, **XXXVII**(1-2), 94–98.
Maity, T.K., Kumar, R. & Misra, A.K. (2008) Development of healthy whey drink with *Lactobacillus rhamnosus, Bifidobacterium bifidum and Propionibacterium freudenreichii subsp. shermanii. Mljekarstvo*, **58**, 315–325.

Makino, S., Sato, A., Goto, A., Nakamura, M., Ogawa, M., Chiba, Y., Hemmi, J., Kano, H. Takeda, K., Okumura, K. & Asami, Y. (2016) Enhanced natural killer cell activation by exopolysaccharides derived from yogurt fermented with *Lactobacillus delbrueckii* ssp. *bulgaricus* OLL1073R-1. *Journal of Dairy Science*, **99**, 915–923.

Malcata, F.X., Gomes, A.M. & Pintado, M.E. (2005) Functional dairy foods - An overview. *Egyptian Journal of Dairy Science*, **33**, 1–12.

Mani-Lopez, E., Palou, E. & Lopez-Malo, A. (2014) Probiotic viability and storage stability of yogurts and fermented milks prepared with several mixtures of lactic acid bacteria. *Journal of Dairy Science*, **97**, 2578–2590.

Mantere-Alhonen, S. (1995) Propionibacteria used as probiotics – a review. *Lait*, **75**, 447–452.

Mantere-Alhonen, S. & Forsen, R. (1990) Microbes in fermented milk products. *Microbes in Milk and Milk Products* (in Finnish), (ed. S. Mantere-Alhonen), pp. 129–148, VAPK-Kustannus, Helsinki.

Marafon, A.P., Sumi, A., Granato, D., Alcantara, M.R., Tamime, A.Y. & de Oliveira, M.N. (2011a) Effects of partially replacing skimmed milk powder with dairy ingredients on rheology, sensory profiling, and microstructure of probiotic stirred-type yogurt during cold storage. *Journal of Dairy Science*, **94**, 5330–5340.

Marafon, A.P., Sumi, A., Alcantara, M.R., Tamime, A.Y. & de Oliveira, M.N. (2011b) Optimization of the rheological properties of probiotic yoghurt supplemented with milk proteins. *LWT - Food Science and Technology*, **44**, 511–519.

Maragkoudakis, P.A., Miaris, C., Rojez, P., Manalis, N., Magkanari, F., Kalantzopoulos, G. & Tsakalidou, E. (2006a) Production of traditional Greek yoghurt using *Lactobacillus* strains with probiotic potential as starter adjuncts. *International Dairy Journal*, **16**, 52–60.

Maragkoudakis, P.A., Zoumpopoulou, G., Miaris, C., Kalantzopoulos, G., Pot, B. & Tsakalidou, E. (2006b) Probiotic potential of *Lactobacillus* strains isolated from dairy products. *International Dairy Journal*, **16**, 189–199.

Marhamatizadeh, M.H., Ehsandoost, E., Gholami, P., Moshiri, H. & Nazemi, M. (2012a) Effect of permeate on growth and survival of *Lactobacillus acidophilus* and *Bifidobacterium bifidum* for production of probiotic nutritive beverages. *World Applied Sciences Journal*, **18**, 1389–1393.

Marhamatizadeh, M.H., Nikbakht, M., Rezazadeh, S., Marhamati, Z., Hosseini, M. & Yakarim, M. (2012b) Effect of oregano on the growth of *Lactobacillus acidophilus* and *Bifidobacterium bifidum* in probiotic dairy products. *World Applied Sciences Journal*, **18**, 1394–1399.

Marhamatizadeh, M.H., Mohammadi, M., Rezazadeh, S. & Jafari, F. (2012c) Effects of garlic on the growth of *Lactobacillus acidophilus* and *Bifidobacterium bifidum* in probiotic milk and yoghurt. *Middle-East Journal of Scientific Research*, **11**(7), 894–899.

Marhamatizadeh, M.H., Ehsandoost, E. & Gholami, P. (2013) The influence of green tea (*Camellia sinensis* L.) extract on characteristic of probiotic bacteria in milk and yoghurt during fermentation and refrigerated storage. *International Journal of Farming and Allied Sciences*, **2**(17), 599–606.

Marhamatizadeh, M.H., Ehsandoost, E. & Gholami, P. (2014) The effect of coffee extract on the growth and viability of *Lactobacillus acidophilus* and *Bifidobacterium bifidum* in probiotic milk and yoghurt. *Journal of Food Biosciences and Technology*, **4**(1), 37–48.

Marsh, A.J., O'Sullivan, O., Hill, C., Ross R.P. & Cotter, P.D. (2013) Sequencing-based analysis of the bacterial and fungal composition of Kefir grains and milks from multiple sources. *PLOS One*, **8**, e69371.

Marshall, V.M. (1987) Fermented milks and their future trends - I. microbiological aspects. *Journal of Dairy Research*, **54**, 559–574.

Marshall, V.M.E. & Tamime, A.Y. (1997) Physiology and biochemistry of fermented milks. *Microbiology and Biochemistry of Cheese and Fermented Milk*, (ed. B.A. Law), 2nd edn., pp. 153–192, Blackie Academic & Professional, London.

Martin-Diana, A.B., Janer, C., Pelaez, C. & Requena, T. (2003) Development of a fermented goat's milk containing probiotic bacteria. *International Dairy Journal*, **13**, 827–833.

Martinez-Villaluenga, C., Frias, J., Gomez, R. & Vidal-Valverde, C. (2006) Influence of addition of raffinose family oligosaccharides on probiotic survival in fermented milk during refrigerated storage. *International Dairy Journal*, **16**, 768–774.

Maruyama, L.Y., Cardarelli, H.R., Buriti, F.C.A. & Saad, S.M.I. (2006) Textura instrumental de queigo Petit-Suisse potencialmente probiotico: Influência de differentes combinações de gomas. *Ciencia e Tecnologia de Alimentos*, **26**, 386–393.

Masco, L., Ventura, M., Zink, R., Huys, G. & Swings, J. (2004) Polyphasic taxonomic analysis of *Bifidobacterium animalis* and *Bifidobacterium lactis* reveals relatedness at the subspecies level: reclassification *Bifidobacterium animalis* as *Bifidobacterium animalis* subsp. *animalis* subsp. nov. and *Bifidobacterium lactis* as *Bifidobacterium animalis* subsp. *lactis* subsp. nov. *International Journal of Systematic and Evolutionary Microbiology*, **54**, 1137–1143.

Masco, L., Huys, G., de Brandt, E., Temmerman, R. & Swings, J. (2005) Culture-dependent and culture-independent qualitative analysis of probiotic products claimed to contain bifidobacteria. *International Journal of Food Microbiology*, **102**, 221–230.

Massoud, R., Massoudi, R., Khosravi-Darani, K. & Reza, N. H. (2015) Improving the viability of probiotic bacteria in yoghurt by homogenization. *Journal of Food Processing and Preservation*, **39**, 2984–2990.

Masuda, T., Ohkawa, Y., Kaneko, D., Umezu, S. & Itoh, T. (2005) Trial production of fermented goat milk containing probiotic *Lactobacillus acidophilus*. *Journal of the Japanese Society for Food Science and Technology (Nippon Shokuhin Kagaku Kogaku Kaishi)*, **52**(3), 131–134.

Matias, N.S., Bedani, R., Castro, I.A. & Saad, S.M.I. (2014) A probiotic soy-based innovative product as an alternative to petit-suisse cheese. *LWT - Food Science and Technology*, **59**, 411–417.

Matijevic, B., Lisak, K., Bozanic, R. & Tratnik, L. (2008) The influence of the different initial probiotic bacteria concentration on sweet whey fermentation. *Mljekarstvo*, **58**, 387–401.

Matijevic, B., Lisak, K., Bozanic, R. & Tratnik, L. (2011) Impact of enzymatic hydrolyzed lactose on fermentation and growth of probiotic bacteria in whey. *Mljekarstvo*, **61**, 154–160.

Mattarelli, P., Bonaparte, C., Pot, B. & Biavati, B. (2008) Proposal to reclassify the three biotypes of *Bifidobacterium longum* as three subspecies: *Bifidobacterium longum* subsp. *longum* subsp. nov., *Bifidobacterium longum* subsp. *infantis* comb. nov. and *Bifidobacterium longum* subsp. *suis* comb. nov. *International Journal of Systematic and Evolutionary Microbiology*, **58**, 767–772.

Mattila-Sandholm, T., Myllarinen, P., Crittenden, R., Mogensen, G., Fonden, R. & Saarela, M. (2002) Technological challenges for future probiotic foods. *International Dairy Journal*, **12**, 173–182.

Matto, J., Malinen, E., Suihko, M.-L., Alander, M., Palva, A. & Saarela, M. (2004) Genetic heterogeneity and technological properties of intestinal bifidobacteria. *Journal of Applied Microbiology*, **97**, 459–470.

Mayo, B., Ammor, M.S., Delgado, S. & Alegria, A. (2010) Fermented Milk Products. *Fermented Foods and Beverages of the World* (Eds. J.P. Tamang & K. Kailasapathy), pp. 263–289, CRC Press (Taylor & Francis Group), Boca Raton.

McBrearty, S., Ross, R.P., Fitzgerald, G.F., Collins, J.K., Wallace, J.M. & Stanton, C. (2001) Influence of two commercially available bifidobacteria cultures on Cheddar cheese quality. *International Dairy Journal*, **11**, 599–610.

de Medeiros, A.C.L., Thomazini, M., Urbano, A., Correia, R.T.P. & Favaro-Trindade, C.S. (2014) Structural characterisation and cell viability of a spray dried probiotic yoghurt produced with goats' milk and *Bifidobacterium animalis* subsp. *lactis* (BI-07). *International Dairy Journal*, **39**, 71–77.

de Medeiros Burkert, J.F., dos Santos da Fonseca, R.A., Oliveira de Moraes, J., Sganzerla, J., Kalil, S. J. & Veiga Burkert, C.A. (2012) Sensory acceptance of potentially symbiotic dairy beverages. *Brazilian Journal of Food Technology*, **15**, 325–332.

Mehanna, N.S., Sharaf, O.M., Ibrahim, G.A. & Tawfik, N.F. (2002) Incorporation and viability of some probiotic bacteria in functional dairy food – 1. Soft cheese. *Egyptian Journal of Dairy Science*, **30**, 217–229.

Meile, L., le Blay, G. & Thierry, A. (2008) Safety assessment of dairy microorganisms: *Propionibacterium* and *Bifidobacterium*. *International Journal of Food Microbiology*, **126**, 316–320.

de Menezes, C.R., Barin, J.S., Chicoski, A.J., Zepka, L.Q., Jacob-Lopes, E., Fries, L.L.M. & Terra, N.N. (2013) Microencapsulation of probiotics: progress and prospects. *Ciencia Rural*, **43**, 1309–1316.

Meng, X.C., Stanton,C., Fitzgerald, G.F., Daly, C. & Ross, R.P. (2008) Anhydrobiotics: The challenges of drying probiotic cultures. *Food Chemistry*, **106**, 1406–1416.

Mercanti, D.J., Rousseau, G.M., Capra, M.L., Quiberoni, A., Tremblay, D.M., Labrie, S.J. & Moineau, S. (2015) Genomic diversity of phages infecting probiotic strains of *Lactobacillus paracasei*. *Applied and Environmental Microbiology*, doi:10.1128/AEM.02723-15. Accessed 2nd July 2016.

Mercenier, M.A., Prioult, G. & Nutten, S. (2012a) Drinking yoghurt preparations containing non-replicating probiotic micro-organisms. *European Patent Application*, **EP 2 449 891 A1**.

Mercenier, A., Nutten, S. & Prioult, G. (2012b) Dry whole milk preparations containing probiotic microorganisms. *European Patent Application*, **EP 2 429 551 A1**.

Mercenier, A. Prioult, G. & Nutten, S. (2013) Drinking yoghurt preparations containing non-replicating probiotic micro-organisms. *European Patent Application*, **EP 2 635 139 A1**.

Milanvic, S.D., Panic, M.D. & Caric, M.D. (2004) Quality of quarg produced by probiotics application. *Acta Periodica Technologica*, **35**, 37–48.

Minelli, E.B., Benini, A, Marzotto, M., Sbarbati, A., Ruzzenete, O., Ferrario, R., Hendriks, H. & Dellaglio, F. (2004) Assessment of a novel probiotic *Lactobacillus casei* strain for production of functional dairy foods. *International Dairy Journal*, **14**, 723–736.

Minervini, F., Siragusa, S., Faccia, M., Dal Bello, F., Gobbetti, M. & de Angelis, M. (2012) Manufacture of Fior di Latte cheese by incorporation of probiotic lactobacilli. *Journal of Dairy Science*, **95**, 508–520.

Milesi, M.M., Vinderola, C.G., Sabbag, N., Meinardi, C.A. & Hynes, E. (2009) Influence on cheese proteolysis and sensory characteristics of non-starter lactobacilli strains with probiotic potential. *Food Res International*, **42**, 1186–1196.

Mirlohi, M., Soleimanian-Zad, S., Dokhani, S. & Sheikh-Zeinodin, M. (2014) Microbial and physicochemical changes in yoghurts containing different *Lactobacillus delbrueckii* subsp. *bulgaricus* strains in association with *Lactobacillus plantarum* as an adjunct culture. *International Journal of Dairy Technology*, **67**, 246–254.

Mituniewicz-Malek, A., Ziarno, M. & Dmytrow, I. (2014) Incorporation of inulin and transglutaminase in fermented goat milk containing probiotic bacteria. *Journal of Dairy Science*, **97**, 3332–3338.

Miyazaki, K & Matsuzaki, T. (2008) Health properties of milk fermented with *Lactobacillus casei* strain Shirota (LcS). *Handbook of Fermented Functional Foods* (ed. E.R. Farnworth), 2nd Edition, pp. 165–208, CRC Press (Taylor & Francis Group), Boca Raton.

Modzelewska-Kapitula, M., Klebukowska, L. & Kornacki, K. (2008) Evaluation of the possible use of potentially probiotic *Lactobacillus* strains in dairy products. *International Journal of Dairy Technology*, **61**, 165–169.

Moeller, C. & de Vrese, M. (2004) Review: Probiotic effects of selected lactic acid bacteria. *Milchwissenschaft*, **59**, 597–601.

Mogensen, G. (1980) Production and properties of yoghurt and ymer made from ultrafiltrated milk. *Desalination*, **35**, 213–222.

Mohammadi, R., Mortazavian, A.M., Khosrokhavar, R. & da Cruz, A.G. (2011) Probiotic ice cream: viability of probiotic bacteria and sensory properties. *Annals of Microbiology*, **61**, 411–424.

Mohd Redzwan, S., Abd Mutalib, M.S., Wang, J.-S., Ahmad, Z., Kang, M.-S., Abdul Rahman, N.A., Nikbakht Nasrabadi, E. & Jamaluddin, R. (2016) Effect of supplementation of fermented milk drink containing probiotic *Lactobacillus casei* Shirota on the concentrations of aflatoxin biomarkers among employees of Universiti Putra Malaysia: A randomised,double-blind, cross-over, placebo-controlled study. *British Journal of Nutrition*, **115**, 39–54.

Mocanu, G.D., Botez, E., Nistor, O.V. & Andronoiu, D.G. (2011) Characterization of probiotic yoghurt obtained with medicinal plant extracts and modelling of bacteria cell growth during its production. *Journal of Agroalimentary Processes and Technologies*, **17**, 65–71.

Moeller, C. & de Vrese, M. (2004) Review: Probiotic effects of selected acid bacteria. *Milchwissenschaft*, **59**, 11–12.

Morelli, L. (2000) *In vitro* selection of probiotic lactobacilli: A critical appraisal. *Current Issues in Intestinal Microbiology*, **1**, 59–67.

Mortazavian, A.M., Ehsani, M.R., Azizi, A., Razavi, S.H., Mousavi, S.M. & Reinheimer, J.A. (2008a) Effect of microencapsulation of probiotic bacteria with calcium alginate on cell stability during the refrigerated storage period in the Iranian yogurt drink (doogh). *Milchwissenschaft*, **63**, 262–265.

Mortazavian, A.M., Azizi, A., Ehsani, M.R., Razavi, S.H., Mousavi, S.M., Sohrabvandi, S. & Reinheimer, J.A. (2008b) Survival of encapsulated probiotic bacteria in Iranian yogurt drink (doogh) after the product exposure to simulated gastrointestinal conditions. *Milchwissenschaft*, **63**, 427–429.

Mortazavian, A.M., Khosrokhavar, R., Rastegar, H. & Mortazaei, G.R. (2010) Effects of dry matter standardization order on biochemical and microbiological characteristics of freshly made probiotic doogh (Iranian fermented milk drink). *Italian Journal of Food Science*, **22**, 98–104.

Mpofu, A., Linnemann, A.R., Sybesma, W., Kort, R., Nout, M.J.R. & Smid, E.J. (2014) Development of a locally sustainable functional food based on mutandabota, a traditional food in Southern Africa. *Journal of Dairy Science*, **97**, 2591–2599.

Mpofu, A., Anita R.L., Nout, M.J.R., Zwietering, M.H., Smid, E. J. & den Besten, H.M.W. (2016) Inactivation of bacterial pathogens in yoba mutandabota, a dairy product fermented with the probiotic *Lactobacillus rhamnosus* strain Yoba. *International Journal of Food Microbiology*, **217**, 42–48.

Muir, D.D., Tamime, A.Y. & Wszolek, M. (1999) Comparison of the sensory profiles of kefir, buttermilk and yogurt. *International Journal of Dairy Technology*, **52**, 129–134.

Murad, H.A., Sadek, Z.I. & Fathy, F.A. (1998) Production of bifidus Kariesh cheese. *Deutsche Lebensmittle-Rhudschau*, **94**, 409–412.

Muramalla, T. & Aryana, K.J. (2011) Some low homogenization pressures improve certain probiotic characteristics of yogurt culture bacteria and *Lactobacillus acidophilus* LA-K. *Journal of Dairy Science*, **94**, 3725–3738.

Mushtaq, M., Gani, A., Masoodi, F.A. & Ahmad, M. (2016) Himalayan cheese (*Kalari/Kradi*) – Effect of different probiotic strains on oxidative stability, microbiological, sensory and nutraceutical properties during storage. *LWT - Food Science and Technology*, **67**, 74–81.

Naeemi, H., Mortazavi, S.A., Milani, E. & Koochaki, A. (2013) The influence of adding inulin and encapsulation on survivability *Lactobacillus casei* storage of synbiotic yoghurt. *Iranian Journal of Food Science and Technology*, **10**(40), 27–36.

Nalbantoglu, U., Cakar, A., Dogan, H., Abaci, N., Ustek, D., Sayood, K. & Can, H. (2014). Metagenomic analysis of the microbial community in kefir grains. *Food Microbiology*, **41**, 42–51.

Nemeckova, I., Kejmarova, M., Chramostova, J., Zikan, V., Laknerova, I. & Hartman, I. (2013) Fermented dairy products with addition of malts. *Mlekarske Listy*, **24**(129 - Suppl. Veda Vyzkum), V–VIII.

Nielsen, B., Gurakan, G.C. & Unlu, G. (2014) Kefir: A multifaceted fermented dairy product. *Probiotics and Antimicrobial Proteins*, **6**(3-4), 123–135.

van den Nieuwboer, M., van de Burgwal, L.H.M. & Claassen, E. (2016) A quantitative key-opinion-leader analysis of innovation barriers in probiotic research and development: valorisation and improving the tech transfer cycle. *Pharma Nutrition*, **4**, 9–18.

Ng, E.W., Yeung, M. & Tong, P.S. (2011) Effects of yogurt starter cultures on the survival of *Lactobacillus acidophilus*. *International Journal of Food Microbiology*, **145**, 169–175.

Nemeckova, I., Kejmarova, M., Chramostova, J., Zikan, V., Laknerova, I. Hartman, I. (2013) Fermented dairy products with addition of malts. *Mlekarske Listy*, **24**(129, Supplement Veda, Vyzkum), V–VIII.

Nguyen, H.T.H., Ong, L., Lefevre, C., Kentish, S.E. & Gras, S.L. (2014) The microstructure and physicochemical properties of probiotic buffalo yoghurt during fermentation and storage: A comparison with bovine yoghurt. *Food and Bioprocess Technology*, **7**, 937–953.

van den Nieuwboer, M., van Hemert, S., Claassen, E. & de Vos, W.M. (2016) *Lactobacillus plantarum* WCFS1 and its host interaction: A dozen years after the genome. *Microbial Biotechnology*, **9**, 452–465.

Nielsen, B., Gurakan, G.C. & Unlu, G. (2014) Kefir: a multifaceted fermented dairy product. *Probiotics and Antimicrobial Proteins*, **6**(3-4), 123–135.

Nose, A., Nozaki, D., Ishikawa, F., Mizusawa, S. & Akahoshi, R. (2007) Novel bacterium belonging to the genus *Bifidobacterium* and utilization of the same. *PCT International Patent Application*, **WO 010977 A1**.

Nousia, F.G., Androulakis, P.I. & Fletouris, D.J. (2011) Survival of *Lactobacillus acidophilus* LMGP-21381 in probiotic ice cream and its influence on sensory acceptability. *International Journal of Dairy Technology*, **64**, 130–136.

Oberg, C.J., Moyes, L.V., Domek, M.J., Brothersen, C. & McMahon, D.J. (2011) Survival of probiotic adjunct cultures in cheese and challenges in their enumeration using selective media. *Journal of Dairy Science*, **94**, 2220–2230.

Oberman, H. & Libudzisk, Z. (1998) Fermented milks. *Microbiology of Fermented Products*, (ed. B.J.B. Wood), Vol. **1**, 2nd Edition, pp. 308–350, Blackie Academic & Professional, London.

Oh, N.S., Joung, J.Y., Lee, J.Y., Kim, S.H. & Kim, Y. (2016) Characterization of the microbial diversity and chemical composition of Gouda cheese made by potential probiotic strains as an adjunct starter culture. *Journal of Agriculture and Food Chemistry*, **64**, 7357–7366.

Olivares, M. & Xaus, J. (2007) Probiotics for the XXI's century: It's time for breast milk probiotics. *Innovations in Food Technology*, **35**, 71–73.

Oliviera, M.N. & Damin, M.R. (2003) Efeito do teor de solidos e da concentrção de sacarose na acidificaca, fermeza e viabilidade de bactérias do iogurte e probióticas em leite fermentado. *Ciência e Technologia de Alimentos*, **23** (Supplement), 172–176.

Oliveira, R.P.de S., Florence, A.C.R., Silva, R.C., Perego, P., Converti, A., Gioielli, L.A. & Oliveira, M.N. (2009) Effect of different prebiotics on the fermentation kinetics, probiotic survival and fatty acids profiles in nonfat symbiotic fermented milk. *International Journal of Food Microbiology*, **128**, 467–472.

Oliveira, R.P.deS., Florence, A.C.R., Perego, P., de Oliveira, M.N. & Converti, A. (2011a) Use of lactulose as prebiotic and its influence on the growth, acidification profile and viable counts of different probiotics in fermented skim milk. *International Journal of Food Microbiology*, **145**, 22–27.

Oliveira, R.P.deS., Perego, P., de Oliveira, M.N. & Converti, A. (2011b) Effect of inulin as a prebiotic to improve growth and counts of a probiotic cocktail in fermented skim milk. *LWT – Food Science and Technology*, **45**, 358–363.

Oliveira, R.P.deS., Perego, P., de Oliveira, M.N. & Converti, A. (2011c) Effect of inulin as prebiotic and synbiotic interactions between probiotics to improve fermented milk firmness. *Journal of Food Engineering*, **107**, 36–40.

Oliveira, R.P.deS., Perego, P., de Oliveira, M.N. & Converti, A. (2012) Effect of inulin on the growth and metabolism of a probiotic strain of *Lactobacillus rhamnosus* in co-culture with *Streptococcus thermophilus*. *LWT – Food Science and Technology*, **44**, 520–523.

de Oliveira Gaino, V., Voltarelli, V.P., de Rensis, C.M.V.B. & Vianna, P.C.B. (2012) Probiotic "requeijao cremoso": evaluation of *Lactobacillus casei* viability, physical-chemical composition and sensory acceptance. *Semina: Ciencias Agrarias*, **33**(Suppl. 2), 3133–3142.

de Oliveira, M.E.G., Garcia, E.F., de Oliveira, C.E.V., Gomes, A.M.P., Pintado, M.M.E., Madureira, A.R.M.F., da Conceicao, M.L., do EgyptoQueiroga, R. de C. R. & de Souza, E.L. (2014) Addition of probiotic bacteria in a semi-hard goat cheese (coalho): Survival to simulated gastrointestinal conditions and inhibitory effect against pathogenic bacteria. *Food Research International*, **64**, 241–247.

Ong, L. & Shah, N.P. (2009) Probiotic Cheddar cheese: Influence of ripening temperatures on survival of probiotic microorganisms, cheese composition and organic acid profiles. *LWT – Food Science and Technology*, **42**, 1260–1268.

Ong, L., Henriksson, A. & Shah, N.P. (2007) Proteolytic pattern and organic acid profiles of probiotic Cheddar cheese as influenced by probiotic strains of *Lactobacillus acidophilus*, *Lb. paracasei*, *Lb. casei* or *Bifidobacterium* sp. *International Dairy Journal*, **17**, 67–78.

Ordonez, G.A., Fung, D.Y.C. & Jeon, I.J. (2000a) Effect of Oxyrase™ on the metabolic process of lactic acid bacteria in frozen yogurt mix. *Journal of Rapid Methods and Automation in Microbiology*, **8**, 71–81.

Ordonez, G.A., Jeon, I.J. & Roberts, H.A. (2000b) Manufacture of frozen yogurt with ultrafiltered milk and probiotic lactic acid bacteria. *Journal of Food Processing and Preservation*, **24**, 163–176.

O'Riordan, K. & Fitzgerald, G.F. (1998) Evaluation of bifidobacteria for the production of antimicrobial compounds and assessment of performance in Cottage cheese at refrigerated temperature. *Journal of Applied Microbiology*, **85**, 103–114.

Ortakci, F., Broadbent, J.R., McManus, W.R. & McMahon, D.J. (2012) Survival of microencapsulated probiotic *Lactobacillus paracasei* LBC-1e during manufacture of Mozzarella cheese and simulated gastric digestion. *Journal of Dairy Science*, **95**, 6274–6281.

Ortu, S., Felis, G.E., Marzotto, M., Deriu, A., Molicotti, P., Sechi, L.A., Dellaglio, F. & Zanetti, S. (2007) Identification and functional characterization of *Lactobacillus* strains isolated from milk and Gioddu, a traditional Sardinian fermented milk. *International Dairy Journal*, **17**, 1312–1320.

Osman, M.M., & Abbas, F.M. (2001) Fate of *Lactobacillus acidophilus* La-5 and *Bifidobacterium lactis* Bb-12 in 'probiotic' Ras cheese. *8th Egyptian Conference for Dairy Science and Technology – Cairo 3–5 November 2001, Research Papers II*, 653–664.

Ostlie, H.M., Helland, M.H. & Narvhus, J.A. (2003) Growth and metabolism of selected strains of probiotic bacteria in milk. *International Journal of Food Microbiology*, **87**, 17–27.

O'Sullivan, A., Farver, M. & Smilowitz, J.T. (2015) The influence of early infant-feeding practices on the intestinal microbiome and body composition in infants. *Nutrition and Metabolic Insight*, **8**(suppl. 1), 1–9. doi: 10.4137/NMI.S29530

O'Sullivan, D.J., McSweeney, P.L.H., Cotter, P.D., Giblin, L. & Jeremiah J. Sheehan, J.J. (2016) Compromised *Lactobacillus helveticus* starter activity in the presence of facultative heterofermentative *Lactobacillus casei* DPC6987 results in atypical eye formation in Swiss-type cheese. *Journal of Dairy Science*, **99**, 2625–2640.

Ouwehand, A.C., Isolauri, E., Kirjavainen, P.V., Tolkko, S. & Salminen, S.J. (2000) The mucus binding of *Bifidobacterium lactis* Bb12 is enhanced in the presence of *Lactobacillus* GG and *Lactobacillus delbrueckii* subsp. *bulgaricus*. *Letters in Applied Microbiology*, **30**, 10–13.

Ozer, B. & Kirmaci, H.S. (2010) Functional milks and dairy beverages. *International Journal of Dairy Technology*, **63**, 1–15.

Ozer, B. & Tamime, A.Y. (2013) Membrane processing of fermented milks. *Membrane Processing – Dairy and Beverage Applications* (ed. A.Y. Tamime), pp. 143–175, Wiley-Blackwell, Oxford.

Ozer, B., Uzun, Y.S. & Kirmaci, H.S. (2008) Effect of microencapsulation on viability of *Lactobacillus acidophilus* LA-5 and *Bifidobacterium bifidum* BB-12 during Kasar cheese ripening. *International Journal of Dairy Technology*, **61**,237–244.

Ozer, B., Kirmaci, H.A., Ebru Senel, H., Atamer, M. & Hayaloglu, A. (2009) Improving the viability of *Bifidobacterium bifidum* BB-12 and *Lactobacillus acidophilus* LA-5 in white-brined cheese by microencapsulation. *International Dairy Journal*, **19**, 22–29.

Paez, R. Lavari, L., Audero, G., Cuatrin, A., Zaritzky, N., Reinheimer, J. & Vinderola, G. (2013) Study of the effects of spray-drying on the functionality of probiotic lactobacilli. *International Journal of Dairy Technology*, **66**, 155–161.

Papadimitriou, K., Alegria, A., Bron, P.A., de Angelis, M., Gobbetti, M., Kleerebezem, M., Lemos, J.A., Linares, D.M., Ross, P., Stanton, C., Turroni, F., van Sinderen, D., Varmanen, P, Ventura, M., Zuniga, M., Tsakalidou, E. & Kok, J. (2016) Stress physiology of lactic acid bacteria. *Microbiology and Molecular Biology Reviews*, **80**, 837–890.

Parsa, P., Alizadeh, M., Bari, M.R. & Moghar, A.A. (2015) Optimisation of probiotic yoghurt enriched with phytosterols. *International Journal of Dairy Technology*, **68**, 557–564.

Patrignani, F., Serrazanetti, D.I., Mathara, J.M., Siroli, L., Gardini, F., Holzapfel, W.H. & Lanciotti, R. (2016) Use of homogenisation pressure to improve quality and functionality of probiotic fermented milk containing *Lactobacillus rhamnosus* BFE 5264. *International Journal of Dairy Technology*, **69**, 262–271.

Pavlovic, H., Hardi, J., Slacanac, V., Halt, M. & Kocevski, D. (2006) Inhibitory effect of goat and cow milk fermented by *Bifidobacterium longum* on *Serratia marcescens* and *Campylobacter jejuni*. *Czech Journal of Food Sciences*, **24**(4), 164–171.

Peneva, D. & Aleksandrov, N. (2013) Probiotics for dietary dairy product. *United States of America Patent Application*, **0236600 A1**.

Peneva, D. & Aleksandrov, N. (2015) Probiotics for dietary dairy product. *United States of America Patent Application*, **9 131 708 B2**.

Penhasi, A. (2013a) Probiotic liquid food products. *United States of America Patent Application*, **0136826 A**.

Penhasi, A. (2013b) Heat resistant probiotic compositions and healthy food comprising them. *United States of America Patent Application Publication*, **0115334 A1**.

Penna, A.L.B, Subbarao-Gurram, A. & Barbosa-Canovas, G.V. (2007) High hydrostatic pressure processing on microstructure of probiotic low-fat yogurt. *Food Research International*, **40**, 510–519.

Pereira, E.P.R., Cavalcanti, R.N., Esmerino, E.A., Silva, R., Guerreiro, L.R.M., Cunha, R.L., Bolini, H.M.A., Meireles, M.A., Faria, J.A.F. & Cruz, A.G. (2016) Effect of incorporation of antioxidants on the chemical, rheological, and sensory properties of probiotic petit suisse cheese. *Journal of Dairy Science*, **99**, 1762–1772.

Perko, B., Matijasic, B.B. & Rogelj, I. (2002) Production of probiotic cheese with addition of *Lactobacillus gasseri* LF221 (Rif^T) and K7 (Rif^T). *Kmetijstvo, Zootehnika*, **80**, 61–70.

Perotti, M.C., Mercanti, D.J., Bernal, S.M. & Zalazar, C.A. (2009) Characterization of the free fatty acids profile of Pategras cheese during ripening. *International Journal of Dairy Technology*, **62**, 331–338.

Phillips, M., Kailasapathy, K. & Tran, L. (2006) Viability of commercial probiotic cultures (*L. acidophilus*, *Bifidobacterium* sp., *L. casei*, *L. paracasei* and *L. rhamnosus*) in cheddar cheese. *International Journal of Food Microbiology*, **108**, 276–280.

Picot, A. & Lacroix, C. (2003) Optimization of dynamic loop mixer operating conditions for production of o/w emulsion for cell microencapsulation. *Lait*, **83**, 237–250.

Picot, A. & Lacroix, C. (2004) Encapsulation of bifidobacteria in whey protein-based microcapsules and survival in simulated gastrointestinal conditions and in yoghurt. *International Dairy Journal*, **14**, 505–515.

Pinto Kempka, A., Leticia Krueger, R., Valduga, E., di Luccio, M., Treichel, H., Cansian, R. & de Oliveira, D. (2009) Formulation of a peach-flavored dairy drink using alternative substrates and probiotic culture. *Ciencia e Tecnologia de Alimentos*, **28**(Suppl.), 170–177.

Pisano, M.B., Patrignani, F., Cosentino, S., Guerzoni, M.E., Franz, C.M.A.P. & Holzapfel, W.H. (2011) Diversity and functional properties *Lactobacillus plantarum*-group strains isolated from Italian cheese products. *Dairy Science & Technology*, **91**, 65–76.

Playne, M.J., Bennet, L.E. & Smithers, G.W. (2003) Functional dairy foods and ingredients. *Australian Journal of Dairy Technology*, **58**, 242–264.
Plessas, S., Bosnea, L., Alexopoulos, A. & Bezirtzoglou, E. (2012) Potential effects of probiotics in cheese and yogurt production: A review. *Engineering in Life Sciences*, **12**, 433–440.
Prado, M.R., Blandon, L.M., Vandenberghe, L.P.S., Rodrigues, C., Castro, G.R., Thomaz-Soccol, V. & Soccol, C.R. (2015) Milk kefir: Composition, microbial cultures, biological activities, and related products. *Frontiers in Microbiology*, **6**, 1–10.
Prajapati, J.B., Shah, R.K. & Dave, J.M. (1986) Nutritional and therapeutic benefits of a blended spray dried acidophilus preparation. *Australian Journal of Dairy Technology*, **42**, 17–21.
Prasanna, P.H.P., Grandison, A.S. & Charalampopoulos, D. (2014) Bifidobacteria in milk products: An overview of physiological and biochemical properties, exopolysaccharide production, selection criteria of milk products and health benefits. *Food Research International*, **55**, 247–262.
de Prisco, A., Maresca, D., Ongeng, D. & Mauriello, G. (2015) Microencapsulation by vibrating technology of the probiotic strain *Lactobacillus reuteri* DSM 17938 to enhance its survival in foods and in gastrointestinal environment. *LWT - Food Science and Technology*, **61**, 452–462.
Psomas, E., Andrighetto, C., Litopoulou-Tzanetaki, E., Lombardi, A. & Tzanetakis, N. (2001) Some probiotic properties of yeast isolates from infant faeces and Feta cheese. *International Journal of Food Microbiology*, **69**, 125–133.
Raeisi, S.N., Ouoba, L.I.I., Farahmand, N. Sutherland, J. & Ghoddusi, H.B. (2013) Variation, viability and validity of bifidobacteria in fermented milk products. *Food Control*, **34**, 691–697.
Rajkowska, K. & Kunicka-Styczynska, A. (2009) Phenotypic and genotypic characterization of probiotic yeasts. *Biotechnology & Biotechnological Equipment*, **23**, 662–665.
Ramakrishnan, V., Goveas, L.C., Prakash, M., Halami, P.M. & Narayan, B. (2014) Optimization of conditions for probiotic curd formulation by *Enterococcus faecium* MTCC 5695 with probiotic properties using response surface methodology. *Journal of Food Science and Technology*, **51**, 3050–3060.
Rao, H.G.R. & Prakash, A.S. (2004) Development of probiotic *kulfi* (Indian ice cream). *Indian Dairyman*, **56**(6), 57–64.
Rao, A.V., Shivnarain, N. & Maharaj, I. (1989) Survival of microencapsulated *Bifidobacterium pseudolongum* in simulated gastric and intestinal juices. *Journal of Canadian Institute of Food Science and Technology*, **22**, 345–349.
Rasic, J.Li. & Kurmann, J.A. (1978) *Yoghurt – Scientific Grounds, Technology, Manufacture and Preparations*, pp. 117–119, Technical Dairy Publishing House, Copenhagen.
Rasic, J.Li. & Kurmann, J.A. (1983) *Bifidobacteria and their Role*, Experienta Supplementum Vol. **39**, pp. 102–133, Birkhäuser Verlag, Basel.
Ravula, R.R., & Shah, N.P. (1998) Viability of probiotic bacteria in fermented frozen dairy desserts. *Food Australia*, **50**, 136–139.
Real, E., Ortega, O., Reyneri, P., Garcias, H. & Gonzalez, J. (2005) Coagulated fermented milk product made with *Streptococcus thermophilus*, *Bifidobacterium bifidum*, and *Lactobacillus casei*. *Alimentaria*, **361**, 65–67.
Reid, G. (2015) The growth potential for dairy probiotics. *International Dairy Journal*, **49**, 16–22.
Riaz, Q.U.A. & Masud, T. (2013) Recent trends and applications of encapsulating materials for probiotic stability. *Critical Reviews in Food Science and Nutrition*, **53**, 231–244.
Ristagno, D., Hannon, J.A., Beresford, T.P. & McSweeney, P.L.H. (2012) Effect of bacteriocins-producing strain of *Lactobacillus paracasei* on the nonstarter microflora of Cheddar cheese. *International Journal of Dairy Technology*, **65**, 523–530.
Robinson, R.K., Tamime, A.Y. & Wszolek, M. (2002) Microbiology of fermented milks. *Dairy Microbiology Handbook* (ed. R.K. Robinson), 3rd Edition, The Microbiology of Milk and Milk Products, pp. 367–430. John Wiley & Sons, New York.
Rodriguez-Alcala, L.M., Braga, T., Malcata, F.X., Gomes, A. & Fontecha, J. (2011) Quantitative and qualitative determination of CLA produced by *Bifidobacterium* and lactic acid bacteria by combining spectrophotometric and Ag^+ HPLC techniques. *Food Chemistry*, **125**, 1373–1378.

Rodrigues, F., Sarmento, B., Andrade, J. & Oliveira, B. (2012) Review: Can microencapsulation be a means to increase survival of probiotics in cheese? *International Journal of Probiotics & Prebiotics*, **7**(2), 65–79.

Rogelj, I., Matijasic, B.B., Majhenic, A.C. & Stojkovic, S. (2002) The survival and persistence of *Lactobacillus acidophilus* LF221 in different ecosystems. *International Journal of Food Microbiology*, **76**, 83–91.

Ross, R.P., Fitzgerald, G., Collins, K., Stanton, C., Deeth, H.C., Smithers, G.W. & Hourigan, J. (2002) Cheese delivering biocultures – probiotic cheese. *Australian Journal of Dairy Technology*, **57**, 71–78.

Rothschild, P. (1995) Internal defences. *Dairy Industries International*, **60**(2), 24–25.

Roy, D. (2005) Technological aspects related to the use of bifidobacteria in dairy products. *Lait*, **85**, 39–56.

Roy, D., Mainville, I. & Mondou, F. (1997) Selective enumeration of bifidobacteria for use under cheesemaking condition. *Milchwissenschaft*, **50**, 139142.

Ruas-Madiedo, P. Gueimonde, M. Reyes-Gavilan, C. G. de los & Salminen, S. (2006) Short communication: effect of exopolysaccharide isolated from "viili" on the adhesion of probiotics and pathogens to intestinal mucus. *Journal of Dairy Science*, **89**, 2355–2358.

Ryan, P.M., Burdikova, Z., Beresford, T., Auty, M.A.E., Fitzgerald, G.F., Ross, R.P., Sheehan, J.J & Stanton, C. (2015) Reduced-fat Cheddar and Swiss-type cheeses harboring exopolysaccharide-producing probiotic *Lactobacillus mucosae* DPC 6426. *Journal of Dairy Science*, **98**, 8531–8544.

Ryhanen, E.L., Pihlanto-Leppala, A. & Pahkala, E. (2001) A new type of ripened, low-fat cheese with bioactive properties. *International Dairy Journal*, **11**, 441–447.

Saad, N., Delattre, C., Urdaci, M., Schmitter, J.M. & Bressollier, P. (2013) An overview of the last advances in probiotic and prebiotic field. *LWT - Food Science and Technology*, **50**, 1–16.

Saarela, M.H. (2011) Probiotic functional foods. *Functional foods: Concept to Product* (ed. M. Saarela), pp. 425–448, Woodhead Publishing, Campbridge.

Saarela, M., Mogensen, G., Fonden, G., Matto, J. & Mattila-Sandholm, T. (2000) Probiotic bacteria: safety, functional and technological properties. *Journal of Biotechnology*, **84**, 197–215.

Saavedra, J.M., Abi-Hanna, A., Moore, N. & Yolken, R.H. (2004) Long term consumption of infant formulas containing live probiotic bacteria: tolerance and safety. *Americam Journal of Clinical Nutritiun*, **79**, 261–267.

Sabikhi, L. & Mathur, B.N. (2000) Selection of suitable strain of *Bifidobacterium bifidum* for the manufacture of probiotic Edam cheese. *Indian Journal of Dairy Science*, **53**, 112–122.

Sabikhi, L. & Mathur, B.N. (2002) Technology and quality attributes of probiotic Edam cheese containing *Bifidobacterium bifidum*. *Indian Journal of Dairy Science*, **55**, 331–337.

Sabikhi, L., Kumar, M.H. & Mathur, B.N. (2014) *Bifidobacterium bifidum* in probiotic Edam cheese: Influence on cheese ripening. *Journal of Food Science and Technology*, **51**, 3902–3909.

Sabir, F., Beyatli, Y., Cokmus, C. & Onal-Darilmaz, D. (2010) Assessment of potential properties of *Lactobacillus* spp., *Lactococcus* spp. and *Pediococcus* spp. strains isolated from kefir. *Journal of Food Science*, **75**, M568–M573.

Saccaro, D.M., Tamime, A.Y., Pilleggi, A.L.O.P.S. & Oliveira, M. N. (2009) The viability of three probiotic organisms grown with yoghurt starter cultures during storage for 21 days at 4 °C. *International Journal of Dairy Technology*, **62**, 397–404.

Sadek, Z.I., Fathi, F.A. & Salem, M.M.E. (2006) Incidence, survival and biocontrol of psychrotrophic *Bacillus cereus* and its potential for toxin production in milk and Tallaga cheese. *Polish Journal of Food and Nutrition Sciences*, **15**/56, 419–425.

Sah, B.N.P, Vasiljevic, T., McKechnie, S. & O.N. Donkor, O.N. (2015) Effect of refrigerated storage on probiotic viability and the production and stability of antimutagenic and antioxidant peptides in yogurt supplemented with pineapple peel. *Journal of Dairy Science*, **98**, 5905–5916.

Sah, B.N.P., Vasiljevic, T., McKechnie, S. & Donkor, O.N. (2016) Physicochemical, textural and rheological properties of probiotic yogurt fortified with fibre-rich pineapple peel powder during refrigerated storage. *LWT – Food Science and Technology*, **65**, 978–986.

Saito, T. (2004) Selection of useful probiotic lactic acid bacteria from *Lactobacillus acidophilus* group and their application to functional foods. *Animal Science Journal*, **75**, 1–13.

Salama, F.M.M. (2002) Production of therapeutic and diabetic stirred yoghurt-like fermented milk products. *Egyptian Journal of Dairy Science*, **30**, 177–190.

Salem, A.S., Gafour, W.A. & Eassawy, E.A.Y. (2006) Probiotic milk beverage fortified with antioxidants as functional ingredients. *Egyptian Journal of Dairy Science*, **34**, 23–32.

Salji, J. (1992) Acidophilus milk products: foods with a third dimension. *Food Science and Technology Today*, **6**, 142–147.

Salva, S., Nunez, M., Villena, J., Ramon, A., Font, G. & Alvarez, S. (2011) Development of a fermented goats' milk containing *Lactobacillus rhamnosus*: *in vivo* study of health benefits. *Journal of the Science of Food and Agriculture*, **91**, 2355–2362.

Sanchez, B., de los Reyes-Gavilan, C.G., Margolles, A. & Gueimonde, M. (2009) Probiotic fermented milks: Present and future. *International Journal of Dairy Technology*, **62**, 472–483.

Santiago-Lopez, L., Hernandez-Mendoza, A., Garcia, H.S., Mata-Haro, V., Vallejo-Cordoba, B. & Gonzalez-Cordova, A.F. (2015) The effects of consuming probiotic-fermented milk on the immune system: A review of scientific evidence. *International Journal of Dairy Technology*, **68**, 153–165.

Santillo, A. & Albenzio, M. (2008) Influence of lamb rennet paste containing probiotic on proteolysis and rheological properties of Pecorino cheese. *Journal of Dairy Science*, **91**, 1733–1742.

Santillo, A., Albenzio, M., Bevilacqua, A., Corbo, M.R. & Sevi, A. (2012) Encapsulation of probiotic bacteria in lamb rennet paste: Effects on the quality of Pecorino cheese. *Journal of Dairy Science*, **95**, 3489–3500.

Santos, A., San Mauro, M., Sanchez, A., Torres, J.M. & Marquina, D. (2003) The antimicrobial properties of different strains of *Lactobacillus* spp. isolated from kefir. *Systematic and Applied Microbiology*, **26**, 434–437.

Savoie, S., Champagne, C.P., Chiasson, S. & P. Audet (2007) Media and process parameters affecting the growth, strain ratios and specific acidifying activities of a mixed lactic starter containing aroma-producing and probiotic strains. *Journal of Applied Microbiology*, **103**, 163–174.

Saxelin, M., Grenov, B., Svensson, U., Fonden, R., Reniero, R. & Mattila-Sandholm, T. (1999) The technology of probiotics. *Trends in Food Science and Technology*, **10**, 387–392.

Schillinger, U. (1999) Isolation and Identification of lactobacilli from novel-type probiotic and mild yoghurts and their stability during refrigerated storage. *International Journal of Food Microbiology*, **47**, 79–87.

Schoevers, A. & Britz, T.J. (2003) Influence of different culturing conditions on kefir grains increase. *International Journal of Dairy Technology*, **56**, 183–187.

Schoina, V., Terpou, A., Angelika-Ioanna, G., Koutinas, A., Kanellaki, M. & Bosnea, L. (2015) Use of *Pistacia terebinthus* resin as immobilization support for *Lactobacillus casei* cells and application in selected dairy products. *Journal of Food Science and Technology*, **52**, 5700–5708.

Seleet, F.L., El-Kholy, W.I. & Abd-Rabou, N.S. (2011) Evaluation of milk drinks fermented by probiotic bacteria and fortified with zinc salts. *Polish Journal of Food and Nutrition Sciences*, **61**, 55–60.

Seyhan, E., Yaman, H. & Ozer, B. (2016) Production of a whey-based functional beverage supplemented with soy isoflavones and phytosterols. *International Journal of Dairy Technology*, **69**, 114–121.

Shah, N.P. (2000) Probiotic bacteria: Selective enumeration and survival in dairy foods. *Journal of Dairy Science*, **83**, 894–907.

Shah, N.P. (2004) Probiotics and prebiotics. *Agro Food Industry Hi-Tech*, **15**, 13–16.
Shah, N.P. (2007) Functional cultures and health benefits. *International Dairy Journal*, **17**, 1262–1277.
Shah, N.P. & Ravula, R. (2000a) Influence of water activity on fermentation, organic acids production and viability of yoghurt and probiotic bacteria. *Australian Journal of Dairy Technology*, **55**, 127–131.
Shah, N.P. & Ravula, R. (2000b) Microencapsulation of probiotic bacteria and their survival in frozen fermented dairy desserts. *Australian Journal of Dairy Technology*, **55**, 139–144.
Shah, N.P. & Ravula, R. (2004) Selling the cells in desserts. *Dairy Industries International*, **69**(1), 31–32.
Shah, N. P., Lankaputhra, W. E. V., Britz, M. L. & Kyle, W.S.A. (1995) Survival of *Lactobacillus acidophilus* and *Bifidobacterium bifidum* in commercial yoghurt during refrigerated storage. *International Dairy Journal*, **5**, 515–521.
Shahabbaspour, Z., Mortazavian, A.M., Pourahmad, R., Moghimi, A. & Sohrabvandi, S. (2013) The effects of ratio of cow's milk to soymilk, probiotic strain and fruit concentrate on qualitative aspects of probiotic flavoured fermented drinks. *International Journal of Dairy Technology*, **66**, 135–144.
Sharafi, H., Derakhshan, V., Paknejad, M., Alidoust, L., Tohidi, A., Pornour, M., Hajfarajollah, H., Shahbani Zahiri, H., Akbari Noghabi, K. (2015) *Lactobacillus crustorum* KH: novel prospective probiotic strain isolated from Iranian traditional dairy products. *Applied Biochemistry and Biotechnology*, **175**, 2178–2194.
Sharma, M. & Devi, M. (2014) Probiotics: A comprehensive approach toward health foods. *Critical Reviews in Food Science and Nutrition*, **54**, 537–552.
Sharma, G. & Ghosh, B.C. (2006) Probiotic dairy foods and prebiotics for health benefit. *Indian Food Industry*, **25**(1), 68–73.
Sharma, P., Tomar, S.K., Goswami, P., Sangwan, V. & Singh, R. (2014) Antibiotic resistance among commercially available probiotics. *Food Research International*, **57**, 176–195.
Shehata, A.E., El-Nawawy, M.A., El-Kenany, Y.M. & Aumara, I.E.M. (2001) Production of soft cheese with health benefits. *8th Egyptian Conference for Dairy Science and Technology – Cairo 3–5 November 2001, Research Papers II*, 635–651.
Shehata, A.E., El-Nawawy, M.A., El-Kenany, Y.M. & Aumara, I.E.M. (2004a) Use of bifidobacteria in Ras cheese production: I Chemical and organoleptic properties. *9th Egyptian Conference for Dairy Science and Technology – Cairo 9–11 October 2004, Research Papers II*, 533–562.
Shehata, A.E., El-Nawawy, M.A., El-Kenany, Y.M. & Aumara, I.E.M. (2004b) Use of bifidobacteria in Ras cheese production: II Microbiological properties. *9th Egyptian Conference for Dairy Science and Technology – Cairo 9–11 October 2004, Research Papers II*, 563–585.
Shehata, A.E., El-Nawawy, M.A., El-Kenany, Y.M. & Aumara, I.E.M. (2004c) Use of bifidobacteria in Ras cheese production: III Changes in protein and microstructure during ripening. *9th Egyptian Conference for Dairy Science and Technology – Cairo 9–11 October 2004, Research Papers II*, 587–594.
Sheu, T.Y. & Marshall, R.T. (1993) Microentrapment of lactobacilli in calcium alginate gels. *Journal of Food Science*, **54**, 557–561.
Shewale, R.N., Sawale, P.D., Khedkar, C.D. & Singh, A. (2014) Selection criteria for probiotics: A review. *International Journal of Probiotics and Prebiotics*, **9**(1/2), 17–22.
Shiby, V.K. & Mishra, H.N. (2013) Fermented milks and milk products as functional foods - A review. *Critical Reviews in Food Science and Nutrition*, **53**, 482–496.
Shihata, A. & Shah, N.P. (2000) Proteolytic profiles of yoghurt and probiotic bacteria. *International Dairy Journal*, **10**, 401–408.
Shihata, A. & Shah, N.P. (2002) Influence of addition of proteolytic strains of *Lactobacillus delbrueckii* subsp. *bulgaricus* to commercial ABT starter cultures on texture of yoghurt, exopolysaccharides production and survival of bacteria. *International Dairy Journal*, **12**, 765–772.

Shortt, C. (1999) The probiotic century: Historical and current prospective. *Trends in Food Science & Technology*, **10**, 411–417.

Shu, G., Wang, S., Chen, Z., Chen, H., Wang, C. & Ma, Y. (2015) Effect of bacteria proportion on the fermentation of goat yoghurt with probiotic culture. *Acta Scientarum Polonorum Technologia Alimentaria*, **14**, 407–414.

da Silveira, E.O., Neto, J.H.L., Silva, L.A., da Raposo, A.E.S., Magnani, M. & Cardarelli, H.R. (2015) The effects of inulin combined with oligofructose and goat cheese whey on the physicochemical properties and sensory acceptance of a probiotic chocolate goat dairy beverage. *LWT - Food Science and Technology*, **62**, 445–451.

Simova, E., Beshkova, D., Angelov, A., Hristozova, T., Frengova, G. & Spasov, Z. (2002) Lactic acid bacteria and yeasts in kefir grains and kefir made from them. *Journal of Industrial Microbiology and Biotechnology*, **28**, 1–6.

Simunek, M. & Evacic, S. (2009) Effect of inulin on the growth and survival of *Bifidobacterium longum* BB536 in fermented goat's and cow's milk. *Mljekarstvo*, **59**, 209–216.

Skrzypczak, K. & Gustaw, W. (2012) Effect of prebiotics and whey proteins on physicochemical properties of bio-yoghurt. *Zywnosc*, **19**(5), 155–165.

Soccol, C.R., Porto de Souza Vandenberghe, L., Rigon Spier, M., Bianchi Pedroni Medeiros, A., Yamaguishi, C.T., de Dea Lindner, J., Pandey, A. & Thomaz-Soccol, V. (2010) The potential of probiotics: A review. *Food Technology and Biotechnology*, **48**, 413–434.

Sodini, I., Lucas, A., Oliveira, M.N., Remeuf, F. & Corrieu, G. (2002) Effect of milk base and starter culture on acidification, texture, and probiotic cell counts in fermented milk processing. *Journal of Dairy Science*, **85**, 2479–2488.

Songisepp, E., Mikelsaar, M., Ratsep, M., Zilmer, M., Hutt, P., Utt, M., Zilmer, K., Uksti, J. & Kojalg, S. (2009) Isolated microorganism strain *Lactobacillus plantarum* Tensia DSM 21380 as antimicrobial and antihypertensive probiotic, food product and composition comprising said microorganism and use of said microorganism for preparation of antihypertensive medicine and method for suppressing pathogens and non-starter lactobacilli in food product. *PCT International Patent Application*, **WO 138091 A2**.

Songisepp, E., Mikelsaar, M., Ratsep, M., Zilmer, M., Hutt, P., Utt, M., Zilmer, K., Uksti, J. & Kojalg, S. (2014) Antimicrobial and antihypertensive probiotic composition, food product and dietary supplement comprising microorganism strain *Lactobacillus plantarum* Tensia DSM 21380 and method for suppressing contaminating microbes in a food product. *United States of America Patent Application*, **0328814 A1**.

Soukoulis, C., Lyroni, E. & Tzia, C. (2010) Sensory profiling and hedonic judgement of probiotic ice cream as a function of hydrocolloids, yogurt and milk fat content. *LWT - Food Science and Technology*, **43**, 1351–1358.

Sparo, M.D., Corso, A., Gagetti, P., Delpech, G., Ceci, M., Confalonieri, A., Urbizu, L. & Bruni, S.F.S. (2012) *Enterococcus faecalis* CECT712: Biopreservation of crafted goat cheese. *International Journal of Probiotics and Prebiotics*, **7**(3-4), 145–152.

Sridar, R., Nguyen, M. & Kailasapathy, K. (2003) Studies on the effect of encapsulation on the survival of probiotic microorganisms under high acid and bile conditions. *Journal of Food Science and Technology (Mysore)*, **40**, 458–460.

Srisuvor, N., Chinprahast, N., Prakitchaiwattana, C. & Subhimaros, S. (2013) Effects of inulin and polydextrose on physicochemical and sensory properties of low-fat set yoghurt with probiotic-cultured banana purée. *LWT - Food Science and Technology*, **51**, 30–36.

Staliano, C.D., Martinez, R.C.R. & Saad, S.M.I. (2015) Beneficial microorganisms viability and sensory acceptance of a potentially synbiotic dairy-based tomato spread. *LWT - Food Science and Technology*, **62** (1 – Part 2), 682–688.

Stanton, C., Gardiner, G., Lynch, P.B., Collins, J.K., Fitzgerald, G. & Ross, R.P. (1998) Probiotic cheese. *International Dairy Journal*, **8**, 491–496.

Stanton, C., Coakley, M., Murphy, J.J., Fitzgerald, G.F., Devery, R. & Ross, R.P. (2002) Development of dairy-based functional foods. *Sciences des Aliments*, **22**, 439–447.

Stroehlein, J.R. (2004) Treatment of *Clostridium difficile* infection. *Current Treatment Options in Gastroenterology*, **7**, 235–239.

Strohmaier, W. (1997) Lactulose, an innovative food ingredient – physiological aspects. *Proceedings of Food Ingredients Europe Conference*, pp. 69–72, Porte de Versailles, Paris.

Su, L.-C., Lin, C.-W. & Chen, M.-J. (2005) Studies of the microbial and physical properties of oriental style dairy product kou woan lao with probiotics. *Asian-Australasian Journal of Animal Sciences*, **18**, 409–413.

Su, L.-C., Lin, C.-W. & Chen, M.-J. (2007) Development of an Oriental-style dairy product coagulated by microcapsules containing probiotics and filtrates from fermented rice. *International Journal of Dairy Technology*, **60**, 49–54.

Suarez-Solis, V., Cardoso, F., Nunez de Villavicencio, M., Fernandez, M. & Fargoso, L. (2002) Probiotic fresh cheese. *Alimentaria* **39**, 83–86.

Surawicz, C.M. (2003) Probiotics, antibiotic-associated diarrhoea and *Clostridium difficile* diarrhoea in humans. *Best Practice and Research in Clinical Gastroenterology*, **17**, 775–83.

Sultana, K., Godward, G., Reynolds, N., Arumugaswamy, R., Peiris, P. & Kailasapathy, K. (2000) Encapsulation of probiotic bacteria with alginate-starch and evaluation of survival in simulated gastrointestinal conditions and yoghurt. *International Journal of Food Microbiology*, **62**, 47–55.

Taha, S.H., Girgis, E.S., Amer, S.N. & Abed-El-Moeety, S.H. (1997) Effect of milk fat substitution with vegetable oils on the quality attributes of Labneh. *Egyptian Journal of Dairy Science*, **25**, 193–203.

Taha, S.H., Abd-El-Fattah, A.M., El-Dairy, S.Y., Assous, M.T. & Attalla, N.R. (2007) Antioxidant activity of flavoured stirred yoghurt like products. *Egyptian Journal of Dairy Science*, **35**, 31–44.

Takahashi, N., Xiao, J.-Z., Miyaji, K., Yaeshiima, T., Hiramatsu, A., Iwatsuki, K., Kokubo, S. & Hosono, A. (2004) Selection of acid tolerant bifdobacteria and evidence of a low-pH-inducible acid tolerance response in *Bifidobacterium longum*. *Journal of Dairy Research*, **71**, 340–345.

Takeda, S., Yamasaki, K., Takeshita, M., Kikuchi, Y., Tsend-Ayush, C., Dashnyam, B., Ahhmed, A.M., Kawahara, S. & Muguruma, M. (2011) The investigation of probiotic potential of lactic acid bacteria isolated from traditional Mongolian dairy products. *Animal Science Journal*, **82**, 571–579.

Takeda, S., Takeshita, M., Matsusaki, T., Kikuchi, Y., Tsend-Ayush, C., Oyunsuren, T., Miyata, M., Maeda, K., Yasuda, S., Aiba, Y., Koga, Y. & Igoshi, K. (2015) *In vitro* and *in vivo* anti-*Helicobacter pylori* activity of probiotics isolated from Mongolian dairy products. *Food Science and Technology Research*, **21**, 3, 399–406.

Talwalkar, A. & Kailasapathy, K. (2003a) Effect of microencapsulation on oxygen toxicity in probiotic bacteria. *Australian Journal of Dairy Technology*, **58**, 36–39.

Talwalkar, A. & Kailasapathy, K. (2003b) Metabolic and biochemical responses of probiotic bacteria to oxygen. *Journal of Dairy Science*, **86**, 2537–2546.

Talwalkar, A. & Kailasapathy, K. (2004) The role of oxygen in the viability of probiotic bacteria with reference to *L. acidophilus* and *Bifidobacterium* spp. *Current Issues in Intestinal Microbiology*, **5**, 1–8.

Talwalker, A., Miller, C.W., Kailasapathy, K. &Nguyen, M.H. (2004) Effect of packaging materials and dissolved oxygen on the survival of probiotic bacteria in yoghurt. *International Journal of Food Science & Technology*, **39**, 605–611.

Tamime, A. Y. (2002) Fermented milks: a historical food with modern applications – a review. *European Journal of Clinical Nutrition*, **56**, S2–S15.

Tamime, A.Y. (ed.) (2006a) *Fermented Milks*, Blackwell Publishing, Oxford.

Tamime, A.Y. (ed.) (2006b) *Brined Chesses*, Blackwell Publishing, Oxford.

Tamime, A.Y. & Marshall, V.M.E. (1997) Microbiology and technology of fermented milks. *Microbiology and Biochemistry of Cheese and Fermented Milk*, (ed. B.A. Law), 2nd edn., pp. 57–152, Blackie Academic & Professional, London.

Tamime, A.Y. & Robinson, R.K. (1999) *Yoghurt, Science and Technology*. Woodhead Publishing Ltd., Cambridge.

Tamime, A.Y. & Robinson, R.K. (2007) Fermented milks popular in Europe and North America. *Handbook of Food Products Manufacturing* (ed. Y.H. Hui), pp. 411–429, J. Wiley & Sons, Hoboken.

Tamime, A.Y., Kalab, M. & Davies, G. (1984) Microstructure of set-style yoghurt manufactured from cow's milk fortified by different methods. *Food Microstructure*, **3**, 83–92.

Tamime, A.Y., Marshall, V.M.E. & Robinson, R.K. (1995) Microbiological and technological aspects of fermented milks by bifidobacteria. *Journal of Dairy Research*, **62**, 151–187.

Tamime, A.Y., Skriver, A. & Nilsson, L.-E. (2006) Starter cultures. *Fermented Milks*, (ed. A.Y. Tamime), pp. 11–52, Blackwell Publishing, Oxford.

Tamime, A.Y., Hickey, M. & Muir, D.D. (2014) Strained fermented milks - A review of existing legislative provisions, survey of nutritional labelling of commercial products in selected markets and terminology of products in some selected countries. *International Journal of Dairy Technology*, **67**, 305–333.

Tawfek, M.A., El-Deeb, A.M. & Hassan, N.S.Y. (2016) Production of reduced calorie frozen yoghurt-like product supplemented soy milk fermented by bifidobacteria. *Egyptian Journal of Dairy Science*, **44**, 55–65.

Thakur, N., Rokana, N. & Panwar, H. (2016) Probiotics: Selection criteria, safety and role in health and disease. *Journal of Innovative Biology*, **3**, 259–270.

Tharmaraj, N. & Shah, N.P. (2004) Survival of *Lactobacillus acidophilus*, *Lactobacillus paracasei* subsp. *paracasei*, *Lactobacillus rhamnosus*, *Bifidobacteria animalis* and *Propionibacterium* in cheese-based dips and the suitability of dips as effective carriers of probiotic bacteria. *International Dairy Journal*, **14**, 1055–1066.

Thorball, J., Skolling, O. & Casas, I.A. (2001) Dispensing tube. *Patent Cooperation Treaty (PCT) / World Intellectual Property Organization*, **WO 01/15985 A1**.

Tratnik, L., Suskovic, J., Bozanic, R. & Kos, B. (2000) Creamed Cottage cheese enriched with *Lactobacillus* GG. *Mljekarstvo*, **50**, 113–123.

Tropcheva, R., Georgieva, R., Paskov, V., Karsheva, M. & Danova, S. (2014) Sensory properties of Bulgarian yogurts, supplemented with lactobacilli as probiotic adjuncts. *Journal of Texture Studies*, **45**(3), 187–194.

Trujillo-de Santiago, G., Saenz-Collins, C.P. & Rojias-de Gante, C. (2012) Elaboration of a probiotic oblea from whey fermented using *Lactobacillus acidophilus* or *Bifidobacterium infantis*. *Journal of Dairy Science*, **95**, 6897–6904.

Tungjaroenchai, W., White, C.H., Holmes, W.E. & Drake, M.A. (2004) Influence of adjunct cultures on volatile free fatty acids in reduced-fat edam cheeses. *Journal of Dairy Science*, **87**, 3224–3234.

Tunick, M.H. & van Hekken, D.L. (2015) Dairy products and health: Recent insights. *Journal of Agricultural and Food Chemistry*, **63**, 9381–9388.

Ulrich, P. (1980) Manufacture of ymer by means of ultrafiltration (UF) (in Danish). *Maelkeritidende*, **93**, 67–73.

Underwood, M.A., Bruce German, J., Lebrilla, C.B. & David A. Mills, D.A. (2015) *Bifidobacterium longum* subspecies *infantis*: champion colonizer of the infant gut. *Pediatr Resarch*, **77**, 229–235. doi:10.1038/pr.2014.156.

Unno, T., Choi, J.-H., Hur, H.-G., Sadowsky, M.J., Ahn, Y.-T., Huh, C.-S., Kim, G.-B. & Cha, C.-J. (2015) Changes in human gut microbiota influenced by probiotic fermented milk ingestion. *Journal of Dairy Science*, **98**, 3568–3576.

Urkek, B., Sengul, M. & Erkaya, T. (2014) Use of *Saccharomyces boulardii* in fermented dairy products. *Akademik Gida*, **12**(2), 108–113.

Uysal-Pala, C., Karagul-Yuceer, Y., Pala, A. & Savas, T. (2006) Sensory properties of drinkable yoghurt made from milk of different goat breeds. *Journal of Sensory Studies*, **21**, 520–533.

van Brandt, L., Coudijzer, K., Herman, L., Michiels, C., Hendrickx, M. & Vlaemynck, G. (2011) Survival of *Mycobacterium avium* ssp. *paratuberculosis* in yoghurt and in commercial fermented milk products containing probiotic cultures. *Journal of Applied Microbiology*, **110**, 1252–1261.

Van de Water, J. (2003) Yogurt and immunity: The health benefits of fermented milk products that contain lactic acid bacteria. *Handbook of Fermented Functional Foods*, (ed. E.R. Farnworth), pp. 113–144, CRC Press LLC, Boca Raton.

Varga, L., Szigeti, J. & Csengeri, E. (2003) Effect of oligofructose on the microflora of an ABT-type fermented milk during refrigerated storage. *Milchwissenschaft*, **58**, 55–58.

Varga, L., Sule, J. & Nagy, P. (2014a) Viability of culture organisms in honey enriched Acidophilus – Bifidus - Thermophilus (ABT) - type fermented camel milk. *Journal of Dairy Science*, **97**, 6814–6818.

Varga, L. Sule, J. & Nagy, P. (2014b) Survival of the characteristic microbiota in probiotic fermented camel, cow, goat, and sheep milks during refrigerated storage. *Journal of Dairy Science*, **97**, 2039–2044.

Vasiljevic, T. & Shah, N.P. (2008) Probiotics - from Metchnikoff to bioactives. *International Dairy Journal*, **18**, 714–728.

Vijayendra, S.V.N. & Gupta, R.C. (2014) Performance evaluation of bulk freeze dried starter cultures of dahi and yoghurt along with probiotic strains in standardized milk of cow and buffalo. *Journal of Food Science and Technology*, **51**, 4114–4119.

Vinderola, C.G, Bailo, N. & Reinheimer, J.A. (2000a) Survival of probiotic microflora in Argentinian yoghurts during refrigerated storage. *Food Research International*, **33**, 97–102.

Vinderola, C.G., Prosello, W., Ghiberto, D. & Reinheimer, J.A. (2000b) Viability of probiotic (*Bifidobacterium* spp., *Lactobacillus acidophilus* and *Lactobacillus casei*) and nonprobiotic microflora in Argentinian Fresco cheese. *Journal of Dairy Science*, **83**, 1905–1911.

Vinderola, C.G., Mocchiutti, P. & Reinheimer, J.A. (2002) Interactions among lactic acid starter and probiotic bacteria used for fermented dairy products. *Journal of Dairy Science*, **85**, 721–729.

Vinderola, G., Prosello, W., Molinari, F., Ghiberto, D. & Reinheimer, J. (2009) Growth of *Lactobacillus paracasei* A13 in Argentinian probiotic cheese and its impact on the characteristics of the product. *International Journal of Food Microbiology*, **135**, 171–174.

de Vos, P., Faas, M.M., Spasojevic, M. & Sikkema, J. (2010) Encapsulation for preservation of functionality and targeted delivery of bioactive food components. *International Dairy Journal*, **20**, 292–302.

Walsh, H., Cheng, J. & Guo, M. (2014) Effects of carbonation on probiotic survivability, physico-chemical, and sensory properties of milk-based symbiotic beverages. *Journal of Food Science*, **79**, M604–M613.

Wang, L.-Q. (2014) Study on the formula of the function milk beverage of yacon and probiotic. *Storage and Process*, **14**(3), 33–36.

Wang, J., Guo, Z., Zhang, Q., Yan, L., Chen, W., Liu, X.M. & Zhang, H.P. (2009) Fermentation characteristics and transit tolerance of probiotic *Lactobacillus casei* Zhang in soymilk and bovine milk during storage. *Journal of Dairy Science*, **92**, 2468–2476.

Wang, J., Guo, Z., Zhang, Q., Yan, L., Chen, Y., Chen, X., Liu, X.-M., Chen, W. & Zhang, H.-P. (2010) Effect of probiotic *Lactobacillus casei* Zhang on fermentation characteristics of set yoghurt. *International Journal of Dairy Technology*, **63** 105–112.

Wang, J.-C., Guo, Z., Qi, M.G.S.D., Zhang, H.-P. & Meng, H.B.L.G. (2013) Effects of probiotic *Lactobacillus casei* Zhang on sensory properties of set fermented milk. *Scientia Agricultura Sinica*, **46**, 575–585.

Wang, X.-P., Lv, J.-L., Yan, Y.-M., Lt, W.-J. & Du, B. (2015a) Research on probiotic properties and lipid-lowering capability of five strains of *Lactobacillus*. *China Dairy Industry*, **43**(9), 11–14.

Wang, G., Xiong, Y., Xu, Q., Yin, J. & Hao, Y. (2015b) Complete genome sequence of *Lactobacillus paracasei* CAUH35, a new strain isolated from traditional dairy product koumiss in China. *Journal of Biotechnology*, **214**, 75–76.

Weinrichter, B., Ginzinger, W., Sollberger, H. & Rohm, H. (2004a) Production of Emmental cheese with adjunct starters with varying degree of autolysis. *Milchwissenschaft*, **59**, 515–519.

Weinrichter, B., Sollberger, H., Ginzinger, W., Jaros, D. & Rohm, H. (2004b) Adjunct starter properties affect characteristic features of Swiss-type cheeses. *Nahrung*, **48**, 73–79.

Weizman, Z. & Alsheikh, A. (2006) Safety and tolerance of a probiotic formula in early infancy comparing two probiotic agents: A pilot study. *Journal of the American College of Nutrition*, **25**(5) 415–419.

Weizman, Z., Asli, G. & Alsheikh, A. (2005) Effect of a probiotic infant formula on infections in child care centers: Comparison of two probiotic agents. *Pediatrics*, **115**, 5–9.

WenChian, L., HungChi, H. & ChengChun, C. (2003) Viability of microencapsulated bifidobacteria in simulated gastric and bile solution. *International Journal of Food Microbiology*, **86**, 293–301.

WenRong, S. & Griffiths, M.W. (2000) Survival of bifidobacteria in yogurt and simulated gastric juice following immobilization in gellan-xanthan beads. *International Journal of Food Microbiology*, **61**, 17–25.

Witthuhn, R.C., Schoeman, T. & Britz, T.J. (2004) Isolation and characterization of the microbial population of different South African kefir grains. *International Journal of Dairy Technology*, **57**, 33–37.

Wolpert, V.M. (1988) Skyr and skyrwhey. *Dairy Industries International*, **90**(3), 29. (Note to publisher - one page only).

Woo, S.I., Kim, J.Y., Lee, Y.J., Kim, N.S. & Hahn, Y.S. (2010) Effect of *Lactobacillus sakei* supplementation in children with atopic eczema-dermatitis syndrome. *Annals of Allergy, Asthma, and Immunology*, **104**, 343–348.

Wszolek, M., Tamime, A.Y., Muir, D.D. & Barclay, M.N.I. (2001) Properties of kefir made in Scotland and Poland using bovine, caprine and ovine milk with different starter cultures. *Lebensmittel-Wissenschaft und –Technologie*, **34**, 251–261.

Wszolek, M., Kupiec-Tehan, B., Skov Guldager, H. & Tamime, A.Y. (2006) Production of kefir, koumiss and other related products. *Fermented Milks* (ed. A.Y. Tamime), pp. 174–216, Blackwell Publishing, Oxford.

Wu, R., Sun, Z., Wu, J., Meng, H. & Zhang, H. (2010) Effect of bile salts stress on protein synthesis of *Lactobacillus casei* Zhang revealed by 2-dimensional gel electrophoresis. *Journal of Dairy Science*, **93**, 3858–3868.

Xanthopoulos, V., Hatzikamari, M., Adamidis, T., Tsakalidou, E., Tzanetakis, N. & Litopoulou-Tzanetaki, E. (2000) Heterogeneity of *Lactobacillus plantarum* isolates from Feta cheese throughout ripening. *Journal of Applied Microbiology*, **88**, 1056–1064.

Xie, N., Zhou, T. & Li, B. (2012) Kefir yeasts enhance probiotic potential of *Lactobacillus casei* H9: The positive effects of co-aggregation between the two strains. *Food Research International*, **45**, 394–401.

Yadav, R. & Shukla, P. (2015) An overview of advanced technologies for selection of probiotics and their expediency: A review. *Critical Reviews in Food Science and Nutrition*, **55**, 3233–3242.

Yamani, M.I., Al-Nabulsi, A.A., Haddadin, M.S. & Robinson, R.K. (1998) The isolation of salt tolerant lactic acid bacteria from ovine and bovine milks for use in the production of Nabulsi cheese. *International Journal of Dairy Technology*, **51**, 86–89.

Yang, Y., An, H., Zhai, Z., Wang, G., Li, J. & Hao, Y. (2016) Complete genomic sequence of *Lactobacillus helveticus* CAUH18, a potential probiotic strain originated from koumiss. *Journal of Biotechnology*, **224**, 18–19.

Yerlikaya, O. (2014) Starter cultures used in probiotic dairy product preparation and popular probiotic dairy drinks. *Food Science and Technology - Campinas*, **34**, 221–229.

Yerlikaya, O. & Ozer, E. (2014) Production of probiotic fresh white cheese using co-culture with *Streptococcus thermophilus*. *Food Science and Technology - Campinas*, **34**, 471–477.

Yerlikaya, O., Ender, G., Torunoglu, F.A. & Akbulut, N. (2013) Production of probiotic milk drink containing *Lactobacillus acidophilus, Bifidobacterium animalis subsp. lactis and Lactobacillus casei. Agro Food Industry Hi-Tech*, **24**(2), 49–52.

Yi, W.-Z., Tang, W.-G. & Liu, C.-G. (2014) Effect of compound oligosaccharides on the stability of brown probiotic milk beverage. *China Dairy Industry*, **42**(7), 34–38.

Yildiz, F. (ed.) (2010) *Development and Manufacture of Yogurt and Other Functional Dairy Products*, CRS Press (Taylor & Francis Group), Boca Raton.

Yilmaz-Ersan, L. (2013) Fatty acid composition of cream fermented by probiotic bacteria. *Mljekarstvo*, **63**, 132–139.

Yilmaztekin, M., Ozer, B.H. & Atasoy, F. (2004) Survival of *Lactobacillus acidophilus* LA-5 and *Bifidobacterium bifidum* BB-02 in white-brined cheese. *International Journal of Food Science and Nutrition*, **55**, 53–60.

Young, C.K. & Nelson, F.E. (1978) Survival of *Lactobacillus acidophilus* in "sweet acidophilus milk" during refrigerated storage. *Journal of Food Protection*, **41**, 248–250.

Yuksel, A.K. & Bakirci, I. (2014) Determination of certain quality characteristics of probiotic yoghurts produced with different prebiotic combinations during storage. *Akademik Gida – Academic Food Journal*, **12**(2), 26–33.

Yuksel, A.K. & Bakirci, I. (2015) Investigation of the volatile compound profiles of probiotic yogurts using different inulin and demineralised whey powder combinations. *Food Science and Biotechnology*, **24**, 807–816.

Zacarias, M.F., Binetti, A., Laco, M., Reinheimer, J. & Vinderola, G. (2011) Preliminary technological and potential probiotic characterisation of bifidobacteria isolated from breast milk for use in dairy products. *International Dairy Journal*, **21**, 548–555.

Zadernowska, A., Chajecka-Wierzchowska, W. & Ogryzek, M.P. (2015) Growth potential of *Yersinia enterocolitica* in blue cheese and in blue cheese with probiotic - *Lactobacillus acidophilus* LA-5. *Journal of Food Science and Technology*, **52**, 7540–7544.

Zafari, S., Marhamatizadeh, M.H. & Azadneya, P. (2013) The effect of *Dorema aucheri* on the activities of *Bifidobacterium bifidum* and *Lactobacillus acidophilus* in probiotic milk and yoghurt. *International Journal of Advanced Biological and Biomedical Research*, **1**, 1033–1039.

Zanirati, D.F., Abatemarco Jr., M., de CiccoSandes, S.H., Nicoli, J.R., Nunes, A.C. & Neumann, E. (2015) Selection of lactic acid bacteria from Brazilian kefir grains for potential use as starter or probiotic cultures. *Anaerobe*, **32**, 70–76.

Zare, F., Champagne, C.P., Simpson, B.K., Orsat, V. & Boye, J.I. (2012) Effect of the addition of pulse ingredients to milk on acid production by probiotic and yoghurt starter cultures. *LWT - Food Science and Technology*, **45**, 155–160.

Zhang, H., Hao, Y., Zhang, D. & Luo, Y. (2011) Characterization of the cyptic plasmid pTXW from *Lactobacillus paracasei* TXW. *Plasmid*, **65**, 1–7.

Zhang, L., Zhang, X., Liu, C., Li, C., Li, S., Li, T., Li, D., Zhao, Y. & Yang, Z. (2013) Manufacture of Cheddar cheese using probiotic *Lactobacillus plantarum* K25 and its cholesterol-lowering effects in a mice model. *World Journal of Microbiology and Biotechnology*, (2013) **29**, 127–135.

Zhihua, H., Xinxi, G., Jining, T., Jie, Y., Hongtao, T. & Yunbo, L. (2013) Optimization of mungbean milk medium for *Lactobacillus helveticus* and *Lactobacillus casei*. *Journal of Chinese Institute of Food Science and Technology*, **13**(6), 117–122.

Zomorodi, S., Khosrowshahi, A., Rohani, S.M.R. & Miraghaei, S. (2011) Survival of *Lactobacillus casei*, *Lactobacillus plantarum* and *Bifdobacterium bifidum* in free and microencapsulated forms on Iranian White cheese produced by ultrafiltration. *International Journal of Dairy Technology*, **64**, 84–91.

5 Current Legislation of Probiotic Products

M. Hickey

5.1 Introduction and background

The belief that certain fermented milks provide health benefits above and beyond their nutritional value has been recognised for centuries by various cultures in many countries. However, it was not until the beginning of the twentieth century that the first scientific claims for food providing health benefits were made by the Nobel laureate Élie (Ilya) Metchnikoff (1845–1914). Metchnikoff was born in the village of Ivanovka, in what is now the province of Kharkiv in Ukraine, then called Kharkoff, Little Russia. He was a Professor of Zoology at the University of Odessa for about 16 years, but following difficulties with the Tsarist regime there, in 1888 he was invited by Louis Pasteur to work in the Pasteur Institute in Paris, where he remained for the rest of his life (Metchnikoff, 1921). While there, he became interested in the study of micro-organisms and especially their roles in the immune system, and learnt of the observations of a young Bulgarian student in Geneva called Stamen Grigoroff (1878–1945) regarding the number of centenarians to be found in Bulgaria where yoghurt was a stable food (Metchnikoff, 1908). In 1908, in 'The Prolongation of Life: Optimistic Studies', Metchnikoff proposed that the longevity of Bulgarian peasant farmers was related to their ingestion of fermented milks. In addition, Metchnikoff believed that there was potential to replace harmful bacteria in our bodies with beneficial bacteria. It is claimed he drank sour milk every day throughout his life. In 1908, he shared the Nobel Prize for Physiology Medicine with Paul Erlich for their work in the field of immunology. Stamen Grigoroff was born in the village of Studen Izvor, Tran Region, Bulgaria. In 1905, aged 27, working in the laboratory of Professor Masole in Geneva, he identified the micro-organisms in yoghurt, which he called *Bacterium bulgaricum* (Grigoroff, 1905). Professor Masole wrote to Mechnikoff telling him of his young assistant's findings. Metchnikoff invited Grigorov to visit the Institut Pasteur, where he read a paper on the *Lactobacillus* strain he had discovered. Soon after Coendi and Mikelson, assistants to Mechnikoff, named the micro-organism *Bacillus bulgaricus* (Grigoroff) in his honour. This micro-organism is now called *Lb. delbrueckii* subsp. *bulgaricus*.

At about the same time a French paediatrician, Henry Tissier, who also worked at the Pasteur Institute, observed that the stools of young children with diarrhoea were

Probiotic Dairy Products, Second Edition. Edited by Adnan Y. Tamime and Linda V. Thomas.

characterised by low numbers of particular *y*- or bifid-shaped bacteria, while those of healthy children had high numbers of the same type of organisms (Tissier, 1906; Anukam & Reid, 2007). Tissier suggested that there was a possibility of administering such bacteria to the ill children. Very little else is known about Tissier; he is not included in the long list of biographies of famous scientists on the Pasteur Institute's website.

Alfred Nissle (1874–1965) was born in Köpenick district in the southeast of Berlin. In 1912, he joined the Institute of Hygiene of the University of Freiburg. From 1915 to 1938, he was head of the Institute for Infectious Diseases in Freiburg. In 1917, he isolated a strain of non-pathogenic *Escherichia coli* (*E. coli* Nissle 1917) from the faeces of a World War 1 soldier who did not develop enterocolitis during a severe outbreak of shigellosis. He used the identified strain to treat intestinal diseases, such as shigellosis and salmonellosis, with a considerable amount of success. *E. coli* Nissle 1917 (EcN) has many features in common with probiotic lactic acid bacteria but was the first non-lactic acid bacteria (LAB) probiotic identified (Sonnenborn & Schulze, 2009).

Leo Rettger (1874–1954) was born in Huntington, Indiana, on 17 March 1874. He taught at Yale University from 1902 to 1942. He was Professor of Bacteriology there and became the first proponent of probiotics in the United States. In 1920, he showed '*Lactobacillus bulgaricus*' (presumed to be *Lactobacillus delbrueckii* subsp. *bulgaricus*) could not survive in the human intestine – this seemed to contradict Metchinikoff's theory, and as a result the concept of the benefits of fermented food waned for some years. However, in 1935, Rettger published a paper that identified that certain strains of *Lactobacillus acidophilus* were very active, when introduced to the human digestive tract. Tests were carried out, and it was found to be helpful in relieving chronic constipation (Rettger *et al.*, 1935).

Minoru Shirota (1899–1982) was born in Inadani, a village in Western Nagano, Japan. In 1921, he chose to study medicine in Kyoto University, at a time when a number of children died in his village due to infectious diseases and malnutrition. Inspired by the ideas of Mechnikoff, he sought to develop a stronger strain of lactic acid bacteria that would help destroy the harmful bacteria living in the intestines, and thus help maintain or improve human health. In 1930, he succeeded in culturing a strain of lactic acid bacteria, *Lactobacillus casei* Shirota. Five years later, he succeeded in incorporating this strain into a drink he called Yakult. This product was sold only in Japan until 1964, when Yakult expanded to markets outside Japan; it is now sold in 35 countries worldwide (Anonymous, 2011c).

Since that time, many researchers have isolated other micro-organisms that have health benefits associated with humans, and many of these have been commercialised. In 2003, a scientific paper listed 17 strains of lactobacilli and five strains of bifidobacteria that were used by internationally known food manufacturers, such as Nestlé, Danone, Arla, Valio, Yakult and Fonterra, and major starter culture suppliers, such as Chr. Hansen and Danisco (Sanders, 2003).

Total safety of any product cannot be guaranteed, but some of the *Lactobacillus* species have been used in the manufacture of fermented milk products and cheese for centuries, and have had a very good safety record during that time. On very rare occasions, there are reports of infections linked with consumption of commercial probiotic fermented products by individuals with underlying medical conditions, but the lactobacilli used were not necessarily causal (Mackay *et al.*, 1999; Rautio *et al.*, 1999). A review on the safety of certain micro-organisms, such as lactobacilli and bifidobacteria, concluded that there were essentially no risks involved in their oral consumption by healthy individuals (Salminen, 1996). Nonetheless, the increasing trend of incorporating into foods specific strains, that have been isolated from humans but not traditionally used as starter cultures, is likely to increase the need, requirements and rigour of safety assessment. Also, the increasing numbers of immunocompromised individuals in the population can only contribute to this need. Furthermore, *Enterococcus*, *Bacillus* and *Saccharomyces* species are also used in fermented food and dietary supplements throughout the world. Safety assessments of such strains need to be quite rigorous as some pathogenic strains are also found in these genera.

While the scientific concept of using beneficial bacteria had its origins early in the twentieth century, the term 'probiotic' only appeared in the 1960s. There seems to be general agreement that the term 'probiotic' was first used in a 1965 paper. However, in this paper, the term was used in a different context to describe substances secreted by one organism that stimulate the growth of another (i.e. associative growth behaviour). It would appear that the term probiotic was used in that instance as an antonym to the term antibiotic. It was not until 1974 that the term probiotic was actually used to describe a feed or food supplement by Parker (1974), who defined it as 'organisms and substances which contribute to intestinal microbial balance'. However, this definition also includes what we now call prebiotics. In 1989, Fuller (1989), an expert in gut microbial ecology at the Agriculture and Food Research Council (AFRC), which was then based in Reading University, modified Parker's definition to: 'a live microbial feed supplement which beneficially affects the host animal by improving its intestinal microbial balance'.

In 2002, a Working Group of a Food and Agriculture Organisation of the United Nations and World Health Organisation (FAO/WHO) Expert Consultation proposed the following definition: 'Live micro-organisms which when administered in adequate amounts confer a health benefit on the host' (FAO/WHO, 2002). The 2002 definition, although widely accepted at least in the scientific community, has not been adopted into any international standard (at least to date). In 2014, a similar panel of scientific experts organised by the International Scientific Association for Probiotics and Prebiotics (ISAPP) agreed that the FAO/WHO (2002) definition for probiotics was still relevant, but advised a minor grammatical correction as follows: 'Live micro-organisms that, when administered in adequate amounts, confer a health benefit on the host' (Hill *et al.*, 2014).

From the 1990s, an increasing number of scientific reports appear in the literature detailing the specific health benefits of probiotic micro-organisms; this led to an increasing clamour to allow the use of specific claims in the labelling, marketing and advertising of products containing probiotic micro-organisms. These considerations not only applied to probiotic products, but also encompassed the overall area of functional foods, which fall into the grey area between foods and medicines. As a reaction to this, some countries developed their own legislative systems to ensure food safety and consumer protection and to address the issues raised by this relatively new category of products.

Inevitably, differences have arisen between the approaches used in different jurisdictions, and this chapter shall cover the approaches in a number of regions of the world, starting with Japan, and going on to discuss the European Union (EU), the United States

(USA), Canada and China. It shall then go on to describe the progress that has taken place at an international level; in particular, the role and status of Codex Alimentarius will be examined. Finally, some conclusions and possible future developments will also be considered.

5.2 The situation in Japan

Whenever legislation on the subject of functional foods is discussed, reference is made to the Japanese system of regulation for foods with health claims and foods for special dietary use. Many will be familiar with the category of Foods for Specified Health Use (FOSHU), although, as we shall see, this is but one specific category of such products in Japan. The Japanese concept of functional foods and the consequential legislation on foods with health claims, including FOSHU, have their origins in two main factors: (a) the severe malnutrition in certain sections of the Japanese population in the aftermath of World War II, leading to the schools' lunch programme and permitting the addition of certain nutrients to certain staple foods, such as bread, and (b) concerns regarding the ageing of the Japanese population and the burden that this was likely to place on the state's finances in the future.

In 1952, the Nutrition Improvement Law [Law No. 248, enacted by the Diet (国会), the Japanese Parliament, on 31 July 1952] created a food category for special dietary uses. Article 1 of the law stated, 'This law aims to contribute to promote national welfare by means of striving for the elevation of ideas of nutritional improvement of the people, the inquiry into the nutritional conditions of the people and the maintenance and the development of the health and the physical strength of the people devising necessary steps about the betterment of the nutritional conditions of the people' (Mitsuda, 1958). This law, which paved the way for later legislation in this area, mainly addressed the implementation of the outcomes of the National Nutrition Surveys that had been undertaken annually since 1946, the development of a national nutrition policy and the provision of, and enrichment with, nutrients in foods for specific groups such as infants, pregnant women and medical patients and for the elderly with certain eating difficulties. FOSHU came somewhat later (Tomita, 2007; Yamada *et al.*, 2008).

In 1984, the Ministry of Education, Science and Culture carried out a study on statistical analysis and an outlook on food nutrition (Ichikawa, 1994) and, in 1986, a similar report of this study identified three main functions of food:

- The food's nutritive value;
- The food's organoleptic appeal; and
- The food's physiological factors, which include the regulation of bodily functions, disease prevention, promotion of recovery and good health.

The concept of functional foods grew out of the third function in this list, and a Functional Foods Forum was established comprising experts from food and nutrition departments of Japanese universities. In addition, the then Ministry of Health and Welfare was promoting functional foods, and the industry was keen to market such products using health claims. In 1989, the Nutrition Law was enacted, and this included the manufacture of functional foods, which could make health claims. A Functional Foods Liaison Board was established to work with the industry, and various Functional Food Sub-Committees were set up. Since 1991, FOSHU became the official term for such foods, replacing the original term functional foods, and they are defined as 'processed foods containing ingredients that aid specific bodily functions in addition to being nutritious'. The law also aimed at preventing misleading and ill-defined health claims and listed five categories of Food for Special Dietary Uses:

- Milk powder for pregnant and lactating women;
- Infant formula;
- Food for elderly people with difficulty in chewing or swallowing;
- Foods for medical patients; and
- FOSHU.

In 2003, an additional category of Foods with Nutrient Function Claims (FNFC) was added to the existing category of FOSHU. In 2005, following expert consultations, the relevant Ministry, then named the Ministry for Health Labour and Welfare (MHLW), changed the existing FOSHU system to include the new subsystems, which were entitled standardised FOSHU, qualified FOSHU and disease risk reduction claims for FOSHU, based on the level of scientific evidence supporting the relationship between the food or ingredient in the food and the health condition concerned (Figure 5.1) (Yamada *et al.*, 2008).

In 2015, a new category of Foods with Function Claims was introduced in order to make more products available, clearly labelled with certain nutritional or health functions, and to enable consumers to make more informed choices. Before being placed on the market, the food manufacturer is required to submit information to Japan's Consumer Affairs Agency (CAA) on food safety, the function claim that is based on scientific evidence, and the system they have put in place to collect information on any adverse health effects. The submitted information is disclosed on the website of the CAA (CAA Japan, 2016b).

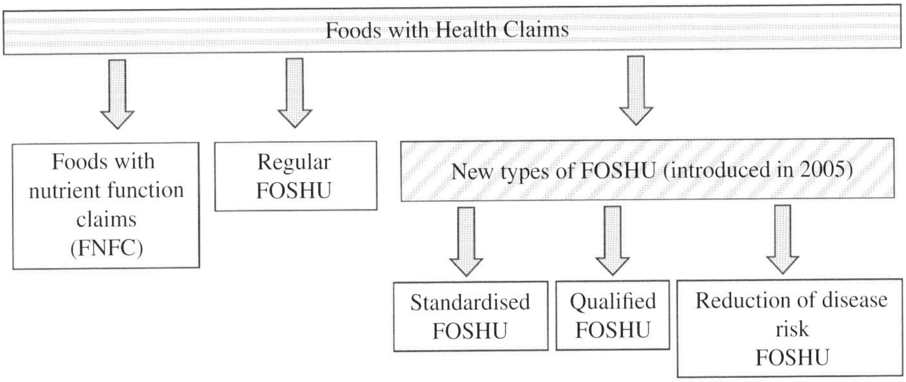

Figure 5.1 Categories of foods with health claims in Japan following the restructuring of the Foods for Special Health Uses (FOSHU) system in 2005 into four groups based on the strength of the evidence behind the claim.

Table 5.1 The seven broad categories of FOSHU claims in Japan.

Gastrointestinal health (most probiotics are included in this category)
Hypertension control
Blood sugar control
Body fat control
Dental health
Cholesterol and triacylglycerol (TAG) level control
Bone formation

The CAA was established as a new government agency on 1 September 2009, and one of the responsibilities it was given was the administration of the Japanese food-labelling system, including those provisions covering FOSHU. The CAA began discussions on a review of the health foods systems in November 2009. The report on these discussions was published in late August 2010 and included improving the FOSHU approval process with regard to speed and transparency and also the presentation of the necessary labelling information. The report also included a commitment to regulate health foods, such as dietary supplements and other health-related products outside the FOSHU system, by developing guidelines to cover false and misleading labelling; such foods are not allowed to make health or nutrient function claims. In addition to the CAA, the Consumer Commission also was established on 1 September 2009. It operates as an independent third-party agency to monitor general consumer affairs administration of the government ministries and agencies, including the CAA. It is composed of a panel of independent experts, and its meetings are held in public and open to consumers to attend. Certain expert consultative groups and committees of the Consumer Commission have roles in the review of FOSHU applications, as shall be outlined in this chapter. Approved FOSHU product claims may be grouped under seven broad headings (see Table 5.1). Some examples of health claims for milk-based probiotic products that have been approved for FOSHU labelling on the Japanese market are listed in Table 5.2.

5.2.1 *Subsystems of FOSHU*

Qualified FOSHU

Qualified FOSHU refers to the approval of a food with a health function which is not substantiated with the high level of scientific evidence that is required of regular FOSHU, or to the approval of a food with certain effectiveness but without the mechanism of the effective element for the function being clearly established. Standardised FOSHU claims are approved where a product meets the set standards and specifications. In the case of a reduction of disease risk FOSHU, a claim is permitted when reduction of disease risk is clinically and nutritionally established for an ingredient (Anonymous, 2011a).

Table 5.2 Examples of health claims approved on FOSHU products containing probiotic micro-organisms.

Commercial company	Health claims
Meiji Milk Products	Due to the effects of *Lactobacillus* spp. LB 81 in Bulgarian yoghurt, this fermented milk regulates the balance of intestinal bacteria that lead to and maintain a good intestinal condition.
Yakult	Due to the effects of the Yakult strain (*Lactobacillus casei* Shirota), which can reach the intestine alive, Yakult maintains the intestine in good health by increasing beneficial bacteria, decreasing harmful bacteria and improving the intestinal environment. Manufactured products are known as fermented milk beverage and Sofhul (smooth-textured yoghurt).
Morinaga Milk	Due to the effects of '*Bifidobacterium longum* BB536' (presumed to be *Bifidobacterium longum* subsp. *longum* BB536), which reaches the intestine alive, the bifidobacteria in the intestines increase, improve the intestinal environment and regulate intestinal and stomach conditions. Manufactured products are known as Bifidus BB536 Yoghurt and Caldus milk.
Milk Takanashi Products	Due to the effects of *Lactobacillus rhamnosus* GG, which can reach the intestine alive, this product increases beneficial bacteria and decreases harmful bacteria. It improves the intestinal environment and regulates the stomach conditions. Manufactured product is known as Onaka-He-GG.

Data compiled from CAA Japan (2016a).

Disease reduction risk FOSHU

There are only two approved disease reduction claims for foods (Anonymous, 2011a). These are: (a) calcium and osteoporosis: where the claim statement is that a proper amount of calcium contained in healthy meals, with appropriate exercise, may support healthy bones of young women and reduce the risk of osteoporosis when aged, and (b) folic acid and neural tube defects: where it may be claimed that the intake of a proper amount of folic acid contained in healthy meals may support a woman to bear a healthy baby, by reducing the risk of neural tube defects, such as spondyloschisis, during foetal development.

Foods with nutrient function claims

As mentioned in this chapter, there are also legal provisions for FNFC which can be made for foods that meet specified requirements for any of 12 named vitamins and the minerals calcium, copper, iron, magnesium and zinc (Yamada *et al.*, 2008; Anonymous, 2011b). Unlike FOSHU products, FNFC products do not require permission from or notification to the Japanese authorities, provided that they meet the established standards and specifications that specify the nutrient content required for the use of each claim as set down, and the labelling must include any warnings as required. As these do not affect probiotics directly, they shall not be addressed further in this chapter.

5.2.2 *Essential elements for obtaining FOSHU approval*

The Japanese approval procedure differs from those of other countries in that approval is given for individual products. This is unlike the situation in the USA, where only generic claims can be made on certain foods; while in the EU, at least in the case of general-function claims, all foods that meet the specified requirements may carry the claim.

Prior to September 2009, it was the norm that applications were submitted to the local or regional health authority and passed on to the Office of Health Policy on Newly Developed Foods of the Ministry of Health, Labour and Welfare but, since the formation of the CAA, applications are now submitted to the CAA's Labelling Division.

If the application is deemed adequate and in order, the CAA then forwards the application to the Assessment and Evaluation Group for Novel Food of the Consumer Commission to review its efficacy. It is then forwarded to the Expert Assessment Group of the Food Safety Commission which reviews the food safety aspects. It is then returned to the Assessment Committee for Novel Foods of the Consumer Commission for a comprehensive review of both efficacy and food safety. Requests for additional documentation and information may be made throughout this process.

Following successful completion of the various reviews, the application is passed to the Ministry of Health, Labour and Welfare to check whether the labelling violates the Pharmaceutical Affairs Act. On completion of the entire process, the product receives FOSHU approval and is permitted to use the approved claim (Figure 5.2). The labelling of approved products can also carry the FOSHU logo (Figure 5.3) or the Qualified FOSHU logo (Figure 5.4) as appropriate. The examples shown are those that indicate, in Japanese characters above the logo, 消費者庁許可, which shows that the products are permitted or approved by the CAA. Products approved by the Ministry of Health, Labour and Welfare prior to the establishment of the CAA in September 2009 may indicate this by the use of the Japanese characters 厚生労働省許可 above the logo.

The approval process generally takes about one year from the time of submission to obtaining approval. Initially, the rate of applications was slow; this was attributed to the length of time taken to obtain approval (in the earlier years, it was reputed to take some years), and also to the costs involved. From 1991 to 1998, only 126 products received FOSHU approval; however, amendments to the FOSHU Law in 1998 reduced the requirements for:

- The amount of scientific documentation that manufacturers must submit;
- A certificate confirming that all submitted scientific documentation had been reviewed by outside scientific experts was removed and replaced by a requirement that studies had been published in a scientific journal; an industry-sponsored journal is deemed acceptable in this respect;
- For products which previously were required to be tested by the National Health and Nutrition Laboratory, now the manufacturer's own analytical tests are deemed acceptable.

By the end of 2003, a total of 396 products had been granted FOSHU status; by April 2011, the total had progressed to 995 products; and by mid-October 2016, it progressed to 1325 products (Figure 5.5) (CAA Japan, 2011).

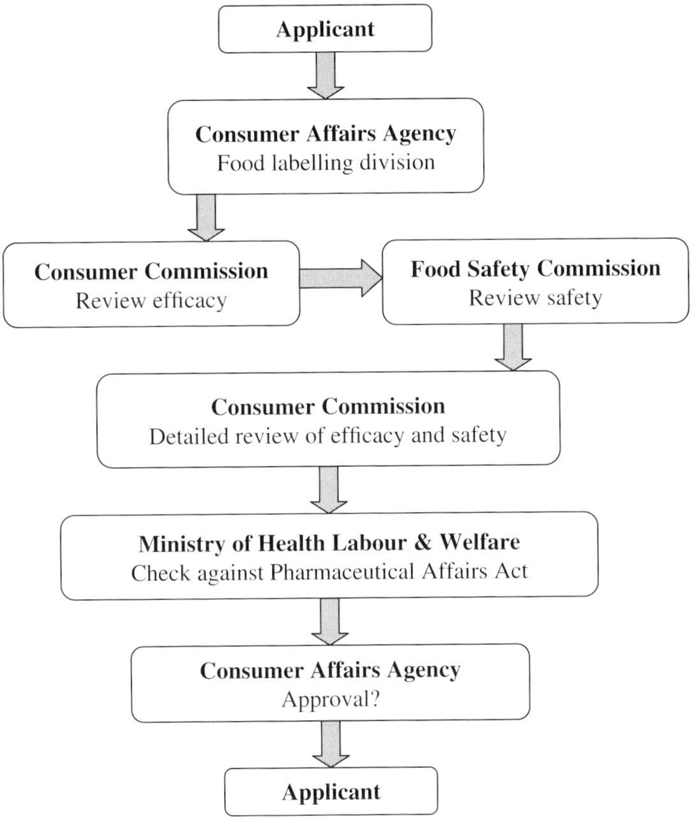

Figure 5.2 The FOSHU approval system.

Figure 5.3 The FOSHU logo.

Figure 5.4 The qualified FOSHU logo.

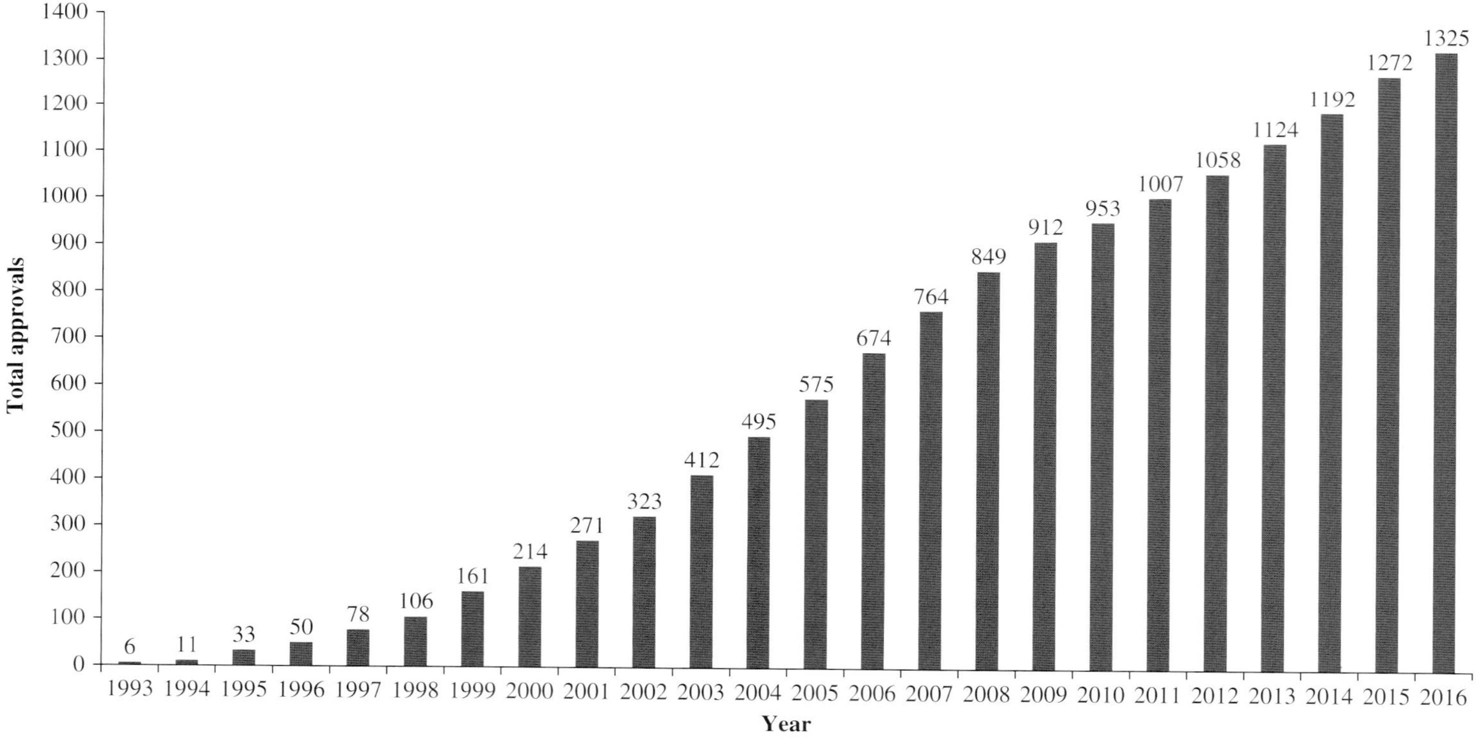

Figure 5.5 Cumulative FOSHU approvals from 1993 to 2010. After CAA Japan (2016a, 2016b).

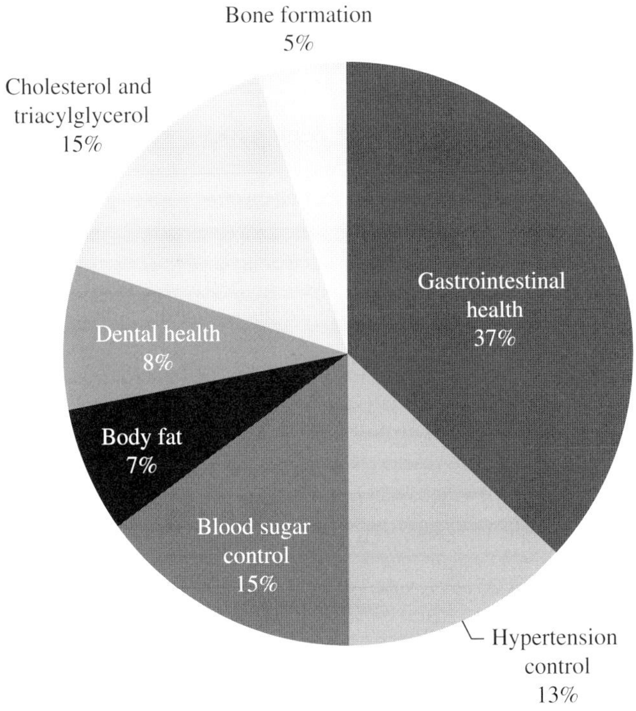

Figure 5.6 FOSHU by health effect, expressed as a percentage of total approvals. After CAA Japan (2011).

A list of all FOSHU products is available in Excel file format but in Japanese only (CAA Japan, 2016a). In terms of product approvals, the largest category is that of gastrointestinal health, which consists mainly of probiotics, prebiotics and dietary fibre. By the end of 2001, such products accounted for about 60% of the 289 approved products (Arai *et al.*, 2002), and probiotics accounted for about one-third of the gastrointestinal health category. In 2008 it was indicated that approvals related to gastrointestinal health still accounted for about half of all FOSHU approvals, and by April 2011 the figure was 36% (Figure 5.6) (Yamada *et al.*, 2008; CAA Japan, 2011). Up-to-date analysis of the categories of product approvals has not been readily available since 2011.

5.2.3 *Features of the new category of foods with function claims*

Under this new category, introduced in 2015, there is no safety assessment or evaluation of functionality by government bodies. The food operators can use functional claims on their own. However, prior notification must be given to the CAA 60 days before launch. The notification number should appear on the packaging. Information on each product, including scientific data, can be seen on the website of the CAA, where any revisions or modifications of the text can also be accessed. According to the website of the CAA, there were 325 notifications up to late March 2016; included are approximately 25 probiotics, including yogurts and probiotic supplements. Examples of some of these

Table 5.3 Examples of notifications of new foods with a function claim in Japan for products containing probiotic micro-organisms, 2015–2016.

Commercial product	Function claim
Megumi yoghurt drink	This product contains *Lactobacillus gasseri* SP that has the function of decreasing visceral fat.
Morinaga bifidus yoghurt and PREMil[1]	This product contains '*Bifidobacterium longum* BB536' (presumed to be *Bifidobacterium longum* subsp. *longum* BB536) that enhances the intestinal environment and adjusts the intestinal condition.
Glico BifiX breakfast fermented milk drink and yoghurt	This product contains Bifidus BifiX ('*Bifidobacterium lactis* GCL 2505' – presumed to be *Bifidobacterium animalis* subsp. *lactis* GCL 2505). This live bacterium has been reported to improve the intestinal environment by reaching and living in the intestine, and by proliferating in the intestine.

[1] Product contains 1.8 g $100\,g^{-1}$ fat, protein-enriched milk with dietary fiber.
Data compiled from CAA Japan (2016b).

products and their claims are Morinaga Milk, Takanashi Yogurt, Calpis Milk, Megmilk Snow and Ekazi Glico (see Table 5.3).

5.2.4 *Unique features of the Japanese FOSHU system*

The unique features of the FOSHU system are summarised as follows:

- It is a voluntary procedure, but it has government approval.
- It is granted to individual products.
- Specific health claims are approved prior to use.
- Approval is based on documented scientific safety and effectiveness considerations.
- Approved products may use the FOSHU logo on their labels.

Being the first regulatory system to address functional foods, the Japanese approach has proved a model and reference point for the sector. It is expensive to get FOSHU approval; the cost has been estimated as being between US$850 000 and US$1.5 million (Yamaguchi, 2004; Patel *et al.*, 2008). This estimate may be outdated, but no recent estimates of current costs are readily available.

5.3 The legislative situation in the European Union

The EU's horizontal legislation considerations addressing aspects of food safety, novel foods and food ingredients, and food labelling shall be addressed before going on to deal with the regulation of nutrition and health claims. The latter have the greatest impact on probiotics products in the EU and shall be discussed in some detail.

5.3.1 *Relevant EU food safety legislation*

Regulation 178/2002 lays down some general principles and requirements on food safety in the EU (EU, 2002). Article 14.1 of this Regulation requires that food put on the market should be safe. The Regulation goes on to specify that food safety in this context encompasses short-term, intermediate and long-term effects on consumers, and also any possible effects on subsequent generations (through teratogenic effects). Prior to its adoption, food manufacturers were required to ensure their products met the specific requirements of EU legislation but, unlike the laws of an increasing number of member states, the law did not include the broad requirement that food put on the market should be safe. While it could be held that the Product Liability Directive 85/374 (EU, 1985, 1999) made it a requirement to have safe products by imposing strict liability on manufacturers whose products caused harm to individuals, it did not allow action by the authorities before the products actually caused a problem. Furthermore, the scope of the Liability Directive excluded primary agricultural products. By way of contrast, the General Food Law Regulation 178/2002 allows pre-emptive action and also removes the exemption for primary agricultural produce.

5.3.2 *Novel food regulation in the European Union*

The Novel Foods Regulation 258/97 (EU, 1997), which came into force on 15 May 1997, requires novel foods and novel food ingredients that have not been previously used to a significant degree within the EU to undergo a safety assessment before being placed on the market. One could have an interesting discussion on the precise legal meaning of the words 'significant degree' in the definition, but doubtless clarity on this shall emerge over time. The categories of foods listed in the scope of the regulation include novel food ingredients and novel food processes, as well as novel foods themselves. Up to the beginning of 2015, about 180 novel food applications were made, with about 80 such products authorised for use in the EU; however, only one of these applications and notifications relates to a probiotic product.

Some milk products containing probiotics were on the market prior to the Novel Foods Regulation coming into force and so were not affected by this regulation, because commercial foods in use in at least one member state before the EU regulation came into force were allowed on the EU market under the 'principle of mutual recognition'. One example of such prior approval of a probiotic is the strain *Lactobacillus rhamnosus* GG, which was approved by the Advisory Committee for Novel Foods and Processes (ACNFP) in 1992 in the United Kingdom (UK) (Anonymous, 1992). It should be noted that its approval was as a novel starter culture and not as a novel probiotic. This was the first micro-organism to undergo a formal novel food evaluation by the UK ACNFP, and there was discussion in the Committee as to whether the proposed use was actually novel. The conclusion was that, while the organism was already present in the UK diet at low levels, the proposed use could significantly increase the level of consumption and that justified its assessment as a novel food. Furthermore, the conclusion of the assessment was that there were no food safety concerns regarding its consumption in foods. In 1996, a scientific paper outlined some of the selection criteria and clinical evaluation applicable to *Lb. rhamnosus* GG (Salminen, 1996), and the specific techniques used in the selection process for the organism are described fully in USA Patent 4839281 (Gorbach & Goldin, 1989).

5.3.3 *Genetically modified organisms*

Genetically modified organisms (GMOs) are organisms, such as plants, animals and micro-organisms, the genetic characteristics of which have been modified artificially in order to give them new or improved properties. To ensure the safety of GMOs, the EU has established a detailed legal framework covering the area.

Given the controversy in recent years regarding GMOs, it is likely to be some time before probiotics developed using genetic modification will appear on the EU market. It has to be borne in mind that, as part of the selection and evaluation of probiotics, improvements of the performance of some of the strains are made; however, to date those that have been developed and are in use have been using non-GMO selection techniques.

5.3.4 *EU food-labelling provisions*

Following much discussion and debate, a new regulation, Regulation 1160/2001 on the provision of food information to consumers, was adopted and published in November 2011 (EU, 2011). Among its purposes, this repeals the earlier food-labelling Directive 2000/13. This regulation entered into force on 13 December 2011 and its provisions applied from 13 December 2014; the provisions on mandatory nutritional declaration applied from 13 December 2016. This new regulation retained the provisions as regards the general principle underpinning food labelling that purchasers must not be misled, including by claims that could mislead. Much more specific provisions on the subject of claims were addressed in the regulation on nutrition and health claims that had been adopted earlier, in late 2006.

5.3.5 *EU nutrition and health claims*

The regulation of nutrition and health claims in the EU has posed a major problem for many existing products, and in particular for probiotics. Efforts had been made to regulate health claims in the late 1980s and early 1990s, at a time when nine of the then 12 member states had either national legislation or guidelines on claims. These national provisions gave rise to a situation that had the potential to create barriers to the completion of the open market. The EU Commission produced draft proposals for a regulation, which was worked on for some years, but this effort came to nought, foundering on the contentious issue of health claims. Some countries wanted a total prohibition on health claims completely, while others were prepared to accept them provided they were subject to strict criteria. Barriers to, and difficulties in, intra-community trade continued to arise as a consequence, while ‘health claims’ continued to appear on foods, and some viewed them as a problem as they were unregulated.

Eventually, in July 2003, the Commission published a new proposal for a regulation on nutrition and health claims. The resulting Regulation 1924/2006 was finally adopted and published on 30 December 2006. The original published version was corrected in early 2007, and two subsequent amendments were adopted in 2008 and another in 2010

(EU, 2006, 2008b, 2008c, 2010). Full implementation of all the provisions in this regulation is not scheduled until 2021, but an implementing Regulation 353/2008 was adopted in April 2008 (EU, 2008a).

The scope of the main Regulation encompasses nutrition and health claims made on all foods marketed within the EU and includes claims made in the labelling, presentation and advertising of foods, including brand names. The scope also includes food supplements, foods for particular nutritional uses, natural mineral waters and water intended for human consumption. A claim is defined as referring to any message or representation which is not mandatory under EU or national legislation, including pictorial, graphical or symbolic representation in any form, which states, suggests or implies that a food has particular characteristics. A nutrition claim is any claim that states, suggests or implies that a food has particular beneficial nutritional properties due to the energy, nutrients or other substances provided, not provided or provided in reduced/increased amounts. A health claim is any claim that states, suggests or implies that a relationship exists between a food category, a food or one of its constituents, and health.

While nutrient content claims had been addressed to some extent already, the regulation of the field of health claims at the community level was new. Health claims are only permitted if the following information is included in the labelling (or if there is no such labelling, in the presentation and advertising):

- A statement indicating the importance of a varied and balanced diet and a healthy lifestyle;
- The quantity of the food and pattern of consumption required to obtain the claimed beneficial effect; and
- Where appropriate, a statement addressed to persons who should avoid using the food; and an appropriate warning for products that are likely to present a health risk if consumed in excess.

5.3.6 *Types of health claims*

Health claims that describe or refer to the role of a nutrient/substance in body functions are called Article 13 claims. These are further subdivided into Article 13.1 health claims and Article 13.5 health claims. Health claims that refer to a reduction of disease risk, or claims referring to children's development and health, are called Article 14 claims. It should be mentioned that medicinal claims, which claim that food can treat, prevent or cure any disease or medical condition, are still prohibited on foods; such medicinal claims can only be made for licensed medicines.

Article 13.1 claims

Article 13.1 claims address health claims other than those referring to the reduction of disease risk and to children's development and health. They refer to:

- The role of a nutrient/substance in growth, development or body functions;
- Psychological and behavioural functions; or
- Slimming or weight control, a reduction in the sense of hunger, an increase in the sense of satiety or a reduction of the available energy from the diet.

These are based on generally accepted scientific evidence and must be well understood by the average consumer. They are sometimes referred to as 'general function' claims.

In mid-2006, the Commission asked member states to submit a list of health claims under Article 13.1 made by the food industry within their jurisdiction, and to this end the competent authorities in each country contacted their food companies. The deadline for the submission of these national lists was the end of January 2008. It is believed the Commission expected to receive about 2000 such claims from member states – they actually received about 40 000. By a process of eliminating duplicates and consolidating, the total number of Article 13.1 claims to be assessed was reduced to 4306. However, when claims on botanicals, numbering 1548, were put on hold by the Commission this left a total of 2758 Article 13.1 claims to be assessed. The European Food Standard Authority (EFSA) opinions on the scientific evidence supporting these claims were published in batches, starting in October 2009 and ending in July 2011, by which time EFSA's panel on Dietetic Products, Nutrition and Allergies (NDA) had completed its assessment of all the remaining 2758 of these claims and issued 341 opinions on them.

An earlier personal review and analysis of 2719 of these EFSA opinions indicated that, of the total, 478 (17.6%) received favourable opinions (i.e. that the claims submitted were borne out by the evidence supplied), while 2212 (81.3%) received unfavourable outcomes, with 30 (1.1%) receiving mixed outcomes. The latter cases are due to the grouping of a number of claims in a single opinion; some of the health claims therein have received favourable outcomes, while others received unfavourable opinions. The remaining 38 health claims relate to the roles of various vitamins, minerals and omega-3 fatty acids on various body functions, which had been evaluated in previously published EFSA opinions; thus, their outcomes cannot be readily analysed without identifying each of the corresponding earlier opinions and their outcomes (Hickey, 2014).

Some types, or categories, of claims received a higher percentage of favourable outcomes than others. However, none of the opinions on 359 probiotic claims were favourable (see Table 5.4).

Table 5.4 Analysis of applications and authorisations of health claims under European Union Regulation 1924/2006, to October 2016.

Claim type	Total submitted	Not authorised	Authorised	Probiotic claims (not authorised)
General health claims (Article 13.1)	2104	1875	229	359
Health claims based on new scientific evidence or where protection of proprietary data was requested (Article 13.5)	112	108	4	17
Reduction of disease risk claims [Article 14.1(a)]	37	23	14	2
Claims referring to children's development and health [Article 14.1(b)]	56	44	12	9
Totals	2309	2050	259	387

EU (2006).

The main reasons for submissions on probiotics receiving unfavourable opinions included:

- Insufficient information to identify or characterise the substance or ingredient on which the claim is based;
- Insufficient scientific evidence to demonstrate that the claimed effect was beneficial to the maintenance or improvement of the functions of the body;
- Lack of precision as regards the wording of the health claim being made (examples include claims using broad terms, such as claims on improved 'gut health'); and
- Lack of sufficient human studies containing the necessary scientific data to demonstrate the claimed health benefit.

Article 13.5 claims

Article 13.5 claims are of a similar nature to the Article 13.1 health claims, but are based on newly developed scientific evidence and/or applications that contain requests for the protection of proprietary data. Up to October 2016, the EFSA panel had evaluated 112 applications; those receiving unfavourable opinions included seven milk-based or enriched-milk products, of which four were probiotic-containing products. Of the total Article 13.5 claims assessed, only four (3.6%) EFSA opinions were favourable for the proposed claims (EFSA, 2016). None of these favourable claims concerned milk-based products or probiotics (see Table 5.4).

Article 14 health claims

As mentioned, Article 14.1 includes two types of health claims: (a) reduction of disease risk claims, and (b) claims referring to children's development and health.

Article 14 claims are assessed individually and are not grouped or consolidated. To October 2016, 37 Article 14.1(a) health claims have been evaluated, including two that are related to probiotics. Fourteen (37.8%) of these were approved, but not those related to the two probiotic claims. A total of 56 Article 14.1(b) claims were evaluated, including nine related to probiotics. Twelve (21.4%) of the claims were approved but, again, none related to probiotics (Table 5.4).

Some Article 14.1(b) claims would appear to be quite similar to some Article 13.1 claims, but because they make reference to effects on children they were dealt with under Article 14. The main reasons for submissions receiving unfavourable opinions were similar to those outlined under Article 13.1 claims.

Overall, therefore, none of the 387 probiotic claims reported on the EU Register of nutrition and health claims were authorised by EFSA (EU, 2016). It is worth noting that these included some of the products which had already received FOSHU approval in Japan, and which had the same, or similar, health claims that were not accepted by the EFSA panel. Although the claims made had been subjected to scientific evaluation by both the Japanese authorities and EFSA, these evaluations had resulted in different outcomes.

It is noted that International Probiotics Association (IPA) Europe, the European chapter of the IPA, established in Brussels in 2015, claim that more than 400 applications were submitted to EFSA (Thomas, 2016). Furthermore, since no probiotic claims have been approved and under a 2007 interpretation of the EU Nutrition & Health Claims Regulation 1924/2006, the use of the term probiotic is regarded as a health claim; therefore, the use of the term probiotic on a food label in the EU has been banned since December 2012. Not surprisingly, the consequences for the market for probiotic dairy products in the EU are quite serious. It is interesting to compare the claims made on the labels of some probiotic fermented milk products on the UK and Irish markets in 2005 (Table 5.5) with the label statements made on the same, or similar, fermented milk products on the same markets in 2016 (Table 5.6). It will be noted that any references to probiotics have been removed in 2016, but the presence of named cultures is still acceptable. Where health claims are made, these now relate to the presence of named vitamins, and use approved wording for the relevant Article 13.1 claims.

Nonetheless, it has been estimated that the probiotic yoghurt industry in the EU, which had grown by an average 5% per annum from 2000 to 2012, declined by 8% in 2013 and may well lose up to €1.5 billion in revenue by 2020 unless the regulatory situation is resolved. This contrasts with continuing steady growth in other regions of the world (Thomas, 2016).

Table 5.5 Examples of claims used on labels of some probiotic fermented milk products on the UK and Irish markets in 2005.

Manufacturer/product	Probiotic micro-organisms	Claim on label
Danone (Actimel)	*Lactobacillus casei* strain Immunitas	Helps support your bodies' natural defenses.
Danone (Activia)	*Bifidobacterium* spp. strain Essensis or Digestivum	*Bifidobacterium* spp. strain Essensis or Digestivum, a natural culture unique to Danone Activia, has been proven to help our digestion work better as it supplements and supports the essential cultures in our intestinal flora. A healthy digestion is essential to a healthy life.
Yoplait (Every Body Probiotic Yoghurt Drink + 15 Vitamins and Minerals)	*Lactobacillus rhamnosus* GG (LGG)	+ LGG, the most clinically researched probiotic in the world proven to enhance your natural resistance and help you maintain a healthy digestive system. + Balance from within LGG + Healthy digestive system LGG
Müller (Vitality)	*Lactobacillus acidophilus* LA-5 and *Bifidobacterium animalis* subsp. *lactis* BB-12 (see Masco *et al.*, 2004)	The good bacteria in Müller Vitality can help maintain the balance of "good" and "bad" bacteria in your digestive system. Vitality is packed with millions of "good" probiotic bacteria *Lb. acidophilus* LA-5 and *Bif. animalis* subsp. *lactis* BB-12. Great at any time of day as often as you like. Contains prebiotic inulin.
Ocean Spray Probiotic Yoghurt	Not listed	Contains millions of friendly bacteria that help to maintain the balance of natural flora in your body, which in turn may aid digestion and general well-being.

Table 5.6 Examples of statements on labels of some fermented milk products on the UK and Irish markets in 2016.

Product	Cultures named on list of ingredients or label statements	Statement on label
Danone Actimel	*Lactobacillus casei* Danone® (This is the registered trademark used for the specific Danone strain *Lactobacillus paracasei* subsp. *paracasei* CNCM I-1518.)	Start your day with Actimel with Vitamins B_6 and D to support your immune system. It also contains L. casei[1] Danone® cultures. Enjoy as part of a healthy balanced diet and lifestyle.
Danone Activia	Bifidus ActiRegularis® (This is the registered trademark name used in the United Kingdom and Ireland for their strain *Bifidobacterium lactis*[2] DN-173 010.)	Yogurt with Bifidus ActiRegularis, and the ingredients list *Bifidobacterium lactis*[2] (Bifidus ActiRegularis®) and *Lactococcus lactis*[3] cultures Every Activia pot contains carefully selected ingredients and 4 billion Bifidus ActiRegularis® cultures to craft our delicious yogurt.
Yakult	*Lactobacillus casei* Shirota	Contains 10^{10} billion *Lactobacillus casei* Shirota per 100 ml when refrigerated (6.5 billion per bottle). Did you know there are billions of unique *Lactobacillus casei* Shirota bacteria in these small bottles? Millions of people enjoy it as part of their daily life. Yakult has come a long way since it was introduced in 1935 by the Japanese scientist Dr. Shirota. Since then Yakult has been chosen by people around the world. With 6.5 billion *Lactobacillus casei* Shirota bacteria you'll want to drink every last drop. You can enjoy this delicious drink every day.
Milbona ProViact Yogurt Drink	*L. casei*	Fat-free yogurt drink with sugars and sweeteners, with L. casei[1] cultures, Vitamin D and Vitamin B_6. Vitamin D and Vitamin B_6 contribute to the normal function of the immune system. As part of a varied, balanced diet and healthy lifestyle.

[1] L. casei stated on the label should read '*L. casei*'.
[2] *Bifidobacterium lactis* stated on the label; presumed to be *Bifidobacterium animalis* subsp. *lactis*.
[3] *Lactococcus lactis* stated on the label should read '*Lactococcus lactis* spp.'.

5.4 The USA's legislative situation on probiotics and related health claims

As is the case with the EU, the USA does not have any specific legislation on probiotics or other functional foods. The US Institute of Medicine of the National Academy of Sciences has defined functional foods as 'any modified food or food ingredient that may provide a health benefit beyond the traditional nutrients it contains'; however, this is not a legal definition. The approach to regulation is again focused on the area of health claims.

The US Food and Drug Administration (FDA) has defined four categories of foods:

- Conventional Foods;
- Dietary Supplements (i.e. intended to supplement the diet and marketed like conventional foods, although they also have to be labelled as dietary supplements);

- Foods for Special Dietary Use (i.e. intended to supply particular dietary needs for physiological conditions, overweight, food allergies and infant formula); and
- Medicinal Foods (i.e. intended for dietary management of a specific disease, under the supervision of a doctor or another health professional).

In theory, probiotics could fit into any of the above-mentioned categories but, to date, none would seem to be used in medical foods, and there are very few applications for their use in foods for special dietary purposes. A number of conventional foods contain probiotics, and these are mainly dairy products, such as yoghurts, cultured milks, milk and Cottage cheese. The USA standard of identity for yoghurt requires the use of the conventional starter culture organisms (*Streptococcus thermophilus* and *Lb. delbrueckii* subsp. *bulgaricus*) to be present, but no levels are proscribed. Certain species of lactobacilli and bifidobacteria may also be added; indeed, at the start of the twenty-first century it was estimated that over 75% of yoghurts sold in the USA market contain *Lb. acidophilus* (Sanders, 2003). However, the National Yogurt Association has introduced a voluntary Live Active Culture seal for products that contain live starter cultures, and this requires refrigerated yoghurt to contain 10^8 colony forming units (cfu) g^{-1} and frozen yoghurt 10^7 cfu g^{-1} at the time of manufacture. These requirements do not apply to the levels of other cultures named on the label (Sanders, 2003).

Now, however, the main market for probiotics in the USA is in dietary supplements, and they are sold in the form of pills, capsules, powders and drinks. The reason that so many probiotics are sold as dietary supplements would appear to be that prior approval of structure/function claims is not required for this category – this aspect will be further discussed in Section 5.4.1.

5.4.1 *Claims and labelling in the USA*

The Nutrition Labelling and Education Act (NLEA) of 1990 allowed health and disease prevention claims on food labels, such as 'any substance that expressly or by implication characterises the relationship of any substance to a disease or health-related condition' (US Congress, 1990). Such claims on foods are subject to prior authorisation by the FDA.

The Dietary Supplement Health Education Act (DSHEA), which was enacted in 1994, created another category of statements, generally referred to as 'structure/function' claims that may be made for dietary supplements (US Congress, 1994). These statements may claim a benefit related to a nutrient deficiency disease or describe the role of a nutrient or dietary ingredient intended to affect a structure or function in humans; they characterise the means by which a nutrient or dietary ingredient acts to maintain such structure or function, or describe general well-being from consumption of a nutrient or dietary ingredient. The manufacturer is responsible for ensuring the accuracy and truthfulness of these claims; the FDA does not approve them. For this reason, the law stipulates that if a dietary supplement label includes such a claim, it must state in a 'disclaimer' that the FDA has not evaluated the claim. The disclaimer must also state that the dietary supplement product is not intended to 'diagnose, treat, cure or prevent any disease' because only a drug can legally make such a claim. Some report that this requirement is not always followed in practice (Berner & O'Donnell, 1998).

In 1997, the Food and Drug Administration Modernisation Act (FDAMA) provided a second way for a health claim to be used on foods (US Congress, 1997). This allows certain health claims to be made as a result of a successful submission of a notification based on an 'authoritative statement' from the National Academy of Sciences or another scientific body of the US government. The government did not include dietary supplements in the provisions for health claims based on authoritative statements. Consequently, this method of managing health claims cannot be used for dietary supplements at this time. Only qualified health claims may be made for dietary supplements, and this came about as a result of the US federal Court of Appeals case of Pearson *v*. Shalala. This court decision requires the FDA to allow appropriately qualified health claims that would be misleading without such qualification. These qualified claims are based on the weight of the scientific evidence (i.e. there is more evidence for than against the relationship, but it falls short of the validity standard required for foods under NLEA).

Nutrition and health claims in the USA fall into three categories:

- *Health claims*: These describe a relationship between a food, food component or dietary supplement ingredient, and reducing risk of a disease or health-related condition.
- *Nutrient content claims*: These describe the level of a nutrient or dietary substance in the product, using terms such as 'free', 'high' or 'low'; or they compare the level in a food to that of another food, using terms such as 'more', 'reduced' or 'lite'.
- *Structure/function claims*: These describe the role of a nutrient or dietary ingredient intended to affect normal structure or function in humans, for example 'Calcium builds strong bones'. They also may characterise the means by which a nutrient or dietary ingredient acts to maintain such structure or function, for example 'Fibre maintains bowel regularity' or 'Antioxidants maintain cell integrity', or they may describe general well-being from consumption of a nutrient or dietary ingredient.

At present, the FDA has approved 12 health claims or nutrient content claims for foods that meet the Significant Scientific Agreement (SSA) standard (see Table 5.7) and five FDAMA (Health Claims Authorised Based on an Authoritative Statement by Federal Scientific Bodies) (Table 5.8); some of these apply to dietary supplements as well as conventional foods. As well as the requirements for approved claims, the FDA has detailed the requirements for the food making the claim, the food claim requirements and model claim statements. Full details of these claims can be found in the Code of Federal Regulations and in Appendix C of the Food Labelling Guide (FDA, 2009).

In USA legislation, there is no explicit recognition of any benefits of functional foods as such. Some contend that statements that a probiotic helps proper digestive function would be structure/function statements and not necessarily health claims; but if the claim was that it helped reduce the risk of cancer, that would be a health-related claim and, thus, subject to FDA approval (Berner & O'Donnell, 1998). Some products sold as dietary supplements make structural/function claims, such as 'when taken daily, helps

Table 5.7 Twelve health claims approved by the US Food and Drug Administration (FDA) that meet the Significant Scientific Agreement (SSA) standard.

Calcium and osteoporosis
Dietary lipids and cancer
Sodium and hypertension
Dietary saturated fat, cholesterol and coronary heart disease
Fibre-containing grain products, fruits and vegetables and cancer
Fruits, vegetables and grain products that contain fibre, particularly soluble fibre, and risk of coronary heart disease
Fruits and vegetables and cancer
Folate and neural tube defects
Dietary sugar alcohols and dental caries
Soluble fibre from certain foods and the risk of coronary heart disease
Soy protein and the risk of coronary heart disease
Plant sterol/stanol esters and the risk of coronary heart disease

After FDA (2015).

fortify your body's natural defences and helps keep your body at its best' (Actimel, Danone), and 'Helps create a favourable environment for the growth of beneficial flora, which dramatically influences metabolism and physical well-being' (Acidophilus, Cell Tech) (Sanders, 2003).

Structure/function claims have historically appeared on the labels of conventional foods and dietary supplements. The Dietary Supplement Health and Education Act of 1994 (DSHEA) established regulatory procedures for such claims for dietary supplement labels (although they can be applied to conventional foods also).

As structure/function claims are not FDA approved, there is no definitive list of such claims. However, such claims may: (a) describe the role of a nutrient or dietary ingredient intended to affect normal structure or function in humans (e.g. 'Calcium builds strong bones'), (b) characterise the means by which a nutrient or dietary ingredient acts to maintain such structure or function (e.g. 'Fibre maintains bowel regularity' or 'Antioxidants maintain cell integrity'), (c) describe general well-being from consumption of a nutrient or dietary ingredient, and (d) describe a benefit related to a nutrient deficiency disease (like vitamin C and scurvy), as long as the statement also tells how widespread (or otherwise) such a disease is in the USA.

Although structure/function claims do not require pre-approval by the FDA, they must be truthful and not misleading – the manufacturer is responsible for ensuring the accuracy and truthfulness of these claims. The FDA must be notified of dietary supplement claims within 30 days of their first use. If a dietary supplement label includes such a claim, it must state in a 'disclaimer' that the FDA has not evaluated the claim – such a disclaimer is not required on conventional foods.

Structure/function claims and disease claims for conventional foods focus on effects derived from nutritive value, while structure/function claims for dietary supplements

Table 5.8 Wording of five US Food and Drug Administration Modernisation Act (FDAMA) health claims[1] – health claims authorised based on an authoritative statement by federal scientific bodies.

Diets rich in whole grain foods and other plant foods and low in total fat, saturated fat, and cholesterol may reduce the risk of heart disease and some cancers.
Diets containing foods that are a good source of potassium and that are low in sodium may reduce the risk of high blood pressure and stroke.
Drinking fluoridated water may reduce the risk of [dental caries or tooth decay].
Diets low in saturated fat and cholesterol, and as low as possible in *trans* fat, may reduce the risk of heart disease.
Replacing saturated fat with similar amounts of unsaturated fats may reduce the risk of heart disease. To achieve this benefit, total daily calories should not increase.

[1]There is also one nutrient content claim authorised under the FDAMA – for choline content of foods.
After FDA (2015).

may focus on nutritive as well as non-nutritive effects. The FDA is likely to interpret the dividing line between structure/function claims and disease claims in a similar manner for conventional foods and for dietary supplements.

If this is not complicated enough, the USA also has a system of qualified health claims. The FDA began considering such claims, under interim procedures, in September 2003. United States of America court decisions had clarified the need to provide for health claims based on less scientific evidence, as long as the claims did not mislead the consumers. As with approved health claims, qualified health claims should also be based on a relationship between a substance and a health-related condition. In common with all health claims, qualified health claims require that a petition (i.e. an application) be submitted to the FDA. An enforcement discretion letter is issued by the FDA if it does not object to the use of the claim specified in the letter, provided that the products that bear the claim are consistent with the stated criteria. The FDA is committed to having all letters of enforcement discretion posted on their website. Once the letter is posted on the website, all manufacturers are informed how the FDA intends to exercise its enforcement discretion on the use of the specific qualified health claim.

The scientific support for qualified health claims does not have to be as strong as that for health claims with SSA. Under its interim guidance, the FDA is tentatively providing for three levels of science below the SSA standard; these are: (a) a good to moderate level of scientific agreement, (b) a low level of scientific agreement, and (c) a very low level of scientific agreement.

The criteria for the scientific review are described in the FDA interim guidance (FDA, 2011). As of late 2016, the USA has 15 qualified health claims (Table 5.8).

5.4.2 *The role of the Federal Trade Commission (FTC) and legal challenges*

The FTC was established under the Federal Trade Commission Act of 1914 and commenced in early 1915. It has two main purposes: to protect consumers and to promote fair competition. Its role is to protect consumers by preventing unfair, deceptive or fraudulent practices. It challenges companies and individuals that break the law. Its remit covers all business activities and is not confined to food trade matters. It also develops rules to ensure a vibrant marketplace, and educates consumers and businesses about their rights and responsibilities. In 2010, it challenged the Dannon company on health claims it was making in the advertising of its DanActive probiotic drink (that it reduced the likelihood of getting colds or flu) and its Activia Yoghurt (that it was clinically proven that if eaten every day, it would help regulate the digestive system in 2 weeks). Dannon agreed to settle FTC charges of deceptive advertising and to drop claims that allegedly exaggerated the health benefits of the two products. Dannon also agreed not to make any other claims about the health benefits, or effectiveness, of any yoghurt, dairy drink, probiotic food or drink, unless the claims are true and backed by competent and reliable scientific evidence. It was noted that while companies usually do not need FDA approval of their health claims to comply with the FTC Act, it strongly recommended that FDA approval of such would help companies avoid such problems. The FTC pointed out that the complaint was a finding that Dannon had actually violated the law and the settlement reached did not constitute admission of a law violation (FTC, 2010).

Another feature of the USA legal system is the use of class actions. A class action is a case in which a group of people, with the same or similar complaint caused by the same product or action, sue the defendant as a group, or the case is taken by an individual on behalf of the group. In 2015, an individual took a case in a US District Court in California against Yakult USA Inc., alleging that the company breached California's Unfair Competition Law by deceptively claiming that its probiotic beverages containing *Lb. casei* Shirota helped balance the digestive system which, as a consequence, supports overall health. Examples from the website and three advertisements of Yakult were also submitted as evidence. In January 2016, the judge in the case ruled that the plaintiff lacked the necessary standing to seek an injunction on behalf of the putative class because he failed to allege or offer evidence of future harm, as he was unlikely to purchase the product again (Anonymous, 2016). Understanding that he could not proceed with his original case, because he had no intention of ever buying Yakult again, the same individual sought to rectify his problem by buying Yakult again some 10 days after the original hearing and stapled a copy of his receipt to the motion. The case was heard by the same judge, who ruled that this newly alleged intent to buy Yakult was nothing more than a barely disguised attempt to manufacture standing and dismissed the case (Nakamura, 2016). It is worth noting that this case was decided on the standing of the plaintiff and not on the validity or otherwise of the health claims made for the product.

So what is the situation as regards labelling and marketing claims for probiotic milk-based products in the USA at this time? The challenges include conveying the benefits of a food or dietary supplement containing probiotic organisms to avoid wording claims in a manner that would be viewed by the FDA as unauthorised health or drug claims; also, in determining if there is sufficient scientific evidence to support petitioning the FDA to permit a health claim or qualified health claim describing the relationship of a food containing a specific species and strain of probiotics to reduction of the risk of disease. To date, the answer to this latter question appears to be no. It would be desirable to encourage the FDA to provide specific guidance on substantiation for structure/function claims and health claims for probiotic foods and dietary supplements. Some believe that the problem is the approach of the FDA Center for Biologics Evaluation and Research (CBER) to probiotic research. Their basic role is to evaluate biological drugs, so when they see the word 'probiotic' on a food product, it seems they automatically think 'drug'. The consequence of this is that probiotic drug development in the USA is alive and well, but probiotic foods, and researchers who want to study them, continue to suffer (Sanders, 2012, 2014).

5.5 The Canadian legislative situation regarding health claims and functional foods

5.5.1 *Background*

It should be noted that Canada has a bilingual policy and all federal legislation is published in French and English. Federal responsibility for the development of the national food-labelling requirements is shared between two federal departments, Health Canada (HC) and the Canadian Food Inspection Agency (CFIA). HC is responsible for the establishment of policies and standards relating to the health, safety and nutritional quality of food sold in Canada under the Food and Drugs Act. The CFIA is responsible for the administration of food-labelling policies related to misrepresentation and fraud in respect to food labelling, packaging and advertising, and the general food and fish-labelling provisions respecting grade, quality and composition. In addition, the CFIA has responsibility for the administration of the food-related provisions of the Consumer Packaging and Labelling Act, including basic food label information, net quantity, metrication and bilingual labelling.

5.5.2 *Health claims on foods in Canada*

The Canadian authorities accept that health claims on foods may assist consumers in making more informed decisions about their food choices, provided that such claims are scientifically valid and not misleading. The Food Directorate of Health Canada (FDHC) elaborates policies, regulations and standards that relate to the use of health claims on foods. It also makes any necessary assessments of these claims by reviewing mandatory and voluntary pre-market submissions. The decisions are based on HC reviews and are available on the FDHC website at http://www.hc-sc.gc.ca/fn-an/consult/index-eng.php. Food products may also be subject to safety assessment if they are considered novel foods. The CFIA enforces policies, regulations and standards, set by HC, governing the safety and nutritional quality of all food sold in Canada.

In 1998, in response to growing consumer and marketing interest in nutrition and health, HC published a Policy Paper on Nutraceuticals/Functional Foods and Health Claims on Foods (Health Canada, 1998). In 2002, an Interim Guidance Document was published that outlined standards of evidence for evaluating foods with health claims (Health Canada, 2002). In 2003, the Canadian Food and Drug Regulations were amended to introduce the first series of authorised health claims in Canada. In 2009, HC updated the Interim Guidance Document, replacing it with the Guidance Document for Preparing a Submission for Food Health Claims (Health Canada, 2009a). Also in 2009, HC posted 'Guidance Document – The Use of Probiotic Micro-organisms in Food' and a new Guidance Document entitled 'Classification of Products at the Food-Natural Health Product Interface: Products in Food Format', which was updated in 2010 (Health Canada, 2009b, 2010).

The term health claim is not defined in the Canadian Food and Drug Regulations, but HC recognises the definition of 'any representation that states, suggests, or implies that a relationship exists between a food or a constituent of that food and health' (L'Abbé *et al*., 2008). This is the definition as included in the Codex Alimentarius, as will be discussed further in this chapter. In Canada, health claims may be stated explicitly with words, or implicitly using symbols, graphics and logos; product names, brand names and trademarks are also included.

Health claims have been classed into three main categories:

- Disease risk reduction and therapeutic claims;
- Function claims; and
- General health claims.

These do not deal with health claims used for probiotics and have been addressed elsewhere (Hickey, 2014).

5.5.3 *Probiotic claims*

Unlike other countries and regions discussed in this chapter, Canada has specifically addressed probiotic claims in its Guide to Food Labelling and Advertising, which contains the definition developed by FAO/WHO (2001) expert consultation on health and nutritional properties of probiotics in food, as discussed earlier (FAO/WHO, 2002; CFIA, 2011).

There are two types of probiotic claims that can be made on food in Canada:

- *Strain-specific claims*: Claims about the health benefits or effects of specific strains of probiotics. As of mid-July 2016, no strain-specific claims have been accepted by HC.
- *Non-strain-specific claims*: Statements about the nature of probiotics. A list of non-strain-specific probiotic claims, acceptable without the need to conduct a detailed review of the scientific literature for the basis of the claim, is given in the Guide.

HC has also prepared a Guidance Document, 'The Use of Probiotic Micro-organisms in Food', that sets out the conditions under which health claims pertaining to probiotics would be considered acceptable (Health Canada, 2009b). This indicates that the term probiotic and other similar terms or representations should be accompanied by specific and validated statements about the benefits or effects of the probiotic. Health claims about the health benefits or effects of probiotics are statements that should be validated and should be supported by strain-specific evidence. When making a probiotic claim, the manufacturer should have the necessary documentation available to support the

Table 5.9 The eligible non-strain-specific species for probiotic claims in the Canadian Guide to Food Labelling and Advertising.

Lactobacilli spp.	Bifidobacteria spp.
Lactobacillus acidophilus	*Bifidobacterium adolescentis*
Lactobacillus casei	*Bifidobacterium animalis* subsp. *animalis*
Lactobacillus fermentum	*Bifidobacterium animalis* subsp. *lactis*[1]
Lactobacillus gasseri	*Bifidobacterium bifidum*
Lactobacillus johnsonii	*Bifidobacterium breve*
Lactobacillus paracasei	*Bifidobacterium longum* subsp. *infantis* comb. nov.
Lactobacillus plantarum	*Bifidobacterium longum* subsp. *longum* subsp. nov.
Lactobacillus rhamnosus	
Lactobacillus salivarius	

[1] The synonym *Bif. lactis* may be used for labelling purposes for this probiotic species.
After Canadian Food Inspection Agency (2011).

identity, safety, viability, concentration and stability of the specific probiotic strain that is used. It is also a requirement that a product should contain a minimum level of 1.0×10^9 cfu per stated serving size of the eligible micro-organism(s) that is (are) the subject of the claim.

Where it could be required, the manufacturer supplier should follow all legal requirements applicable to the sale of food, and this could include any requirements concerning the use and labelling of ingredients used in novel technology. The food should contain the amount of the probiotic micro-organism(s) that would be required to result in the claimed health benefit for the full shelf life of the product. Furthermore, documentation to support the stability and viability of the probiotic strain(s) should be maintained.

Wording of acceptable claims has been developed that may be made on 16 non-strain-specific probiotics (Table 5.9):

- Probiotic that naturally forms part of the gut/digestive tract flora;
- Provides live micro-organisms that naturally form part of the gut/digestive tract flora;
- Probiotic that contributes to healthy gut/digestive tract flora; and
- Provides live micro-organisms that contribute to healthy gut flora.

In the case of advertising, if the probiotic micro-organism is identified or referred to in the advertisement, then the identity of the micro-organism (genus, species and strain) should be declared using acceptable nomenclature.

5.6 Health foods and functional foods in China

5.6.1 *Introduction*

Traditional Chinese medicine (TCM) has a long documented history going back to the West Zhou Dynasty in about 1000 BC that links the use, or the avoidance, of food to prevent, treat or alleviate certain diseases or their symptoms. Examples of such foods or food ingredients include both relatively common products such as honey, oysters, mussels, almonds, garlic, ginger and others, as well as more exotic products such as deer horn antler, musk, snake, bear bile, ant beverage, earthworm, chrysanthemum and ginseng. Foods and medicines came to be seen as coming from the same sources, and many of the herbs used in TCM were used both as medicines and as ingredients of regular foods. Theories on nutrition in TCM came to regard food as having four roles or uses (Weng & Chen, 1996): (a) food as diet, (b) food used as a tonic, (c) food used as a medicine and (d) the avoidance of food in certain cases.

The use of food in TCM is outside the scope of this chapter, except insofar as it shows the historical basis of the situation that evolved over 3 millennia in China.

5.6.2 *Chinese legislative structures*

In China, the highest state body and the only legislative house of government is the National People's Congress (NPC). Delegates who are elected by China's provinces, autonomous regions, municipalities and armed forces hold office for 5 years, and the full congress is convened for one session each year. The 12th and current NPC, which was elected in March 2013, has 2987 members. The NPC enacts the national laws and appoints the prime minister, other ministers and the president. The NPC chooses a Standing Committee from among its members, which is the permanent body of the NPC, and it is the NPC Standing Committee that convenes the parliament, issues regulations, interprets laws and oversees government activities. The NPC also appoints the members of the State Council, which is the equivalent of the government in other countries. The State Council is organised as ministries based on the various sectors of the economy; these include the Ministry of Health and the Ministry of Agriculture (Lähteenmäki-Uutela, 2009a, 2009b).

The Ministry of Health is the most relevant agency in regulating foodstuffs and medicines. However, the China Food and Drug Administration (CFDA), which was founded in 2003, has a major role in drafting various guidelines related to safety and efficacy of foodstuffs, and at this time it operates directly under the State Council. Foods, such as healthy foods and novel foods that require pre-market authorisation, come under the jurisdiction and competence of the CFDA (Lähteenmäki-Uutela, 2009b; CFDA, 2011).

5.6.3 *The healthy (functional) foods sector in China and its regulation*

By the late 1980s, healthy foods in China were regarded as fitting into one of three categories: (a) fortified foods, (b) special nutrition foods and (c) foods for special health use. In 1986, the Chinese Ministry of Health enacted a Hygienic Standard and Regulation on the Use of Nutritional Fortification Substances; later revisions resulted in 40 substances being authorised for such use. These comprised amino acids, vitamins and minerals, but not any of the plants, herbs or products of animal origin used in TCM. In 1992, a National Standard for the Labelling of foods was issued, which covered infant formulae, nutritionally fortified food and modified food such as high-fibre, low-sodium and low-fat foods, where the labels were required to specify the target groups for whom the foods were intended. Because of the TCM link with certain foods, it was also deemed necessary to clearly differentiate between what was a food and what was a drug, and to avoid confusion in their regulation; that was achieved by defining food in Article 54 of the current Food Hygiene Law as 'any finished product or raw materials to be provided for people to eat or drink, as well as any product that has traditionally served as both food and medicament, with the exception of products used solely for medical purposes' (Ministry of Health of China, 1995; Kan, 1996).

Drugs were defined in Article 57 of the Drug Administration Law of 1984, which was revised in 2001, and they may now be found in Article 102 of the current Drug Administration Law (Ministry of Health of China, 2001). This definition is:

> Drugs refer to articles which are used in the prevention, treatment and diagnosis of human diseases and intended for the regulation of the physiological functions of human beings, for which indications, usage and dosage are established, including Chinese crude drugs, prepared slices of Chinese crude drugs, traditional Chinese medicine preparations, chemical drugs substances and their preparations, antibiotics, biochemical drugs, radioactive pharmaceuticals, serum, vaccines, blood products and diagnostic agents.

Therefore, the main factor that decides whether a product is a food or a drug is whether it is used as a medicine or not; however, there was still a certain ambiguity as regards the use of certain traditional herbs, and at that time it came down to how the product was labelled and advertised.

The development of the Chinese healthy (functional) food sector began in the early 1980s, when commercially produced products started to be offered for retail sale. Some of these stated they could improve health and remedy certain diseases or conditions. By 1994 there were estimated to be about 3000 factories producing such foods, estimated to be worth $4 billion at that time. However, there were concerns as regards the identity, name, efficacy and possibly even food safety aspects of these products, and it came to be recognised that a system of evaluation and assessment of these foods was necessary.

In March 1996, the Ministry of Health China developed provisions to cover healthy foods (which began to be regarded as functional foods), which it defined as food that has special health functions. Such foods were intended for consumption by specific population groups, for the regulation of certain functions of the body, but this did not include therapeutic use.

The new Food Safety Law of 2009, which replaced an earlier Food Hygiene Law of 1995, contained some specific provisions relating to health foods in Article 51 requiring the state to stringently supervise foods claimed to have particular effects on human health (National People's Congress of China, 1995, 2009). The relevant supervision and administration departments were obliged to perform their functions and undertake the responsibilities in accordance with the law. The new Food Safety Law required that:

- No food claimed to have particular effects on human health shall cause any acute, sub-acute or chronic harm to human health.
- The labels and instructions of such food shall not involve the effect of prevention or treatment of any disease.

- The contents thereof shall be true and indicate applicable groups of people, inapplicable groups of people, effective ingredients or symbolic ingredients and the contents thereof, and so on.
- The effects and ingredients of a product shall be consistent with the indications in the labels and instructions.

5.6.4 *Types of health claims in China and their approval*

There are 27 broad headings under which health claims may be made for foods or food ingredients in China. The type of scientific data required for each of these categories is specified, and indicates which should be obtained from human trials, animal trials or both (Table 5.10). Prior to submission of the application, reports must be obtained from an authorised Chinese laboratory under the following headings:

- A toxicological safety assessment
- An evaluation of the functionality or efficacy of the active ingredient
- An analysis of the level of the active ingredient present
- The stability or shelf life of the product
- The microbiological quality of the product.

In the case of health products manufactured in China, the required laboratory reports should be from one of a list of laboratories approved by the CFDA throughout China; for imported products, the reports should be from the Institute of Nutrition and Food Safety of the Chinese Centre for Disease Control in Beijing.

Applications for product approval of Chinese-produced health foods should be made through the provincial health administrative department, which carries out the preliminary examination before passing it on to the CFDA. Applications for imported health foods are sent straight to the CFDA. The application should detail the product characteristics; the full formulation or ingredients; the analytical methods required; details of the manufacturing process; the relevant product and raw material specifications; the product label, including directions for the product's use; product samples and evidence from scientific literature regarding the characterisation, safety and efficacy of the food or its active ingredient.

5.6.5 *China's probiotic market size and potential*

Approximately 8200 functional food products were approved in China by the Ministry of Health or the CFDA in the period between 1996 and mid-2007; however, at the end of that period, it was estimated that only about 30% of the approved products were still on the market. The approval rate of applications by the CFDA in the period 2003 to 2007 was between 27% and 33%. Approximations of the top three categories, among the 27 broad headings for which health claims may be made for foods or food supplements, were 33% for enhancement of immune function, 15% for alleviating fatigue and 9% for reduction in blood lipids. Supplements in the form of pills or capsules make up more than 60% of the functional food products on the market, while conventional foods comprise less than 1% (Yang, 2008).

Table 5.10 Health claims permitted in China.

Health claims	Type of supporting scientific test data required
Alleviating physical fatigue	Animal
Assisting in protection against chemical injury of the liver	Animal
Assisting in protection against irradiation	Animal
Enhancement of anoxia endurance	Animal
Enhancement of the immune system	Animal
Improvement of sleep	Animal
Increase in bone density	Animal
Anti-oxidative function	Animal + human
Assisting in blood lipids reduction	Animal + human
Assisting in blood pressure reduction	Animal + human
Assisting in blood sugar reduction	Animal + human
Assisting in memory improvement	Animal + human
Assisting in weight control (overweight or obesity control)	Animal + human
Facilitating bowel movement	Animal + human
Facilitating digestion (regularity)	Animal + human
Facilitating lead excretion	Animal + human
Improving child growth and development	Animal + human
Improving nutritional anaemia	Animal + human
Increasing milk secretion	Animal + human
Moistening and cleaning throat	Animal + human
Protection of gastric mucosa from damage	Animal + human
Regulating gastrointestinal microflora	Animal + human
Alleviating eye fatigue	Human
Eliminating acne	Human
Eliminating skin melasma (dark pigmentation patches)	Human
Improving skin's ability to retain moisture	Human
Improving skin's oil content	Human

Data compiled from Yang (2008).

As of June 2016, the CFDA had approved a total of 16 573 health food products, of which 15 822 (95.5%) were domestically made and 751 (4.5%) were imported. According to estimates, functional foods account for about 65% while nutritional supplements account for the remaining 35% (Hong Kong Trade Development Council, 2016). The top categories involved those regulating the immune system, alleviating physical fatigue, promoting anti-ageing (likely involving multiple categories) and assisting blood lipids reduction.

Earlier estimates for the annual value of functional foods in China in 2007 ranged from US$4 billion to US $6 billion, and were US$14 billion in 2009 (Ford *et al.*, 2007; Patel *et al.*, 2008; Yang, 2008; RedFern Associates, 2010). In 2015, the market for probiotics products in China was estimated as £5.6 billion (€6.5 billion) and is predicted to be £15.5 billion (€18 billion) by 2021 (AskCI Consulting, 2016). The market share of the main probiotic yoghurt companies in China in 2013 were the Wahaha Group (35%), Mengui Dairy (18%), Yakult Honsha (17%), the Yili Group (14%) and the Bright Dairy Group (10%) (Hung, 2015).

While the market size for such products in China has large potential, the requirement for testing in government-approved agencies within China is a major challenge, especially as all the relevant information has to be provided in Chinese. This both increases the approval costs and increases the time taken to get to market. The testing process in China can take from 6 to 12 months – this variation is mainly dependent on whether or not human trials are required. Following the submission to the CFDA, and if everything thereafter progresses smoothly, the application can be approved within a further 3 months. However, if additional information is required, this period will be extended from 9 to 19 months. The application fee is relatively modest (US$1200), and the main cost is in the testing. It has been estimated that the overall approval cost in China can be between US$14 500 and US$34 000, which is considerably cheaper than in Japan, where the total cost has been estimated as being between US$850 000 and US$1.5 million (Yamaguchi, 2004; Patel *et al.*, 2008).

5.7 Codex Alimentarius Commission (CAC)

5.7.1 *Background*

The Codex Alimentarius was established by the FAO and WHO in 1963 to develop harmonised international food standards, protect consumer health and promote fair practices in food trade. The CAC is an international intergovernmental body. Its membership is open to member nations and associate members of the FAO and/or the WHO; and, as of 2016, it has 187 member countries, one member organisation (the EU) and 240 Codex observers, made up of 56 international intergovernmental organisations (IGOs), 168 non-governmental organisations (NGOs) and 16 United Nations (UN) representatives. Nowadays, the CAC meets annually, and the venue alternates between the FAO headquarters in Rome and the WHO headquarters in Geneva. Nominated senior officials represent member governments at Codex meetings. National delegations may also include representatives of the industry, consumers and academia. Codex observers are allowed to contribute to meetings at all stages except in final decisions. This is the exclusive prerogative of member governments.

The CAC has established two types of subsidiary committees: (a) Codex Committees, and (b) Co-ordinating Committees. The former committee type is sub-divided into General Subject Committees (currently nine in number) that are so called because of the horizontal nature of their work, and Commodity Committees (currently 16 in number) which develop the standards for specific foods or classes of foods. There are five

Regional Co-ordinating Committees whose role is to ensure that the CAC is responsive to regional interests and the needs of developing countries. The CAC also establishes *ad hoc* Intergovernmental Task Forces given stated tasks on specific topics. Currently, there is one such Task Force on antimicrobial resistance.

Food standards cover specific commodities and also general issues that have cross-sectoral horizontal application. For example, they encompass food labelling, food additives, food hygiene, contaminants, nutrition and foods for special dietary uses, and methods of analysis and sampling.

The CAC has established a number of principles on a scientific basis for its decision making (Randell & Race, 1996). These principles ensure that the quality and food safety provisions shall be based on sound science and that, in establishing food standards, other legitimate factors relevant to consumers' health and the promotion of fair trade may be considered. The standards and related texts are subject to revision, as and when deemed necessary by the CAC and its subsidiary bodies, to ensure that they are consistent with, and reflect, current scientific knowledge. Any member of the Codex may identify and present new scientific or other information to the relevant body that may warrant a revision.

Following the decisions and text adoptions of the July 2015 meeting of the CAC, the Codex Alimentarius contained (FAO/WHO, 2016):

- 191 Commodity Standards;
- 73 Guideline texts;
- 51 Codes of Practice;
- 17 Maximum Levels (MLs) for contaminants in foods;
- 3770 MLs for 301 food additives;
- 4347 Maximum Residue Limits (MRLs) covering 196 pesticides; and
- 610 MRLs covering 75 veterinary drugs.

5.7.2 *Acceptance of Codex standards and their role in the World Trade Organisation (WTO)*

Codex standards, codes of practice, guidelines and other such texts are not legally binding. However, they are used as the basis for the national legislations of many countries, especially developing countries, and developed countries also take them into account when updating or revising their legislations.

The Uruguay round of multilateral trade negotiations held under the General Agreement on Tariffs and Trade (GATT), which took place between 1986 and 1994, led to the formation of the WTO on 1 January 1995. For the first time, GATT agreements included agriculture and food in their scope; however, the Marrakesh agreement of 1994 also included the agreements on sanitary and phytosanitary (SPS) measures and on technical barriers to trade (TBT). These agreements acknowledge the need for the harmonisation of international standards to minimise the risk of sanitary, phytosanitary and other technical standards becoming barriers to international trade. Thus, the SPS and TBT agreements gave formal recognition to the standards, guidelines and recommendations of international organisations, including the CAC, as reference points for facilitating international trade and resolving disputes. Hence, the role of the CAC in this regard is now well recognised.

It should be noted, however, that consumer groups have expressed some criticism of Codex standards on the basis of the time taken to elaborate standards, and sound science; the latter basis may not necessarily take into account other considerations, such as consumer concerns (O'Rourke, 1999).

5.7.3 *Codex and food-labelling claims*

The Codex General Standard for the Labelling of Pre-packaged Foods (GSLPF; CODEX STAN 1-1985, Rev. 1-1991) is the basic Codex Food Labelling standard (FAO/WHO, 2007c). In addition, the Codex General Standard for the Labelling of and Claims for Pre-packaged Foods for Special Dietary Uses was adopted by the CAC in 1985, and it defines a claim as 'any representation which states, suggests or implies that a food has particular qualities relating to its origin, nutritional properties, nature, processing, composition or any other quality' (FAO/WHO, 2007b).

The Codex General Guidelines on Claims were originally adopted in 1979, with a revised version adopted in 1991 (FAO/WHO, 2007a). The text contains the definition given above under the GSLPF, and also gives a list of claims which should be prohibited:

- Claims which state that a given food will provide an adequate source of all nutrients, except in certain well-defined products where a Codex standard regulates this claim as admissible or appropriate authorities have accepted that the specific food does so.
- Claims that imply that a balanced diet or ordinary foods cannot supply adequate amounts of all ingredients.
- Claims that cannot be substantiated.
- Claims that a food is suitable for use in the prevention, alleviation, treatment or cure of a disease, disorder or particular condition unless those are in accordance with and follow the principles of Codex standards or guidelines on foods for special dietary uses or, in the absence of Codex standards and guidelines, are permitted by the laws of the country of sale (these are usually called medicinal claims).

The task of developing guidelines on the use of nutrition and health claims comes under the remit of the Codex Committee on Food Labelling (CCFL). This has proved a difficult task; for example, in written comments submitted to the CCFL meeting in 1994, some countries (Denmark and Finland) were opposed to both nutrient function and health claims, whilst others (New Zealand and Switzerland) opposed health claims but could accept nutrient function claims. However, Australia, Sweden and the USA were prepared to accept both types of claims provided they were subject to strict criteria (Pascal, 1996). At its meeting in 1996, the CCFL agreed to delete all references to health claims in the guidelines and sent the Guidelines for Use of Nutrition Claims to the CAC; these were adopted in 1997. Work continued addressing the more contentious issue of health claims and finally, at its meeting, the CCFL agreed Guidelines on the use of

Nutrition and Health Claims. These were adopted by CAC in 2004, including provisions for health claims (FAO/WHO, 2007d).

These guidelines are long and detailed and, at the outset, laid down a number of principles concerning health claims, such as:

- They should be consistent with national health policy and nutrition policy, and support same as applicable.
- They should be supported by a sound and sufficient body of scientific evidence to substantiate the claim.
- They should provide truthful and non-misleading information to aid consumers in choosing healthy diets.
- They should be supported by specific consumer information.
- Their impact on consumers' eating habits and buying habits should be monitored.
- The prohibition of claims, as stated in Section 3.4 of the Codex General Guidelines for the use of claims, should remain.

The Codex definition of a health claim is given as 'any representation that states, suggests, or implies that a relationship exists between a food or a constituent of that food and health.' Three types of claims are also defined: (a) Nutrient Function Claims, (b) Other Function Claims, and (c) Reduction of Disease Risk Claims.

There is a requirement that any health claim must be accepted by, or acceptable to, the competent authorities of the country where the product is sold. As regards substantiation of health claims, when the original text was adopted in 2004, reference was made to a parallel text being developed by the CCNFSDU at that time. The resultant text was finally adopted by the CAC 2009, is now entitled Recommendations on the Scientific Substantiation of Health Claims and is included as an Annex to the Guidelines on the Use of Nutrition and Health Claims (FAO/WHO, 2013). A footnote to the Annex states that this document should be read in conjunction with the Working Principles for Risk Analysis for Food Safety for Application by Governments CAC/GL 62-2007 (FAO/WHO, 2007d). They are intended for governments to facilitate their own evaluation of health claims made by industry, and as a reference for industry in preparing dossiers aimed at providing substantiation of such claims. They cover the following aspects:

- Health claims should primarily be based on evidence provided by well-designed human intervention studies. Human observational studies per se are not necessarily sufficient but they may contribute to the totality of the evidence.
- Data from *ex vivo* or *in vitro* animal model studies are not regarded as sufficient, but may be used to provide additional supportive information.
- The totality of the evidence, including appropriate unpublished data, should be identified and reviewed.
- Evidence based on human studies should demonstrate a consistent association between the food or food constituent and the claimed health effect, with little or no evidence to the contrary.
- Substantiation can take into account specific situations or alternate processes; e.g. based on generally accepted statements by recognised expert scientific bodies deemed acceptable over time.

- Health claims involving a relationship between a food category and a health effect, based substantially on observational studies which should provide a consistent body of evidence from a number of well-designed studies.
- Evidence-based dietary guidelines and authoritative statements prepared or endorsed by a competent authoritative body and meeting the required high scientific standards may also be used.

The following is a summary of the labelling requirements under the Codex Guidelines for the use of Nutrition and Health Claims:

- A statement of the quantity of any nutrient or other constituent of the food that is the subject of the claim.
- The target group, if appropriate.
- How to use the food to obtain the claimed benefit and other lifestyle factors or other dietary sources, where appropriate.
- If appropriate, advice to vulnerable groups on how to use the food, and to groups who would need to avoid the food.
- Maximum safe intake of the food or constituent where necessary.
- How the food or food constituent fits within the context of the total diet.
- A statement on the importance of maintaining a healthy diet.

When combined, the above guidelines and recommendations should provide a firmer legal basis for health claims at the international level, but leave the actual approval or acceptance of such claims to individual governments to ensure that they are in line with national dietary policies and guidelines. It should be borne in mind that the Codex does not evaluate health claims – the guidelines are intended for individual governments to facilitate their evaluation of health claims made by industry. They also should provide a reference in preparing dossiers aimed at substantiating such claims. It should also be mentioned that none of these guidelines and recommendations include the definition of a probiotic, as elaborated by the 2002 Working Group of an FAO/WHO Expert Consultation (FAO/WHO, 2002) or the slightly revised definition in 2014 (Hill *et al.*, 2014).

5.7.4 *Codex standard for fermented milks*

At its 26th Session in July 2003, the CAC adopted a new Codex Standard for Fermented Milks, which replaced two former standards: (a) for Yoghurt and Sweetened Yoghurt, and (b) for Flavoured Yoghurt and Products Heat-Treated after Fermentation. However, the new standard is also expanded to encompass a broader range of fermented milks, such as Kefir, Acidophilus Milk and Koumiss (Kymus), and it was also revised to encompass a new category of Drinks Based on Fermented Milk (FAO/WHO, 2011). While not specifically aimed at probiotics, this new standard includes compositional requirements for the minimum level of starter culture organisms (1×10^7 cfu g^{-1}) and, where a content labelling claim is made for a specific micro-organism other than the normal starter cultures, a minimum of 1×10^6 cfu g^{-1} is required. The earlier standards required only that micro-organisms should be viable and abundant, without setting specific minima.

5.8 Some conclusions and possible future legislative prospects for probiotics

While the scientific basis for the benefits of probiotics have been recognised for over 100 years, the regulatory status has not evolved at the same rate as the science. Although the regulatory systems we have considered in this chapter require scientific validation for health claims, the outcomes differ significantly. Probiotic-related claims fared much better in Japan and China than in the USA and the EU. Claims for probiotic milk-based products which are acceptable in Japan, for instance, were not accepted by EFSA, although it is likely the same or similar scientific data were used for certain major brands. This is likely due to different levels or standards being required for approval. In the EU and the USA, it would appear as if the requirements are for a similar level of validation to those demanded of medical drugs. Another factor may be that the Japanese system evolved with the involvement and participation of both government and industry in the development of foods with health claims, while in the EU and the USA, the approach was more of a reactionary one to products already on the markets there. It is understood that industry efforts are continuing to have probiotic claims treated as nutrient function claims and not health claims. The outcome of this is still uncertain.

The size and opportunities of the Chinese market for fermented foods with probiotics are very large and attractive, but it also has challenges in the requirements for claim validation and in market access. It has developed rapidly and is expected to continue to develop for some years to come. It remains to be seen, given the claim validation difficulties in the EU and the USA, whether probiotic research activity will also move to China from the west.

Meanwhile, Canada has adopted the definition of a probiotic given by the FAO/WHO in their joint 2002 report. Canada also recognises 17 probiotic species, permits specified non-strain health claims and allows strain-specific claims, although none of the latter type of claims are approved at this time. It is not clear if applications have been submitted, but no outcomes of evaluations of such applications are available.

Will the future for probiotics be as foods or as supplements? This is likely to be decided for individual country or regional markets. In the markets of developed countries, leading probiotic milk-based brands may continue to grow, albeit on a slower or more limited basis than in Europe.

Some of the bigger questions are: what will happen in developing markets, and how will new products be promoted? In countries where regulatory challenges are greatest, will other novel marketing approaches be developed? For instance, will the use of scientific conferences aimed at health professionals substitute for, or replace, the need for health claims in product labelling?

Acknowledgements

I gratefully acknowledge my thanks and appreciation to two of my International Dairy Federation colleagues and friends for their support and advice in compiling this chapter. Firstly, to Dr Yuki Morita, Director of Regulations and Public Affairs, Yakult Europe, for her generous assistance with the translation and interpretation of certain Japanese texts; for product labels and other documentation, including the Japanese characters used in Section 5.2; and in supplying the images of the FOSHU logos used in this chapter.

Secondly, to Ms Cary Frye, Vice-President of Regulatory and Scientific Affairs, International Dairy Foods Association, Washington, DC, for supplying material and details on US legislation as addressed in Section 5.4, and whose knowledge in such matters is thorough and extensive.

While I have used information and material supplied by these colleagues, the opinions, views and conclusions expressed are mine alone and do not purport to be theirs.

References

Anonymous (1992) *Annual Report 1992: Advisory Committee for Novel Foods and Processes*. UK Ministry of Agriculture Fisheries and Foods Publications, London.

Anonymous (2011a) Food for Specified Health Uses (FOSHU). Japanese Ministry of Health, Labour and Welfare. http://www.mhlw.go.jp/english/topics/foodsafety/fhc/02.html

Anonymous (2011b) Food with Nutrient Function Claims. Japanese Ministry of Health, Labour and Welfare. http://www.mhlw.go.jp/english/topics/foodsafety/fhc/01.html

Anonymous (2011c) The story of Yakult founder. http://usahawan-yakult.blogspot.ie/2011/12/story-of-dr-minoru-shirota-yakult.html

Anonymous (2016) Nicolas Torrent, on Behalf of Himself and All Others Similarly Situated, Plaintiff, v. Yakult U.S.A., Inc., Defendant. *SACV 15-00124-CJC(JCGx)*, 1–11. US District Court, Central District of California, Southern Division.

Anukam, K.C. & Reid, G. (2007) Probiotics: 100 years (1907–2007) after Elie Metchnikoff's observation. In *Communicating Current Research and Educational Topics and Trends in Applied Microbiology* (ed. A. Méndez-Vilas), Vol. **1**, 466–474. Formatex, Badajoz, Spain.

Arai, S.Y., Moringa, T., Yoshikawa, E., Ichiishi, Y., Kiso, M., Yamazaki, M., Morotimo, M., Shimizu, T., Kuwata, T. & Kaminogawa, S. (2002) Recent trends in functional food science and industry in Japan. *Bioscience Biotechnology & Biochemistry*, **66**, 2017–2029.

AskCI Consulting (2016) Market analysis of probiotics industry in China 2016–2021 proposal. In *Market Publishers Report Database*, 1–87. AskCI Consulting Co. Ltd, Birmingham.

Berner, L.A. & O'Donnell, J.A. (1998) Functional foods and health claims legislation: applications to dairy foods. *International Dairy Journal*, **8**, 355–362.

Canadian Food Inspection Agency (CFIA) (2011) Health claims. In *Guide to Food Labelling and Advertising*, 8–46. Canadian Food Inspection Agency, Ottawa.

CAA Japan (2011) Regulatory systems of health claims in Japan. http://www.caa.go.jp/en/pdf/syokuhin338.pdf

CAA Japan (2016a) Foods for Special Health Use (FOSHU) permit (approved) list. http://www.caa.go.jp/foods/pdf/151224_2.pdf

CAA Japan (2016b) List of notifications of Foods with Function Claims. http://www.caa.go.jp/foods/todoke_1-25.html. Accessed in October 2016

China Food and Drug Administration (CFDA) (2011) Main responsibilities. http://eng.sfda.gov.cn/WS03/CL0756/

EFSA (2016) EU Register on nutrition and health claims. http://ec.europa.eu/food/safety/labelling_nutrition/claims/register/public/?event=searh
EU (1985) Council Directive 85/374/EEC of 25 July 1985 on the approximation of the laws, regulations and administrative provisions of the Member States concerning liability for defective products. *Official Journal of the European Commission*, **L210**, 29–33.
EU (1997) Regulation (EC) 258/97/EC of 27 January 1997 concerning novel foods and novel food ingredients. *Official Journal of the European Commission*, **L43**, 1–6.
EU (1999) Directive 1999/34/EC of the European Parliament and of the Council of 10 May 1999 amending Council Directive 85/374/EEC on the approximation of the laws, regulations and administrative provisions of the Member States concerning liability for defective products. *Official Journal of the European Commission*, **L141**, 20–21.
EU (2002) Regulation (EC) No 178/2002 of the European Parliament and of the Council of 28 January 2002 laying down the general principles and requirements of food law, establishing the European Food Safety Authority and laying down procedures in matters of food safety. *Official Journal of the European Commission*, **L31**, 1–24.
EU (2006) Corrigendum to Regulation (EC) No 1924/2006 of the European Parliament and of the Council of 20 December 2006 on nutrition and health claims made on foods. *Official Journal of the European Commission*, **L12**, 3–18.
EU (2008a) Commission Regulation (EC) No 353/2008 of 18 April 2008 establishing implementing rules for applications for authorisation of health claims as provided for in Article 15 of Regulation (EC) No 1924/2006 of the European Parliament and of the Council. *Official Journal of the European Commission*, **L109**, 11–16.
EU (2008b) Regulation (EC) No 107/2008 of the European Parliament and of the Council of 15 January 2008 amending Regulation (EC) No 1924/2006 on nutrition and health claims made on foods as regards the implementing powers conferred on the Commission. *Official Journal of the European Commission*, **L39**, 8–10.
EU (2008c) Regulation (EC) No 109/2008 of the European Parliament and of the Council of 15 January 2008 amending Regulation (EC) No 1924/2006 on nutrition and health claims made on foods. *Official Journal of the European Commission*, **L39**, 14–15.
EU (2010) Commission Regulation (EU) No 116/2010 of 9 February 2010 amending Regulation (EC) No 1924/2006 of the European Parliament and of the Council with regard to the list of nutrition claims. *Official Journal of the European Commission*, **L37**, 16–18.
EU (2011) Regulation (EU) No 1169/2011 of the European Parliament and of the Council of 25 October 2011 on the provision of food information to consumers, amending Regulations (EC) No 1924/2006 and (EC) No 1925/2006 of the European Parliament and of the Council, and repealing Commission Directive 87/250/EEC, Council Directive 90/496/EEC, Commission Directive 1999/10/EC, Directive 2000/13/EC of the European Parliament and of the Council, Commission Directives 2002/67/EC and 2008/5/EC and Commission Regulation (EC) No 608/2004. *Official Journal of the European Commission*, **L304**, 18–63.
EU (2012) Commission Regulation (EU) No 1047/2012 of 8 November 2012 amending Regulation (EC) No 1924/2006 with regard to the list of nutrition claims text with EEA relevance. *Official Journal of the European Commission*, **L310**, 36–37.
EU (2016) EU Register on nutrition and health claims. http://ec.europa.eu/food/safety/labelling_nutrition/claims/register/public/?event=searh
FAO/WHO (2001) *Evaluation of Health and Nutritional Properties of Probiotics in Food Including Powder Milk with Live Lactic Acid Bacteria*. Report of a Joint FAO/WHO Expert Consultation. Food and Agriculture Organization of the United Nations, Rome.
FAO/WHO. (2002) *Guidelines for the Evaluation of Probiotics in Food*. Report of a Joint FAO/WHO Working Group on Drafting Guidelines for the Evaluation of Probiotics in Food. ftp://ftp.fao.org/es/esn/food/wgreport2.pdf
FAO/WHO. (2007a) General guidelines for claims. In *Food Labelling*, 21–23. Secretariat of the Codex Alimentarius Commission, Rome.

FAO/WHO (2007b) General standard for the labelling of and claims for prepackaged foods for special dietary uses (CODEX STAN 146-1985). In *Food Labelling*, 17–20. Secretariat of the Codex Alimentarius Commission, Rome.
FAO/WHO (2007c) General standard for the labelling of prepackaged foods (CODEX STAN 1-1985). In *Food Labelling*, 1–. Secretariat of the Codex Alimentarius Commission, Rome.
FAO/WHO (2007d) Working Principles for Risk Analysis for Food Safety for Application by Governments CAC/GL 62-2007 pp. 3–10. Secretariat of the Codex Alimentarius Commission, Rome.
FAO/WHO (2011) Fermented milks (CODEX STAN 243-2003). In *Milk and Milk Products*, 6–16. Secretariat of the Codex Alimentarius Commission, Rome.
FAO/WHO (2013) Guidelines for the use of nutrition and health claims (CAC/GL 23-1977). Accessible at: http://www.fao.org/fao-who-codexalimentarius/standards/list-standards/en/
FAO/WHO (2016) *Understanding CODEX*. Secretariat of the Joint FAO/WHO Food Standards Programme, Rome.
FDA (2009) Guidance for industry: a food labeling guide (11. Appendix C: health claims). http://www.fda.gov/Food/GuidanceComplianceRegulatoryInformation/GuidanceDocuments/FoodLabelingNutrition/FoodLabelingGuide/default.htm
FDA (2011) Guidance for industry: a food labeling guide (8. Claims). http://www.fda.gov/Food/GuidanceComplianceRegulatoryInformation/GuidanceDocuments/FoodLabelingNutrition/FoodLabelingGuide/ucm064908.htm#qualified
FDA (2015) FDA Modernization Act (FDAMA) claims. http://www.fda.gov/Food/GuidanceRegulation/GuidanceDocumentsRegulatoryInformation/LabelingNutrition/ucm064919.htm
FTC (2010) Dannon agrees to drop exaggerated health claims for Activia Yogurt and DanActive dairy drink (File No. 0823158). https://www.ftc.gov/news-events/press-releases/2010/12/dannon-agrees-drop-exaggerated-health-claims-activia-yogurt
Ford, K., Konishi, Y., Rajalahti, R. & Pehu, E. (2007) *Health Enhancing Foods: Country Case Studies of China and India*. World Bank, Washington, DC. http://documents.worldbank.org/curated/en/732061468771539378/pdf/387150IN0CHA0Health0food01PUBLIC1.pdf
Fuller, R. (1989) Probiotics in man and animals. *Journal of Applied Bacteriology*, **66**, 365–378.
Gorbach, S.L. & Goldin, B.R. (1989) *Lactobacillus* strains and methods of selection. New England Medical Center Hospitals. USA Patent No. 4839281.
Grigoroff, S. (1905) Etude sur le lait fermenté comestible: le "Kissélo mléko" de Bulgarie. *Revue Médicale de la Suisse Romande*, **25**, 714–720.
Health Canada (1998) Policy paper: nutraceuticals/functional foods and health claims on foods. http://www.hc-sc.gc.ca/fn-an/label-etiquet/claims-reclam/nutra-funct_foods-nutra-fonct_aliment-eng.php
Health Canada (2002) Interim guidance document – preparing a submission for foods with health claims: incorporating standards of evidence for evaluating foods with health claims. http://www.hc-sc.gc.ca/fn-an/label-etiquet/claims-reclam/abstract_guidance-orientation_resume-eng.php
Health Canada (2009a) Guidance document for preparing a submission for food health claims. http://www.hc-sc.gc.ca/fn-an/alt_formats/hpfb-dgpsa/pdf/legislation/health-claims_guidance-orientation_allegations-sante-eng.pdf
Health Canada (2009b) Guidance document – the use of probiotic micro-organisms in food. http://www.hc-sc.gc.ca/fn-an/alt_formats/hpfb-dgpsa/pdf/legislation/probiotics_guidance-orientation_probiotiques-eng.pdf
Health Canada (2010) Classification of products at the food – natural health product interface: products in food formats, Version 2.0. http://www.hc-sc.gc.ca/dhp-mps/prodnatur/legislation/docs/food-nhp-aliments-psn-guide-eng.php
Hill, C., Guarner, F., Reid, G., Gibson, R., Merenstein, D.J., Pot, B., Morelli, L., Canani, R.B., Flint, H.J., Salminen, S., Calder, P.C. & Sanders, M.E. (2014) Expert consensus document: the

International Scientific Association for Probiotics and Prebiotics consensus statement on the scope and appropriate use of the term probiotic. *Nature Review Gastroenterology Hepatology*, **11**, 506–514.

Hong Kong Trade Development Council (2016) *China's Health Food Market*. Hong Kong Trade Development Council, Hong Kong.

Hung, W.L. (2015) Consumerism of probiotics in China. In *Beneficial Microorganisms in Food and Nutraceuticals* (ed. M.-T. Liong), Vol. **27**, 183–201. Springer International Publishing, Cham, Switzerland.

Hickey, M. (2014) Legislations and relevant regulations. In *Milk and Dairy Products as Functional Foods* (ed. A. Kanekanian), 314–372. John Wiley & Sons Ltd, Chichester.

Ichikawa, T. (1994) Functional foods in Japan. In *Functional Foods: Designer Foods, Pharmafoods, Nutraceuticals* (ed. I. Goldberg), 453–467. Chapman & Hall, London.

Kan, X. (1996) Legal aspects of functional foods: the situation in China. *Nutrition Reviews*, **54**, S162–S166.

L'Abbé, M.R., Dumais, L., Chao, E. & Junkins, E. (2008) Health claims on foods in Canada. *Journal of Nutrition*, **138**, 1221S–1227S.

Lähteenmäki-Uutela, A. (2009a) Chinese law on foods and medicines. *Nordic Journal of Commercial Law*, **2**, 1–41.

Lähteenmäki-Uutela, A. (2009b) *Foodstuffs and Medicines as Legal Categories in the EU and China – Functional Foods as a Borderline Case*. PhD thesis, University of Turku, Turku, Finland. http://doria32-kk.lib.helsinki.fi/bitstream/handle/10024/52491/diss2009lahteenmaki-uutela.pdf?sequence=1

Mackay, A.D., Taylor, M.B., Kibbler, C.C. & Hamilton-Miller, J.M.T. (1999) *Lactobacillus endocarditis* caused by a probiotic organism. *Clinical Microbiology Infectious Diseases*, **5**, 290–292.

Masco, L., Ventura, M., Zink, R., Huys, G. & Swings, J. (2004) Polyphasic taxonomic analysis of *Bifidobacterium animalis* and *Bifidobacterium lactis* reveals relatedness at the subspecies level: reclassification *Bifidobacterium animalis* as *Bifidobacterium animalis* subsp. *animalis* subsp. nov. and *Bifidobacterium lactis* as *Bifidobacterium animalis* subsp. *lactis* subsp. nov. *International Journal of Systematic and Evolutionary Microbiology*, **54**, 1137–1143.

Metchnikoff, É. (1908) Lactic acid as inhibiting intestinal putrefaction. In *The Prolongation of Life: Optimistic Studies* (ed. P.C. Mitchell), 161–183. G.P. Putnam's Sons, London.

Metchnikoff, É. (1921) *The Life of Elie Metchnikoff 1845–1916*. Houghton Mifflin Company, New York.

Ministry of Health of China (1995) Food Hygiene Law of the People's Republic of China. http://www.chinafdc-law.com/laws/detail_156.html

Ministry of Health of China (2001) Drug Administration Law of the People's Republic of China (2001 revision). http://www.chinafdc-law.com/laws/detail_80.html

Mitsuda, H. (1958) Studies on enriched food. *Bulletin of the Institute for Chemical Research. Kyoto University*, **35**, 140–148.

Nakamura, A. (2016) Class cert denial redux: plaintiff's "manufactured" standing falls short in Yakultyogurtaction.http://classdismissed.mofo.com/food-misbranding/class-cert-denial-redux-plaintiffs-manufactured-standing-falls-short-in-yakult-yogurt-action/

National People's Congress of China (1995) Food Hygiene Law of the People's Republic of China. http://www.chinafdc-law.com/laws/detail_156.html

National People's Congress of China (2009) Food Safety Law of the People's Republic of China. http://www.usdachina.org/info_details1.asp?id=2431

O'Rourke, R. (1999) The international context. In *European Food Law with 1999 Update*, 148–160. Palladian Law Publishing Ltd, Bembridge.

Parker, R.B. (1974) Probiotics: the other half of the antibiotic story. *Animal Nutrition Health*, **29**, 4–8.

Patel, D., Dufour, Y. & Domigan, N. (2008) Functional food and nutraceutical registration processes in Japan and China: similarities and differences. *Journal of Pharmacy & Pharmaceutical Sciences*, **11**, 1–11.

Randell, A. & Race, J. (1996) Regulatory and legal aspects of functional foods: an international perspective. *Nutrition Reviews*, **54**, 152–155.

Rautio, M., Jousimes-Somer, H., Kauma, H., Pirtarinen, I., Saxelin, M., Tynkkynen, S. & Koskela, M. (1999) Liver abscess due to a *Lactobacillus rhamnosus* strain indistinguishable from *Lactobacillus rhamnosus* strain GG. *Clinical Infectious Diseases*, **29**, 1159–1160.

RedFern Associates (2010) The China health food market overview 2010. *New Zealand Trade & Enterprise*, 1–34. RedFern Associates, Shanghai.

Rettger, L.F., Levy, W.N., Weinstein, L. & Weiss, J.E. (1935) *Lactobacillus acidophilus and its Therapeutic Application*. Yale University Press, New Haven, CT.

Salminen, S. (1996) Functional dairy foods with *Lactobacillus* strain GG. *Nutrition Reviews*, **54**(11), 99–101.

Sanders, M.E. (2003) Probiotics: considerations for human health. *Nutrition Reviews*, **61**, 91–99.

Sanders, M.E. (2012) How FDA's actions are guaranteeing research on probiotic foods is not conducted in the USA. http://cdrf.org/2012/10/13/how-fdas-actions-are-guaranteeing-research-on-probiotic-foods-is-not-conducted-in-the-usa/

Sanders, M.E. (2014) Probiotics: achieving a better regulatory fit. http://cdrf.org/2014/07/24/probiotics-achieving-better-regulatory-fit/

Sonnenborn, U. & Schulze, J. (2009) The non-pathogenic *Escherichia coli* strain Nissle 1917 – features of a versatile probiotic. *Microbial Ecology in Health and Disease*, **21**, 122–158.

Thomas, L.V. (2016) Probiotics – the journey continues. *International Journal of Dairy Technology*, **49**, 469–480.

Tissier, H. (1906) Traitement des infections intestinales par la méthode de la transformation de la flore bactérienne de l'intestin. *Comptes Rendus des Séances et Mémoires de la Société de Biologie*, **60**, 359–361.

Tomita, F. (2007) Japanese legislation on nutrition and health products: the regulatory issues and products on Japanese market. http://congress.utu.fi/NHC2007/presentations/Fusao_Tomita.pdf

US Congress (1990) Nutrition Labeling and Education Act (NLEA) of 1990 (Pub. L 101-535). http://thomas.loc.gov/cgi-bin/query/C?c101:./temp/~c101Xru6IX

US Congress (1994) Dietary Supplement Health and Education Act of 1994. http://thomas.loc.gov/cgi-bin/query/C?c103:./temp/~c103iJhsSt

US Congress (1997) Food and Drug Administration Modernization Act of 1997 (Pub. L 105-115). http://www.fda.gov/RegulatoryInformation/Legislation/FederalFoodDrugandCosmeticActFDCAct/SignificantAmendmentstotheFDCAct/FDAMA/FullTextofFDAMAlaw/default.htm

Weng, W. & Chen, J. (1996) The Eastern perspective on functional foods based on traditional Chinese medicine. *Nutrition Reviews*, **54**, S11–S16.

Yamada, K., Sato-Mito, N., Nagata, J. & Umegaki, K. (2008) Health claim evidence requirements in Japan. *Journal of Nutrition*, **138**, 1192S–1198S.

Yamaguchi, P. (2004) FOSHU approval – is it worth the price? http://newhope360.com/foshu-approval-it-worth-price

Yang, Y. (2008) Scientific substantiation of functional food health claims in China. *Journal of Nutrition*, **138**, 1199S–1205S.

6 Enumeration and Identification of Mixed Probiotic and Lactic Acid Bacteria Starter Cultures

A.Č. Majhenič, P.M. Lorbeg and P. Treven

6.1 Introduction

Lactic acid bacteria (LAB) have been used to ferment or culture foods for thousands of years. Without people being aware of their presence and fermentation power, LAB have been accidently exploited since ancient times to produce cultured foods with improved preservation properties and with characteristic flavours and textures different from the raw material. LAB are widespread in nature, associated with plants, meat and dairy, and are also found in human and animal gastrointestinal (GI) tracts. Although LAB refer to a large and diverse group of beneficial microbes, they have similar properties and they all produce lactic acid as a major end product of the fermentation process. They are best known for their role in the preparation of fermented dairy products, but they are equally important for pickling of vegetables, baking, and curing fish, meats and sausages. Modern industrial processes utilise specially prepared LAB as starter cultures that can be defined as microbial preparations containing large numbers of bacteria of at least one strain to be added to a raw material to produce a fermented food, by accelerating and steering its fermentation process (Hati *et al.*, 2013). The food industry benefits from the use of the LAB that contribute to the formation of desired chemical, physical and organoleptic properties of the final food products.

This chapter reviews the importance of proper and reliable enumeration and classification of probiotic and LAB starter cultures by the use of different approaches based on either phenotypic and/or genotypic methods. Particular attention will be given to modern molecular methods.

6.2 Classification

Methods for the identification and enumeration of probiotics and LAB starter cultures can be systematically classified in various ways on the basis of properties inherent to each method. Methods are most often classified based on the following: (a) culture dependency, (b) cellular vitality, (c) taxonomic discriminative power, and (d) analytical power. All these features have to be considered especially when selecting a new method

Probiotic Dairy Products, Second Edition. Edited by Adnan Y. Tamime and Linda V. Thomas.

for a specific application. In a broader sense, however, all methods can be divided in two groups: methods that rely on phenotypic properties of bacteria (phenotyping) and methods that rely on their genetic properties (genotyping).

Despite the well-known limitations of culture-dependent methods, the cultivation of microbes is routinely used in all microbiology laboratories and cannot be avoided even when opting for the genetic approach. Although a few International Standards Organisation (ISO) methods are also available to quantify probiotics and fermenting microbes employed in the dairy industry (Boyer & Combrisson, 2013), the range of selective media available to identify and enumerate probiotic strains or starter cultures is relatively limited (Vinderola & Reinheimer, 1999). When choosing the method for enumeration and/or identification of probiotics and starter cultures, the culture dependency of the method is one of the first factors to take into account.

The essential criterion for both probiotics and starter cultures is cellular viability, and some methods do not possess the ability to differentiate between dead and live bacterial cells. The physiological state of probiotic bacteria and starter cultures is an important parameter since many probiotic and technological effects depend on their metabolic activity.

Probiotic ability is a strain-specific feature as well as certain key phenotypic traits, and the interest for dairy applications is for particular strains within a given species. Specificity describes the degree to which a certain method can detect members of a target taxon (Bokulich & Mills, 2012). The term strain-specific is very strict, since a method should able to exclusively detect a specific particular strain. In reality, the term strain-specific is relative and is largely dependent on what tests are performed that prove the method's strain specificity. A common example of strain-specificity assessment is the demonstration of the presence or the absence of specific polymerase chain reaction (PCR) products on a certain set of more or less closely related strains, using isolated colonies from faeces or deoxyribonucleic acid (DNA) isolated from faecal samples (Treven, 2015).

Another important characteristic of the method of choice is analytical power (i.e. its quantification ability). Some methods offer limited or no quantitative information about the target strain, and these are used only for typing purposes. The classification of methods available for the identification and enumeration of probiotics and LAB starter cultures is shown in Table 6.1.

6.3 Phenotypic methods

6.3.1 *Differential plating*

Plate counting remains the most frequently used method for enumeration of LAB in dairy products and starter cultures to ensure the quality control of products. It should be noted that different organisms have different requirements for nutrients and microelements; therefore, no single medium or combination of defined media is applicable to all dairy products. Until now, only a few protocols for the enumeration of dairy starters and probiotic bacteria have been validated and published by the European Food Safety

Table 6.1 Classification of methods for identification and enumeration of probiotics and lactic acid bacteria starter cultures.

Analytical method	Culture dependency[1]	Viability discrimination[2]	Maximal taxonomic discriminative power	Quantification[3]
Differential plating	+	–	Species	+
API 50 CHL	+	–	Species	–
BIOLOG	+	–	Species	–
FTIRS	+	–	Strain	–
MALDI-TOF MS	+	–	Species	–
Flow cytometry	–	+	NA	+
Fluorescence microscopy	–	+	NA	+
PCR	–/+	–	Strain	–
qPCR	–	–	Strain	+
RT-PCR	–	–	Strain	–
PMA/EMA PCR	–	+	Strain	+
PCR-DGGE	–	–	Species	–
RAPD-PCR	+	–	Strain	–
SSCP	–	–	Species	–
PFGE	+	–	Strain	–
AFLP	–	–	Species	–
RFLP	–	–	Species	–
T-RFLP	–	–	Species	–
ARDRA	+	–	Species	–
MLST	+	–	Strain	–
Sequencing of specific genes	–/+	–	Strain	–
Sequencing of repeats and noncoding regions	+	–	Strain	–
WGS	+	–	Strain	+
Metagenomics	–	–	Strain	+
DNA/cDNA microarrays	–	+	Strain	+
FISH	–	+	Strain	+

FTIRS = Fourier transform infrared spectroscopy; MALDI-TOF MS = matrix-assisted laser desorption/ionisation–time of flight mass spectrometry; PCR = polymerase chain reaction; qPCR = quantitative PCR; RT-PCR = reverse transcription PCR; PMA/EMA-PCR = propidium monoazide–PCR/ethidium monoazide–PCR; PCR-DGGE = PCR–denaturing gradient gel electrophoresis; RAPD = randomly amplified polymorphic DNA–PCR; SSCP = single-strand conformation polymorphism; AFLP = amplified fragment length polymorphism; RFLP = restriction fragment length polymorphism; T-RFLP = terminal-restriction fragment length polymorphism; ARDRA = amplified ribosomal DNA restriction analysis; MLST = multilocus sequence typing; WGS = whole genome sequencing; FISH = fluorescence in situ hybridisation; NA = not applicable.

[1] + = Culture dependent; –/+ = can be performed also on culture; – = culture independent.

[2] – = The method cannot discriminate between viable and nonviable bacterial cells.

[3] + = Quantification is possible; – = quantification is not possible.

Authority (EFSA) and ISO in co-operation with the International Dairy Federation (IDF); the selection of suitable selective media still remains a challenge.

According to ISO/IDF (2003), the enumeration of yoghurt bacteria should be performed on M17 agar supplemented with lactose (LM17) for *Streptococcus thermophilus*, and on de Man, Rogosa and Sharpe (MRS) agar acidified with acetic acid to pH 5.4 for *Lactobacillus delbrueckii* subsp. *bulgaricus*. However, this standard is usually not applicable to probiotic cultures in yoghurt, since probiotic strains and yoghurt starter cultures often have similar growth requirements and therefore selection of the probiotic strain is not guaranteed. Two ISO standardised methods are available for the selective enumeration of probiotic bacteria: the method for the enumeration of *Lactobacillus acidophilus* using MRS agar supplemented with clindamycin and ciprofloxacin (ISO/IDF, 2006), and the method for bifidobacteria enumeration using transgalactosylated oligosaccharide (TOS)–propionate agar supplemented with mupirocin (ISO/IDF, 2010b). Since TOS–propionate agar does not support growth of all strains of bifidobacteria, the use of other media like MRS agar or modified Wilkins Chalgren agar (WCA) supplemented with mupirocin may be needed (Simpson *et al.*, 2004; Bunesova *et al.*, 2015). Suggestions for the enumeration of other bacterial groups are also stated in the standard, which specifies the characteristics of starter cultures used for the production of fermented milk (ISO/IDF, 2010a).

Several media have been used for selective enumeration of starter cultures and probiotic bacteria when they are combined in the dairy products. Only a few media are useful for selective enumeration of total lactobacilli, since several other micro-organisms can grow on media used for their cultivation (Coeuret *et al.*, 2004). Lactobacilli are most frequently cultivated in anaerobic conditions using MRS agar (Tharmaraj & Shah, 2003; Leverrier *et al.*, 2005; Aureli *et al.*, 2010; Vardjan *et al.*, 2013; Succi *et al.*, 2014) or Lactobacilli Selective Agar (LBS), also known as Rogosa agar (Čanžek Majhenič *et al.*, 2007; Ong & Shah, 2009). Unlike MRS, LBS does not support the growth of bifidobacteria; therefore, it may be more appropriate for analysis of products containing both lactobacilli and bifidobacteria (Ong & Shah, 2009). Nevertheless, MRS remains the most frequently used agar for enumeration of lactobacilli. For differential enumeration of heterofermentative lactobacilli (e.g. *Lactobacillus casei*, *Lactobacillus paracasei* spp., *Lactobacillus rhamnosus*, *Lactobacillus plantarum*, *Lactobacillus fermentum* etc.), MRS-V agar (MRS supplemented with 1 mg L^{-1} of vancomycin) is appropriate, since the majority of homofermentative lactobacilli (e.g. *Lb. delbrueckii* spp., *Lb. acidophilus* and *Lactobacillus salivarius*), bifidobacteria, lactococci and enterococci are susceptible to vancomycin (Aureli *et al.*, 2010; Karimi *et al.*, 2012). Lactobacilli are usually incubated at 37 °C, but the ability to grow at different temperatures can sometimes be used as an additional selective factor. When both *Lb. casei* and *Lb. rhamnosus* are present in the sample, the counts of *Lb. rhamnosus* on MRS-V at 45 °C can be subtracted from the total count on MRS-V incubated at 37 °C to obtain the *Lb. casei* count (Tharmaraj & Shah, 2003; Karimi *et al.*, 2012).

Since MRS-V also supports the growth of *Pediococcus* (Simpson *et al.*, 2004) and *Leuconostoc* (Mathot *et al.*, 1994; Hemme & Foucaud-Scheunemann, 2004) strains, it can be used for their selective enumeration at 37 °C and 30 °C, respectively. The addition of 1 mg L^{-1} of ampicillin to MRS-V improves the selectivity for enumeration of pediococci (Simpson *et al.*, 2004) because heterofermentative lactobacilli are usually susceptible to ampicillin. The selective enumeration of *Leuconostoc* spp. can be improved by the addition of tetracycline and tomato juice (Hemme & Foucaud-Scheunemann, 2004).

The LM17 culture medium is proposed for *Str. thermophilus* enumeration (Leverrier *et al.*, 2005; Aureli *et al.*, 2010; Succi *et al.*, 2014) as well as for the enumeration of lactococci (Ong & Shah, 2009; Oberg *et al.*, 2011). Anaerobic or aerobic incubation at 37 °C or 45 °C is used for *Str. thermophilus*, while lactococci are incubated aerobically at 30 °C. For both bacterial groups, a short incubation time (24 h) improves selectivity, since some lactobacilli can grow on LM17 but much more slowly than lactococci or *Str. thermophilus* (Dave & Shah, 1996; Oberg *et al.*, 2011). For the enumeration of streptococci in yoghurt, *Str. thermophilus* (ST) agar can also be successfully used (Dave & Shah, 1996; Vinderola *et al.*, 2000; Tharmaraj & Shah, 2003; Leverrier *et al.*, 2005).

Sodium lactate and yeast extract lactate (YELA) agar can be used for selective enumeration of *Propionibacterium* strains. While 7 days of incubation in anaerobic conditions at 30 °C is required for propionibacterial growth, some other bacteria can already be observed after 3 d; therefore, the total count at day 3 should be subtracted from the total counts at day 7 in order to obtain the appropriate propionibacterial count.

6.3.2 *Carbohydrate fermentation-based methods*

The conventional microbiological methods for bacterial identification are based on morphological and physiological characteristics, such as Gram staining, cell shape, spore formation, enzyme production and the fermentation of different carbohydrates. With regard to the latter approach, the API® system and Biolog are widely used. Both methods differentiate bacterial isolates according to their fermentation patterns (Moraes *et al.*, 2013).

API 50 CHL

The API test has become a well-established method for the identification of microorganisms to the species level. It is used for infectious disease diagnosis and identification of industrially important micro-organisms. There are several types of API identification tests for healthcare and food safety applications, provided by the manufacturer BioMerieux. Among them, the API 50 CH system is used in conjunction with API 50 CHL Medium for the identification of *Lactobacillus* and related genera. The API 50 CH is a standardised system using biochemical tests to analyse carbohydrate metabolism and in this way to identify micro-organisms. It is composed of 50 microtubes, with 49 of them containing different carbohydrates and their derivatives; the first microtube is a negative control that does not contain any active ingredient. During incubation, fermentation is revealed by a colour change in the microtube, caused by the anaerobic production of acid and detected by the pH indicator present in the medium. According to the manufacturer's instructions, results are read after 48 h of incubation and further analysed with the API web database offered by BioMerieux (Herbel *et al.*, 2013). There are many papers reporting the use of the API system for biochemical fingerprinting of LAB and probiotics, but reports on the specificity of the system are rather ambiguous; discrepancies between the results of phenotypic and genotypic identification are evident. When API identification of LAB was compared to results obtained by species-specific PCR (Čanžek Majhenič *et al.*, 2007; Brolazo *et al.*, 2011), 16S rRNA (ribosomal ribonucleic acid) sequencing (Moraes *et al.*, 2013; Ni *et al.*, 2015) or the use of whole chromosomal DNA probes (Boyd *et al.*, 2005), the agreement between the methods was usually low.

Nevertheless, the API 50 CHL system can serve as a preliminary taxonomic identification method, but due to the high level of phenotypic variability among LAB, and especially lactobacilli, this time-consuming and lab-intensive method should not be solely used. The misidentification and non-interpretable results are clear drawbacks of this method (Herbel *et al.*, 2013).

Biolog

According to the manufacturer's introduction (Anonymous, 2007), the Biolog AN MicroPlate™ system is designed for identification of a very wide range of anaerobic bacteria, including the genera *Bifidobacterium*, *Clostridium*, *Lactobacillus*, *Lactococcus*, *Pediococcus*, *Propionibacterium* and *Weissella*. These genera are important in industrial and environmental applications, especially in the food industry where they are responsible for both food production and food control. The AN MicroPlate employs the same redox chemistry used in the Biolog GP2 and GN2 MicroPlate. Based on reduction of tetrazolium, the test responds to the process of metabolism (i.e. oxidation of substrates) rather than to metabolic by-products (e.g. acid). Biolog's universal chemistry works with any carbon source and greatly simplifies the testing process, as no colour-developing chemicals need to be added after incubation.

The Biolog system tests a micro-organism's ability to utilise or oxidise a panel of 95 carbon sources, where tetrazolium violet is incorporated into each of the substrates contained in a 96-well microtitre plate. As a bacterium begins to use the carbon sources, it respires, which reduces the tetrazolium redox dye and thus changes those wells to a purple colour. The end result is a unique biochemical pattern or fingerprint of coloured wells on the microplate that is characteristic of that bacterial species. The fingerprint data are analysed and compared to a database, and an identification result is generated. The Biolog system was originally created for the identification of Gram-negative bacteria, but since the introduction of this system in 1989, the identification capability of the system has broadened to include Gram-positive bacteria (Tshikhudo *et al.*, 2013). There are some studies reporting use of the Biolog system for LAB and/or probiotic bacteria identification. Comparison of results obtained by the Biolog system with genotypic identification (e.g. species-specific PCR or 16S rDNA sequence) showed that the reliability of commercial phenotypic identification systems was inadequate when analysing LAB isolates from natural, spontaneous fermentations, and these need to be confirmed with genotypic identification methods (Morgan *et al.*, 2009; Paveljšek *et al.*, 2014; Čitar *et al.*, 2015).

The Biolog system is fairly simple to use, requiring little technical expertise to operate and interpret results. A downside is that the Biolog system requires pure cultures and subsequent growth of the bacteria. Pure culture and growth are frequently problematic when it comes to slow-growing, fastidious, unusual, nonviable or non-culturable bacteria.

As many LAB share similar nutritional and growth requirements, biochemical-based methodologies for identification are not conclusive in many cases; therefore, there is a need for any phenotypic approach to be supported or combined with a molecular approach. Proper LAB and probiotic identification can only be achieved by a multi-stage approach. Nevertheless, the API and Biolog systems are both culture-dependent methods that are time-consuming, since they demand approximately 4 or 3 d, respectively, to be completed. Their discriminatory power for LAB and probiotic bacteria identification, which is based on physiological properties, is proposed to be at genus and species levels (Herbel *et al.*, 2013), but at both levels identification can be questionable on many occasions (Boyd *et al.*, 2005; Brolazo *et al.*, 2011; Moraes *et al.*, 2013). Therefore, the species name cannot be assigned on the basis of such a test; there is a need to confirm Biolog results with other methods, such as whole genome sequencing (WGS), or at least 16S rRNA gene sequencing, matrix-assisted laser desorption/ionisation–time-of-flight mass spectrometry (MALDI-TOF MS) or multi-locus sequence typing (MLST) (Tshikhudo *et al.*, 2013).

6.3.3 *Spectroscopic methods*

Spectroscopic methods enable identification of bacteria on the basis of recording spectra of whole bacterial cells, which reflects their phenotypic fingerprint. Fourier transform infrared spectroscopy (FTIRS) and MALDI-TOF MS are both suitable methods for the routine identification of food-related micro-organisms. FTIRS generates a biochemical fingerprint of cell components such as proteins, lipids, sugars, lipopolysaccharides and nucleic acids, while MALDI-TOF MS records more specific protein mass spectra (Wenning *et al.*, 2014).

Fourier transform infrared spectroscopy (FTIRS)

Fourier transform infrared spectroscopy is based on measurement of the interaction of mid-infrared light with different chemical components in the sample. Chemical bonds present in the sample can absorb infrared (IR) radiation of specific wavelengths resulting in different vibrations of a molecule, such as stretching, contraction and bending. When IR radiation is passed through the sample, some wavelengths are absorbed by functional groups present, regardless of other structures in the sample, and the detector records absorbed/transmitted light as a fingerprint that reflects the chemical composition of the sample. There is a correlation between the band position and chemical structures in the sample (Davis & Mauer, 2010).

Different bacteria vary considerably in their chemical composition, particularly in cell wall or membrane composition, resulting in unique and characteristic IR fingerprints. These differences are particularly evident at the strain level, but not at the genus or species level; therefore, FTIRS can be used to differentiate bacterial strains and species, based on the FTIRS fingerprints database (Davis & Mauer, 2010).

Fourier transform infrared spectroscopy is a relatively fast, simple and sensitive technique, requiring only a small amount of sample. Furthermore, almost no sample preparation is needed before the measurement (Davis & Mauer, 2010). Prior to the analysis, the micro-organisms need to be cultivated either in liquid or on solid medium. Cells grown on solid media can be harvested directly from the media, suspended in the water and subjected to the analysis, while cultures from the liquid media first need to be centrifuged and washed to remove the medium. The most frequently employed FTIRS technique is transmission, where the sample is placed on an IR transparent ZnSe crystal (Amiel *et al.*, 2000; Oust *et al.*, 2004; Bosch *et al.*, 2006; Luginbühl *et al.*, 2006; Dziuba *et al.*, 2007; Nicolaou *et al.*, 2011); the use of reflectance on different optical plates has also been reported (Savić *et al.*, 2008; Foca *et al.*, 2016). Although spectroscopic equipment is relatively expensive, no additional costs are needed for analysis. Furthermore, FTIRS can also be used for process monitoring, quality control and authenticity determination of dairy products (Karoui & De Baerdemaeker, 2007; Woodcock *et al.*, 2008).

For correct differentiation of LAB species, it is crucial to use an appropriate spectral range and the appropriate multivariate statistical methods. Usually between 4000 and 400 cm^{-1} of bacterial spectra are recorded, including the several repetitions of the measurements of the same strain that are required in order to verify repeatability (Santos *et al.*, 2015). Comparison of spectra derived from different LAB revealed that the region between 1500 and 700 cm^{-1} is the best suited for discrimination and identification of LAB (Oust *et al.*, 2004; Dziuba *et al.*, 2007; Savić *et al.*, 2008; Prabhakar *et al.*, 2011). Multivariate statistical methods are of two types: supervised or unsupervised analysis. For unsupervised methods, such as principal component analysis (PCA) and hierarchical cluster analysis (HCA), no prior knowledge about the test bacteria is needed. On the other hand, supervised methods like discriminant analysis (DA), partial least squares regression (PLSR) and artificial neural network (ANN) require prior knowledge about the sample identity (Davis & Mauer, 2010).

All protocols, including cultivation, harvesting, drying and registration of spectra, should be strictly standardised to guarantee reproducibility of results. Alterations in the media used for cultivation and/or variation in incubation time have a strong influence on identification accuracy (Bosch *et al.*, 2006; Wenning & Scherer, 2013). For identification purposes, a spectrum database with a sufficient number of reference strains should be created, preferably from a different origin, to cover intra-species diversity (Luginbühl *et al.*, 2006; Savić *et al.*, 2008; Santos *et al.*, 2015). Several libraries created for identification of LAB have mostly focused on the narrow range of bacterial species included.

Several authors have demonstrated the potential of FTIRS for the discrimination and classification of bacterial species used as dairy starter cultures or probiotics (Amiel *et al.*, 2001; Oust *et al.*, 2004; Bosch *et al.*, 2006). However, the use of FTIRS for classification and identification of lactobacilli isolated from Kefir (Luginbühl *et al.*, 2006) and cheese samples (Savić *et al.*, 2008) revealed difficulties in differentiating between closely related bacterial species. Although FTIRS spectroscopy is a phenotypic method, the typing of yeast and bacterial starters for Limburger cheese correlated well with genotypic methods (Goerges *et al.*, 2008). Moreover, identification of cheese isolates by FTIRS was in agreement with the results of sodium dodecyl sulphate–polyacrylamide gel electrophoresis (SDS-PAGE) identification (Weinrichter *et al.*, 2001). Although

FTIRS is mostly considered as a qualitative method, Nicolaou *et al.* (2011) successfully used it for enumeration of bacterial strains of *Staphylococcus aureus* and *Lactococcus lactis* subsp. *cremoris* both in pure culture and as co-cultures grown in ultra-high-temperature (UHT) milk.

Matrix-assisted laser desorption/ionisation–time-of-flight mass spectrometry

The MALDI-TOF MS method generates the protein profiles of whole bacterial cells. The laser beam desorbs and ionises the sample (previously spotted onto the sample target plate), which leads to formation of mainly singly protonated ions. The ions are separated on the basis of their mass-to-charge ratio (m:z) and detected using the time-of-flight (TOF) analyser. Based on this TOF information, a characteristic spectrum called a peptide mass fingerprint is created (Singhal *et al.*, 2015). The recorded protein mass spectra can be used for identification of bacteria at the genus and species levels and, in some cases, even to the subspecies level (Sauer & Kliem, 2010). Although purchase costs are extremely high, the method is a rapid, sensitive and automated system for bacterial identification.

For microbial identification, spectra from 2 to 20 kDa proteins are typically collected, which represent mainly ribosomal proteins along with a few housekeeping proteins. The protein pattern of the tested microbe is compared to the spectra of reference bacteria in the library. Some commercial libraries are already available, such as the MALDI Biotyper (Bruker Daltonics), the SARAMIS™ (BioMerieux) and the Andromas (Andromas SAS) database. In addition, several research groups have created open-source software and databases that are freely available (Singhal *et al.*, 2015). At first, such databases focused on pathogens and clinically relevant micro-organisms, but now several food-related micro-organisms are included. The MALDI Biotyper database already contains more than 200 *Lactobacillus* species and therefore is commonly used for the analysis of LAB from milk and dairy products (Albesharat *et al.*, 2011; Angelakis *et al.*, 2011; Dušková *et al.*, 2012; Delavenne *et al.*, 2013; Bunesova *et al.*, 2014; Nacef *et al.*, 2016). The MALDI-TOF MS identification at the species level, presented as either highly probable or probable identification, was confirmed by the use of molecular methods, such as 16S rDNA sequencing or species-specific PCR, while strains with only probable identification at the genus level have, in some cases, been misidentified (Delavenne *et al.*, 2013). Commercial databases are continually increasing in size and are regularly updated, which will certainly improve the identification of different micro-organisms used in the dairy industry. As well as identification, MALDI-TOF MS has been shown to be an appropriate method for typing closely related *Lactococcus* spp. (Tanigawa *et al.*, 2010) and *Bifidobacterium* spp. (Sato *et al.*, 2011).

Usually one colony is enough for analysis. Intact bacterial cells from agar plates or after harvesting by centrifugation from liquid media can be transferred directly onto the MALDI target plate; previous extraction of bacterial proteins from the cells, however, may improve the quality of the spectra obtained (Sedo *et al.*, 2011). Different protocols have been used for protein extraction from LAB, such as solvent extraction with formic acid in acetonitrile (Albesharat *et al.*, 2011; Dušková *et al.*, 2012; Delavenne *et al.*, 2013; Bunesova *et al.*, 2014; Wenning *et al.*, 2014; Španová *et al.*, 2015), cell disruption using bead beating (Teramoto *et al.*, 2007; Tanigawa *et al.*, 2010) and treatment with trypsin (Schmidt *et al.*, 2009). Although culture conditions might have some impact on the phenotypic appearance of microbes, they do not have significant impact on ribosomal proteins; therefore, changes in culture conditions have only minor effects on the identification of food-related bacteria (Wenning *et al.*, 2014).

6.3.4 *Fluorescence dyes-based methods*

Traditionally, plate counting is used for the enumeration of starter and probiotic bacteria. The method is time-consuming and often provides an underestimate of microbial count due to the presence of damaged or viable but non-culturable (VBNC) cells. Since starter and probiotic bacteria in food products can often suffer chemical or physical stresses that result in the temporary loss of culturability, different fluorescence-based techniques have been proposed as alternatives to plate counting (Zotta *et al.*, 2012). A variety of fluorescence probes can be used to examine physiological characteristics of living cells, such as membrane integrity, intracellular enzyme activity, membrane potential or cytoplasmic pH (Davis, 2014).

Staining by fluorescein diacetate (FDA), carboxifluorescein diacetate (cFDA) and - cyano-2,3-ditolyl tetrazolium chloride (CTC) may be used for detection of enzyme activity that indicates viable bacteria. These fluorochromes are non-fluorescent until intracellular enzymes (usually esterase) cleave them.

The evaluation of membrane integrity, which is the most definitive proof of cell viability, can be detected by dye exclusion or dye retention methods. Exclusion or cell-impermeable dyes, like propidium iodide (PI) or TOTO stains, are excluded by intact cells and can stain only membrane-compromised (dead) cells. On the other hand, cell-permeable dyes (like DAPI – 4′,6-diamidino-2-phenylindole – and most SYTO® dyes) are able to stain also the bacteria with intact cell membranes. These dyes usually bind to nucleic acids. Combinations of cell-permeable and cell-impermeable dyes can be used to distinguish between live and dead cells.

Only live cells are able to maintain membrane potential, which can be measured by means of membrane potential-sensitive dyes, such as carbocianines (DiOCn) or rhodamine (Díaz *et al.*, 2010).

Fluorescence microscopy

Bacteria in samples can be directly visualised microscopically, but assessment of viability requires differentiation between live and dead bacteria (Davis, 2014). The LIVE/DEAD *Bac*Light™ kit, consisting of the fluorescence nucleic acid stains SYTO 9 and PI, is generally used for direct fluorescence microscopy of bacteria in milk and dairy products (Auty *et al.*, 2001; Bunthof *et al.*, 2001; Gatti *et al.*, 2006; Moreno *et al.*, 2006; Olszewska *et al.*, 2012). While the green-fluorescent SYTO 9 stain penetrates both viable and nonviable bacteria, the red-fluorescent PI penetrates only bacteria with a damaged membrane. Thus, live bacteria fluoresce green while dead bacteria fluoresce red. Several other fluorescent dyes have also been used for LAB staining. Corich *et al.*

(2004), for example, successfully used FDA, CTC and DAPI for the enumeration of bacteria in whey starter cultures, while the use of acridine orange gave good results only with pure cultures but not whey cultures. Moreover, Zotta *et al.* (2012) tested different combinations of SYTO 9; PI; 5,(6)-carboxyfluorescein diacetate (cFDA) and DAPI stains for viability assessment of LAB subjected to oxidative or heat stress. For most species, cFDA–DAPI, DAPI–PI and cFDA–PI combinations provided better results compared to SYTO 9–PI. The use of cFDA–PI gave satisfactory results for almost all LAB strains tested.

Comparison of analysis of whey starters (Gatti *et al.*, 2006) and fermented milk samples (Moreno *et al.*, 2006) by direct microscopic enumeration and plate counting indicated that plate counting may lead to an underestimation of bacterial numbers, which can be related not only to the presence of VBNC but also to bacterial clumping. In contrast, the numbers of bacteria in cheese and spray-dried probiotic milk powder obtained by direct confocal scanning laser microscopy were lower than the numbers obtained by plate counting (Auty *et al.*, 2001). Direct epifluorescence microscopy may also be used for checking the viability of starter and probiotic bacteria after drying (Perdana *et al.*, 2012) or freezing (Passot *et al.*, 2015).

Conventional epifluorescence microscopy can be used for liquid samples, while the use of confocal scanning laser microscopy enables the observation of the sub-surface of foods (Auty *et al.*, 2001). Nevertheless, methods based on fluorescent detection in combination with microscopic detection are usually very useful, but they are not convenient for routine use for quality control in the dairy industry because they are too labour-intensive.

Flow cytometry (FC)

Flow cytometry can be described as automated microscopy with the advantages of automation, objectivity and speed, as many thousands of cells can be analysed in a second (Veal *et al.*, 2000). The basic principle of FC is measurement of the optical characteristics of a single microbial cell. The sample in liquid form is introduced to a fast-flowing fluid stream that forces the cells to pass in single file through a laser beam. Single microbial cells are illuminated with the laser beam, and the intensity of the optical signals generated is collected using a combination of optical filters and light detectors. A combination of light-scattering and fluorescence signals provides information on cell size, morphology and granularity. The fluorescence signal can also provide additional information about cell structure and functionality, depending on the fluorochromes used (Comas-Riu & Rius, 2009; Díaz *et al.*, 2010).

The FC technique is used as a routine method for measuring the total bacterial count in milk. However, the BactoScan™ (manufactured by FOSS) used for raw milk testing is unsuitable for analysis of starter cultures and dairy products, as it counts total (live and dead) bacteria. For the analysis of dairy-manufacturing processes, where the greater part of the natural microbiota is killed during pasteurisation, a system that counts only viable bacteria is required (Flint *et al.*, 2007). Kramer *et al.* (2009) evaluated the possibility of using SYTO 9 and PI in combination with FC for the determination of strains of *Lb. acidophilus* LA-5 and *Bifidobacterium animalis* subsp. *lactis* BB-12 in lyophilised probiotic product. They concluded that FC can complement plate counting as it detects a ratio of intact versus total bacteria, but one thing to be considered is that this method does not enable the differential counting of single strains. Bunthof *et al.* (2001) showed that a combination of cFDA and TOTO-1 is more appropriate for FC enumeration of viable LAB than a cFDA and PI combination. Double staining with cFDA and TOTO-1 was later used for enumeration of bacteria in dairy starters and probiotic products, as well as for pasteurised milk. Milk samples with low bacterial concentration needed a prior clearing procedure to reduce milk background, which was performed by the use of a commercially available clearing solution. The same solution was successfully used for clearing of probiotic products, although in the samples with high bacterial concentration the sample background could already be sufficiently reduced by diluting. The numbers of intact bacterial cells found with FC were higher compared to those obtained by plate count technique, revealing that some cells were not culturable. Since non-culturable bacteria can contribute to fermentation processes carried out by starter cultures, or to health effects in the case of probiotic bacteria, FC may be used for fast (1 h) and accurate viability assessment of starter and probiotic products (Bunthof & Abee, 2002). Another challenge is the counting microencapsulated bacteria introduced into different matrices. Doherty *et al.* (2010) successfully extracted protein-encapsulated *Lb. rhamnosus* cells by homogenisation of the sample followed by enzymatic protein digestion. Moreover, differences in preparation of dried cultures may affect bacterial counts. Adjustment of reconstitution solution, pH, time, addition of sugars and dilution should be optimised for the bacteria present in the sample, since different conditions are optimal for different bacteria (Muller *et al.*, 2010).

Besides the sample preparation, the optimisation of the instrument is also an important step in FC analysis; this can be difficult and tedious (Kramer *et al.*, 2009). Furthermore, in some cases FC can underestimate the number of viable LAB and bifidobacteria due to the presence of cell clumps, since one clump is counted as one cell in FC. The use of microscopy (Maukonen *et al.*, 2006) or propidium monoazide–quantitative PCR (PMA-qPCR) (Kramer *et al.*, 2009) has the advantage that these methods can account for clumped bacterial cells.

Flow cytometry was used successfully for the assessment of viability and physiological activity of LAB and bifidobacteria subjected to different stress conditions, such as heat, drying or freezing, osmotic stress and the presence of bile salts, as well as for stability testing (Amor *et al.*, 2002; Rault *et al.*, 2007; Sunny-Roberts & Knorr, 2008; Ananta & Knorr, 2009; Doherty *et al.*, 2010; Leandro *et al.*, 2014). Comparisons of FC with the results of traditional plate counting revealed that some of the stressed cells had lost cultivability.

In 2015, the ISO standardised method for the enumeration of LAB in starter cultures, probiotics and fermented milk by FC was published (ISO/IDF, 2015). This contains a detailed protocol for the enumeration of bacteria in freeze-dried or frozen cultures and fermented milk products. Three different combinations of fluorescent dyes can be used, cFDA and PI, PI and SYTO 24 or 3,3′-diethyloxacarbocyanine iodide ($DiOC_2$). Since FC enables the differentiation of active versus total bacteria, it can be used for stability assessment of both bacterial cultures and milk products during the entire shelf life. Only total viable bacteria can be assessed using the above-mentioned protocol, while the use of selective probes in combination with FC also enables enumeration of defined bacterial populations. Geng *et al.* (2014) used a two-step immune-labelling protocol for the enumeration of *Bif. animalis* subsp. *lactis* in fermented dairy products, using polyclonal antibody specific for *Bif. animalis* subsp. *lactis* as the primary antibody.

6.4 Genetic methods

One of the most commonly used approaches to quantify probiotic strains or LAB starter cultures is still based on standard cultivation techniques, but cell culture-based methods only measure replicating cells. Alternative methods, so-called culture-independent methods – such as fluorescent *in situ* hybridisation (FISH), nucleic acid amplification techniques such as real-time quantitative PCR (RT-qPCR), reverse transcriptase PCR (RT-PCR) and PMA-PCR, and cell-sorting techniques such as FC – offer the potential to enumerate both culturable and VBNC bacteria (Davis, 2014).

As in other fields of microbiology, species identification in dairy or probiotic products can be assessed through the use of either culture-dependent or culture-independent methods, but culture-independent methods offer a number of advantages over culture-dependent methods.

In culture-independent methods, micro-organisms are studied not because they are able to grow on a specific microbiological medium, but because they possess DNA, RNA and proteins, which are the preferred targets for such approaches. Moreover, the physiological status of the microbial cell does not affect the outcome of the investigation. Populations that are numerically less important are also not detected by means of traditional methods because they are masked on the plates (Cocolin *et al.*, 2013).

Despite all the advantages listed here, culture-independent methods are not perfect; their limitations and pitfalls were critically and constructively discussed in the review of Jany & Barbier (2008). One of the limitations is the difficulty of accessing every genotype from the community as a result of poor DNA extraction yield, or PCR inhibition by various extraction by-products or by substances coming from the food–probiotic product matrix itself. Furthermore, techniques such as denaturing gradient gel electrophoresis (DGGE), temperature gradient gel electrophoresis (TGGE), single-strand conformation polymorphism (SSCP) and terminal-restriction fragment length polymorphism (T-RFLP) have limitations in terms of resolution as they can generate patterns in which different genotypes group together due to co-migration. Another limitation of gel migration-based methods is the difficulty in obtaining profiles in which the less commonly amplified sequences cannot be distinguished from background noise. This problem increases with the diversity of the community. Finally, culture-independent methods regularly fail to identify species obtained using culture-dependent methods. These two different types of methods reveal different profiles of the same community; therefore, researchers suggest that using a polyphasic approach, combining culture-dependent and culture-independent methods, may be best in order to obtain a more accurate view of the structure of a microbial community.

6.4.1 *Polymerase chain reaction-based methods*

Most of the techniques used in food microbiology for culture-independent analysis are based on PCR. After amplification of the nucleic acids extracted directly from the food matrix, the PCR product is subjected to specific analyses that are able to highlight differences in the amplified DNA sequences (Cocolin *et al.*, 2013). Since its development by Kary Mullis in the 1980s, PCR has become fundamental to the work of biological and medical research laboratories. When there is a need to copy, sequence or quantify DNA, PCR is the starting point. Basically, this biochemical technique combines thermos-cycling and heat-labile enzymes that enable the quick and reliable multiplication of DNA. Thus, PCR exploits the ability of DNA polymerase to synthesise new strands of DNA complementary to a targeted template strand. Therefore, nucleotides are needed but DNA polymerase can add a nucleotide only onto a pre-existing 3′-OH group, which is provided by a primer, to which it can add the first nucleotide. This requirement makes it possible to delineate a specific region of template sequence that the researcher wants to amplify. At the end of the PCR reaction, the specific sequence is accumulated as billions of copies. The most commonly used PCR-based methods for LAB and probiotic bacteria identification and/or enumeration are PCR, RT-PCR, qPCR and PMA–ethidium monoazide (EMA) qPCR.

Polymerase chain reaction (PCR)

Widely employed for descriptive purposes such as the detection of microbes and analyses of ecosystem composition in combination with other technologies, PCR is now routinely used for the detection of pathogenic and spoilage microbes as well as technologically relevant LAB and probiotics in food products (Sohier *et al.*, 2014).

The simplest culture-independent PCR approach for the genus-, species- or strain-specific detection of LAB in dairy or probiotic products is the use of specific primers for PCR-based detection of the target organisms in the total bacterial DNA extracted from a sample. Such approaches, however, have not been widely used in culture-independent community studies of dairy products since a specific primer pair is needed for every bacterial species, yet this can be a helpful approach for confirming the presence and unequivocal identification of targeted species (Pogačić *et al.*, 2010).

An extensive and detailed chapter about application of PCR-based methods to dairy products and to non-dairy probiotic products has been published by Monnet & Bogovič Matijašić (2012). The review describes many interesting applications of PCR-based methods for dairy products that can be used to detect, identify and quantify either unwanted or beneficial micro-organisms. Bagheripoor-Fallah *et al.* (2015) also discussed the most commonly used molecular approaches to identify and/or quantify probiotic bacteria in fermented dairy products. They agreed that PCR-based techniques equipped with species-specific primers targeting 16S rRNA genes are rapid and reliable detection methods for species. When 16S rDNA sequencing failed in discrimination of analogous and intra-species strains due to their high similarity, species-specific primers were used (Moraes *et al.*, 2013). PCR with specific 16S rDNA-based oligonucleotide primers is a powerful method for the detection of target bacteria within complex ecosystems, such as human faeces or dairy products. In their review, Matsuki *et al.* (2003) described the use of genus- and species-specific PCR primers for the detection and identification of bifidobacteria that colonise the human gut or occur in dairy products. Genus- or species-specific primers are not too difficult to construct, but problems arise when intending to confirm different strains of the same species in the product. A variety of PCR-based genotyping techniques have been reviewed, such as random amplified polymorphic DNA (RAPD) analysis, repetitive sequence-based PCR (rep-PCR), pulsed-field gel electrophoresis (PFGE), amplified fragment length polymorphism (AFLP) ribotyping and so on; these are successfully used to distinguish different strains of probiotic bacteria, even closely related ones (Monnet & Bogovič Matijašić, 2012).

Brolazo *et al.* (2011) used another PCR approach. They analysed vaginal lactobacilli from 135 healthy women by using a multiplex PCR technique that applies primers based on specific sequences of bacterial DNA that codify the regions of the 16S and 23S rRNA which are exclusive for each species and produce fragments of DNA of different sizes for each species.

Often, after finding a suitable phylogenetic marker, the PCR strain-specific assay can be developed to target specific parts of the targeted gene. A multiplex PCR assay was designed, utilising two novel strain-specific primer sets, which enabled the identification of *Lb. casei* ATCC 393 by targeting polymorphic sites within the *hsp60* gene (Karapetsas *et al.*, 2010). The quantitative detection of *Lb. acidophilus* LAB20 was based on targeting the variable region of a novel S-layer protein, allowing the specific enumeration of the probiotic in dog faeces (Tang & Saris, 2013). Bacteriocin-specific primers were found useful for the detection and quantification of probiotic strain *Lactobacillus gasseri* K7 in biological samples (Treven, 2015; Treven *et al.*, 2015).

Ni *et al.* (2015) identified LAB from forage paddy rice silage by biochemical characterisation that was followed by genotypic characterisation based on sequence analyses of their 16S rRNA and *recA* genes. The latter was used for further discrimination of strains in the *Lb. plantarum* group where a multiplex PCR assay was performed with *recA* gene-based primers.

Reverse transcription polymerase chain reaction (RT-PCR)

The RT-PCR technique is a two-stage process, in which a target messenger RNA (mRNA) sequence is first transcribed into a complementary DNA (cDNA) sequence, using either random hexanucleotide primers or sequence-specific primers. The cDNA sequence may then be used to generate a second-strand cDNA or serve directly as a template for a PCR or qPCR (Keer & Birch, 2003). RT-PCR is, therefore, a PCR that is preceded by conversion of sample RNA into cDNA with enzyme reverse transcriptase, and it is one of the many variants of PCR. One such application of RT-PCR was reported by Trmčić *et al.* (2011), who evaluated expression of all 11 genes involved in the biosynthesis of the bacteriocin nisin during cheese production using real-time RT-PCR. Special attention has been given to the use of mRNA as a marker of viability. This marker is a highly labile molecule with a very short half-life (measured in seconds) in bacteria (Davis, 2014), and it therefore should provide a more closely correlated indication of viability status than DNA-based methods (Keer & Birch, 2003). For example, Saito *et al.* (2004) successfully demonstrated the viability of *Lb. helveticus* GCL1001 in human faeces as they detected the mRNA from this strain in the faeces of volunteers using nested RT-PCR. Sheridan *et al.* (1998) also used an RT-PCR method to investigate the relationship between detection of mRNA and cellular viability in *Escherichia coli*. Nevertheless, rRNA has also been investigated as an indicator of viability and has been found to positively correlate with viability under some bacterial-stress regimes. Lahtinen *et al.* (2008) assessed the stability of 16S rRNA of VBNC probiotic bifidobacteria during storage. They concluded that cells that gradually lost culturability in fermented products retained high levels of rRNA, whereas the rRNA of acid-killed control cells decreased at a faster rate. However, the longer half-life of rRNA species and their variable retention following a variety of bacterial stress treatments make rRNA, under many conditions, a less accurate indicator of viability than mRNA targets (Keer & Birch, 2003).

Real-time PCR (RT-PCR), or quantitative PCR (qPCR)

Since they are abbreviated similarly, reverse transcription PCR and real-time PCR are often mistakenly interchanged. To avoid confusion, real-time PCR is labelled as quantitative PCR (qPCR), whereas reverse transcription PCR is abbreviated as RT-PCR. Besides real-time PCR, there are also other quantitative PCR approaches that will not be described here. The qPCR technique differs greatly from PCR because qPCR measures the amplification in real time, not just at the end point.

The principle of qPCR involves monitoring the progress of DNA amplification using fluorescent reagents, which bind with the amplicon at the end of each cycle without disrupting the amplification of the template DNA. The quantification strategy is based on the threshold cycle number (C_T) that is inversely proportional to the cell number corresponding to the template DNA concentration. Absolute quantification of microbial populations is achieved by plotting the C_T against the cell number corresponding to the template DNA concentration used for qPCR (Nagarajan & Loh, 2014).

Quantitative PCR enables the discrimination of different taxa and the quantification of bacteria in a sample. In qPCR analysis, it is possible to measure the amplification process using genus or species-specific primers. As reviewed by Nagarajan and Loh (2014), the fluorescent reagents used to follow PCR amplification of target DNA sequences include non-specific DNA-binding dyes, hydrolysis probes, hybridisation probes, light-up probes, molecular beacons, sunrise primers and scorpion primers. Among them, SYBR Green I as a non-specific DNA-binding dye, and TaqMan probes as hybridisation probes, are the most commonly used detection chemistries in qPCR. SYBR Green I binds non-specifically to double-stranded DNA (dsDNA), and emits fluorescence as amplicons accumulate during the PCR.

Bogovič Matijašić *et al.* (2010) used species-specific primers and reaction conditions for conventional PCR and SYBR Green I qPCR to quantify *Lb. gasseri*, *Enterococcus faecium* and *Bifidobacteriun longum* subsp. *infantis* in an over-the-counter probiotic drug. Ladero *et al.* (2012) employed multiplex qPCR for the detection and quantification of putrescine-producing LAB based on the detection of the agmatine deaminase gene (*aguA*) in dairy products. The authors proposed a multiplex qPCR method for the quantitative detection and identification of putrescine-producing lactobacilli, lactococci and enterococci present in dairy products, which could also serve as a screening method for putrescine producers in starter culture collections. Furet *et al.* (2004) developed qPCR assays for the quantification of LAB in fermented milk products, designing specific primers for the detection of bacterial species or groups of species. With this method, authors successfully identified *Str. thermophilus*, *Lb. delbrueckii* spp., *Lb. casei* and the *Lb. casei* group, *Lb. paracasei* subsp. *paracasei*, *Lb. rhamnosus*, *Lb. acidophilus* and

Lb. johnsonii as initially present in commercial fermented milk products, with a detection threshold of 10^3 cells mL^{-1} of product.

The qPCR technique is often combined with reverse transcription to quantify mRNA. This gives another visible difference between the two methods; RT-PCR can be used for the amplification process, but it needs to be combined with qPCR for quantification purposes. The obvious advantages of qPCR are its high specificity, better sensitivity and wide identification range, but optimisation of primers and probe design, nucleic acid extraction and PCR biases are the main pitfalls of the method (Nagarajan & Loh, 2014).

Propidium monoazide-PCR (PMA-PCR) and ethidium monoazide-PCR (EMA-PCR)

The PMA-PCR and EMA-PCR are emerging techniques that limit enumeration to cells with intact membranes; these methods can also be termed as viability PCR (vPCR) (Davis, 2014) because bacterial cells with intact membranes are assumed to be viable. Prior to genomic DNA extraction and qPCR analysis, bacterial cells are treated with intercalating agents, such as PMA or EMA, which penetrate only cells with compromised membranes and subsequently prevent amplification of DNA by PCR. Quantification of viable cells in probiotic products or viable LAB in fermented milk was proposed by using EMA-PCR and PMA-PCR, and these methods showed good correlations with plate counts. The PMA-qPCR has been successfully implemented to mitigate overestimation of the microbial community due to nonviable cells.

Kramer *et al.* (2009) evaluated the possibility to use PMA in combination with qPCR using SYBR Green I chemistry for selective enumeration of viable probiotic bacteria of the strains *Lb. acidophilus* LA-5 and *Bif. animalis* subsp. *lactis* BB-12 in lyophilised products. The results obtained by FC (LIVE/DEAD) analysis were comparable with those by PMA-qPCR. The authors concluded that PMA-qPCR and FC determination of the viability of probiotic bacteria could complement the plate count method that considers only the culturable part of the bacterial population.

In another study, the PMA-qPCR approach was evaluated for enumeration of probiotic strains *Lb. gasseri* K7 and *Lb. delbrueckii* subsp. *bulgaricus* IM414 microencapsulated in calcium alginate beads (Oketič *et al.*, 2015). The lactobacilli were analysed by plate counting and PMA-qPCR by species-specific primers during storage at 4 °C for 90 d. The authors concluded that cell injury interferes with colony formation without affecting membrane integrity; therefore, they agreed that the culture-independent PMA-qPCR method could not be an alternative for the plate count method, but that it might complement the latter well-established method, providing useful information about the ratio of compromised bacteria in the samples.

When Weber *et al.* (2014) analysed the composition of a bacterial community in bulk tank milk, they compared culture-dependent and culture-independent methods, including the plate count method, chemotaxonomic differentiation of isolates, subsequent identification by 16S rRNA gene sequencing, and the PMA treatment of milk before DNA extraction and construction of 16S rRNA gene clone libraries. Since they observed certain discrepancies in bacterial community composition in raw milk based on either culture-dependent or culture-independent methods, they strongly recommended a combination of both approaches, as this would enable the detection of the highest bacterial diversity in the raw milk samples analysed.

Despite its slight advantage in more strongly suppressing dead cell signals, the usefulness of EMA can be severely hindered by the penetration of intact cells in a species-dependent manner. PMA, on the other hand, proves more selection for live cells but can show greater suppression of dead cell signals. Different approaches have been used to overcome these deficiencies (Fittipaldi *et al.*, 2012). In their extensive review, Fittipaldi *et al.* (2012) discussed in detail current knowledge and present aspects that are important when designing experiments employing viability dyes. Moreover, the crucial importance of viability determination of probiotics by PCR-based methods, as well as new high-throughput molecular technologies such as microarray technology and next-generation sequencing (NGS), was reviewed by Monnet & Bogovič Matijašić (2012).

6.4.2 *DNA banding pattern-based methods*

Denaturing gradient gel electrophoresis (DGGE)

The DGGE technique is based on the separation of PCR amplicons of the same size but with different sequences (Jany & Barbier, 2008) where the 16S rRNA is the most frequent target gene because it exists in all bacteria and can easily be amplified without prior knowledge of studied strains (Li *et al.*, 2009). In addition, DGGE also is a two-step technique composed of: (a) PCR amplification of the genes encoding the 16S rRNA, and (b) separation of PCR amplicons based on the decreased electrophoretic mobility of PCR-amplified, partially melted, dsDNA molecules in polyacrylamide gels containing a linear gradient of DNA denaturants (Bagheripoor-Fallah *et al.*, 2015), most often chemical (urea and formamide in DGGE) or, very rarely, physical TGGE (Cocolin *et al.*, 2013). As a result, a mixture of amplified PCR products will form a banding pattern after staining that reflects the different denaturing behaviour of the various sequences that represent components of the microbiota. The resulting bands in the gel are analysed by comparing them to the control DNA ladder run on the same gel (Figure 6.1). Subsequent identification of specific bacterial groups or species present in the sample can be achieved either by cloning and sequencing the excised bands or by hybridisation of the profile using phylogenetic probes (Amor *et al.*, 2007).

The DGGE technique has been reported as being successful in a discrimination and stability study of lactobacilli and yeast microbiota in kefir grains (Vardjan *et al.*, 2013), and in an analysis of the microbial consortium of kefir grains with a focus on *Lactobacillus kefiranofaciens* (Hamet *et al.*, 2013). Lorbeg *et al.* (2009) evaluated six primer pairs amplifying different variable regions of 16S rDNA that were selected and applied in DGGE analysis of 12 species belonging to genus *Enterococcus* and eight other bacterial species often found in cheeses, namely seven lactobacilli and one *Lactococcus lactis* spp. For differentiation and identification of pure enterococcal isolates, primer pair P1V1/P2V1 showed the most promising results: all 12 enterococcal isolates gave distinctive DGGE fingerprints but with multiple bands patterns, indicating that this primer pair is not appropriate for identification of enterococcal species in mixed cultures. Much better potential for detection and identification of enterococci in mixed

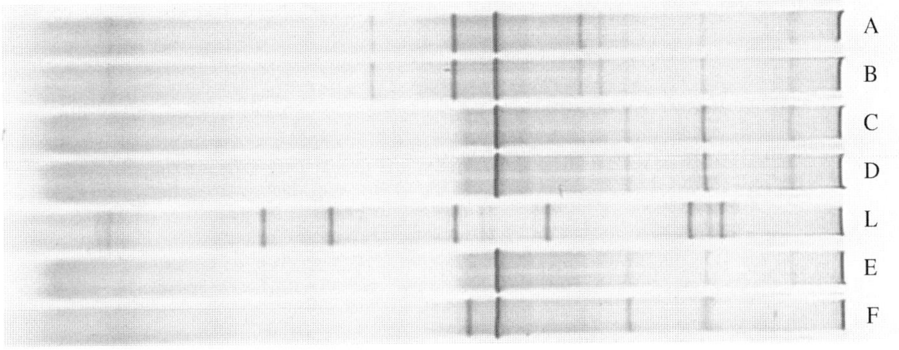

Figure 6.1 DGGE comparison of different Kefirs from Slovenian market using primers HDA1-GC and HDA2.
Note: Lanes A–F: DNA isolated directly from different Kefirs; lane L: DNA ladder.
DGGE = Denaturing gradient gel electrophoresis.

communities was shown by primers HDA1/HDA2 and V3f/V3 amplifying the V3 region but, since some bacterial species showed the same fingerprint, for clear identification the authors suggested a combination of DGGE with another method. Appropriate primer selection is essential, but unfortunately a universal primer set does not exist for community profiling.

One potential of DGGE is that it offers the possibility of performing ecological studies that target both nucleic acids (i.e. DNA and RNA). In order to reveal the metabolically active microbiota of artisanal cheeses, some researchers have performed DGGE analyses on reverse-transcribed (RT) RNA (Randazzo *et al.*, 2002; Leite *et al.*, 2013). By combining RT-PCR-DGGE (RNA-based) and PCR-DGGE (DNA-based), it is possible to differentiate metabolically active (RNA-derived) microbiota from the total diversity (DNA-derived) of microbiota. Therefore, the RT-PCR-DGGE approach might be very useful in studies of long-matured cheeses since different microbial groups might be active during different periods of maturation. Regardless of whether bacteria are viable or nonviable, their DNA is always present in the cheese matrix. Since RNA is less stable than DNA, RNA will degrade quicker in dead organisms (Pogačić *et al.*, 2010).

Finally, as with other methods, there are some limitations of the DGGE fingerprint technique, for example its low sensitivity of detection of rare members of the community (<1%); however, with group- or species-specific primers, the sensitivity of detecting less frequent bacteria has been significantly improved (Amor *et al.*, 2007). In their review, Cocolin *et al.* (2013) extensively summarised studies from all around the world that exploit the DGGE technique for analysing the microbial ecology of various food products where the limit of detection was about 10^3 colony-forming units (cfu) mL^{-1} or g^{-1} of product (Cocolin *et al.*, 2013). The DGGE approach also suffers from a weakness in differentiating between species with high phylogenetic relationships, so that sequencing of the bands in the DGGE profile is necessary (Lorbeg *et al.*, 2009; Vardjan *et al.*, 2013). Two suggestions have been proposed to fill this gap: (a) applying a narrower denaturing gradient, based on an increase in the band position, and (b) employment of other primers that might lead to products with easy separation on DGGE gels. However, every change in the gradient, primer set or electrophoresis conditions will result in new parameters so that a new database has to be constructed. This indicates that DGGE is only applicable to microbial ecosystems that are simple in nature, such as probiotic products, and the increasing complexity of a microbial population means that more parameters would need to be modified, which would make this technique time-consuming (Bagheripoor-Fallah *et al.*, 2015).

Randomly amplified polymorphic DNA-PCR (RAPD-PCR) and repetitive genomic element PCR (rep-PCR)

The RAPD-PCR and rep-PCR techniques are DNA fingerprinting methods that rely only on PCR (Temmerman *et al.*, 2004). Both methods are based on the presence of repetitive elements present in the bacterial genome. The use of single primers complementary to these sequences in the PCR reaction enables the amplification of different-sized DNA fragments lying between these elements (Masco *et al.*, 2003). Isolated DNA from a pure strain is used as a template in the PCR reaction, and its DNA fragments are separated using agarose gel electrophoresis to produce a specific fingerprint (Figure 6.2). Since these fingerprints are strain specific, these methods can be successfully used for strain typing.

The RAPD-PCR technique uses short arbitrary (10 bp) primer and low-stringency conditions. The primer anneals to a number of partial or complete complementary sequences in the genome of an organism to produce randomly sized DNA fragments (Mohania *et al.*, 2008).

The RAPD technique is a simple and fast method, and therefore it is one of the most frequently used methods for typing LAB. It has been successfully used for typing LAB isolated from dairy products (Fitzsimons *et al.*, 1999; De Angelis *et al.*, 2001; Rossetti &

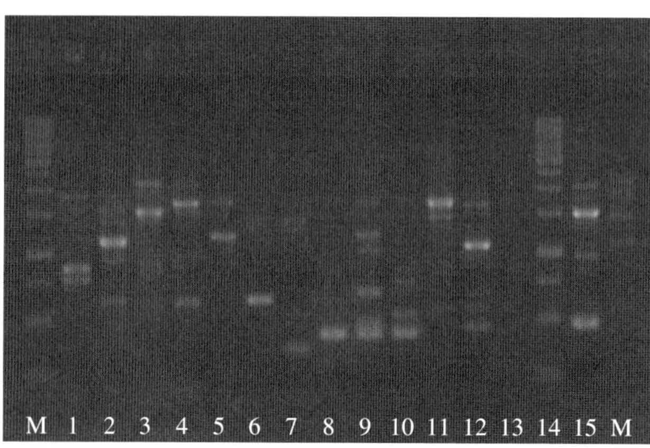

Figure 6.2 Comparison of RAPD-PCR profiles of selected LAB strains.
Note: Lanes 1–6: *Lactobacillus gasseri* strains; lanes 7 and 8: *Lactobacillus acidophilus* strains; lanes 9 and 10: *Lactobacillus johnsonii* strains; lane 11: *Lactobacillus plantarum*; lane 12: *Lactobacillus rhamnosus*; lane 13: negative control; lane 14: *Enterococcus faecium*; lane 15: *Bifidobacterium longum* subsp. *infantis*; lane M: 1 kb ladder.
LAB = Lactic acid bacteria; RAPD-PCR = randomly amplified polymorphic DNA–polymerase chain reaction.

Giraffa, 2005; Golowczyc *et al.*, 2008; Martín-Platero *et al.*, 2008; Nieto-Arribas *et al.*, 2009; Vardjan *et al.*, 2013), natural whey starters (Cocconcelli *et al.*, 1997; Andrighetto *et al.*, 2004; de Candia *et al.*, 2007), commercial dairy starters (Tailliez *et al.*, 1998; Vincent *et al.*, 1998; Giraffa & Rossetti, 2004; Ma *et al.*, 2013) and probiotic products (Gardiner *et al.*, 1998; Reuter *et al.*, 2002; Schillinger *et al.*, 2003). It has also been used for identification of dairy LAB strains at the species level (Fitzsimons *et al.*, 1999; Rossetti & Giraffa, 2005). Although several different primers may be used for RAPD analysis, primer M13 has most often been employed for typing dairy LAB (Giraffa & Rossetti, 2004; Rossetti & Giraffa, 2005; Martín-Platero *et al.*, 2008; Nieto-Arribas *et al.*, 2009; Bove *et al.*, 2011).

In addition, RAPD primers are not directed against a specific sequence; therefore, the reproducibility of the method is poor (Temmerman *et al.*, 2004). Other factors (e.g. annealing temperature, DNA purity and DNA concentration) may also affect reproducibility. The use of strictly controlled PCR conditions and several single-primer reactions may improve both the reproducibility and discrimination of the method (Vincent *et al.*, 1998).

In contrast to RAPD-PCR, rep-PCR is more reproducible due to the use of primers specific against known repetitive element sequences in the bacterial genome (Singh *et al.*, 2009). Examples of repetitive sequences are the repetitive extragenic palindromic (REP) sequences, the polytrinucleotide $(GTG)_5$ sequences, the enterobacterial repetitive intergenic consensus (ERIC) sequences and BOX sequences.

Bove *et al.* (2011) used a combination of RAPD and rep-PCR for characterisation of *Lb. rhamnosus* strains isolated from the same batch of Parmigiano Reggiano cheese during its maturation period. Analysis of different LAB isolated from Grana Padano, using a combination of RAPD and $(GTG)_5$-PCR in comparison with restriction fragment length polymorphism (RFLP), demonstrated that RFLP is a more reliable method for identification at the species level (Mancini *et al.*, 2012). Moreover, rep-PCR alone was observed to be adequate for the accurate species identification of LAB isolates from Salers cheese (Callon *et al.*, 2004). The suitability of rep-PCR using primer $(GTG)_5$ was confirmed for clustering of different LAB from cheeses and other fermented dairy products (Ouadghiri *et al.*, 2005; Zamfir *et al.*, 2006; Nikolic *et al.*, 2008; Van Hoorde *et al.*, 2008).

Masco *et al.* (2003) evaluated the applicability of several rep-PCR techniques for the differentiation of *Bifidobacterium* strains from infants' faeces. The highest discriminatory power was obtained using BOX-PCR with primer BOXA1R, compared to $(GTG)_5$ and several ERIC and REP primers. In contrast, RAPD-PCR was more suitable for the typing of LAB from various sources compared to ERIC-PCR or $(GTG)_5$-PCR (Ruiz *et al.*, 2014).

Single-stranded conformation polymorphism (SSCP)

The SSCP analysis detects sequence variations between different DNA fragments, which are usually PCR-amplified from variable regions of the 16S rRNA gene. This technique is essentially based on the sequence-dependent differential intra-molecular folding of single-stranded DNA (ssDNA), which alters the migration speed of the molecules. The SSCP analysis requires uniform, low-temperature, non-denaturing electrophoresis to maintain ssDNA secondary structure. The discriminatory ability and reproducibility of

SSCP analysis, which is generally most effective for fragments up to 400 bp in size, are also dependent on the position of the sequence variations in the gene studied (Giraffa & Neviani, 2001).

The SSCP analysis involves the following four steps: (a) PCR amplification of the DNA sequence of interest, (b) denaturation of the double-stranded PCR products, (c) cooling of the denatured ssDNA to maximise self-annealing, and (d) detection of mobility differences between the ssDNAs using electrophoresis under non-denaturing conditions. Several methods have been developed to visualise the SSCP mobility shifts. These include the incorporation of radioisotope labelling, silver staining, fluorescent dye-labelled PCR primers and, more recently, capillary-based electrophoresis (Dong & Zhu, 2005).

The SSCP method also was adapted for the rapid identification of bacteria to the genus and species levels. It is a culture-independent tool evaluating LAB communities in food such as cheese (Duthoit *et al.*, 2003; Samelis *et al.*, 2011; Csikos *et al.*, 2016) and pickled vegetables (Wu *et al.*, 2011). As reviewed by Jany and Barbier (2008), SSCP-PCR is the second most-used method for the study of microbial communities of cheese.

When SSCP is used to profile a complex microbial ecosystem, a robust database needs to be created in order to identify each single component, by comparing the retention time of each signal with a reference time in the database. If matching does not occur, identification cannot be obtained (Cocolin *et al.*, 2013).

Diagnostics using SSCP are less time-consuming and expensive than establishing species-specific primers for PCR. In addition, it is a DNA sequence-based method that does not need any sequence analysis software (Herbel *et al.*, 2013). However, when using an automated sequencer, one of the disadvantages of this technique lies in the difficulty of appending new data to an existing database: samples presenting unknown profiles cannot be directly sequenced because they are labelled (Jany & Barbier, 2008). Similarly to DGGE, however, SSCP provides community fingerprints that cannot be phylogenetically assigned directly (Giraffa & Neviani, 2001).

Pulsed-field gel electrophoresis (PFGE)

The PFGE technique has been verified as an efficient means for differentiating strains, and it is an electrophoretic technique used to separate large DNA molecules (10 kb–10 Mb). In a conventional constant electric field, DNA molecules >20 kb show the same mobility, making it impossible to differentiate between them. By applying alternating electric fields at different angles, however, PFGE can separate large DNA molecules in a flat agarose gel. Restriction enzymes (REs) with uncommon recognition motifs, so-called rare cutters, are used in PFGE to generate large DNA fragments, and the banding patterns of PFGE in a group of strains reflect DNA polymorphism at the RE recognition sites. The PFGE technique also provides high-resolution, macro-restriction analysis at the genome level, leading it to be considered as the 'gold standard' for typing bacteria (Li *et al.*, 2009).

The choice of RE is one of the most important factors in determining the PFGE banding pattern because the cleavage site of each RE is unique. The REs with long, infrequently occurring recognition motifs may provide higher resolution in PFGE; this is because the generated DNA fingerprint depends on the specificity of the restriction enzyme used and the sequence of the bacterial genome, and is therefore characteristic of a particular species or strain of bacteria (Amor *et al.*, 2007).

A high discriminatory power of PFGE has been reported for the differentiation of different probiotic strains in commercial products (Coeuret *et al.*, 2004; Yeung *et al.*, 2004) or for strain differentiation of LAB in fermented foods, such as olives (Doulgeraki *et al.*, 2013), viili (Kahala *et al.*, 2008), fermented sausage (Tran *et al.*, 2011), cheese (Vernile *et al.*, 2008) and different fermented dairy products (Xu *et al.*, 2012).

Although widely used, PFGE has several limitations. This method is time and labour consuming, as well as lacking in reproducibility and inter-laboratory comparability. It also requires high-quality DNA, is poorly applicable to human or environmental samples and may lack the resolution needed to distinguish bands of nearly identical size. Other drawbacks include the risk of laboratory-acquired infection due to prolonged handling of bacterial strains before treating with proteases and REs, and many other factors such as concentration of DNA in the agarose plugs, amount of agarose in the gel, electrophoresis voltage, gel temperature and buffer strength, which may also influence patterns (Li *et al.*, 2009).

Amplified fragment length polymorphism (AFLP), restriction fragment length polymorphism (RFLP) and terminal-restriction fragment length polymorphism (T-RFLP)

The AFLP technique combines the power of RFLP with the flexibility of PCR-based methods by ligating primer-recognition sequences (adaptors) to the digested DNA (Amor *et al.*, 2007). The AFLP technique involves restriction of total bacterial DNA with two endonucleases of different cutting frequencies, one with an average cutting frequency and a second with a higher cutting frequency, followed by ligation of the fragments to oligonucleotide adapters complementary to the sequences of the restriction site (Bagheripoor-Fallah *et al.*, 2015). The use of PCR primers complementary to the adapter and the restriction site sequence yields strain-specific amplification patterns (Amor *et al.*, 2007). The AFLP method has mostly been employed in clinical studies, but its successful application in identifying LAB from traditionally produced sauerkraut (Beganovic *et al.*, 2014) and in two Flemish artisanal raw-milk Gouda-type cheeses (Van Hoorde *et al.*, 2008) has been reported. A strain-discriminative differentiation of *Bifidobacterium* strains by AFLP with satisfactory discriminative power and reproducibility was developed by Dimitrov (2012). When three molecular typing methods, RAPD, PFGE and AFLP, were evaluated for their ability to differentiate *Lactobacillus* strains of human origin, AFLP proved to be the most discriminatory (Dimitrov *et al.*, 2008). The high resolution of AFLP has also enabled the delineation of closely related *Lac. lactis* strains from different ecological niches (Kutahya *et al.*, 2011). According to Li *et al.* (2009), the requirement for automated analysis equipment due to a huge quantity of information generated, as well as the fact that it is impossible to use the AFLP on DNA taken directly from human specimens and environmental samples, are presumed to be two disadvantages of AFLP.

The RFLP technique includes whole genome DNA extraction, its digestion with RE and separation of the resulting array of DNA fragments by conventional agarose gel electrophoresis (Bagheripoor-Fallah *et al.*, 2015). If frequently cutting REs are used, they may produce hundreds of short fragments that are difficult to be clearly separated by gel electrophoresis. This difficulty can be simplified by subjecting the partial restriction fragments to Southern blotting with labelled probes (Li *et al.*, 2009).

The T-RFLP is a PCR-fingerprinting method that is commonly used for comparative microbial community analysis and phylogenetic affiliation of consortium members (Nagarajan & Loh, 2014). Its analysis is based on the digestion of fluorescently end-labelled PCR product. Primers are usually designed to anneal at consensus sequences in the bacterial 16S rRNA genes, and either one or both 5′ and 3′ ends of the amplicon can be labelled by incorporating a dye on either one or both PCR primers. The digested products are separated by electrophoresis using either gel- or capillary-based systems, with laser detection of the labelled fragments using an automated analyser (Giraffa & Neviani, 2001). This system only detects the end-labelled terminal restriction fragments (TRFs) of the digested PCR products, and their size can be calculated based on the use of DNA size standards that are run simultaneously with the samples (Jany & Barbier, 2008). The T-RFLP analysis was used to characterise the members of the complex bacterial microbiota of different Croatian raw ewe's milk cheeses (Fuka *et al.*, 2013), while the use of $tRNA^{Ala}$–23S rDNA-RFLP identification of micro-organisms was reported for LAB, associated with dairy ecosystems and 75 isolates from Grana Padano cheese (Mancini *et al.*, 2012), and for LAB and yeast strains in Ragusana donkey's milk (Randazzo *et al.*, 2016).

Amplified ribosomal DNA restriction analysis (ARDRA)

The ARDRA technique has been another useful technique for identifying various micro-organisms. The PCR-amplified 16S rRNA fragments are digested at specific sites with restriction enzymes, and the resulting digest is separated by gel electrophoresis. Different DNA sequences will be cut in different locations and will result in a profile unique to the community being analysed. Collado & Hernandez (2007) showed that the ARDRA technique can be a simple, rapid and useful method for routine identification of lactobacilli, streptococci and bifidobacteria in fermented milk products. The ARDRA was also proven to be applicable for differentiation of LAB species from different habitats (Rodas *et al.*, 2003; Ksicova *et al.*, 2013). Although accurate identification of lactobacilli and other co-isolated bacteria in probiotic prospective studies of human, animal or food origin can be a difficult task, Moreira *et al.* (2005) successfully typed lactobacilli isolates at the strain level by PCR amplification of 16S-23S rRNA intergenic spacers using universal primers that anneal within 16S and 23S genes, followed by restriction digestion analyses of PCR products. Some other studies report the satisfactory molecular identification of probiotic lactobacilli isolates using ARDRA (Shehata, 2012; Ozturk & Meterelliyoz, 2015).

6.4.3 *DNA sequencing-based methods*

The DNA sequencing-based methods can be divided in three clusters: (a) sequencing of coding genes like MLST, (b) sequencing of repeats and noncoding regions such as internally transcribed spacers (ITS), and (c) WGS and metagenomics. All DNA sequencing-based methods can be classified as culture-dependent or independent, depending on whether bacterial isolates or clones are used. These are conditionally strain-specific methods (depending on the specificity of the target) which are not suitable for quantification or the assessment of viability of the strain of interest. The rapid development and fall in costs of NGS methods resulted in a rise in the use of WGS and metagenomic approaches, not only for phylogenetic analysis but also for the detection of probiotics and starter cultures.

Sequencing of specific genes

The already known genotypic characteristics of the target strain can be used in order to identify or profile probiotic and LAB starter cultures. The target gene should be stable or only partially susceptible to spontaneous mutations that result in a low degree of similarity among strains within the same species. Several protein-encoding genes have been found suitable for detecting or profiling probiotics and LAB by sequencing.

Sequencing of variable regions of 16S rRNA genes is most commonly used for bacterial identification at the species level; however, the flaw in this approach is that as the evolutionary distance decreases, the diversity level in the 16S rRNA often becomes insufficient, and thus genetic relationships of closely related species cannot be accurately defined on the basis of a 16S rRNA gene sequence (Singh *et al.*, 2009). In order to achieve better phylogenetic resolution, several alternative phylogenetic markers can be used as targets, such as a gene-encoding S-layer protein, *tuf* for elongation factor Tu, *rec A* coding for Rec A protein and *hsp60* for heat shock protein 60 (Yu *et al.*, 2012). For example, Huang & Lee (2011) and Huang *et al.* (2014, 2015) suggested several genes (*yych*, *dnak* and *dnaj*) as good phylogenetic markers for discrimination of the *Lb. casei* group. Claesson *et al.* (2008) concluded that the ubiquitous chaperonin GroEL is a more robust single-gene phylogenetic marker for the genus *Lactobacillus* compared to the 16S rRNA gene.

Multilocus sequence typing (MLST)

The MLST is a method that uses DNA sequencing to characterise bacterial isolates at the molecular level. The principle behind the MLST method is the analysis of differences in the sequences (approx. 400 to 500 bp) of multiple housekeeping genes (6–10, usually 7) (Maiden, 2006). Traditionally, unique sequences (alleles) are assigned a random integer number, and a unique combination of alleles at each locus is given an ‘allelic profile’, which specifies the sequence type (ST). In practice, MLST starts with a PCR amplification step using primers that are specific for the loci of the MLST scheme, followed by Sanger sequencing. Deoxyribonucleic acid sequences are stored in online databases, which allow convenient exchange of strain-typing data both within and between laboratories. The MLST method is suitable for long-term investigation of bacterial population structures, particularly when subtyping bacterial species with a high rate of genetic recombination (Li *et al.*, 2009). However, MLST is a laborious method, and the use of highly conserved housekeeping genes in MLST often fails to detect the variability of closely related strains. Moreover, the cost of traditional MLST is nowadays even higher than WGS, since the costs of WGS continue to decline (Larsen *et al.*, 2012).

Several MLST schemes for *Bifidobacterium* and LAB species have been developed mostly for genetic profiling and evolution studies of isolated strains from different sources. The MLST method was used for characterisation of type strains of the *Bifidobacterium* genus (Ventura *et al.*, 2006), strains of the *Lb. acidophilus* complex (Ramachandran *et al.*, 2013) and strains of the *Lb. delbrueckii* spp. (Tanigawa & Watanabe, 2011), *Lb. casei* (Cai *et al.*, 2007), *Lb. plantarum* (de las Rivas *et al.*, 2006; Gosiewski *et al.*, 2012), *Lb. fermentum* (Dan *et al.*, 2015) and *Lac. lactis* spp. (Fernandez *et al.*, 2011; Xu *et al.*, 2014). Most of the developed schemes used seven housekeeping genes; however, Dan *et al.* (2015) used 11 genes to differentiate *Lb. fermentum* strains, and Xu *et al.* (2014) used 12 to determine the genetic diversity and phylogenetic relationships among *Lac. lactis* spp. isolates.

In addition to genetic profiling, MLST has been used to assess the degree and persistence of the intestinal colonisation of the probiotic strain (*Lb. plantarum* 57B) after oral administration of a mixture of lactobacilli, including the *Lb. plantarum* 57B strain (Strus *et al.*, 2012). Recently, Liu *et al.* (2016) applied functional gene sequence MLST technology to predict the fermentation and flavour-producing characteristics of yoghurt-producing bacteria (strains of *Lb. delbrueckii* subsp. *bulgaricus*). The fact that groups of strains established on the basis of genotype data were consistent with groups identified based on their phenotypic traits indicates that MLST has the potential to replace time-consuming conventional methods based on direct measurement of phenotypic traits (Liu *et al.*, 2016).

In order to compare WGS data with experimentally gained MLST data, a web-based method for MLST based on WGS data has been developed (Larsen *et al.*, 2012). In a similar manner using comparative genomics of *Bifidobacterium* spp., *Lactobacillus* spp. and related probiotic genera, Lukjancenko *et al.* (2012) performed *in silico* MLST and compared the variable gene content of genomes within the genera. Although this is an improvement on the basic technique with modern WGS, it should be noted that an MLST tree should be interpreted with caution, as it represents only a tiny fraction of the complete core genome of a strain.

Sequencing of repeats and noncoding regions

Microsatellites or short sequence repeats (SSRs) are widespread genetic markers consisting of hypermutable short nucleotide motifs (1–6 bp long), tandemly repeated from two or three up to a few dozen times at a locus. Their applications in genetics are extensive due to their ceaseless mutational degree, widespread length variations and hypermutability skills. These properties make them an excellent tool for a number of approaches like genotyping, mapping and positional cloning of genes (Saeed *et al.*, 2016). The diversity of compound SSRs in *Lactobacillus* genomes may be useful for better understanding their genetic diversity, evolutionary biology and strain/genotype differentiations (Saeed *et al.*, 2016). Buhnik-Rosenblau *et al.* (2012) successfully used the SSR technique in combination with MLST to infer the genetic relationships among *Lb. johnsonii* isolates from different sources.

A genomic region separating 16S and 23S rRNA genes in prokaryotic micro-organisms consists mainly of a noncoding sequence which is transcribed together with the ribosomal genes, and this is called an internal transcribed spacer (ITS) (Gurtler & Stanisich, 1996).

In comparison with 16S rRNA genes, an ITS is more variable and exhibits greater resolution for the subtyping of bacteria at the strain level. ITSs vary not only in sequence and length but also in the number of alleles and their positions on the chromosome (Garcia-Martinez *et al.*, 1999); ITS sequencing is mainly used for species or subspecies identification, and less often for strain typing (Li *et al.*, 2009). Usually the sequencing of ITS is the basis for construction of strain-specific primers (Treven, 2015) or RFLP analysis (Sandes *et al.*, 2014). In any case, the use of ITS sequencing has proved to be useful in distinguishing between strains of bifidobacteria from human intestinal mucosal and faecal samples (Turroni *et al.*, 2009) and for phylogenetic analysis of *Str. thermophilus* from yoghurt or cheeses (Galia *et al.*, 2009).

Clustered regularly interspaced short palindromic repeats (CRISPRs) and their CRISPR-associated (Cas) proteins are part of the CRISPR-Cas immune system in bacteria and archaea (Barrangou, 2015). The discovery of CRISPRs opened new possibilities for their use as high-resolution genetic-fingerprinting tools for the assessment of diversity of bacteria. According to the CRISPRdb database (Grissa *et al.*, 2007), 45% of bacteria contain convincing CRISPRs and the occurrence of CRISPR loci varies among different genera and species. Among 102 genomes of LAB, CRISPR loci were identified in 47 genomes (Horvath *et al.*, 2009). This ratio (46.1%) is much lower than the ratio found in *Bifidobacterium* spp., which have a very high frequency of CRISPR-Cas occurrence (77%, or 37 of 48) (Briner *et al.*, 2015). Sanozky-Dawes *et al.* (2015) reported that six of 17 (35.2%) investigated *Lb. gasseri* strains harboured a CRISPR-Cas system, with considerable diversity in array size and spacer content. Although type II-A CRISPR-Cas systems are valuable for genotyping of *Lactobacillus buchneri*, this system is not ubiquitous in *Lb. buchneri* genomes, limiting its potential as a universal target for typing purposes within this species (Briner & Barrangou, 2014). Overall, the absence of CRISPR in some strains or species is the major drawback of this technique, and therefore the potential of CRISPR-based genotyping must be assessed on an individual basis for each candidate species. The sole presence of a CRISPR array in a draft genome is only a starting point; however, several studies have shown potential for CRISPR-based typing of industrial starter cultures, probiotic strains, animal commensal species and pathogens (Barrangou & Dudley, 2016).

Whole genome sequencing and metagenomics

The extensive development of NGS technologies in recent years has resulted in a huge reduction of sequencing cost for a typical bacterial genome (Koser *et al.*, 2012). As well as this, metagenomic approaches have also become widely available. It is interesting to note that more than half of the complete or draft bacterial genome sequences available in the National Center for Biotechnology Information (NCBI) Genome database (http://www.ncbi.nlm.nih.gov/genome/browse/) were submitted from 2015 to September 2016. NGS denotes high-throughput DNA-sequencing technologies permitting the sequencing of millions of DNA strands in parallel and thus generating large amounts of sequence data in a relatively short period of time (Pettersson *et al.*, 2009). Several NGS methods with different characteristics in terms of accuracy, average read length, reads per run and time of analysis have been developed (Loman *et al.*, 2012; Kelleher *et al.*, 2015).

The WGS analysis of bacterial strains offers a tremendous range of applications in various fields. In the food industry, key genetic markers derived from genomic analysis can be the basis for starter strain selection, with particular emphasis on phage resistance and flavour development (Kelleher *et al.*, 2015). With WGS, we can identify the regulatory mechanisms of secondary metabolite overproduction and subsequently improve the fermentation processes, which could result in the reduction of manufacturers' costs (Pettersson *et al.*, 2009).

Also, the WGS technique offers an in-depth insight into the evolutionary background of specific strains of particular species, and it can be used for the identification of probiotics and LAB starter cultures. In fact, with appropriate bioinformatics analysis, WGS can replace any other sequencing-based method. Several groups have used complete genome sequences of bifidobacteria and lactobacilli strains to develop new genotyping methods based on differences in insertion sequences and single nucleotide polymorphisms (Briczinski *et al.*, 2009; Kaleta *et al.*, 2009; Lomonaco *et al.*, 2015).

With the increased availability of genome sequences of bacterial strains from the same species and improved bioinformatics computing, the concept of the pangenome has emerged. Pangenomics is the analysis of the genome sequences of a number of members of the same species (Medini *et al.*, 2005), and the pangenome represents (or tries to represent) 'the genetic information of a bacterial species' (Tettelin *et al.*, 2005). The pangenome can be divided into three parts: (a) a core genome (shared by all strains), (b) a set of dispensable genes (shared by some but not all genomes), and (c) a set of strain-specific genes (present in one genome only) (Medini *et al.*, 2005). The pangenome concept can be used for the identification of strain-specific features, which can be targeted by strain-specific primers. Several tools and pipelines needing various levels of programming skills are freely available online to analyse pangenomes (Treven, 2015). Lv *et al.* (2015) designed strain-specific primers based on the available genetic information of *Bif. longum* subsp. *longum* BBMN68, and used them to monitor this probiotic strain's distribution in the rat gut. In the case of the recently deposited whole-genome shotgun project of probiotic strain *Lb. gasseri* K7 (Treven *et al.*, 2014), five unique putative genes were assumed to be the best candidates for strain-specific PCR when *Lb. gasseri* K7 genome was compared to 299 publicly available *Lactobacillus* genomes.

One of the goals of omic studies is to identify key biomarkers that could be used to screen for new probiotic or technologically interesting strains. In order to improve functionality in industrial processes, the omic approach is also suitable for the evaluation of the physiological state of targeted probiotics (Sohier *et al.*, 2014). In addition, the metagenomic approach could be used to detect specific strains. The utilisation of sequencing coverage, in combination with the application of high-throughput sequencing methods, has enabled the metagenomic 'binning' of assemblies. Sequences that originate from the same genome are grouped into bins by the similarity of their coverage vectors; this helps to differentiate strains (Turaev & Rattei, 2016). In order for microbial strains to be characterised at high resolution, several bioinformatics methods have been developed, such as MathPhlan2, ConStrains and latent strain analysis (LSA). Johansen *et al.* (2014) applied metagenomic-based approaches to quantify seven *Lac. lactis* subsp. *cremoris* strains in a defined mixed-strain starter culture. Recently, Tu *et al.* (2014) suggested a novel k-mer-based approach that identifies genome-specific markers in currently sequenced microbial genomes. These markers could then be used for strain/species-level identification in metagenomes. Taking into account that the price of metagenomic sequencing is falling, it can be speculated that in future, strain-specific identification (especially in complex environments) will be performed mostly by using metagenomic sequencing and appropriate bioinformatics analysis (Treven, 2015).

6.4.4 *Probe hybridisation methods*

DNA and cDNA microarrays

As mentioned elsewhere, NGS methods produce a substantial volume of sequence information. It is now possible to take advantage of such sequencing data to develop comprehensive microarrays using modern probe design strategies. DNA microarrays are composed of microscopic DNA spots (oligonucleotide probes) immobilised on a two-dimensional solid support, forming an array of thousands of probes in a single chip (Nagarajan & Loh, 2014). The first step of the analysis is labelling the community DNA in the sample. A fluorescent signal is recorded after hybridisation with oligonucleotide probes onto the chip (Talbot *et al.*, 2008). The quantification of the targeted bacteria is based on the assumption that the intensity of the hybridisation signal is proportional to abundance. The success of a microarray experiment strongly depends on the selection of the probe set while considering the biological question. Probe design strategies for oligonucleotide microarrays were recently reported by Parisot *et al.* (2016). Oligonucleotide probes are designed based on either: (a) conserved marker genes, such as 16S rRNA genes (phylogenetic oligonucleotide array), (b) key functional genes involved in the physiological processes (functional gene array), (c) whole genomic DNA isolated from pure cultures (community genome array), or (d) DNA cloned directly from the environmental samples (metagenomics array) (Nagarajan & Loh, 2014).

Microarrays can be used for parallel detection of complex microbial communities in many environments, including food matrices. Rungrassamee *et al.* (2012) designed the microarray with a total of 164 bacteria-specific probes from 16S rRNA gene sequences to target 12 bacteria species, including LAB and selected food pathogens, in Thai fermented sausage. The taxonomic discriminative power of microarrays largely depends on the selection of strain/species-specific probes from huge sequence databases. Tu *et al.* (2013) developed a k-mer-based approach that can quickly and comprehensively select 50-mer strain/species-specific probes for microbial strains and species, which can be used to construct microarrays for strain/species-level identification of micro-organisms in complex microbial communities. Patro *et al.* (2015) recently demonstrated the applicability of microarrays in the field of probiotics. They developed a custom microarray (FDA GutProbe) that included genomes and plasmids representative of the most common bacteria in the human gut and food products, to verify the accuracy of labelling in commercial probiotic supplements. The microarray proved to be suitable for the identification of various species found in dietary supplements; therefore, it could be used for the quality control of probiotic products in terms of labelling information and batch variation. Furthermore, the GutProbe microarray enables identification to the species level, and can even discriminate between closely related strains. This method, however, is limited to species that are represented on the current version of the array.

There are several concerns regarding the usage of DNA microarrays for detection and quantification of bacterial communities. For instance, the specificity of probe–target interactions may be compromised by non-specific target binding and cross-hybridisation. Another drawback is the time-consuming and complex process of validation and designing specific oligonucleotides for the robust identification and characterisation of the microbial consortia. However, the microarray is much more cost-effective than sequencing and can reduce time and cost when monitoring bacteria populations, especially when handling a large number of food samples. This microarray method would be preferred for rapid surveillance and analysis of initial products (Patro *et al.*, 2015).

Fluorescence in situ *hybridisation (FISH)*

Fluorescence *in situ* hybridisation is based on the specific hybridisation of a fluorescently labelled probe to a complementary target sequence within the cell. This enables the identification of bacteria *in situ*, without the isolation of community DNA (Pogačić *et al.*, 2010). Fluorescently labelled probes target specific DNA or RNA sequences in order to detect or confirm genes within chromosomes or gene expression. The protocol for species-specific bacterial enumeration in mixed culture consists of four major steps. First, the cells are fixed and permeabilised, and then fluorescently labelled probes are incubated with the permeabilised cells to allow the probes to hybridise with a targeted sequence. The last step is the quantification of hybridised cells using FC (FLOW-FISH) or epifluorescence microscopy (Nagarajan & Loh, 2014). For milk and other dairy products, sample preparation may include homogenisation, simple or multiple centrifugation steps, the addition of sodium citrate buffer or unspecific proteases to reduce the background and auto-fluorescence of the sample (Rohde *et al.*, 2015). The FISH experiments often employ several probes of different specificity that are labelled with different fluorophores. Bottari *et al.* (2006) reviewed the most commonly employed dyes for FISH for microbiological analysis. Domain-, group- and strain-specific probes are the most commonly used for the analysis of mixed bacterial communities (Nagarajan & Loh, 2014).

Fluorescence *in situ* hybridisation is a popular technique for research into probiotics and dairy starter cultures, although few publications have described its use for the enumeration of dairy microbes (Sohier *et al.*, 2014). Babot *et al.* (2011) successfully designed oligonucleotide probes targeting the 16S rRNA of dairy propionibacteria and optimised the FISH protocol to enumerate these bacteria in Gruyère cheese. Similarly, the use of fluorescently labelled oligonucleotide probes specific to *Lac. lactis* spp., *Lb. plantarum* and *Leuconostoc mesenteroides* spp. in FISH experiments enabled an assessment of the spatial distribution of the different microbial species on Stilton cheese sections (Ercolini *et al.*, 2003). Mounier *et al.* (2009) employed four FISH probes to enumerate the main yeasts in Livarot cheese. Fluorescence *in situ* hybridisation with probes specific for bifidobacteria was also used for the enumeration of these bacteria in a fermented oat drink during storage (Lahtinen *et al.*, 2006) and for examination of the probiotic strain *Bif. animalis* subsp. *lactis* BB-12 in microcapsules (Lisova *et al.*, 2013). Olsen *et al.* (2007) quantified *Leuconostoc* spp. populations in five different mixed starter cultures using a whole cell *in situ* hybridisation assay with 16S rRNA-targeted oligonucleotide probes. Recently, Lu *et al.* (2014) used FISH to determine the distribution of yeasts in Tibetan kefir grains for the following genera: *Kluyveromyces*, *Saccharomyces* and *Yarrowia*.

Several research groups have used modified versions of FISH to quantify specific bacteria in different matrices. Garcia-Hernandez *et al.* (2012) developed a direct viable count (DVC)–FISH procedure for quick and easy discrimination between viable and nonviable cells of traditional yoghurt bacteria. Friedrich & Lenke (2006) showed flow cytometric FISH (FLOW-FISH) to be an effective and accurate tool for the bacterial community analysis of complex starter cultures. Another approach is the use of fluorescently labelled rRNA-targeted peptide nucleic acid (PNA)–FISH probes. Matte-Tailliez *et al.* (2001) and Machado *et al.* (2013) used PNA-FISH for the detection and identification of growing LAB cells in industrial starter cultures or in milk.

The popularity of FISH is mainly due to its ability to provide information about the distribution of micro-organisms in a specific matrix and the possibility of performing highly specific detection, in some cases to the strain level. However, it has several disadvantages such as interferences with food matrices, artefacts, low repeatability, problems with limited permeabilisation, poor limit of detection and consequently a laborious optimisation procedure (Nagarajan & Loh, 2014; Sohier *et al.*, 2014).

6.5 Conclusions

Although laborious and time-consuming, traditional plate counting still remains the most commonly used technique for bacterial enumeration. The method is still appropriate for monitoring the quality of final products where, in addition to total plate count, the method also enables a partial idea of the composition of the bacterial community. For monitoring of fermentation processes, faster methods (e.g. FC) are crucial that, above all, allow counting of the total number of bacteria. The advantage is that it counts the VBNC bacteria. Perhaps even better is the PMA/EMA qPCR that counts viable bacteria whilst also distinguishing between species.

Simple phenotypic methods, such as API or Biolog, are insufficient for adequate and conclusive identification of bacterial strains to the species level. The use of genetic methods, such as 16S rDNA sequencing or species-specific PCR, can significantly increase the reliability of species identification. Moreover, with the depositing of more and more protein sequences of different LAB in available databases, even MALDI-TOF MS may eventually become a more reliable method since it is very rapid, and sample pre-treatment is fast and easy.

To monitor the stability of strains present in a product or starter culture, the preferred methods are RAPD and rep-PCR as they allow easy and quick screening of strains based on their genomes. Lately, whole genome sequencing has become more accessible, and application of DNA sequencing has become more realistic due to recent cost reduction. Nevertheless, a new research era is called for where the omics/metagenomic approach is given priority so that, with the support of modern bioinformatic methods, we can monitor and quantify a specific strain, and, through deep sequencing of 16S rRNA, we can determine its physiological state in a complex microbial community without prior cultivation.

References

Albesharat, R., Ehrmann, M.A., Korakli, M., Yazaji, S. & Vogel, R.F. (2011) Phenotypic and genotypic analyses of lactic acid bacteria in local fermented food, breast milk and faeces of mothers and their babies. *Systematic and Applied Microbiology*, **34**, 148–155.

Amiel, C., Mariey, L., Curk-Daubié, M.C., Pichon, P. & Travert, J. (2000) Potentiality of Fourier transform infrared spectroscopy (FTIR) for discrimination and identification of dairy lactic acid bacteria. *Lait*, **80**, 445–459.

Amiel, C., Mariey, L., Denis, C., Pichon, P. & Travert, J. (2001) FTIR spectroscopy and taxonomic purpose: contribution to the classification of lactic acid bacteria. *Lait*, **81**, 249–255.

Amor, K.B., Breeuwer, P., Verbaarschot, P., Rombouts, F.M., Akkermans, A.D.L., De Vos, W.M. & Abee, T. (2002) Multiparametric flow cytometry and cell sorting for the assessment of viable, injured, and dead *Bifidobacterium* cells during bile salt stress. *Applied and Environmental Microbiology*, **68**, 5209–5216.

Amor, K.B., Vaughan, E.E. & de Vos, W.M. (2007) Advanced molecular tools for the identification of lactic acid bacteria. *The Journal of Nutrition*, **137**, 741S–747S.

Ananta, E. & Knorr, D. (2009) Comparison of inactivation pathways of thermal or high pressure inactivated *Lactobacillus rhamnosus* ATCC 53103 by flow cytometry analysis. *Food Microbiology*, **26**, 542–546.

Andrighetto, C., Marcazzan, G. & Lombardi, A. (2004) Use of RAPD-PCR and TTGE for the evaluation of biodiversity of whey cultures for Grana Padano cheese. *Letters in Applied Microbiology*, **38**, 400–405.

Angelakis, E., Million, M., Henry, M. & Raoult, D. (2011) Rapid and accurate bacterial identification in probiotics and yoghurts by MALDI-TOF mass spectrometry. *Journal of Food Science*, **76**, M568–M572.

Anonymous (2007) *Anaerobe Identification Test Panel*. AN MicroPlate™. Biolog Inc., Hayward, CA.

Aureli, P., Fiore, A., Scalfaro, C., Casale, M. & Franciosa, G. (2010) National survey outcomes on commercial probiotic food supplements in Italy. *International Journal of Food Microbiology*, **137**, 265–273.

Auty, M.A.E., Gardiner, G.E., McBrearty, S.J., O'Sullivan, E.O., Mulvihill, D.M., Collins, J.K., Fitzgerald, G.F., Stanton, C. & Ross, R.P. (2001) Direct *in situ* viability assessment of bacteria in probiotic dairy products using viability staining in conjunction with confocal scanning laser microscopy. *Applied and Environmental Microbiology*, **67**, 420–425.

Babot, J.D., Hidalgo, M., Arganaraz-Martinez, E., Apella, M.C. & Chaia, A.P. (2011) Fluorescence *in situ* hybridization for detection of classical propionibacteria with specific 16S rRNA-targeted probes and its application to enumeration in Gruyere cheese. *International Journal of Food Microbiology*, **145**, 221–228.

Bagheripoor-Fallah, N., Mortazavian, A., Hosseini, H., Khoshgozaran-Abras, S. & Homayouni Rad, A. (2015) Comparison of molecular techniques with other methods for identification and enumeration of probiotics in fermented milk products. *Critical Reviews in Food Science and Nutrition*, **55**, 396–413.

Barrangou, R. (2015) Diversity of CRISPR-Cas immune systems and molecular machines. *Genome Biology*, **16**, 1–11.

Barrangou, R. & Dudley, E.G. (2016) CRISPR-based typing and next-generation tracking technologies. *Annual Review of Food Science and Technology*, **7**, 395–411.

Beganovic, J., Kos, B., Pavunc, A.L., Uroic, K., Jokic, M. & Suskovic, J. (2014) Traditionally produced sauerkraut as source of autochthonous functional starter cultures. *Microbiological Research*, **169**, 623–632.

Bogovič Matijašić, B., Obermajer, T. & Rogelj, I. (2010) Quantification of *Lactobacillus gasseri*, *Enterococcus faecium* and *Bifidobacterium infantis* in a probiotic OTC drug by real-time PCR. *Food Control*, **21**, 419–425.

Bokulich, N.A. & Mills, D.A. (2012) Next-generation approaches to the microbial ecology of food fermentations. *Bmb Reports*, **45**, 377–389.

Bosch, A., Golowczyc, M.A., Abraham, A.G., Garrote, G.L., De Antoni, G.L. & Yantorno, O. (2006) Rapid discrimination of lactobacilli isolated from kefir grains by FT-IR spectroscopy. *International Journal of Food Microbiology*, **111**, 280–287.

Bottari, B., Ercolini, D., Gatti, M. & Neviani, E. (2006) Application of FISH technology for microbiological analysis: current state and prospects. *Applied Microbiology and Biotechnology*, **73**, 485–494.

Bove, C.G., Lindner, J.D., Lazzi, C., Gatti, M. & Neviani, E. (2011) Evaluation of genetic polymorphism among *Lactobacillus rhamnosus* non-starter Parmigiano Reggiano cheese strains. *International Journal of Food Microbiology*, **144**, 569–572.

Boyd, M.A., Antonio, M.A.D. & Hillier, S.L. (2005) Comparison of API 50 CH strips to whole-chromosomal DNA probes for identification of *Lactobacillus* species. *Journal of Clinical Microbiology*, **43**, 5309–5311.

Boyer, M. & Combrisson, J. (2013) Analytical opportunities of quantitative polymerase chain reaction in dairy microbiology. *International Dairy Journal*, **30**, 45–52.

Briczinski, E.P., Loquasto, J.R., Barrangou, R., Dudley, E.G., Roberts, A.M. & Roberts, R.F. (2009) Strain-specific genotyping of *Bifidobacterium animalis* subsp. *lactis* by using single-nucleotide polymorphisms, insertions, and deletions. *Applied and Environmental Microbiology*, **75**, 7501–7508.

Briner, A.E. & Barrangou, R. (2014) *Lactobacillus buchneri* genotyping on the basis of clustered regularly interspaced short palindromic repeat (CRISPR) locus diversity. *Applied and Environmental Microbiology*, **80**, 994–1001.

Briner, A.E., Lugli, G.A., Milani, C., Duranti, S., Turroni, F., Gueimonde, M., Margolles, A., van Sinderen, D., Ventura, M. & Barrangou, R. (2015) Occurrence and diversity of CRISPR-Cas systems in the genus *Bifidobacterium*. *PLoS One*, **10**, 1–16.

Brolazo, E.M., Leite, D.S., Iba, M.R., Villarroel, M., Marconi, C. & Simoes, J.A. (2011) Correlation between API 50 CH and multiplex polymerase chain reaction for the identification of vaginal lactobacilli in isolates. *Brazilian Journal of Microbiology*, **42**, 225–232.

Buhnik-Rosenblau, K., Matsko-Efimov, V., Jung, M., Shin, H., Danin-Poleg, Y. & Kashi, Y. (2012) Indication for co-evolution of *Lactobacillus johnsonii* with its hosts. *BMC Microbiology*, **12**, 1–10.

Bunesova, V., Killer, J., Vlkova, E., Musilova, S., Tomaska, M., Rada, V. & Kmet, V. (2014) Isolation and characterization of bifidobacteria from ovine cheese. *International Journal of Food Microbiology*, **188**, 26–30.

Bunesova, V., Musilova, S., Geigerova, M., Pechar, R. & Rada, V. (2015) Comparison of mupirocin-based media for selective enumeration of bifidobacteria in probiotic supplements. *Journal of Microbiological Methods*, **109**, 106–109.

Bunthof, C.J., van Schalkwijk, S., Meijer, W., Abee, T. & Hugenholtz, J. (2001) Fluorescent method for monitoring cheese starter permeabilization and lysis. *Applied and Environmental Microbiology*, **67**, 4264–4271.

Bunthof, C.J. & Abee, T. (2002) Development of a flow cytometric method to analyze subpopulations of bacteria in probiotic products and dairy starters. *Applied and Environmental Microbiology*, **68**, 2934–2942.

Cai, H., Rodriguez, B.T., Zhang, W., Broadbent, J.R. & Steele, J.L. (2007) Genotypic and phenotypic characterization of *Lactobacillus casei* strains isolated from different ecological niches suggests frequent recombination and niche specificity. *Microbiology (SGM)*, **153**, 2655–2665.

Callon, C., Millet, L. & Montel, M.C. (2004) Diversity of lactic acid bacteria isolated from AOC Salers cheese. *Journal of Dairy Research*, **71**, 231–244.

Claesson, M.J., van Sinderen, D. & O'Toole, P.W. (2008) *Lactobacillus* phylogenomics – towards a reclassification of the genus. *International Journal of Systematic and Evolutionary Microbiology*, **58**, 2945–2954.

Cocconcelli, P.S., Parisi, M.G., Senini, L. & Bottazzi, V. (1997) Use of RAPD and 16S rDNA sequencing for the study of *Lactobacillus* population dynamics in natural whey culture. *Letters in Applied Microbiology*, **25**, 8–12.

Cocolin, L., Alessandria, V., Dolci, P., Gorra, R. & Rantsiou, K. (2013) Culture independent methods to assess the diversity and dynamics of microbiota during food fermentation. *International Journal of Food Microbiology*, **167**, 29–43.

Coeuret, V., Gueguen, M. & Vernoux, J.P. (2004) Numbers and strains of lactobacilli in some probiotic products. *International Journal of Food Microbiology*, **97**, 147–156.

Collado, M.C. & Hernandez, M. (2007) Identification and differentiation of *Lactobacillus*, *Streptococcus* and *Bifidobacterium* species in fermented milk products with bifidobacteria. *Microbiological Research*, **162**, 86–92.

Comas-Riu, J. & Rius, N. (2009) Flow cytometry applications in the food industry. *Journal of Industrial Microbiology & Biotechnology*, **36**, 999–1011.

Corich, V., Soldati, E. & Giacomini, A. (2004) Optimization of fluorescence microscopy techniques for the detection of total and viable lactic acid bacteria in whey starter cultures. *Annals of Microbiology*, **54**, 335–342.

Csikos, A., Hodzic, A., Pasic-Juhas, E., Javor, A., Hrkovic-Porobija, A., Goletic, T., Gulyas, G. & Czegledi, L. (2016) Applicability and sensitivity of PCR-SSCP method for milk species identification in cheese. *Acta Alimentaria*, **45**, 69–76.

Čanžek Majhenič, A., Mohar Lorberg, P. & Rogelj, I. (2007) Characterisation of the *Lactobacillus* community in traditional Karst ewe's cheese. *International Journal of Dairy Technology*, **60**, 182–190.

Čitar, M., Hacin, B., Tompa, G., Štempelj, M., Rogelj, I., Dolinšek, J., Narat, M. & Bogovič Matijašić, B. (2015) Human intestinal mucosa-associated *Lactobacillus* and *Bifidobacterium* strains with probiotic properties modulate IL-10, IL-6 and IL-12 gene expression in THP-1 cells. *Beneficial Microbes*, **6**, 325–336.

Dan, T., Liu, W.J., Song, Y.Q., Xu, H.Y., Menghe, B., Zhang, H.P. & Sun, Z.H. (2015) The evolution and population structure of *Lactobacillus fermentum* from different naturally fermented products as determined by multilocus sequence typing (MLST). *BMC Microbiology*, **15**, 1–11.

Dave, R.I. & Shah, N.P. (1996) Evaluation of media for selective enumeration of *Streptococcus thermophilus*, *Lactobacillus delbrueckii* ssp. *bulgaricus*, *Lactobacillus acidophilus*, and bifidobacteria. *Journal of Dairy Science*, **79**, 1529–1536.

Davis, C. (2014) Enumeration of probiotic strains: review of culture-dependent and alternative techniques to quantify viable bacteria. *Journal of Microbiological Methods*, **103**, 9–17.

Davis, R. & Mauer, L.J. (2010) Fourier tansform infrared (FT-IR) spectroscopy: a rapid tool for detection and analysis of foodborne pathogenic bacteria. In *Current Research, Technology and Education Topics in Applied Microbiology and Microbial Biotechnology* (ed. A. Mendez-Vilas), 1582–1594. Formatex Research Center, Badajoz, Spain.

De Angelis, M., Corsetti, A., Tosti, N., Rossi, J., Corbo, M.R. & Gobbetti, M. (2001) Characterization of non-starter lactic acid bacteria from Italian ewe cheeses based on phenotypic, genotypic, and cell wall protein analyses. *Applied and Environmental Microbiology*, **67**, 2011–2020.

de Candia, S., De Angelis, M., Dunlea, E., Minervini, F., McSweeney, P.L.H., Faccia, M. & Gobbetti, M. (2007) Molecular identification and typing of natural whey starter cultures and microbiological and compositional properties of related traditional Mozzarella cheeses. *International Journal of Food Microbiology*, **119**, 182–191.

de las Rivas, B., Marcobal, A. & Munoz, R. (2006) Development of a multilocus sequence typing method for analysis of *Lactobacillus plantarum* strains. *Microbiology (SGM)*, **152**, 85–93.

Delavenne, E., Ismail, R., Pawtowski, A., Mounier, J., Barbier, G. & Le Blay, G. (2013) Assessment of lactobacilli strains as yogurt bioprotective cultures. *Food Control*, **30**, 206–213.

Díaz, M., Herrero, M., García, L.A. & Quirós, C. (2010) Application of flow cytometry to industrial microbial bioprocesses. *Biochemical Engineering Journal*, **48**, 385–407.

Dimitrov, Z. (2012) Development of strain discriminative amplified fragment length polymorpfic DNA for bifidobacteria: design of strain-specific markers. *Biotechnology & Biotechnological Equipment*, **26**, 35–38.

Dimitrov, Z.P., Minkova, S. & Michaylova, M. (2008) Comparative evaluation of three molecular typing methods in their applicability to differentiate *Lactobacillus* strains with human origin. *World Journal of Microbiology & Biotechnology*, **24**, 1305–1312.

Doherty, S.B., Wang, L., Ross, R.P., Stanton, C., Fitzgerald, G.F. & Brodkorb, A. (2010) Use of viability staining in combination with flow cytometry for rapid viability assessment of *Lactobacillus rhamnosus* GG in complex protein matrices. *Journal of Microbiological Methods*, **82**, 301–310.

Dong, Y. & Zhu, H. (2005) Single-strand conformational polymorphism analysis: basic principles and routine practice. *Methods in Molecular Medicine*, **108**, 149–157.

Doulgeraki, A.I., Pramateftaki, P., Argyri, A.A., Nychas, G.J.E., Tassou, C.C. & Panagou, E.Z. (2013) Molecular characterization of lactic acid bacteria isolated from industrially fermented Greek table olives. *LWT – Food Science and Technology*, **50**, 353–356.

Dušková, M., Šedo, O., Kšicová, K., Zdráhal, Z. & Karpíšková, R. (2012) Identification of lactobacilli isolated from food by genotypic methods and MALDI-TOF MS. *International Journal of Food Microbiology*, **159**, 107–114.

Duthoit, F., Godon, J.J. & Montel, M.C. (2003) Bacterial community dynamics during production of registered designation of origin Salers cheese as evaluated by 16S rRNA gene single-strand conformation polymorphism analysis. *Applied and Environmental Microbiology*, **69**, 3840–3848.

Dziuba, B., Babuchowski, A., Nałęcz, D. & Niklewicz, M. (2007) Identification of lactic acid bacteria using FTIR spectroscopy and cluster analysis. *International Dairy Journal*, **17**, 183–189.

Ercolini, D., Hill, P.J. & Dodd, C.E.R. (2003) Bacterial community structure and location in Stilton cheese. *Applied and Environmental Microbiology*, **69**, 3540–3548.

Fernandez, E., Alegria, A., Delgado, S., Martin, M.C. & Mayo, B. (2011) Comparative phenotypic and molecular genetic profiling of wild *Lactococcus lactis* subsp. *lactis* strains of the *L. lactis* subsp. *lactis* and *L. lactis* subsp. *cremoris* genotypes, isolated from starter-free cheeses made of raw milk. *Applied and Environmental Microbiology*, **77**, 5324–5335.

Fittipaldi, M., Nocker, A. & Codony, F. (2012) Progress in understanding preferential detection of live cells using viability dyes in combination with DNA amplification. *Journal of Microbiological Methods*, **91**, 276–289.

Fitzsimons, N.A., Cogan, T.M., Condon, S. & Beresford, T. (1999) Phenotypic and genotypic characterization of non-starter lactic acid bacteria in mature Cheddar cheese. *Applied and Environmental Microbiology*, **65**, 3418–3426.

Flint, S., Walker, K., Waters, B. & Crawford, R. (2007) Description and validation of a rapid (1 h) flow cytometry test for enumerating thermophilic bacteria in milk powders. *Journal of Applied Microbiology*, **102**, 909–915.

Foca, G., Ferrari, C., Ulrici, A., Sciutto, G., Prati, S., Morandi, S., Brasca, M., Lavermicocca, P., Lanteri, S. & Oliveri, P. (2016) The potential of spectral and hyperspectral-imaging techniques for bacterial detection in food: a case study on lactic acid bacteria. *Talanta*, **153**, 111–119.

Friedrich, U. & Lenke, J. (2006) Improved enumeration of lactic acid bacteria in mesophilic dairy starter cultures by using multiplex quantitative real-time PCR and flow cytometry-fluorescence in situ hybridization. *Applied and Environmental Microbiology*, **72**, 4163–4171.

Fuka, M.M., Wallisch, S., Engel, M., Welzl, G., Havranek, J. & Schloter, M. (2013) Dynamics of bacterial communities during the ripening process of different Croatian cheese types derived from raw ewe's milk cheeses. *PLoS One*, **8**, 1–10.

Furet, J.P., Quenee, P. & Tailliez, P. (2004) Molecular quantification of lactic acid bacteria in fermented milk products using real-time quantitative PCR. *International Journal of Food Microbiology*, **97**, 197–207.

Galia, W., Perrin, C., Genay, M. & Dary, A. (2009) Variability and molecular typing of *Streptococcus thermophilus* strains displaying different proteolytic and acidifying properties. *International Dairy Journal*, **19**, 89–95.

Garcia-Hernandez, J., Moreno, Y., Amorocho, C.M. & Hernandez, M. (2012) A combination of direct viable count and fluorescence in situ hybridization for specific enumeration of viable *Lactobacillus delbrueckii* subsp. *bulgaricus* and *Streptococcus thermophilus*. *Letters in Applied Microbiology*, **54**, 247–254.

Garcia-Martinez, J., Acinas, S.G., Anton, A.I. & Rodriguez-Valera, F. (1999) Use of the 16S-23S ribosomal genes spacer region in studies of prokaryotic diversity. *Journal of Microbiological Methods*, **36**, 55–64.

Gardiner, G., Ross, R.P., Collins, J.K., Fitzgerald, G. & Stanton, C. (1998) Development of a probiotic Cheddar cheese containing human-derived *Lactobacillus paracasei* strains. *Applied and Environmental Microbiology*, **64**, 2192–2199.

Gatti, M., Bernini, V., Lazzi, C. & Neviani, E. (2006) Fluorescence microscopy for studying the viability of micro-organisms in natural whey starters. *Letters in Applied Microbiology*, **42**, 338–343.

Geng, J., Chiron, C. & Combrisson, J. (2014) Rapid and specific enumeration of viable bifidobacteria in dairy products based on flow cytometry technology: a proof of concept study. *International Dairy Journal*, **37**, 1–4.

Giraffa, G. & Neviani, E. (2001) DNA-based, culture-independent strategies for evaluating microbial communities in food-associated ecosystems. *International Journal of Food Microbiology*, **67**, 19–34.

Giraffa, G. & Rossetti, L. (2004) Monitoring of the bacterial composition of dairy starter cultures by RAPD-PCR. *FEMS Microbiology Letters*, **237**, 133–138.

Goerges, S., Mounier, J., Rea, M.C., Gelsomino, R., Heise, V., Beduhn, R., Cogan, T.M., Vancanneyt, M. & Scherer, S. (2008) Commercial ripening starter microorganisms inoculated into cheese milk do not successfully establish themselves in the resident microbial ripening consortia of a South German red smear cheese. *Applied and Environmental Microbiology*, **74**, 2210–2217.

Golowczyc, M.A., Gugliada, M.J., Hollmann, A., Delfederico, L., Garrote, G.L., Abraham, A.G., Semorile, L. & De Antoni, G. (2008) Characterization of homofermentative lactobacilli isolated from kefir grains: potential use as probiotic. *Journal of Dairy Research*, **75**, 211–217.

Gosiewski, T., Chmielarczyk, A., Strus, M., Brzychczy-Wloch, M. & Heczko, P.B. (2012) The application of genetics methods to differentiation of three *Lactobacillus* species of human origin. *Annals of Microbiology*, **62**, 1437–1445.

Grissa, I., Vergnaud, G. & Pourcel, C. (2007) The CRISPRdb database and tools to display CRISPRs and to generate dictionaries of spacers and repeats. *BMC Bioinformatics*, **8**, 1–10.

Gurtler, V. & Stanisich, V.A. (1996) New approaches to typing and identification of bacteria using the 16S-23S rDNA spacer region. *Microbiology*, **142**, 3–16.

Hamet, M.F., Londero, A., Medrano, M., Vercammen, E., Van Hoorde, K., Garrote, G.L., Huys, G., Vandamme, P. & Abraham, A.G. (2013) Application of culture-dependent and culture-independent methods for the identification of *Lactobacillus kefiranofaciens* in microbial consortia present in kefir grains. *Food Microbiology*, **36**, 327–334.

Hati, S., Mandal, S. & Prajapati, J.B. (2013) Novel starters for value added fermented dairy products. *Current Research in Nutrition and Food Science*, **1**, 83–91.

Hemme, D. & Foucaud-Scheunemann, C. (2004) *Leuconostoc*, characteristics, use in dairy technology and prospects in functional foods. *International Dairy Journal*, **14**, 467–494.

Herbel, S.R., Vahjen, W., Wieler, L.H. & Guenther, S. (2013) Timely approaches to identify probiotic species of the genus *Lactobacillus*. *Gut Pathogens*, **5**, 1–13.

Horvath, P., Coute-Monvoisin, A.C., Romero, D.A., Boyaval, P., Fremaux, C. & Barrangou, R. (2009) Comparative analysis of CRISPR loci in lactic acid bacteria genomes. *International Journal of Food Microbiology*, **131**, 62–70.

Huang, C.H. & Lee, F.L. (2011) The dnaK gene as a molecular marker for the classification and discrimination of the *Lactobacillus casei* group. *Antonie Van Leeuwenhoek International Journal of General and Molecular Microbiology*, **99**, 319–327.

Huang, C.H., Chang, M.T. & Huang, L.N. (2014) Use of highly variable gene (yycH) as DNA marker to resolve interspecific relationships within the *Lactobacillus casei* group and a target for developing novel species-specific PCR primers. *European Food Research and Technology*, **239**, 719–724.

Huang, C.H., Chang, M.T., Huang, L.N. & Chu, W.S. (2015) The dnaJ gene as a molecular discriminator to differentiate among species and strain within the *Lactobacillus casei* group. *Molecular and Cellular Probes*, **29**, 479–484.

ISO/IDF (2003) *Yogurt – Enumeration of Characteristic Microorganisms – Colony-Count Technique at 37 Degrees C*. ISO 7889/IDF 117. International Organization for Standardization, Geneva, and International Dairy Federation, Brussles.

ISO/IDF (2006) *Milk Products – Enumeration of Presumptive* Lactobacillus acidophilus *on a Selective Medium – Colony-Count Technique at 37 Degrees C*. ISO 20128/IDF 192. International Standard Organization, Geneva, and International Dairy Federation, Brussles.

ISO/IDF (2010a) *Fermented Milk Products – Bacterial Starter Cultures – Standard of Identity*. ISO 27205/IDF 149. International Standard Organization, Geneva, and International Dairy Federation, Brussles.

ISO/IDF (2010b) *Milk Products – Enumeration of Presumptive Bifidobacteria – Colony Count Technique at 37 Degrees C*. ISO 29981/IDF 220. International Standard Organization, Geneva, and International Dairy Federation, Brussles.

ISO/IDF (2015) *Milk and Milk Products – Starter Cultures, Probiotics and Fermented Products – Quantification of Lactic Acid Bacteria by Flow Cytometry*. ISO 19344/IDF 232. International Standard Organization, Geneva, and International Dairy Federation, Brussles.

Jany, J.L. & Barbier, G. (2008) Culture-independent methods for identifying microbial communities in cheese. *Food Microbiology*, **25**, 839–848.

Johansen, P., Vindelov, J., Arneborg, N. & Brockmann, E. (2014) Development of quantitative PCR and metagenomics-based approaches for strain quantification of a defined mixed-strain starter culture. *Systematic and Applied Microbiology*, **37**, 186–193.

Kahala, M., Maki, M., Lehtovaara, A., Tapanainen, J.M., Katiska, R., Juuruskorpi, M., Juhola, J. & Joutsjoki, V. (2008) Characterization of starter lactic acid bacteria from the Finnish fermented milk product viili. *Journal of Applied Microbiology*, **105**, 1929–1938.

Kaleta, P., Callanan, M.J., O'Callaghan, J., Fitzgerald, G.F., Beresford, T.P. & Ross, R.P. (2009) Exploitation of the diverse insertion sequence element content of dairy *Lactobacillus helveticus* starters as a rapid method to identify different strains. *Journal of Microbiological Methods*, **79**, 32–36.

Karapetsas, A., Vavoulidis, E., Galanis, A., Sandaltzopoulos, R. & Kourkoutas, Y. (2010) Rapid detection and identification of probiotic *Lactobacillus casei* ATCC 393 by multiplex PCR. *Journal of Molecular Microbiology and Biotechnology*, **18**, 156–161.

Karimi, R., Mortazavian, A.M. & Amiri-Rigi, A. (2012) Selective enumeration of probiotic microorganisms in cheese. *Food Microbiology*, **29**, 1–9.

Karoui, R. & De Baerdemaeker, J. (2007) A review of the analytical methods coupled with chemometric tools for the determination of the quality and identity of dairy products. *Food Chemistry*, **102**, 621–640.

Keer, J.T. & Birch, L. (2003) Molecular methods for the assessment of bacterial viability. *Journal of Microbiological Methods*, **53**, 175–183.

Kelleher, P., Murphy, J., Mahony, J. & van Sinderen, D. (2015) Next-generation sequencing as an approach to dairy starter selection. *Dairy Science & Technology*, **95**, 545–568.

Koser, C.U., Ellington, M.J., Cartwright, E.J.P., Gillespie, S.H., Brown, N.M., Farrington, M., Holden, M.T.G., Dougan, G., Bentley, S.D., Parkhill, J. & Peacock, S.J. (2012) Routine use of microbial whole genome sequencing in diagnostic and public health microbiology. *PLoS Pathogens*, **8**, 1–9.

Kramer, M., Obermajer, N., Bogovič Matijašić, B., Rogelj, I. & Kmetec, V. (2009) Quantification of live and dead probiotic bacteria in lyophilised product by real-time PCR and by flow cytometry. *Applied Microbiology and Biotechnology*, **84**, 1137–1147.

Ksicova, K., Duskova, M. & Karpiskova, R. (2013) Differentiation of *Lactobacillus* species by ARDRA. *Czech Journal of Food Sciences*, **31**, 180–188.

Kutahya, O.E., Starrenburg, M.J.C., Rademaker, J.L.W., Klaassen, C.H.W., Vlieg, J.E.T.V., Smid, E.J. & Kleerebezem, M. (2011) High-resolution amplified fragment length polymorphism typing of *Lactococcus lactis* strains enables identification of genetic markers for subspecies-related phenotypes. *Applied and Environmental Microbiology*, **77**, 5192–5198.

Ladero, V., Canedo, E., Perez, M., Martin, M.C., Fernandez, M. & Alvarez, M.A. (2012) Multiplex qPCR for the detection and quantification of putrescine-producing lactic acid bacteria in dairy products. *Food Control*, **27**, 307–313.

Lahtinen, S.J., Gueimonde, M., Ouwehand, A.C., Reinikainen, J.P. & Salminen, S.J. (2006) Comparison of four methods to enumerate probiotic bifidobacteria in a fermented food product. *Food Microbiology*, **23**, 571–577.

Lahtinen, S.J., Ahokoski, H., Reinikainen, J.P., Gueimonde, M., Nurmi, J., Ouwehand, A.C. & Salminen, S.J. (2008) Degradation of 16S rRNA and attributes of viability of viable but non-culturable probiotic bacteria. *Letters in Applied Microbiology*, **46**, 693–698.

Larsen, M.V., Cosentino, S., Rasmussen, S., Friis, C., Hasman, H., Marvig, R.L., Jelsbak, L., Sicheritz-Ponten, T., Ussery, D.W., Aarestrup, F.M. & Lund, O. (2012) Multilocus sequence typing of total-genome-sequenced bacteria. *Journal of Clinical Microbiology*, **50**, 1355–1361.

Leandro, E.d.S., Lima, G., de Carvalho, K.A.F., Pereira, O.G. & de Moraes, C.A. (2014) Flow cytometric assessment of *Lactococcus Lactis* isolates viability after lyophilization. *International Journal of Nutrition and Food Sciences*, **3**, 391–396.

Leite, A.M.O., Leite, D.C.A., Del Aguila, E.M., Alvares, T.S., Peixoto, R.S., Miguel, M.A.L., Silva, J.T. & Paschoalin, V.M.F. (2013) Microbiological and chemical characteristics of Brazilian kefir during fermentation and storage processes. *Journal of Dairy Science*, **96**, 4149–4159.

Leverrier, P., Fremont, Y., Rouault, A., Boyaval, P. & Jan, G. (2005) *In vitro* tolerance to digestive stresses of propionibacteria: influence of food matrices. *Food Microbiology*, **22**, 11–18.

Li, W., Raoult, D. & Fournier, P.E. (2009) Bacterial strain typing in the genomic era. *FEMS Microbiology Reviews*, **33**, 892–916.

Lisova, I., Horackova, S., Kovacova, R., Rada, V. & Plockova, M. (2013) Emulsion encapsulation of *Bifidobacterium animalis* subsp. *lactis* BB-12 with the addition of lecithin. *Czech Journal of Food Sciences*, **31**, 270–274.

Liu, W.J., Yu, J., Sun, Z.H., Song, Y.Q., Wang, X.N., Wang, H.M., Wuren, T.Y., Zha, M.S., Menghe, B. & Zhang, H.P. (2016) Relationships between functional genes in *Lactobacillus delbrueckii* ssp *bulgaricus* isolates and phenotypic characteristics associated with fermentation time and flavor production in yogurt elucidated using multilocus sequence typing. *Journal of Dairy Science*, **99**, 89–103.

Loman, N.J., Misra, R.V., Dallman, T.J., Constantinidou, C., Gharbia, S.E., Wain, J. & Pallen, M.J. (2012) Performance comparison of benchtop high-throughput sequencing platforms. *Nature Biotechnology*, **30**, 434–439.

Lomonaco, S., Furumoto, E.J., Loquasto, J.R., Morra, P., Grassi, A. & Roberts, R.F. (2015) Development of a rapid SNP-typing assay to differentiate *Bifidobacterium animalis* ssp. *lactis* strains used in probiotic-supplemented dairy products. *Journal of Dairy Science*, **98**, 804–812.

Lorbeg, P.M., Majhenic, A.C. & Rogelj, I. (2009) Evaluation of different primers for PCR-DGGE analysis of cheese-associated enterococci. *Journal of Dairy Research*, **76**, 265–271.

Lu, M., Wang, X.X., Sun, G.W., Qin, B., Xiao, J.Z., Yan, S.L., Pan, Y.J. & Wang, Y.J. (2014) Fine structure of Tibetan kefir grains and their yeast distribution, diversity, and shift. *PLoS One*, **9**, 1–10.

Luginbühl, W., Jimeno, J. & Zehntner, U. (2006) Identification of seven species of the *Lactobacillus acidophilus* group by FT–IR spectroscopy. *LWT – Food Science and Technology*, **39**, 152–158.

Lukjancenko, O., Ussery, D.W. & Wassenaar, T.M. (2012) Comparative genomics of *Bifidobacterium*, *Lactobacillus* and related probiotic genera. *Microbial Ecology*, **63**, 651–673.

Lv, Y., Qiao, X., Zhao, L. & Ren, F. (2015) Biodistribution of a promising probiotic, *Bifidobacterium longum* subsp. *longum* strain BBMN68, in the rat gut. *Journal of Microbiology and Biotechnology*, **25**, 863–871.

Ma, C., Wu, Z., Chen, Z., Du, Z., Sun, K. & Ma, A. (2013) Differentiation of *Streptococcus thermophilus* strains in commercial Direct Vat Set yoghurt starter. *Food Science and Biotechnology*, **22**, 987–991.

Machado, A., Almeida, C., Carvalho, A., Boyen, F., Haesebrouck, F., Rodrigues, L., Cerca, N. & Azevedo, N.F. (2013) Fluorescence in situ hybridization method using a peptide nucleic acid probe for identification of *Lactobacillus* spp. in milk samples. *International Journal of Food Microbiology*, **162**, 64–70.

Maiden, M.C.J. (2006) Multilocus sequence typing of bacteria. *Annual Review of Microbiology*, **60**, 561–588.

Mancini, A., Lazzi, C., Bernini, V., Neviani, E. & Gatti, M. (2012) Identification of dairy lactic acid bacteria by tRNAAla-23S rDNA-RFLP. *Journal of Microbiological Methods*, **91**, 380–390.

Martín-Platero, A.M., Valdivia, E., Maqueda, M., Martín-Sánchez, I. & Martínez-Bueno, M. (2008) Polyphasic approach to bacterial dynamics during the ripening of Spanish farmhouse cheese, using culture-dependent and -independent methods. *Applied and Environmental Microbiology*, **74**, 5662–5673.

Masco, L., Huys, G., Gevers, D., Verbrugghen, L. & Swings, J. (2003) Identification of *Bifidobacterium* species using rep-PCR fingerprinting. *Systematic and Applied Microbiology*, **26**, 557–563.

Mathot, A.G., Kihal, M., Prevost, H. & Divies, C. (1994) Selective enumeration of *Leuconostoc* on vancomycin agar media. *International Dairy Journal*, **4**, 459–469.

Matsuki, T., Watanabe, K. & Tanaka, R. (2003) Genus- and species-specific PCR primers for the detection and identification of bifidobacteria. *Current Issues in Intestinal Microbiology*, **4**, 61–69.

Matte-Tailliez, O., Quenee, P., Cibik, R., van Opstal, J., Dessevre, F., Firmesse, O. & Tailliez, P. (2001) Detection and identification of lactic acid bacteria in milk and industrial starter culture with fluorescently labeled rRNA-targeted peptide nucleic acid probes. *Lait*, **81**, 237–248.

Maukonen, J., Alakomi, H.-L., Nohynek, L., Hallamaa, K., Leppämäki, S., Mättö, J. & Saarela, M. (2006) Suitability of the fluorescent techniques for the enumeration of probiotic bacteria in commercial non-dairy drinks and in pharmaceutical products. *Food Research International*, **39**, 22–32.

Medini, D., Donati, C., Tettelin, H., Masignani, V. & Rappuoli, R. (2005) The microbial pan-genome. *Current Opinion in Genetics & Development*, **15**, 589–594.

Mohania, D., Nagpal, R., Kumar, M., Bhardwaj, A., Yadav, M., Jain, S., Marotta, F., Singh, V., Parkash, O. & Yadav, H. (2008) Molecular approaches for identification and characterization of lactic acid bacteria. *Journal of Digestive Diseases*, **9**, 190–198.

Monnet, C. & Bogovič Matijašić, B. (2012) Application of PCR-based methods to dairy products and to non-dairy probiotic products. In *Polymerase Chain Reaction* (ed. P. Hernandez-Rodriguez & A.P. Ramirez Gomez), 11–51. InTech, Rijeka, Croatia.

Moraes, P.M., Perin, L.M., Silva, A. & Nero, L.A. (2013) Comparison of phenotypic and molecular tests to identify lactic acid bacteria. *Brazilian Journal of Microbiology*, **44**, 109–112.

Moreira, J.L.S., Mota, R.M., Horta, M.F., Teixeira, S.M.R., Neumann, E., Nicoli, J.R. & Nunes, A.C. (2005) Identification to the species level of *Lactobacillus* isolated in probiotic prospecting studies of human, animal or food origin by 16S-23S rRNA restriction profiling. *BMC Microbiology*, **5**, 1–9.

Moreno, Y., Collado, M.C., Ferrús, M.A., Cobo, J.M., Hernández, E. & Hernández, M. (2006) Viability assessment of lactic acid bacteria in commercial dairy products stored at 4°C using LIVE/DEAD® BacLight™ staining and conventional plate counts. *International Journal of Food Science & Technology*, **41**, 275–280.

Morgan, M.C., Boyette, M., Goforth, C., Sperry, K.V. & Greene, S.R. (2009) Comparison of the Biolog OmniLog Identification System and 16S ribosomal RNA gene sequencing for accuracy in identification of atypical bacteria of clinical origin. *Journal of Microbiological Methods*, **79**, 336–343.

Mounier, J., Monnet, C., Jacques, N., Antoinette, A. & Irlinger, F. (2009) Assessment of the microbial diversity at the surface of Livarot cheese using culture-dependent and independent approaches. *International Journal of Food Microbiology*, **133**, 31–37.

Muller, J.A., Stanton, C., Sybesma, W., Fitzgerald, G.F. & Ross, R.P. (2010) Reconstitution conditions for dried probiotic powders represent a critical step in determining cell viability. *Journal of Applied Microbiology*, **108**, 1369–1379.

Nacef, M., Chevalier, M., Chollet, S., Drider, D. & Flahaut, C. (2016) MALDI-TOF mass spectrometry for the identification of lactic acid bacteria isolated from a French cheese: the Maroilles. *International Journal of Food Microbiology*, **247**, 2–8.

Nagarajan, K. & Loh, K.C. (2014) Molecular biology-based methods for quantification of bacteria in mixed culture: perspectives and limitations. *Applied Microbiology and Biotechnology*, **98**, 6907–6919.

Ni, K., Wang, Y., Li, D., Cai, Y. & Pang, H. (2015) Characterization, identification and application of lactic acid bacteria isolated from forage paddy rice silage. *PLoS One*, **10**, 1–14.

Nicolaou, N., Xu, Y. & Goodacre, R. (2011) Fourier transform infrared and Raman spectroscopies for the rapid detection, enumeration, and growth interaction of the bacteria *Staphylococcus aureus* and *Lactococcus lactis* ssp. *cremoris* in milk. *Analytical Chemistry*, **83**, 5681–5687.

Nieto-Arribas, P., Sesena, S., Poveda, J.M., Palop, L. & Cabezas, L. (2009) Genotypic and technological characterization of *Lactococcus lactis* isolates involved in processing of artisanal Manchego cheese. *Journal of Applied Microbiology*, **107**, 1505–1517.

Nikolic, M., Terzic-Vidojevic, A., Jovcic, B., Begovic, J., Golic, N. & Topisirovic, L. (2008) Characterization of lactic acid bacteria isolated from Bukuljac, a homemade goat's milk cheese. *International Journal of Food Microbiology*, **122**, 162–170.

Oberg, C.J., Moyes, L.V., Domek, M.J., Brothersen, C. & McMahon, D.J. (2011) Survival of probiotic adjunct cultures in cheese and challenges in their enumeration using selective media. *Journal of Dairy Science*, **94**, 2220–2230.

Oketič, K., Bogovič Matijašić, B., Obermajer, T., Radulovič, Z., Lević, S., Mirković, N. & Nedović, V. (2015) Evaluation of propidium monoazide real-time PCR for enumeration of probiotic lactobacilli microencapsulated in calcium alginate beads. *Beneficial Microbes*, **6**, 573–581.

Olsen, K.N., Brockmann, E. & Molin, S. (2007) Quantification of *Leuconostoc* populations in mixed dairy starter cultures using fluorescence in situ hybridization. *Journal of Applied Microbiology*, **103**, 855–863.

Olszewska, M., Staniewski, B. & Łaniewska-Trokenheim, Ł. (2012) Cell viability of *Bifidobacterium lactis* strain in long-term storage butter assessed with the plate count and fluorescence techniques. *Czech Journal of Food Sciences*, **30**, 421–428.

Ong, L. & Shah, N.P. (2009) Probiotic Cheddar cheese: influence of ripening temperatures on survival of probiotic microorganisms, cheese composition and organic acid profiles. *LWT – Food Science and Technology*, **42**, 1260–1268.

Ouadghiri, M., Amar, M., Vancanneyt, M. & Swings, J. (2005) Biodiversity of lactic acid bacteria in Moroccan soft white cheese (Jben). *FEMS Microbiology Letters*, **251**, 267–271.

Oust, A., Møretrø, T., Kirschner, C., Narvhus, J.A. & Kohler, A. (2004) FT-IR spectroscopy for identification of closely related lactobacilli. *Journal of Microbiological Methods*, **59**, 149–162.

Ozturk, M. & Meterelliyoz, M. (2015) Practical identification of human originated *Lactobacillus* species by amplified ribosomal DNA restriction analysis (ARDRA) for probiotic use. *Molecular Biology Reports*, **42**, 1323–1332.

Parisot, N., Peyretaillade, E., Dugat-Bony, E., Denonfoux, J., Mahul, A. & Peyret, P. (2016) Probe design strategies for oligonucleotide microarrays. *Methods in Molecular Biology*, **1368**, 67–82.

Passot, S., Gautier, J., Jamme, F., Cenard, S., Dumas, P. & Fonseca, F. (2015) Understanding the cryotolerance of lactic acid bacteria using combined synchrotron infrared and fluorescence microscopies. *Analyst*, **140**, 5920–5928.

Patro, J.N., Ramachandran, P., Lewis, J.L., Mammel, M.K., Barnaba, T., Pfeiler, E.A. & Elkins, C.A. (2015) Development and utility of the FDA 'GutProbe' DNA microarray for identification, genotyping and metagenomic analysis of commercially available probiotics. *Journal of Applied Microbiology*, **118**, 1478–1488.

Paveljšek, D., Trmčić, A., Hacin, B. & Rogelj, I. (2014) PCR verification of microplate phenotypic system identification for LAB from traditional Western Balkan raw milk cheeses. *Mljekarstvo*, **64**, 245–253.

Perdana, J., Bereschenko, L., Roghair, M., Fox, M.B., Boom, R.M., Kleerebezem, M. & Schutyser, M.A.I. (2012) Novel method for enumeration of viable *Lactobacillus plantarum* WCFS1 cells after single-droplet drying. *Applied and Environmental Microbiology*, **78**, 8082–8088.

Pettersson, E., Lundeberg, J. & Ahmadian, A. (2009) Generations of sequencing technologies. *Genomics*, **93**, 105–111.

Pogačić, T., Kelava, N., Zamberlin, Š., Dolenčić-Špehar, I. & Samaržija, D. (2010) Methods for culture-independent identification of lactic acid bacteria in dairy products. *Food Technology and Biotechnology*, **48**, 3–10.

Prabhakar, V., Kocaoglu-Vurma, N., Harper, J. & Rodriguez-Saona, L. (2011) Classification of Swiss cheese starter and adjunct cultures using Fourier transform infrared microspectroscopy. *Journal of Dairy Science*, **94**, 4374–4382.

Ramachandran, P., Lacher, D.W., Pfeiler, E.A. & Elkins, C.A. (2013) Development of a tiered multilocus sequence typing scheme for members of the *Lactobacillus acidophilus* complex. *Applied and Environmental Microbiology*, **79**, 7220–7228.

Randazzo, C.L., Torriani, S., Akkermans, A.D.L., de Vos, W.M. & Vaughan, E.E. (2002) Diversity, dynamics, and activity of bacterial communities during production of an artisanal Sicilian cheese as evaluated by 16S rRNA analysis. *Applied and Environmental Microbiology*, **68**, 1882–1892.

Randazzo, C.L., Restuccia, C., Mancini, A., Muccilli, S., Gatti, M. & Caggia, C. (2016) Ragusana donkey milk as a source of lactic acid bacteria and yeast strains of dairy technological interest. *International Journal of Dairy Science & Processing*, **3**, 38–46.

Rault, A., Béal, C., Ghorbal, S., Ogier, J.C. & Bouix, M. (2007) Multiparametric flow cytometry allows rapid assessment and comparison of lactic acid bacteria viability after freezing and during frozen storage. *Cryobiology*, **55**, 35–43.

Reuter, G., Klein, G. & Goldberg, M. (2002) Identification of probiotic cultures in food samples. *Food Research International*, **35**, 117–124.

Rodas, A.M., Ferrer, S. & Pardo, I. (2003) 16S-ARDRA, a tool for identification of lactic acid bacteria isolated from grape must and wine. *Systematic and Applied Microbiology*, **26**, 412–422.

Rohde, A., Hammerl, J.A., Appel, B., Dieckmann, R. & Al Dahouk, S. (2015) FISHing for bacteria in food – a promising tool for the reliable detection of pathogenic bacteria? *Food Microbiology*, **46**, 395–407.

Rossetti, L. & Giraffa, G. (2005) Rapid identification of dairy lactic acid bacteria by M13-generated, RAPD-PCR fingerprint databases. *Journal of Microbiological Methods*, **63**, 135–144.

Ruiz, P., Seseña, S. & Palop, M.L. (2014) A comparative study of different PCR-based DNA fingerprinting techniques for typing of lactic acid bacteria. *European Food Research and Technology*, **239**, 87–98.

Rungrassamee, W., Tosukhowong, A., Klanchui, A., Maibunkaew, S., Plengvidhya, V. & Karoonuthaisiri, N. (2012) Development of bacteria identification array to detect lactobacilli in Thai fermented sausage. *Journal of Microbiological Methods*, **91**, 341–353.

Saeed, A.F., Wang, R.Z. & Wang, S.H. (2016) Microsatellites in pursuit of microbial genome evolution. *Frontiers in Microbiology*, **6**, 1–15.

Saito, Y., Sakamoto, M., Takizawa, S. & Benno, Y. (2004) Monitoring the cell number and viability of *Lactobacillus helveticus* GCL1001 in human feces by PCR methods. *FEMS Microbiology Letters*, **231**, 125–130.

Samelis, J., Bleicher, A., Delbes-Paus, C., Kakouri, A., Neuhaus, K. & Montel, M.C. (2011) FTIR-based polyphasic identification of lactic acid bacteria isolated from traditional Greek Graviera cheese. *Food Microbiology*, **28**, 76–83.

Sandes, S.H.C., Alvim, L.B., Silva, B.C., Zanirati, D.F., Jung, L.R.C., Nicoli, J.R., Neumann, E. & Nunes, A.C. (2014) *Lactobacillus* species identification by amplified ribosomal 16S–23S rRNA restriction fragment length polymorphism analysis. *Beneficial Microbes*, **5**, 471–481.

Sanozky-Dawes, R., Selle, K., O'Flaherty, S., Klaenhammer, T. & Barrangou, R. (2015) Occurrence and activity of a type II CRISPR-Cas system in *Lactobacillus gasseri*. *Microbiology (SGM)*, **161**, 1752–1761.

Santos, M., Gerbino, E., Tymczyszyn, E. & Gomez-Zavaglia, A. (2015) Applications of infrared and Raman spectroscopies to probiotic investigation. *Foods*, **4**, 283–305.

Sato, H., Teramoto, K., Ishii, Y., Watanabe, K. & Benno, Y. (2011) Ribosomal protein profiling by matrix-assisted laser desorption/ionization time-of-flight mass spectrometry for phylogenety-based subspecies resolution of *Bifidobacterium longum*. *Systematic and Applied Microbiology*, **34**, 76–80.

Sauer, S. & Kliem, M. (2010) Mass spectrometry tools for the classification and identification of bacteria. *Nature Reviews: Microbiology*, **8**, 74–82.

Savić, D., Joković, N. & Topisirović, L. (2008) Multivariate statistical methods for discrimination of lactobacilli based on their FTIR spectra. *Dairy Science & Technology*, **88**, 273–290.

Schillinger, U., Yousif, N.M.K., Sesar, L. & Franz, C.M.A.P. (2003) Use of group-specific and RAPD-PCR analyses for rapid differentiation of *Lactobacillus* strains from probiotic yogurts. *Current Microbiology*, **47**, 453–456.

Schmidt, F., Fiege, T., Hustoft, H.K., Kneist, S. & Thiede, B. (2009) Shotgun mass mapping of *Lactobacillus* species and subspecies from caries related isolates by MALDI-MS. *Proteomics*, **9**, 1994–2003.

Sedo, O., Sedlacek, I. & Zdrahal, Z. (2011) Sample preparation methods for MALDI-MS profiling of bacteria. *Mass Spectrometry Reviews*, **30**, 417–434.

Shehata, A.I. (2012) Molecular identification of probiotics *Lactobacillus* strain isolates by amplified ribosomal DNA restriction analysis (ARDRA). *African Journal of Microbiology Research*, **6**, 3034–3041.

Sheridan, G.E.C., Masters, C.I., Shallcross, J.A. & Mackey, B.M. (1998) Detection of mRNA by reverse transcription PCR as an indicator of viability in *Escherichia coli* cells. *Applied and Environmental Microbiology*, **64**, 1313–1318.

Simpson, P.J., Fitzgerald, G.F., Stanton, C. & Ross, R.P. (2004) The evaluation of a mupirocin-based selective medium for the enumeration of bifidobacteria from probiotic animal feed. *Journal of Microbiological Methods*, **57**, 9–16.

Singh, S., Goswami, P., Singh, R. & Heller, K.J. (2009) Application of molecular identification tools for *Lactobacillus*, with a focus on discrimination between closely related species: a review. *LWT – Food Science and Technology*, **42**, 448–457.

Singhal, N., Kumar, M., Kanaujia, P.K. & Virdi, J.S. (2015) MALDI-TOF mass spectrometry: an emerging technology for microbial identification and diagnosis. *Frontiers in Microbiology*, **6**, 1–16.

Sohier, D., Pavan, S., Riou, A., Combrisson, J. & Postollec, F. (2014) Evolution of microbiological analytical methods for dairy industry needs. *Frontiers in Microbiology*, **5**, 1–10.

Strus, M., Chmielarczyk, A., Kochan, P., Adamski, P., Chelmicki, Z., Chelmicki, A., Palucha, A. & Heczko, P.B. (2012) Studies on the effects of probiotic *Lactobacillus* mixture given orally on vaginal and rectal colonization and on parameters of vaginal health in women with intermediate vaginal flora. *European Journal of Obstetrics & Gynecology and Reproductive Biology*, **163**, 210–215.

Succi, M., Sorrentino, E., Di Renzo, T., Tremonte, P., Reale, A., Tipaldi, L., Pannella, G., Russo, A. & Coppola, R. (2014) Lactic acid bacteria in pharmaceutical formulations: presence and viability of "healthy microorganisms". *Journal of Pharmacy and Nutrition Sciences*, **4**, 66–75.

Sunny-Roberts, E.O. & Knorr, D. (2008) Evaluation of the response of *Lactobacillus rhamnosus* VTT E-97800 to sucrose-induced osmotic stress. *Food Microbiology*, **25**, 183–189.

Španová, A., Dráb, V., Turková, K., Špano, M., Burdychová, R., Šedo, O., Šrůtková, D., Rada, V. & Rittich, B. (2015) Selection of potential probiotic *Lactobacillus* strains of human origin for use in dairy industry. *European Food Research and Technology*, **241**, 861–869.

Tailliez, P., Tremblay, J., Ehrlich, S.D. & Chopin, A. (1998) Molecular diversity and relationship within *Lactococcus lactis*, as revealed by randomly amplified polymorphic DNA (RAPD). *Systematic and Applied Microbiology*, **21**, 530–538.

Talbot, G., Topp, E., Palin, M.F. & Masse, D.I. (2008) Evaluation of molecular methods used for establishing the interactions and functions of microorganisms in anaerobic bioreactors. *Water Research*, **42**, 513–537.

Tang, Y. & Saris, P.E.J. (2013) Strain-specific detection of orally administered canine jejunum-dominated *Lactobacillus acidophilus* LAB20 in dog faeces by real-time PCR targeted to the novel surface layer protein. *Letters in Applied Microbiology*, **57**, 330–335.

Tanigawa, K., Kawabata, H. & Watanabe, K. (2010) Identification and typing of *Lactococcus lactis* by matrix-assisted laser desorption ionization-time of flight mass spectrometry. *Applied and Environmental Microbiology*, **76**, 4055–4062.

Tanigawa, K. & Watanabe, K. (2011) Multilocus sequence typing reveals a novel subspeciation of *Lactobacillus delbrueckii*. *Microbiology (SGM)*, **157**, 727–738.

Temmerman, R., Huys, G. & Swings, J. (2004) Identification of lactic acid bacteria: culture-dependent and culture-independent methods. *Trends in Food Science & Technology*, **15**, 348–359.

Teramoto, K., Sato, H., Sun, L., Torimura, M. & Tao, H. (2007) A simple intact protein analysis by MALDI-MS for characterization of ribosomal proteins of two genome-sequenced lactic acid bacteria and verification of their amino acid sequences. *Journal of Proteome Research*, **6**, 3899–3907.

Tettelin, H., Masignani, V., Cieslewicz, M.J., Donati, C., Medini, D., Ward, N.L., Angiuoli, S.V., Crabtree, J., Jones, A.L., Durkin, A.S., DeBoy, R.T., Davidsen, T.M., Mora, M., Scarselli, M., Ros, I.M.Y., Peterson, J.D., Hauser, C.R., Sundaram, J.P., Nelson, W.C., Madupu, R., Brinkac, L.M., Dodson, R.J., Rosovitz, M.J., Sullivan, S.A., Daugherty, S.C., Haft, D.H., Selengut, J., Gwinn, M.L., Zhou, L.W., Zafar, N., Khouri, H., Radune, D., Dimitrov, G., Watkins, K.,

O'Connor, K.J.B., Smith, S., Utterback, T.R., White, O., Rubens, C.E., Grandi, G., Madoff, L.C., Kasper, D.L., Telford, J.L., Wessels, M.R., Rappuoli, R. & Fraser, C.M. (2005) Genome analysis of multiple pathogenic isolates of *Streptococcus agalactiae*: implications for the microbial "pan-genome". *Proceedings of the National Academy of Sciences of the United States of America*, **102**, 13950–13955.

Tharmaraj, N. & Shah, N.P. (2003) Selective enumeration of *Lactobacillus delbrueckii* ssp. *bulgaricus*, *Streptococcus thermophilus*, *Lactobacillus acidophilus*, bifidobacteria, *Lactobacillus casei*, *Lactobacillus rhamnosus*, and propionibacteria. *Journal of Dairy Science*, **86**, 2288–2296.

Tran, K.T.M., May, B.K., Smooker, P.M., Van, T.T.H. & Coloe, P.J. (2011) Distribution and genetic diversity of lactic acid bacteria from traditional fermented sausage. *Food Research International*, **44**, 338–344.

Treven, P. (2015) Strategies to develop strain-specific PCR based assays for probiotics. *Beneficial Microbes*, **6**, 887–898.

Treven, P., Trmčić, A., Bogovič Matijašić, B. & Rogelj, I. (2014) Improved draft genome sequence of probiotic strain *Lactobacillus gasseri* K7. *Genome Announcements*, **2**, e00725-14.

Treven, P., Mrak, V., Bogovič Matijašić, B., Horvat, S. & Rogelj, I. (2015) Administration of probiotics *Lactobacillus rhamnosus* GG and *Lactobacillus gasseri* K7 during pregnancy and lactation changes mouse mesenteric lymph nodes and mammary gland microbiota. *Journal of Dairy Science*, **98**, 2114–2128.

Trmčić, A., Monnet, C., Rogelj, I. & Bogovič Matijašić, B. (2011) Expression of nisin genes in cheese: a quantitative real-time polymerase chain reaction approach. *Journal of Dairy Science*, **94**, 77–85.

Tshikhudo, P., Nnzeru, R., Ntushelo, K. & Mudau, F. (2013) Bacterial species identification getting easier. *African Journal of Biotechnology*, **12**, 5975–5982.

Tu, Q.C., He, Z.L., Deng, Y. & Zhou, J.H. (2013) Strain/species-specific probe design for microbial identification microarrays. *Applied and Environmental Microbiology*, **79**, 5085–5088.

Tu, Q.C., He, Z.L. & Zhou, J.Z. (2014) Strain/species identification in metagenomes using genome-specific markers. *Nucleic Acids Research*, **42**, e67.

Turaev, D. & Rattei, T. (2016) High definition for systems biology of microbial communities: metagenomics gets genome-centric and strain-resolved. *Current Opinion in Biotechnology*, **39**, 174–181.

Turroni, F., Foroni, E., Pizzetti, P., Giubellini, V., Ribbera, A., Merusi, P., Cagnasso, P., Bizzarri, B., de'Angelis, G.L., Shanahan, F., van Sinderen, D. & Ventura, M. (2009) Exploring the diversity of the bifidobacterial population in the human intestinal tract. *Applied and Environmental Microbiology*, **75**, 1534–1545.

Van Hoorde, K., Vandamme, P. & Huys, G. (2008) Molecular identification and typing of lactic acid bacteria associated with the production of two artisanal raw milk cheeses. *Dairy Science & Technology*, **88**, 445–455.

Vardjan, T., Mohar Lorbeg, P., Rogelj, I. & Čanžek Majhenič, A. (2013) Characterization and stability of lactobacilli and yeast microbiota in kefir grains. *Journal of Dairy Science*, **96**, 2729–2736.

Veal, D.A., Deere, D., Ferrari, B., Piper, J. & Attfield, P.V. (2000) Fluorescence staining and flow cytometry for monitoring microbial cells. *Journal of Immunological Methods*, **243**, 191–210.

Ventura, M., Canchaya, C., Del Casale, A., Dellaglio, F., Neviani, E., Fitzgerald, G.F. & van Sinderen, D. (2006) Analysis of bifidobacterial evolution using a multilocus approach. *International Journal of Systematic and Evolutionary Microbiology*, **56**, 2783–2792.

Vernile, A., Giammanco, G., Spano, G., Beresford, T.P., Fox, P.F. & Massa, S. (2008) Genotypic characterization of lactic acid bacteria isolated from traditional Pecorino Siciliano cheese. *Dairy Science & Technology*, **88**, 619–629.

Vincent, D., Roy, D., Mondou, F. & Dery, C. (1998) Characterization of bifidobacteria by random DNA amplification. *International Journal of Food Microbiology*, **43**, 185–193.

Vinderola, C.G. & Reinheimer, J.A. (1999) Culture media for the enumeration of *Bifidobacterium bifidum* and *Lactobacillus acidophilus* in the presence of yoghurt bacteria. *International Dairy Journal*, **9**, 497–505.

Vinderola, C.G., Prosello, W., Ghiberto, D. & Reinheimer, J.A. (2000) Viability of probiotic (*Bifidobacterium*, *Lactobacillus acidophilus* and *Lactobacillus casei*) and nonprobiotic microflora in Argentinian Fresco cheese. *Journal of Dairy Science*, **83**, 1905–1911.

Weber, M., Geissert, J., Kruse, M. & Lipski, A. (2014) Comparative analysis of bacterial community composition in bulk tank raw milk by culture-dependent and culture-independent methods using the viability dye propidium monoazide. *Journal of Dairy Science*, **97**, 6761–6776.

Weinrichter, B., Luginbühl, W., Rohm, H. & Jimeno, J. (2001) Differentiation of facultatively heterofermentative lactobacilli from plants, milk, and hard type cheeses by SDS-PAGE, RAPD, FTIR, energy source utilization and autolysis type. *LWT – Food Science and Technology*, **34**, 556–566.

Wenning, M. & Scherer, S. (2013) Identification of microorganisms by FTIR spectroscopy: perspectives and limitations of the method. *Applied Microbiology and Biotechnology*, **97**, 7111–7120.

Wenning, M., Breitenwieser, F., Konrad, R., Huber, I., Busch, U. & Scherer, S. (2014) Identification and differentiation of food-related bacteria: a comparison of FTIR spectroscopy and MALDI-TOF mass spectrometry. *Journal of Microbiological Methods*, **103**, 44–52.

Woodcock, T., Fagan, C.C., O'Donnell, C.P. & Downey, G. (2008) Application of near and mid-infrared spectroscopy to determine cheese quality and authenticity. *Food and Bioprocess Technology*, **1**, 117–129.

Wu, Z., Weng, P., Zhang, R. & Shenb, X. (2011) Microbial community structure and its dynamic analysis of pickled mustard tuber during a low-salt processing. In *Remote Sensing, Environment and Transportation Engineering (RSETE), International Conference, Nanjing, 2011*, 8339–8345.

Xu, F.L., Guo, Y.C., Wang, H.X., Fu, P., Zeng, H.W., Li, Z.G., Pei, X.Y. & Liu, X.M. (2012) PFGE genotyping and antibiotic resistance of *Lactobacillus* distributed strains in the fermented dairy products. *Annals of Microbiology*, **62**, 255–262.

Xu, H.Y., Sun, Z.H., Liu, W.J., Yu, J., Song, Y.Q., Lv, Q., Zhang, J.C., Shao, Y.Y., Menghe, B. & Zhang, H.P. (2014) Multilocus sequence typing of *Lactococcus lactis* from naturally fermented milk foods in ethnic minority areas of China. *Journal of Dairy Science*, **97**, 2633–2645.

Yeung, P.S.M., Kitts, C.L., Cano, R., Tong, P.S. & Sanders, M.E. (2004) Application of genotypic and phenotypic analyses to commercial probiotic strain identity and relatedness. *Journal of Applied Microbiology*, **97**, 1095–1104.

Yu, J., Sun, Z.H., Liu, W.J., Bao, Q.H., Zhang, J.C. & Zhang, H.P. (2012) Phylogenetic study of *Lactobacillus acidophilus* group, *L. casei* group and *L. plantarum* group based on partial *hsp*60, *pheS* and *tuf* gene sequences. *European Food Research and Technology*, **234**, 927–934.

Zamfir, M., Vancanneyt, M., Makras, L., Vaningelgem, F., Lefebvre, K., Pot, B., Swings, J. & De Vuyst, L. (2006) Biodiversity of lactic acid bacteria in Romanian dairy products. *Systematic and Applied Microbiology*, **29**, 487–495.

Zotta, T., Guidone, A., Tremonte, P., Parente, E. & Ricciardi, A. (2012) A comparison of fluorescent stains for the assessment of viability and metabolic activity of lactic acid bacteria. *World Journal of Microbiology and Biotechnology*, **28**, 919–927.

7 Prebiotic Ingredients in Probiotic Dairy Products

X. Wang and R.A. Rastall

7.1 Introduction

The human body functions as a complex ecosystem with more micro-organisms being present than human cells. The gastrointestinal (GI) tract is the biggest and most important habitat for micro-organisms due to the abundance of nutrients in the form of digesta that flow through the lumen without being absorbed by the small intestine. According to several human intestinal metagenomic studies, the most widely represented phyla of micro-organisms in the human gut are Firmicutes, Bacteroidetes, Actinobacteria, Proteobacteria, Verrucomicrobia, Fusobacteria and Euryarchaeota methanogens (Eckburg *et al.*, 2005; Arumugam *et al.*, 2011). The GI tract's physiochemical status affects the microbiota to a large extent. Due to the low pH of the stomach, only a few microorganisms can survive there. Although pH fluctuates in the small intestine as a result of the secretion of alkaline pancreatic juices and acid bile, the numbers of bacteria increase gradually and reach a high number in the large intestine. On the other hand, the microbiota composition and metabolism affect host health in various ways, for instance by influencing immunity, mineral absorption, energy intake regulation and lipid metabolism. A good symbiosis between the human body and its microbiota is essential for human health, whereas dysbiosis, which can be caused by antibiotic therapy, drugs, diseases, injury, surgery, stress or ageing, is problematic. Diet plays an important role in the gut microbiota; therefore, food ingredients and supplements that can promote beneficial bacteria can confer health benefits.

Prebiotics are defined as 'selectively fermented ingredients that result in specific changes, in the composition and/or activity of the GI microbiota, thus conferring benefit upon host health' (Gibson *et al.*, 2010). Prebiotics not only boost the growth of beneficial bacteria in the GI tract, but also have the potential to inhibit pathogens, improve mineral absorption, increase satiety and improve well-being.

Probiotic Dairy Products, Second Edition. Edited by Adnan Y. Tamime and Linda V. Thomas.

7.2 Criteria for an ingredient to be classified as a prebiotic

Prebiotics were originally defined by Gibson and Roberfroid (1995) as 'a non-digestible food ingredient that beneficially affects the host by selectively stimulating the growth and/or activity of one or a limited number of bacteria in the colon, and thus improves host health'. Since then, the definition has been revisited several times and criteria have been proposed to qualify a compound as prebiotic. Three criteria are proposed by Gibson *et al.* (2004):

- *Non-digestibility*: resistant to gastric acid, hydrolysate enzymes and GI absorption.
- *Fermentable* by gut microbiota.
- *Selectivity*: selectively boosts the growth or/and activity of beneficial colonic bacteria.

Current recognised prebiotics are all carbohydrates: non-digestible oligosaccharides that contain different oligomers of saccharides with different degrees of polymerisation.

Inulin-type fructans are oligosaccharides or polysaccharides composed principally of fructose. D-fructose molecules are linked by β $(2 \rightarrow 1)$ linkages, and the chain is terminated by a D-glucose molecule bonded with fructose by an $\alpha(1 \leftrightarrow 2)$ linkage; inulin often refers to molecules with more than 10 degrees of polymerisation (DP), whereas fructooligosaccharide (FOS) or oligofructose (OF) often refers to molecules with 2 ~ 10 DP. Inulin is naturally present in many plants such as chicory, artichoke, leek, banana, asparagus and onion. FOS also occurs in natural food as well as being produced from sucrose or inulin. These molecules can stay intact until reaching the colon, where they principally simulate the growth of bifidobacteria (Kelly, 2008).

Galactooligosaccharides (GOS) are oligosaccharides with fewer than 10 DP; they consist of one or more galactose residues and a terminal glucose, or they are disaccharides of galactose (Jose Gosalbes *et al.*, 2011). GOS can be produced from lactose, and in the GI tract can enhance the growth of bifidobacteria while supressing the growth of clostridia; this has been shown both *in vitro* and *in vivo* (Rycroft *et al.*, 2001; Vulevic *et al.*, 2013; Giovannini *et al.*, 2014).

Other carbohydrates that have been investigated as candidate prebiotics include human milk oligosaccharides, resistant starch (RS), isomaltooligosaccharides, lactosucrose, xylooligosaccharides (XOS), lactulose, soya/soybean oligosaccharides, glucooligosaccharides, arabinoxylan, arabinogalactans, pectin/pectic oligosaccharides, seaweeds/microalgae and β-glucans.

7.3 Health benefits of prebiotics and their mechanisms of action

Commensal bacteria in the human gut can utilise undigested carbohydrates or proteins as energy sources. Microbial breakdown of carbohydrates produces short-chain fatty acids (SCFAs) (e.g. acetate, propionate and butyrate), ethanol, formate, lactate, succinate, carbon dioxide, methane and hydrogen through various pathways. Protein fermentation also generates SCFAs, ethanol, gases and other organic acids together with branched-chain fatty acids (BCFAs), such as iso-butyrate and iso-valerate, ammonia (NH_3) and amines. Fermentation of aromatic amino acids leads to the production of phenolic and indolic compounds, whereas fermentation of sulphur-containing amino acids often generates hydrogen sulphide (H_2S). In general, SCFAs are believed to be involved in various health benefits (Russell *et al.*, 2013); however, NH_3, phenolic and indolic compounds, and H_2S are believed to have a negative impact on human cells, as indicated by various *in vitro* cell studies (Windey *et al.*, 2012). Commensal bacteria in the human gut can be saccharolytic, proteolytic or both. Consumption of prebiotics usually promotes increased growth of beneficial saccharolytic bacteria and an increase in saccharolytic activity of the gut microbiota, resulting in increased SCFA production.

7.3.1 *Short-chain fatty acids and human metabolism*

Bacteria are present throughout the GI tract; however, they are more abundant in the large intestine than the small intestine. This coincides with the SCFA concentration within the human body: SCFAs are at low concentrations in the terminal ileum and high in the colon, which confirms the role of colonic bacteria in SCFA production (Cummings *et al.*, 1987). Within the colon, SCFAs are produced primarily in the ascending colon where substrates are abundant and to a lesser extent in the descending colon where most substrates have been utilised by the bacteria in the early part of the colon and are scarce. Although the concentration of SCFAs in the human gut varies between individuals, the ratio of acetate, propionate, butyrate and BCFAs is generally around 50:20:20:10 (Cummings *et al.*, 1987; Macfarlane & Macfarlane, 2003). Whether considering the amount of substrates reaching the lower gut and then being fermented, or the requirement to sustain the survival of colonic bacteria, the amount of SCFAs produced in the human colon is considerable; however, human faecal output contains only low amounts of SCFAs (Cummings, 1981). This is due to SCFA absorption in the colon, which happens not only in humans but also in other mammalian species such as rat, horse and pig. McNeil *et al.* (1978) studied 46 human subjects' rectal SCFA absorption by connecting dialysis bags filled with different testing solutions to the volunteers' rectum. The fluid in the dialysis bags was measured before and one hour after connecting: this showed that absorption of SCFAs occurs together with the excretion of bicarbonate and is not related to the pH level in the lumen.

After absorption by the colon, butyrate is present in low concentrations in portal blood due to utilisation of butyrate as an energy source by colonic epithelial cells. The remaining SCFAs are transported to the liver (Cummings *et al.*, 1987). Colonocytes oxidise acetate, propionate, butyrate, glucose and glutamine to generate energy; however, butyrate is the preferred substrate for colonocytes. The presence of butyrate inhibits the oxidation of other energy sources, with the preference order being: butyrate > acetate > propionate > glucose > glutamine (Clausen & Mortensen, 1995). After being absorbed by the colon, SCFAs are transported to the liver via portal blood. Most propionate and butyrate are metabolised by the liver; however, acetate is further transported by the venous blood to peripheral tissues (Bloemen *et al.*, 2009). In venous blood, more than 90% of SCFAs are acetate, which can join the tricarboxylic acid cycle in peripheral tissues by forming acetyl-CoA and then providing energy (Cummings *et al.*, 1987). Around one-third of the acetate absorbed from the colon will be taken by the cells in peripheral tissues providing energy (Cummings *et al.*, 1987).

7.3.2 *Mineral absorption*

Prebiotic consumption can directly lead to an increase in mineral absorption. Feeding rats with FOS can improve the absorption by the gut of multiple minerals, including calcium, magnesium and iron (Delzenne *et al.*, 1995; Ohta *et al.*, 1995). Supplementation with prebiotics, such as GOS, lactulose and other resistant carbohydrates, to rats also revealed similar findings: calcium absorption by the gut was enhanced (Brommage *et al.*, 1993; Chonan *et al.*, 1995). The GOS study also measured bone ash and found that rats fed with GOS had higher levels, implying that prebiotics can improve bone health (Chonan *et al.*, 1995).

Most studies on prebiotics and mineral absorption have focused on calcium as it is crucial for bone health, especially in children and women. Most calcium in the human body is distributed in bones, and adult calcium levels are maintained by a dynamic equilibrium of calcium deposition and resorption. Bone calcium in elderly people decreases, particularly in postmenopausal women. Calcium deficiency in children causes rickets, while low bone mass in the elderly causes osteoporosis and elevated risk of fracture (Greer *et al.*, 2006; Jackson *et al.*, 2006). Calcium is absorbed both by active absorption, which is vitamin D dependent, and in the small intestine by passive absorption.

Many human trials have been carried out, and these have confirmed that prebiotics stimulate mineral absorption. Such studies have investigated FOS, inulin, GOS and lactulose or have targeted specific groups (male and female adolescents, and postmenopausal women) (van den Heuvel *et al.*, 1999a, 1999b, 2000, 2009; Whisner *et al.*, 2013).

The mechanism underlying prebiotic stimulation of mineral absorption is not fully known; however, it may involve protonation of the minerals which increases passive absorption across cell membranes. A feeding study involving ten ileostomy subjects fed with FOS, inulin and sucrose did not show changes in mineral absorption, indicating that the effect of prebiotics occurs in the large intestine (Ellegard *et al.*, 1997).

7.3.3 *Energy intake and appetite regulation*

There is increasing interest in the potential role of the microbiome in human energy metabolism. In an acute study that used inulin as a fat replacer in a sausage patty, a significant decrease of 1521 kJ energy intake was seen ($P = 0.039$) compared to the full-fat patty; this was similar to results obtained in a study using inulin in yoghurt (Archer *et al.*, 2004; Perrigue *et al.*, 2009). Twenty-one volunteers had two consecutive days of FOS and β-glucan supplement and did not show any difference in satiety compared with the control intervention (Peters *et al.*, 2009). However, five volunteers who were on a supplementation of 16 g of FOS for 2 weeks showed an enhancement of satiation compared with another five volunteers who had placebo (Cani *et al.*, 2006).

Table 7.1 Gastrointestinal hormones.

Name	Secretion site	Produced by	Hormone signals targeting organ	Effect on food intake
Ghrelin	Stomach	Gastric oxyntic cells	Vagus and hypothalamus	Increase
Cholecystokinin (CCK)	Small intestine	I cells	Vagus and brainstem	Decrease
Glucagon-like peptide-1 (GLP-1)	Ileum and colon	L cells	Vagus and brainstem	Decrease
Polypeptide YY (PYY)	Ileum and colon	L cells	Hypothalamus	Decrease
Oxyntomodulin	Ileum and colon	L cells	Hypothalamus	Decrease

Note: Table compiled from Cummings & Overduin (2007).

It is hard to quantify satiety and satiation; however, gut hormones that regulate energy intake can be quantified. There are a number of hormones that are secreted by the GI tract that can affect energy intake and satiety. They are produced by different cells and have various functions on satiety regulation. Table 7.1 shows a list of gut hormones and how they affect satiety.

Bioactive glucagon-like peptide-1 (GLP1) has two forms: GLP1(7-37) and GLP1(7-36), which can both be obtained from the biologically inactive 37-amino-acid peptide GLP1(1-37) by cleavage of the peptide chain. Both forms of bioactive GLP1 have alanine at position 2, so they can be inactivated by dipeptidyl peptidase-4 which breaks down alanine-containing peptides. The molecules can then be cleared through the kidneys. GLP1 is produced in L cells in the gut, but the receptor for GLP1 is present in the pancreatic islets (mainly β cells), kidney, lung, heart and nervous system. In addition, GLP1 can increase gastric emptying time by communicating with the nervous system. GLP1 not only increases insulin secretion, but also enhances the storage of insulin and stimulates insulin gene expression; it can also mediate glucose levels by glucagon secretion inhibition and activation of glucose disposal sensors (Drucker, 2006). Seven days of FOS supplementation significantly increased gastric transit time and higher plasma GLP1 levels in nine gastroesophageal reflux disease patients (Piche *et al.*, 2003). A placebo-controlled study with 48 volunteers on a 12-week intervention confirmed FOS can regulate energy intake, decrease ghrelin and increase plasma peptide YY (PYY) (Parnell & Reimer, 2009).

Gut hormones involved in energy homeostasis can be regulated by prebiotics via metabolites, such as SCFAs. Infusion of SCFAs directly into rat colon increased PYY, and in pigs both PYY and GLP1 reached higher concentrations after SCFA infusion (Cherbut *et al.*, 1998; Cuche *et al.*, 2000). SCFAs also have their own receptors, which can affect energy uptake and storage. G protein receptor-40 (GPR40), GPR41 and GPR43 are also known as, respectively, free fatty acid receptor-1 (FFA1), FFA3 and FFA2. FFA1 is activated by long-chain fatty acids; however, FFA2 and FFA3 can be activated by SCFAs. Both of these SCFA receptors are expressed in a variety of human tissues, including the colon (Karaki *et al.*, 2008; Tazoe *et al.*, 2009). FFA2 is more highly expressed in immune cells, and FFA3 has the highest expression in adipose tissues (Brown *et al.*, 2003). For FFA2, the agonist strength is propionate = butyrate = acetate > valerate, while for FFA3, it is propionate = valerate = butyrate > acetate (Brown *et al.*, 2003). FFA2 is expressed in colon L cells and can activate PYY secretion, while FFA3 stimulates the secretion of leptin in adipose tissues (Xiong *et al.*, 2004; Karaki *et al.*, 2008).

Propionate and its potential for energy regulation were observed in a human study: inulin-propionate ester (10 g d^{-1}) was fed to 60 overweight adults recruited on a randomised, double-blind, placebo-controlled, parallel study. Acute measurements revealed that inulin-propionate ester significantly increased plasma concentration of PYY and GLP1 after 6 h compared to inulin; after 24 weeks of supplementation, the inulin-propionate ester group gained less weight and had lower low-density lipoprotein (LDL) cholesterol and total cholesterol compared to the inulin group (Chambers *et al.*, 2015).

7.3.4 *Lipid metabolism*

High concentrations of LDL cholesterol and triacylglycerol in the blood are risk factors for cardiovascular disease. Triacylglycerol levels can be reduced by prebiotics partly by their regulation of gut hormones and fat intake. SCFA receptor FFA2 is involved in mediating the plasma fatty acid level by inhibiting lipolysis and simulating adipogenesis (Stoddart *et al.*, 2008). Prebiotics can modulate microbiota composition with a bifidogenic effect, and anaerobic growth of lactobacilli and bifidobacteria can assimilate cholesterol with bile salts (Pereira & Gibson, 2002). In rats, prebiotic supplementation can inhibit hepatic lipogenesis by downregulating lipogenesis enzymes (Kok *et al.*, 1996). Intervention with FOS resulted in reduction of blood glucose and LDL cholesterol in 18 diabetic subjects compared to ten diabetic subjects who received sucrose (Yamashita *et al.*, 1984). Other studies, mainly on inulin-type fructans, have produced inconsistent results where reduction of LDL cholesterol or triacylglycerol was not always observed. These studies, however, targeted different groups and used different doses of prebiotics (Canzi *et al.*, 1996; Pedersen *et al.*, 1997; Davidson *et al.*, 1998).

7.3.5 *Immune function modulation of prebiotics*

An immune system that functions properly is able to protect the human body from invasion of pathogens and other antigens; however, immunity disorders can lead to problems, such as allergy and inflammatory disease. Prebiotics can modulate immune function from two perspectives: they enhance the defence against pathogenic infections and reduce unwanted inflammatory events. SCFA receptors, which were discussed in Section 7.3.4, may help to modulate immunity in the human body: FFA2 expression is highest in immune cells, which indicates SCFAs may be involved in host defences (Le Poul *et al.*, 2003).

Prebiotic defence against pathogens can prevent acute gastroenteritis or shorten its duration. Two research groups have studied two different prebiotics for their effect on travellers' diarrhoea: GOS had significant improvement on both the occurrence ($P<0.05$) and the length of travellers' diarrhoea ($P<0.05$); however, a study on FOS improved the general well-being of volunteers but did not show improvement on diarrhoea (Cummings *et al.*, 2001; Drakoularakou *et al.*, 2010). Another study, which used a synbiotic containing FOS and two strains of probiotics, also failed to find any impact on travellers' diarrhoea (Virk *et al.*, 2013). The rationale behind any effect is that prebiotics often shorten GI transit time, resulting in excretion of pathogens before they have had time to grow. Furthermore, supplementation with prebiotics boosts the growth of beneficial bacteria that compete with the pathogens.

Inflammatory bowel disease (IBD) describes two GI disorders, ulcerative colitis (UC) and Crohn's disease (CD), which present as abnormal inflammation. These disorders are believed to be related to gut microbiota dysbiosis; therefore, by modulating gut bacteria composition, prebiotics have the potential to improve well-being or even ease the symptoms of IBD. FOS (15 g daily) reduced CD activity in ten patients, as observed by Lindsay *et al.* (2006); however, there was no significant finding with another study feeding 103 patients 15 g FOS daily (Benjamin *et al.*, 2011). Fourteen UC patients and 17 CD patients received 10 g lactulose every day for 4 months, and although no disease parameter improved, quality of life scores increased with lactulose compared to before the intervention (Hafer *et al.*, 2007). The synbiotic is the probiotic '*Bifidobacterium longum*' (presumed to be *Bifidobacterium longum* subsp. *longum*) and the prebiotic Synergy 1 (a mixture of FOS and inulin) resulted in a significant improvement in disease parameters in a randomised placebo-controlled trial with 18 UC patients (Furrie *et al.*, 2005). There are not many human trials into the effects of prebiotics on IBD, and those trials that had positive results did not have high numbers of patients; therefore, more research needs to be done.

Atopic dermatitis is an allergic skin disease that mostly happens in early infancy and childhood. There is evidence that feeding infants with mixed prebiotics can reduce incidence of atopic dermatitis; however, the severity of ongoing atopic dermatitis is not improved by GOS, as indicated by the results of a randomised control trial with 107 infants (Moro *et al.*, 2006; Gruber *et al.*, 2010; Bozensky *et al.*, 2015).

7.3.6 *Colorectal cancer risk and prebiotics*

Since prebiotics can benefit human gut health in many aspects, their effects on colorectal cancer have also been investigated. Feeding colon cancer patients with prebiotics for 12 weeks did not result in any significant difference in cancer-related biomarkers compared to a control group (Rafter *et al.*, 2007). However, an observational study with a large population and a 9-year follow-up revealed that dietary fibre consumption and colorectal cancer risk were negatively correlated (Bingham *et al.*, 2003). Direct studies of the mechanism by which prebiotics may lower human colorectal cancer risk are few in number, but many *in vitro* or animal studies have been performed. Three groups of mice that had 6 weeks of supplementation with inulin, FOS or cellulose were then challenged with a carcinogen, and abnormal crypt foci were significantly lower in the prebiotic group compared to the control group (Buddington *et al.*, 2002). Prebiotics may not be able to cure colon cancer, but they show potential for reducing the risk of colorectal cancer incidence. Two mouse studies found FOS, long-chain inulin and a mixture of both could affect the onset of colon cancer by inducing apoptosis (Hughes & Rowland, 2001; Femia *et al.*, 2002). Burns and Rowland (2004) found that fermentation by probiotics and faecal genotoxic water with FOS or inulin helped human cells resist genotoxicity.

Prebiotics may reduce the genotoxicity of faecal water by modulating the metabolism of the gut microbiota. Proteolysis by gut bacteria can generate carcinogens or co-carcinogens. Fermentation of aromatic amino acids, for instance, generates indolic or phenolic compounds, which are generally considered as carcinogens. Supplementation of prebiotics can shift the microbiota to a more saccharolytic one by increasing the number of bifidobacteria and lactobacilli, thereby reducing production of such carcinogenic compounds. The primary bile acids are synthesised in the liver from cholesterol, and some of them can escape intestinal absorption to be metabolised by colonic bacteria. Secondary bile acids are produced by bacterial groups, such as bacteroides, clostridia, lactobacilli, bifidobacteria and *Eubacterium*, by deconjugation and dihydroxylation (Ridlon *et al.*, 2006). Secondary bile acids can lead to a loss of mucosal cells in the colon due to disruption of the cell membrane and, furthermore, induce hyper-proliferation of mucosal cells that helps the development of colon cancer (Nagengast *et al.*, 1995). Prebiotics can possibly reduce secondary bile acids production by changing microbial metabolism and reducing colonic transit time. Glucuronidation is one of the most important and widely present detoxification pathways in the human body. β-glucuronidase activity of some bacteria, which can deconjugate toxins, leads to a longer transit time of toxins; this leads to an increase of cytotoxicity and genotoxicity.

Prebiotics possibly exert apoptosis induction by SCFA production. In healthy subjects, colonic epithelial cells, which are derived from stem cells, start from the bottom of crypts. They move to the proliferation region and then move upwards until undergoing programmed cell death at the top of crypts; epithelial cells become increasingly more differentiated during this process towards apoptosis. SCFAs are believed to enhance cell differentiation and cell apoptosis, with butyrate being the most effective (Hague *et al.*, 1995). Butyrate can act as a histone deacetylase inhibitor, which can selectively modulate gene expressions involved with the cell cycle (Boffa *et al.*, 1978; Sambucetti *et al.*, 1999).

7.3.7 *Gut permeability*

The gut forms a semipermeable barrier preventing the translocation of antigens, proinflammatory compounds and toxins from the digestive lumen to the sterile organs and tissues; reduced gut barrier function is related to infection, carcinogenesis and other chronic diseases. Supplementation with FOS for 2 weeks did not result in improvement in gut barrier function among 34 healthy volunteers (Ten Bruggencate *et al.*, 2006). However, a human study feeding volunteers with inulin-enriched pasta for 5 weeks revealed decreased gut permeability (Russo *et al.*, 2012). An *in vitro* study indicated that a prebiotic along with a probiotic exerted better improvement of barrier function compared to the prebiotic alone (Commane *et al.*, 2005).

The mechanism behind this could be SCFA modulation of cell-signalling pathways. The colonic mucosa has a layer of cells joined by tight junctions composed of specific proteins, for instance claudins and occludin, which forms a physical barrier. Direct application of mixed SCFAs to rat intestinal wall resulted in reduced permeability, with higher concentrations of acetate exerting better effects (Suzuki *et al.*, 2008). Butyrate is believed to be an epithelial cell differentiation inducer, and differentiated mature epithelial cells have enhanced expression of tight junction proteins. Bordin *et al.* (2004) found that butyrate treatment of several cell lines resulted in an increase in tight junction protein expression.

7.3.8 *Colon motility and faecal bulking with application to constipation*

Generally, prebiotics increase faecal weight and colon motility and decrease colonic transit time which can exert benefits, such as soothing constipation. SCFAs produced by fermentation may also play crucial roles in colon motility: propionate increased muscle contraction frequency in rats that was possibly linked to the SCFA receptors FFA2 and FFA3 (Tazoe *et al.*, 2008), and reported studies on colon motility in relation to prebiotic food supplementation are shown in Table 7.2.

7.4 Inulin-type fructans as prebiotics

Inulin-type fructans, which include FOS with DP = 2–10 and inulin with DP > 10, are composed of a number of fructose residues sometimes with a glucose at the terminal: D-fructose molecules are linked by β $(2 \rightarrow 1)$ linkages and when there is a glucose, the chain is terminated by a D-glucose molecule bonded with fructose by an α $(1 \leftrightarrow 2)$ linkage (see Figure 7.1). Fructan is a generic term used for all molecules that contain one or more fructosyl–fructose links such as inulin and levan, which mainly has β $(2 \rightarrow 6)$ linkages. In this section, inulin-type fructans are discussed.

Both inulin and FOS are widely present in nature and have been in the human diet for a very long time. Due to variations in dietary habits, people consuming Western diets obtain 1–10 g of inulin-type fructans, while people consuming an American diet obtain 5.1 g on average with 2.6 g of inulin and 2.5 g of FOS (van Loo *et al.*, 1995; Moshfegh *et al.*, 1999). The occurrence of these fructans in the human diet is mainly in plant-based foods, especially onion, Jerusalem artichoke, chicory and asparagus. Table 7.3 shows a list of plant foods that are rich in inulin in the Western diet. In addition, people from India, Japan and other parts of the world also consume dahlia tuber and thistle roots, which also contain inulin-type fructans (van Loo *et al.*, 1995).

The degree of polymerisation of these plant fructans is relatively low with a maximum DP < 200 and they also tend to be less branched, whereas fructans from bacterial origin can be both highly branched and highly polymerised with a maximum DP of up to 100 000 (Roberfroid, 2005). The food industry uses chicory root to produce inulin and its derivatives because the dry weight of chicory root contains >70% inulin (van Loo *et al.*, 1995).

Table 7.2 Human studies of colon motility and faecal bulking with prebiotic supplementation.

Prebiotics and control or placebo	Duration of the treatment	Targeted group and subject numbers	Results	References
Low RS (5 g) and high RS (39 g)	3 weeks – a crossover study	11 healthy human subjects	Increased faecal weight	Phillips *et al.* (1995)
GOS (9 g d^{-1})	2 weeks – a crossover study	14 elderly females suffering constipation	Increased defecation frequency	Teuri & Korpel (1998)
GOS (15 g d^{-1})	2 weeks – before and after treatment	12 healthy human subjects	Increased defecation frequency	Teuri *et al.* (1998)
Control diet, wheat bran diet and wheat bran+RS diet	3 weeks – a crossover study	12 healthy human subjects with family history of colorectal cancer	Increased faecal weight and defecation frequency	Muir *et al.* (2004)
Inulin (13 g d^{-1})	3 weeks – a crossover study	15 wheelchair-bound adults	Increased faecal weight and no difference in defecation frequency	Dahl *et al.* (2005)
Inulin (20 g) and resistant maltodextrin	20 d – a placebo-controlled parallel study	32 constipation sufferers	Increased defecation frequency and improvement of constipation	Roman *et al.* (2008)
Polydextrose (8 g d^{-1})	3 weeks – a placebo-controlled parallel study	45 healthy human subjects	Increased defecation frequency and no difference in faecal weight	Hengst *et al.* (2009)
Inulin (20 g d^{-1})	3 weeks – a crossover study	12 healthy male subjects	No difference in both faecal weight and defecation frequency	Slavin & Feirtag (2011)
Inulin and guar gum mixture (15 g)	3 weeks – a placebo-controlled parallel study	60 female constipation sufferers	Increased defecation frequency but no difference between control and prebiotic groups	Linetzky Waitzberg *et al.* (2012)

GOS = Galactooligosaccharides; RS = resistant starch.

7.4.1 *Determination of inulin-type fructans*

Both capillary gas chromatography (CGC) and high-pressure liquid chromatography (HPLC) can be used for the determination of short-chain fructans, such as FOS. Sample preparation for HPLC is quick and straightforward as diluted samples can be injected into the HPLC. van Loo *et al.* (1995) used two packed Aminex HPX87K columns with water (H_2O; pH 9.5 adjusted by potassium hydroxide, or KOH) as eluent, and the separation was performed at 85 °C. Sample preparation for CGC included adding internal standard, dilution with water, drying, formation of oxime by hydroxylamine-HCl, derivatisation with trimethylsilylimidazole (TSIM), extraction of silylated fructans by addition of isooctane, and centrifugation; separation was performed by a capillary column together with a flame-ionisation detector with an helium gas flow of 9 mL min^{-1} (van Loo *et al.*, 1995). The American Official Association of Chemists (AOAC) validated

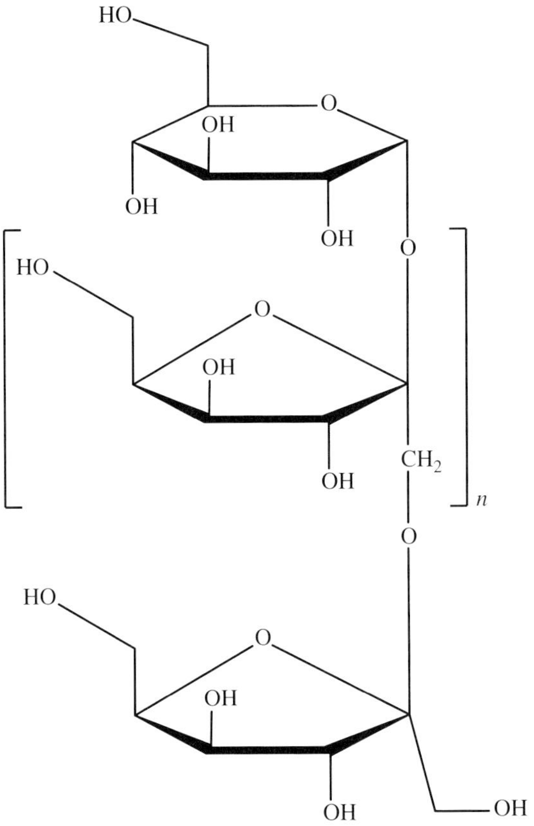

Figure 7.1 The chemical structure of fructooligosaccharides (n = 2–9) and inulin (n > 9).

Table 7.3 Inulin-rich plant foods in the Western diet and their inulin content.

Name	Inulin content (100 g^{-1})	Degrees of polymerisation of its inulin	Characteristics
Onion	1.1–7.5 g	2–12	Depolymerises during storage
Asparagus	2–3 g	–	–
Jerusalem artichoke	16–20 g	2–50	–
Leek	3 g	12	–
Rye flour	0.5–1 g	–	Resistant to baking
Garlic	9.8–16 g	2–50	–
Dandelion	12–15 g	–	–
Artichoke globe	2.5–9.5 g	≥19	High DP range
Barley	0.5–1 g	–	–
Banana	0.3–0.7 g	–	–
Wheat	1–4 g	50% of its inulin ≤5	Resistant to baking
Chicory root	15–20 g	2–65	Resistant to roasting: >70% of inulin still present after roasting

– = Data not reported; DP = degrees of polymerisation.
Data adapted from van Loo *et al*. (1995).

two methods to quantify fructans: AOAC method 997.08 [analysis by high-performance anion exchange chromatography with pulsed amperometric detection (HPAEC-PAD)] and AOAC method 999.03 (spectrophotometry). Both depend on enzymatic treatment to hydrolyse the polysaccharide and release sugars for analysis. Method 997.08 of the AOAC needs three chromatography readings from one sample to quantify fructans: the quantity after direct hot water extraction, the quantity after amyloglucosidase treatment of the first extraction and the quantity after fructozyme treatment of the second extraction (Hoebregs, 1997). AOAC method 999.03 contains two enzymatic treatments: hydrolysis of starch and sucrose by a mixture of enzymes followed by reduction with borohydride, then hydrolysis of fructans by a fructanase mixture followed by spectrophotometric determination of the reducing sugars with *para*-hydroxybenzoic acid hydrazide (McCleary *et al.*, 2000).

7.4.2 *Production of inulin-type fructans*

As discussed, chicory roots are the main raw material for inulin production on an industrial scale. The extraction of inulin from chicory roots is carried out by diffusion in hot water, purification and then spray drying, in a process similar to extraction of sucrose from sugar beet. The inulin produced by simple extraction is native inulin or standard inulin which has DP from 2 to 65 and an average DP of 12, such as Orafti® ST (Beneo, Belgium). Native inulin can be refined by physical separation to obtain high polymerised inulin (inulin HP) with DP from 10 to 60 and an average DP of 25, such as Orafti HP.

The partial enzymatic hydrolysis of inulin can produce FOS with DP 2–7 and an average DP of 4, by means of an endoinulinase (EC 3.2.1.7) or an exo-inulinase together with the endoinulinase (EC 3.2.1.80) (Roberfroid, 2005, 2007). FOS can also be synthesised from sucrose by fructosyltranferases (EC 3.2.1.99 and EC 3.2.1.100), which catalyse the transfer of fructofuranosyl residues from sucrose (Gibson & Rastall, 2006). The transfer of fructofuranosyl residue from sucrose to sucrose, the growing fructan chain or water can produce GF2 (where 'G' represents glucose, 'F' represents fructose and '2' represents the number of fructose moieties), extend the fructan chain or break down sucrose, respectively. The yield of FOS from synthesis is around 55–60%, and separation is needed to obtain purified FOS (Gibson & Rastall, 2006).

Long-chain inulin and short-chain FOS may exert different health benefits; therefore, Beneo provide a product known as Orfati® Synergy 1, which is a mixture of short-chain FOS and long-chain inulin HP.

7.4.3 *Physical and chemical characteristics of inulin-type fructans and application in the food industry*

Chicory root inulins are white odourless powders, while FOS can be a powder obtained by spray drying or a viscous syrup with 75 g 100 g^{-1} dry matter obtained by evaporation (Franck, 2002). Inulin-type fructans do not possess any off-flavour or aftertastes; FOS and native inulin have a slightly sweet taste due to the small amount of glucose, fructose and sucrose present. Water solubility negatively correlates with inulin chain length: at

25 °C, more than 75% of FOS can dissolve in water, whereas only 2.5% of inulin HP can dissolve in water. Although inulin and FOS can resist general food processing, such as heat, the linkage between fructoses can break down at very low pH. Dissolving inulin in water or any other liquid results in a gel with a creamy structure that can be used as a fat replacer (Franck, 2002). Inulin can support most gelling agents, stabilise foam and emulsions and improve the taste and texture of bakery and cereal products (Franck, 2002). It has been used in many foods, such as baked goods, baby food, beverages, breakfast cereals, candy, dairy products, frozen desserts, soups, sauces and table spreads.

7.4.4 *Prebiotic effects of inulin-type fructans*

One of the main characteristics of prebiotics is their resistance to digestion in the GI tract. Humans and other mammals lack the enzyme to break down the β (2 → 1) linkages within inulin and FOS. Human studies have confirmed that most inulin and FOS are not absorbed by the small intestine: 88% of FOS and 89% of inulin were recovered in the ileostomy effluent of ten ileostomy patients (Ellegard *et al.*, 1997). However, gut bacteria do possess enzymes that hydrolyse β (2 → 1) linkages, and McKellar and Modler (1989) found that three strains of bifidobacteria have cell-associated β-fructosidases that hydrolyse inulin-type fructans, although different strains may have different preference for chain length. Enrichment culture of different species of bifidobacteria also revealed their growth by metabolisation of FOS (Gibson & Wang, 1994). Twenty-eight strains of lactobacilli, bifidobacteria and *Streptococcus thermophilus* were examined in MRS agar together with FOS: 12 out of 16 strains of lactobacilli, six out of seven strains of bifidobacteria and none of the *Str. thermophilus* strains were capable of utilising FOS (Kaplan & Hutkins, 2000). Bifidogenic effects of inulin and FOS were observed by Wang and Gibson (1993) in an *in vitro* single-vessel fermentation simulating the human colon, which was achieved by inoculation with human faecal slurries. A number of human studies have been carried out with different dosages, treatment durations, target groups and numbers of volunteers to investigate the microbiota composition changes associated with inulin or FOS supplementation. All of the studies found there was an elevated bifidobacterial count, which in some cases was significant (Hidaka, 1991; Williams *et al.*, 1994; Gibson *et al.*, 1995; Buddington *et al.*, 1996; Kleessen *et al.*, 1997; Bouhnik *et al.*, 1999; Kruse *et al.*, 1999). Gibson *et al.* (1995) put eight volunteers on a controlled diet for 45 d, feeding them with 15 g sucrose for the first 15 d, then 15 d with 15 g FOS. Four volunteers continued to complete another intervention for 15 d with 15 g inulin. The authors did not observe any change in total faecal bacteria; however, both FOS and inulin boosted growth of bifidobacteria significantly. Furthermore, decreased levels of bacteroides, clostridia and fusobacteria were seen in this diet-controlled study.

7.4.5 *Health benefits of inulin-type fructans*

Inulin-type fructans are the best studied prebiotics, and many human studies have been carried out to investigate various health benefits, such as increased mineral absorption, effect on energy regulation, effect on lipid metabolism, improvement of immunity, potential to prevent colorectal cancer and gut function improvement.

Most of the mineral absorption studies have focused on calcium intake with an emphasis on girls and postmenopausal women. Supplementation of inulin or FOS in postmenopausal women resulted in increased absorption and bone health (based on bone turnover markers and other biomarkers) in different studies using various methods, dosages of inulin or FOS and products (Tahiri *et al.*, 2003; Holloway *et al.*, 2007; Adolphi *et al.*, 2009; Slevin *et al.*, 2014; Kruger *et al.*, 2015). Inulin-type fructans have also been shown to increase calcium absorption in adolescents, both male and female (van den Heuvel *et al.*, 1999b; Griffin *et al.*, 2002, 2003; Abrams *et al.*, 2005, 2007a). However, one study showed contradictory results: 10 g d^{-1} of FOS for 36 d did not reveal any improvement in calcium absorption, but subjects in this study were girls with low calcium intake (van den Heuvel *et al.*, 2009). A further two studies have been carried out in healthy adults and these resulted in improvement of calcium absorption (Coudray *et al.*, 1997; Abrams *et al.*, 2007b). Enhancement of magnesium absorption by inulin-type fructans was observed in both postmenopausal women and young girls (Tahiri *et al.*, 2001; van den Heuvel *et al.*, 2009).

As discussed in Section 7.3.4, satiety and energy intake regulation are other possible health benefits from prebiotics. Inulin-type fructans showed improvement of satiation in some human intervention studies, and some studies have measured hormones which are related to energy regulation and revealed enhancement of satiation (Cani *et al.*, 2006; Antal *et al.*, 2008; Parnell & Reimer, 2009; Tarini & Wolever, 2010; Russo *et al.*, 2011). Although inulin and FOS showed potential to regulate energy intake, 8 g d^{-1} of FOS for 12 weeks failed to reduce weight in overweight and obese children (Liber & Szajewska, 2014); therefore, the long-term effects on weight management need further investigation.

Lipid-lowering effects of inulin-type fructans were studied in 18 subjects with diabetes: reductions of blood glucose, total cholesterol and LDL cholesterol were seen in those who received 8 g d^{-1} of FOS for 14 d (Yamashita *et al.*, 1984). Two studies with 4-week interventions of FOS or inulin did not find any difference in the lipid profiles of healthy subjects (Luo *et al.*, 1996; Pedersen *et al.*, 1997). Seventeen healthy volunteers had either placebo or a mixture of inulin and FOS for 6 months, and this resulted in only a trend for total cholesterol and LDL cholesterol reduction in the inulin study group (Forcheron & Beylot, 2007). Many other studies in different target groups, such as healthy subjects, subjects with hypercholesterolemia and individuals with type 2 diabetes, found that inulin significantly improved lipid profile, at different doses, study durations and monitored parameters (Davidson *et al.*, 1998; Brighenti *et al.*, 1999; Jackson *et al.*, 1999; Causey *et al.*, 2000; Letexier *et al.*, 2003; Russo *et al.*, 2008; Dehghan *et al.*, 2013).

Elderly people tend to have weaker immune function, hence there is an interest in improving their health by prebiotic supplementation. Nineteen elderly individuals were recruited in a study supplementing 8 g d^{-1} of FOS for 3 weeks, and their immunity was compared before and after the intervention. There was a reduction of phagocytic activity after the treatment; however, a decrease of interleukin-6 (IL6) was observed with FOS supplementation (Guigoz *et al.*, 2002). Another crossover study had 43 elderly subjects being given a synbiotic comprising the probiotic '*Bif. longum*' (presumed to be *Bif. longum* subsp. *longum*; Mattarelli *et al.*, 2008) and

FOS-enriched inulin. Proinflammatory cytokines were significantly lower in the synbiotic treatment group (Macfarlane *et al.*, 2013). This inhibition of proinflammatory cytokines was also seen in healthy adult volunteers aged 18–24 supplemented with a combination of XOS and inulin (Lecerf *et al.*, 2012). FOS-enriched inulin improved seasonal influenza vaccination in middle-aged humans, as seen by increased antibody titres after vaccination (Lomax *et al.*, 2015). However, supplementation of FOS in infant cereal did not show any change in subjects' immune function (Duggan *et al.*, 2003).

A synbiotic featuring '*Bif. longum*' (presumed to be *Bif. longum* subsp. *longum*) and FOS-enriched inulin not only reduced UC patients' inflammatory cytokines but also improved epithelial tissue regeneration (Furrie *et al.*, 2005). Studies on CD have not shown any change (Lindsay *et al.*, 2006; Benjamin *et al.*, 2011). Apart from these studies on inflammatory bowel diseases, one study showed that intervention with 24 g d^{-1} of inulin for 3 weeks decreased inflammation in patients who had ileal pouch–anal anastomosis (Welters *et al.*, 2002).

Some animal models have suggested that inulin can reduce cancer risk, and Pool-Zobel (2005) reviewed data from animal models that investigated the effects of inulin-type fructans on colorectal cancer risk. Inulin-type fructans can reduce faecal water genotoxicity and reduce secondary bile acid production in humans (Klinder *et al.*, 2004; Boutron-Ruault *et al.*, 2005). However, experimental results with human colorectal cancer patients are controversial: different studies have monitored genotoxicity, proliferation, immune parameters or aberrant crypt foci. Only a limited number of studies with inulin-type fructans have shown significant changes in colorectal cancer patients (Rafter *et al.*, 2007; Roller *et al.*, 2007; Limburg *et al.*, 2011).

7.5 Galactooligosaccharides as prebiotics

Studies of infant faecal microbiota composition found different patterns depending on feeding regime: (a) bifidobacteria tend to be higher in breastfed infants compared to formula-fed or formula–breast mixed-fed infants, (b) clostridia and some facultative microbes are lower in breastfed infants, and (c) proteolytic metabolism is less active in breastfed infants (Stark & Lee, 1982; Benno *et al.*, 1984; Mevissenverhage *et al.*, 1987; Harmsen *et al.*, 2000; Heavey *et al.*, 2003). This led to the identification and isolation of bifidogenic compounds in human milk. Human milk contains more oligosaccharides compared to cow's milk, and some of them have a galactose–glucose structure (Kobata & Ginsburg, 1969, 1972; Yamashita & Kobata, 1974; Yamashita *et al.*, 1976a, 1976b, 1977a, 1977b). The ability of GOS to act as a substitute for these human oligosaccharides has attracted interest in further researching their prebiotic and health-beneficial effects.

Galactooligosaccharide or *trans*-GOS are oligosaccharides composed of a number of galactose monomers with a glucose molecule at the terminus, with DP 3–10 and disaccharides comprising two galactose monomers. Galactoses are often linked by β $(1 \rightarrow 4)$ and β $(1 \rightarrow 6)$ linkages, while β $(1 \rightarrow 2)$ and β $(1 \rightarrow 3)$ linkages occur less frequently in GOS (Figure 7.2).

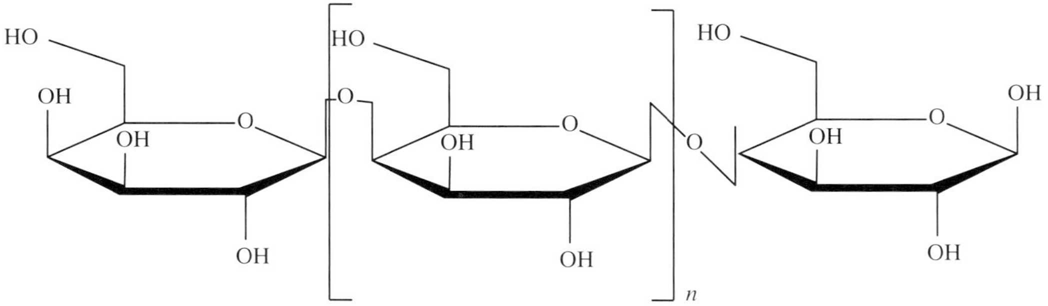

Figure 7.2 Chemical structure of galactooligosaccharides (n = 1–8).

7.5.1 *Production and determination of galactooligosaccharides*

The GOS are synthesised by transgalactosylation in an enzymatic catalysis from lactose by β-galactosidase (EC 3.2.1.23). β-Galactosidase catalyses the transfer of a galactosyl residue from lactose to a sugar chain or water, and can either extend the sugar chain and produce GOS or break down lactose, respectively. During enzymatic transgalactosylation, the GOS production rate is influenced by the enzyme source, substrate concentration and reaction conditions (temperature and pH). Generally, GOS production is favoured by high substrate concentration, although different enzyme sources have different reaction condition preferences and can form different glycoside linkages.

Various organisms can produce β-galactosidase, including: *Aspergillus oryzae*, *Kluyveromyces marxianus* subsp. *lactis* and *Cryptococcus laurentii* in eukaryotic organisms; *Sulfolobus solfataricus* and *Pyrococcus furiosus* in Archaea; and *Bifidobacterium* spp., *Lactobacillus* spp., *Bacillus* spp., *Str. thermophilus*, *Escherichia coli* and *Enterobacter cloacae* in bacteria. Table 7.4 shows the main characteristics of some GOS products in the market.

The AOAC published a validated method (method 2001.02) to determine GOS by HPAEC-PAD. This method requires enzymatic treatment with β-galactosidase to hydrolyse the oligosaccharides and release sugars for analysis. Before the enzymatic reaction, an initial solution is taken and heated to 100 °C for 10 min to deactivate the enzyme. This is then analysed by HPAEC-PAD to determine the concentration of lactose and monosaccharides. The remaining solution with active β-galactosidase is incubated at 60 °C for 30 min, and then sugar analysis is performed by HPAEC-PAD (Slegte, 2002). GOS content can be calculated by subtracting the initial galactose and lactose concentrations from the final galactose concentration. This method needs efficient deactivation of β-galactosidase in the initial solution to avoid underestimating GOS concentration.

7.5.2 *Application of galactooligosaccharides in the food industry*

GOS is categorised as generally regarded as safe (GRAS) (Boudry *et al.*, 2013) in the USA, and it is regarded as a Food for Specific Health Use (FOSHU) in Japan. It has been used in infant formula to enhance bifidobacteria growth in an attempt to provide a functional mimic to human milk oligosaccharides (Boehm *et al.*, 2002). GOS is slightly sweet without aftertaste, is resistant to heat and can retain moisture. It is also not digested or absorbed by the human small intestine; therefore, it can be used as a sugar substitute that has reduced calories. Examples of GOS application in dairy products are fermented milk, lactic acid bacteria beverages, ice cream and milk beverages. Apart from dairy products, GOS can be used in beverages, sweets, dessert, bakery, jams and other food products.

7.5.3 *The prebiotic effect of galactooligosaccharides*

The non-digestibility of GOS was demonstrated by van Loo *et al.* (1999), and they concluded that more than 90% of GOS can be recovered in the colon. However, there is no *in vivo* human study showing non-digestibility, which is normally done by analysing digesta from ileostomy volunteers after feeding the prebiotic.

Table 7.4 Commercially available galactooligosaccharides products.

Name	Format	GOS content (g 100 g^{-1}, dry weight)	Enzyme source	Glycoside bond	References
Cup Oligo H-70	Syrup	≥70	*Cryptococcus laurentii*	β (1→4)	Ohtsuka *et al*. (1990)
Cup Oligo P	Powder	≥70	*Cry. laurentii*	β (1→4)	Ohtsuka *et al*. (1990)
OLIGOMATE® 55N	Syrup	≥55	*Sporobolomyces singularis* and *Kluyveromyces marxianus* subsp. *lactis*	β (1→4) and β (1→6)	Asp *et al*. (1980) and Gorin *et al*. (1964)
OLIGOMATE 55NP	Powder	≥55	*Spo. singularis* and *K. marxianus* subsp. *lactis*	β (1→4) and β (1→6)	Asp *et al*. (1980) and Gorin *et al*. (1964)
Vivinal® GOS	Syrup	≥59	*Bacillus circulans*	β (1→4)	Yanahira *et al*. (1995)
	Syrup (easy-drying syrup)	≥72			
	Powder	≥69			
Bimuno	Syrup	≥57	*Bifidobacterium bifidum*	β (1→3)	Rabiu *et al*. (2001)
	Powder	≥80			
Purimune™ BIOLIGO GL 5700 IMF GOS	Syrup	65	*B. circulans*	β (1→4)	Yanahira *et al*. (1995)
Floraid® GOS Syrup	Syrup	39	*Aspergillus oryzae*	β (1→6)	Toba *et al*. (1985)
Floraid GOS	Powder	39	*A. oryzae*	β (1→6)	Culhane & Tanugraha (2013)

GOS = galactooligosaccharides.

Tanaka *et al.* (1983) carried out a human study to observe the changes in faecal microbiota composition associated with GOS supplementation: *Bifidobacterium* spp. increased while *Bacteroidaceae* spp. decreased. Other human studies have shown similar results: that GOS can modulate human colonic bacteria by increasing bifidobacteria (Ito, 1993; Bouhnik *et al.*, 1997). One study that recruited 37 volunteers (who were more than 50 years old) revealed an increase in faecal *Bifidobacterium* spp. after a 3-week intervention, of the ten bacteria groups that were monitored (Walton *et al.*, 2012). A study with elderly people and another with overweight volunteers both found an increase of bifidobacteria, and decrease of *Bacteroides* spp., *Clostridium histolyticium* and *Desulfovibrio* spp. (Vulevic *et al.*, 2008, 2013). Consuming 2.5 g d^{-1} of GOS can affect microbiota changes within one week. A parallel study with eight volunteers in each group that tested interventions with 2.5 g, 5 g, 7.5 g and 10 g d^{-1} of GOS did not show any dose-dependent effect (Bouhnik *et al.*, 2004). However, another study with 18 volunteers consuming different doses of GOS for 3 weeks showed a dose-dependent change of faecal bifidobacteria (Davis *et al.*, 2010). GOS produced from various enzymatic sources differ in their bifidogenic effect. A crossover study with 59 healthy human subjects found intake of GOS produced from *Bifidobacterium* spp. was more effective for enhancing bifidobacteria growth (Depeint *et al.*, 2008).

7.5.4 *Infant nutrition and galactooligosaccharides*

Infants are free of bacteria before delivery. The infant gut is first inoculated with microbes resulting from contact with the mother during delivery. The maternal microbiota is transferred to the infant during natural delivery; therefore, modulation of the mother's microbiota while pregnant should help infant health by passing on a healthier microbiota. Sixteen expectant mothers received prebiotic treatment with 9 g d^{-1} of 9:1 of GOS–FOS for 15 weeks before their delivery; these women showed higher number of faecal bifidobacteria compared to the placebo group ($n = 17$). However, bifidobacteria of the neonates did not differ significantly between the two groups (Shadid *et al.*, 2007).

Breast and formula feeding shape the infant microbiome in different ways: bifidobacteria tend to be higher in breastfed infants compared to formula-fed infants. After the introduction of solid food, children acquire more microbes and the microbial ecology of the gut starts to become similar to that of adults. Before this, bifidobacteria are the dominant bacterial group in breastfed infants. It is believed that human milk oligosaccharides play an important role in bifidobacterial colonisation in the infant colon. Researchers have investigated the addition of prebiotics to infant formula with the aim of boosting bifidobacteria and exerting health benefits to the infants. The most widely studied prebiotic combination for infant formula is a mixture of 90% GOS and 10% FOS. This combination has both low-molecular-weight GOS and inulin, which is believed to have a similar health benefit as human milk oligosaccharides (Boehm *et al.*, 2002).

As many as 1032 infants have been enrolled in various studies investigating infant tolerance of GOS with diverse targeted groups, such as premature infants, full-term infants and infants born from HIV-positive mothers. Safety and tolerance were evaluated in terms of weight gain, length gain, digestive tolerance and adverse events, and no safety issues were found with GOS or a GOS–FOS combination as supplement in infant formula (Boehm *et al.*, 2002; Chouraqui *et al.*, 2008; da Costa Ribeiro *et al.*, 2015; Fanaro *et al.*, 2009; Holscher *et al.*, 2012; Lee *et al.*, 2015; Ribeiro *et al.*, 2012).

Studies investigating infant formula, prebiotic-supplemented formula and human milk showed a higher number of bifidobacteria, improved stool consistency and lower faecal pH in the infants fed the prebiotic formula or human milk compared to the infants given non-supplemented formula. The organic acid profile of the infant's faecal water was also different, with higher concentrations of acetate and lactate with the prebiotic formula and human milk–fed infants (Bakker-Zierikzee *et al.*, 2005; Ben *et al.*, 2008; Boehm *et al.*, 2002; Knol *et al.*, 2005).

Apart from GI function, other studies have focused on other health benefits of prebiotics given to infants, such as improvement of lipid profile. Alliet *et al.* (2007) investigated lipid profile differences among infants who were breastfed, fed prebiotic-supplemented formula or fed non-supplemented formula; they found that total cholesterol and LDL were higher in breastfed infants compared to the formula-fed infants. There was no difference between two different formulae in terms of the infants' lipid profile.

Many studies have looked at various parameters regarding the prebiotic effects on immune function in infants. Faecal secretory immunoglobulin A (IgA) levels were higher in 19 infants who had 6 g L^{-1} 9:1 GOS–FOS supplemented formula for 16 weeks compared to 19 infants given control formula, indicating a better host defence against pathogens (Bakker-Zierikzee *et al.*, 2006). The feeding of 8 g L^{-1} 9:1 GOS–FOS formula to infants for 6 months did not affect their diphtheria–tetanus–polio (DTP) vaccination response; however, a significant reduction of immunoglobulins relating to cow's milk allergy was observed with prebiotic formula (van Hoffen *et al.*, 2009). Fifty-five new-born premature infants, who consumed prebiotic supplement for 30 d, had the same level of proinflammatory cytokines as the control group (Westerbeek *et al.*, 2011). Without a control formula group, Bocquet *et al.* (2013) failed to find any significant difference between a prebiotic-fed group and a probiotic-fed group (*Bifidobacterium animalis* subsp. *lactis*) in terms of infection incidence. Another recent study did not find any significant difference in IgA secretion, infection incidence and allergic incidence between infants who had GOS until they were one year old, when compared to a control group (Sierra *et al.*, 2015). A summary of human studies of infant health feed containing GOS is shown in Table 7.5.

7.5.5 *Health benefit of galactooligosaccharides*

Rat studies on GOS found enhanced absorption of calcium, iron, magnesium and zinc; furthermore, two studies that monitored bone health parameters confirmed improvement of bone health with GOS supplementation (Chonan *et al.*, 1995, 1996; Chonan & Watanuki, 1996; Takasugi *et al.*, 2013; Weaver *et al.*, 2011). Unlike with inulin, the effect of GOS supplementation on mineral absorption is not well studied in human trials. Healthy males are not the main target group for mineral absorption improvement.

Table 7.5 Some reported human studies of infant health with galactooligosaccharides supplementation.

Treatments and dosage	Formula and subject numbers	Targeted group	Duration	Results	References
10 g L^{-1} 9:1 GOS/FOS	Prebiotic formula ($n=15$) Control formula ($n=15$) Human milk ($n=12$)	Premature infants	28 d	Higher number of bifidobacteria; improved stool consistency	Boehm *et al.* (2002)
6 g L^{-1} 9:1 GOS/FOS or *Bifidobacterium animalis* BB-12[1] (6×10^{10} colony-forming units (cfu) L^{-1})	Prebiotic formula ($n=19$) Probiotic formula ($n=19$) Control formula ($n=19$) Human milk ($n=63$)	New-born infants starting the study at day 5 after delivery	16 weeks	Higher acetate and lactate; lower pH	Bakker-Zierikzee *et al.* (2005)
8 g L^{-1} 9:1 GOS/FOS	Prebiotic formula ($n=15$) Control formula ($n=19$) Human milk ($n=19$)	Infants with average age of 7.7 weeks at enrolment	6 weeks	Higher number of bifidobacteria; higher acetate and lactate; lower pH	Knol *et al.* (2005)
4.5 g d^{-1} 9:1 GOS/FOS	Prebiotic formula ($n=11$) Placebo ($n=9$)	Weaning infants aged 4–6 months	6 weeks	Higher number of bifidobacteria	Scholtens *et al.* (2006)
2.4 g L^{-1} GOS	Prebiotic formula ($n=37$) Prebiotic formula and human milk ($n=58$) Human milk ($n=24$) Control formula ($n=45$)	Term infants	3 months	Higher number of bifidobacteria and lactobacilli; higher acetate; lower pH	Ben *et al.* (2008)

(*Continued*)

Table 7.5 (Continued)

Treatments and dosage	Formula and subject numbers	Targeted group	Duration	Results	References
5 g L^{-1} GOS	Prebiotic formula ($n = 77$) Control formula ($n = 82$)	Infants aged 4–6 months	12 weeks	Higher number of bifidobacteria; improved stool consistency	Fanaro *et al.* (2009)
4 g L^{-1} 1:1 PDX/GOS, 4 g L^{-1} 3:2:1 PDX/GOS/LOS and 8 g L^{-1} 3:2:1 PDX/GOS/LOS	Prebiotic formula ($n = 27$) Prebiotic formula ($n = 27$) Prebiotic formula ($n = 25$) Human milk ($n = 30$) Control formula ($n = 25$)	Term infants	28 d	Human milk group had better stool consistency. No other significant difference was observed.	Nakamura *et al.* (2009)
4 g L^{-1} 9:1 GOS/FOS	Prebiotic formula ($n = 53$) Control formula ($n = 59$) Human milk ($n = 57$)	Infants aged no more than 30 d at enrolment	Until infants reach 4 months old	Higher number of bifidobacteria; improved stool consistency	Vivatvakin *et al.* (2010)
9:1 GOS/FOS	Prebiotic formula ($n = 36$) Control formula ($n = 33$) Human milk ($n = 33$)	Term infants aged 2–8 weeks at enrolment	6 weeks	Higher number of bifidobacteria; lower pH; no difference with stool consistency	Holscher *et al.* (2012)
4 g L^{-1} 1:1 PDX/GOS	Prebiotic formula ($n = 78$) Control formula ($n = 81$) Human milk ($n = 71$)	Infants aged 21–30 days at enrolment	60 d	Higher number of bifidobacteria	Scalabrin *et al.* (2012)

4 g L^{-1} GOS	Prebiotic formula ($n=83$) Control formula ($n=80$) Human milk ($n=199$)	New-born infants starting the study at day 15 after delivery	Before complementary feeding	Higher number of bifidobacteria and lower number of clostridia	Giovannini *et al.* (2014)
3 g L^{-1} GOS	Prebiotic formula ($n=9$) Control formula ($n=13$)	Term infants	2 weeks	Higher number of bifidobacteria; no difference with SCFAs and faecal pH	Matsuki *et al.* (2016)

[1] '*Bifidobacterium animalis* BB-12' (presumed to be *Bifidobacterium animalis* subsp. *lactis* BB-12; Anonymous, 2013).
FOS = fructooligosaccharides; GOS = galactooligosaccharides; LOS = lactulose; PDX = polydextrose; SCFAs = short-chain fatty acids.

Two studies with females of different age ranges found mineral absorption enhancement associated with GOS: ingesting 20 g d^{-1} for 9 days significantly increased true calcium absorption in 12 postmenopausal women in a crossover design study (van den Heuvel *et al.*, 2000). Another crossover study looked at how two different doses of GOS (5 g and 10 g d^{-1} for 3 weeks) affected calcium absorption in adolescent girls compared to the control; GOS improved calcium absorption significantly without any dose-dependent effect (Whisner *et al.*, 2013). Though mineral absorption was analysed in these studies, no further analysis on bone mass density and other bone health parameters was done to confirm the health benefit.

Lipid profile improvement associated with intervention with 5.5 g d^{-1} of GOS for 12 weeks was seen in overweight adult subjects, but not in generally healthy young adults who had 15 g d^{-1} of GOS (van Dokkum *et al.*, 1999; Vulevic *et al.*, 2008). Various studies with healthy subjects, overweight subjects, UC patients and elderly subjects have investigated the effect of GOS on immune function in terms of reduction of pro-inflammatory cytokines, improvement of host defence against pathogens and improvement in the well-being of chronic inflammatory disease patients (see Table 7.6).

7.6 Resistant starch and other glucose-based non-digestible carbohydrates

Resistant starch (RS) refers to those types of starch that are not hydrolysed and absorbed in the small intestine. Type 1 RS is physically surrounded by other material that makes digestion impossible; type 2 RS represents natural uncooked starches, such as potato starch, green banana starch and high-amylose maize starch; type 3 RS is retrograded amylose and starch; and type 4 RS is chemically modified starches.

The non-digestibility of RS was confirmed by Englyst *et al.* (1996) with nine ileostomy subjects: more than 90% of tested RS 2 and RS 3 were recovered in ileostomy effluent. Microbiota changes associated with RS have not been conclusive in human intervention studies, which may be due to the different RS types having different physiological effects. Two recent studies using RS 4 as supplementary treatment did not find any bifidobacterial changes; one of them found significantly higher numbers of *Bacteroides* spp. and *Ruminococcus* spp. with RS 4 (Dahl *et al.*, 2016; Upadhyaya *et al.*, 2016). An increase of *Ruminococcus* spp. was also seen in another two human intervention studies with RS 2 (Abell *et al.*, 2008; Venkataraman *et al.*, 2016). Twenty healthy young adults who had RS 2 for 3 weeks, and 24 volunteers who had RS 3, had higher numbers of faecal bifidobacteria after these treatments (Costabile *et al.*, 2016; Venkataraman *et al.*, 2016). RS has been studied in numerous human intervention trials, with an emphasis on insulin sensitivity, glycaemic homeostasis, appetite, satiety and weight management; however, many of these studies did not investigate if health benefits correlated with any change in the gut microbiota.

Table 7.6 Human studies of adult health with galactooligosaccharides supplementation.

Treatments and dosage	Study design and subject numbers	Targeted group	Duration	Result	References
15 g d^{-1} inulin 15 g d^{-1} fructooligosaccharide (FOS) 15 g d^{-1} GOS	Crossover study ($n = 12$)	Healthy males with an average age of 23	3 weeks	Inulin and GOS increased faecal acetate concentration; inulin increased faecal valerate concentration; inulin and FOS decreased secondary bile acids concentration; GOS and inulin decreased β-glucuronidase activity; no significant difference was found with lipid profile.	van Dokkum *et al.* (1999)
5.6 g d^{-1} GOS	Crossover study ($n = 44$)	Elderly subjects	10 weeks	Higher number of bifidobacteria; increased natural killer (NK) cell activity and anti-inflammatory cytokines; decreased proinflammatory cytokines; no difference with lipid profile	Vulevic *et al.* (2008)
Probiotic strains (*Lactobacillus rhamnosus* GG, *Lb. rhamnosus* LC705, *Propionibacterium freudenreichii* subsp. *shermanii* JS, and *Bifidobacterium breve* BB-99) [2×10^{10} colony-forming units (cfu) d^{-1}] and 3.8 g d^{-1} GOS 120 g d^{-1} Whole grain rye bread	Sequential intervention ($n = 18$)	Healthy males	6 weeks	Probiotic lactobacilli, propionibacteria and GOS increased the counts of bifidobacteria and decreased β-glucuronidase activity.	Kekkonen *et al.* (2011)
Bif. breve (3×10^{9} cfu d^{-1}) and 5.5 g d^{-1} GOS	Parallel study ($n = 44$)	Ulcerative colitis patients	1 year	Colonoscopy showed better condition; decreased myeloperoxidase indicating decreased severity of ulcerative colitis (UC); decreased Bacteroidaceae and faecal pH.	Ishikawa *et al.* (2011)
8 g d^{-1} GOS	Crossover study ($n = 39$)	Healthy subjects aged more than 50	3 weeks	Increased counts of bifidobacteria; no difference of faecal water genotoxicity.	Walton *et al.* (2012)

(*Continued*)

Table 7.6 (Continued)

Treatments and dosage	Study design and subject numbers	Targeted group	Duration	Result	References
5.5 g d^{-1} GOS	Crossover study ($n = 45$)	Overweight adults	12 weeks	Increased number of bifidobacteria and decreased numbers of *Bacteroides* and *Clostridium histolyticum*; no difference with blood cytokines; faecal calprotectin and plasma C-reactive protein decreased, indicating decreased inflammation; increased faecal secretory IgA; decreased plasma insulin; decreased total cholesterol and total cholesterol/high-density lipoprotein (HDL) cholesterol ratio; triglyceride reduction was only significant in males.	Vulevic *et al.* (2013)
7.5 g d^{-1} GOS Amoxicillin (1125 mg d^{-1}) for 5 days	Parallel study ($n = 12$)	Healthy adults	12 d	Restored bifidobacteria from antibiotic treatment and increased butyrate concentration.	Ladirat *et al.* (2014)
6, 12, 18 g d^{-1} α-GOS	Parallel study ($n = 88$)	Overweight adults	14 d	Improvement of appetite with dose-dependent effect; reduced energy intake; lipopolysaccharides reduced dose-dependently; decreased plasma C-reactive protein.	Morel *et al.* (2015)

GOS = galactooligosaccharides.

7.7 Xylooligosaccharides

XOS and xylan are xylose-based oligosaccharides or polysaccharides. They are produced by hydrolysis of hemicellulose followed by purification. Nine species of bifidobacteria were tested in pure culture fermentation with different carbon sources, and one species (*Bifidobacterium catenulatum*) preferred XOS over FOS (Palframan *et al.*, 2003). There are two *in vivo* studies on the effect of XOS in the human colon (Lecerf *et al.*, 2012; Finegold *et al.*, 2014). XOS (4.2 g d^{-1} for 4 weeks) reduced constipation severity in constipated pregnant women (Tateyama *et al.*, 2005).

7.8 Other potential prebiotics candidates and summary

There are many other emerging carbohydrates that may have prebiotic effect, such as human milk oligosaccharides, isomaltooligosaccharides, lactosucrose, lactulose, soya/soybean oligosaccharides, pyrodextrins, polydextrose, arabinogalactans, pectin/pectic oligosaccharides and seaweeds/microalgae (Gibson *et al.*, 2004). Many of them are under investigation by scientists.

Gut bacteria are involved in various metabolic activities, and these activities affect human health in different ways. It is becoming apparent that it is not only gut health that is related to colonic bacteria activity; there is an increasing interest in the gut–brain axis, gut–kidney axis and gut–heart axis (Meijers & Evenepoel, 2011; Cryan & Dinan, 2012; Tang *et al.*, 2013). Serotonin is a key neurotransmitter, and a recent study found that gut bacteria regulate serotonin biosynthesis through signalling by host colon enterochromaffin cells (Yano *et al.*, 2015). Production of *p*-cresol and indole by gut bacteria from protein fermentation contributes to serum *p*-cresyl sulphate and indoxyl sulphate levels, which are risk factors for chronic kidney disease (Meijers *et al.*, 2010). Metabolism of L-carnitine and phosphatidylcholine by some gut bacteria produces trimethylamine, which can be further oxidised to trimethylamine-*N*-oxide; the latter is a promoter of cardiovascular disease, such as atherosclerosis (Wang *et al.*, 2011; Koeth *et al.*, 2013). With more host–microbiome interactions to be elucidated, prebiotics may apply to more health areas by their modulation of the gut bacteria composition and associated health benefits.

References

Abell, G.C.J., Cooke, C.M., Bennett, C.N., Conlon, M.A. & Mcorist, A.L. (2008) Phylotypes related to *Ruminococcus bromii* are abundant in the large bowel of humans and increase in response to a diet high in resistant starch. *FEMS Microbiology Ecology*, **66**, 505–515.

Abrams, S.A., Griffin, I.J., Hawthorne, K.M., Liang, L., Gunn, S.K., Darlington, G. & Ellis, K.J. (2005) A combination of prebiotic short- and long-chain inulin-type fructans enhances calcium absorption and bone mineralization in young adolescents. *American Journal of Clinical Nutrition*, **82**, 471–476.

Abrams, S.A., Griffin, I.J. & Hawthorne, K.M. (2007a) Young adolescents who respond to an inulin-type fructan substantially increase total absorbed calcium and daily calcium accretion to the skeleton. *Journal of Nutrition*, **137**, 2524s–2526s.

Abrams, S.A., Hawthorne, K.M., Aliu, O., Hicks, P.D., Chen, Z. & Griffin, I.J. (2007b) An inulin-type fructan enhances calcium absorption primarily via an effect on colonic absorption in humans. *Journal of Nutrition*, **137**, 2208–2212.

Adolphi, B., Scholz-Ahrens, K.E., de Vrese, M., Acil, Y., Laue, C. & Schrezenmeir, J. (2009) Short-term effect of bedtime consumption of fermented milk supplemented with calcium, inulin-type fructans and caseinphosphopeptides on bone metabolism in healthy, postmenopausal women. *European Journal of Nutrition*, **48**, 45–53.

Alliet, P., Scholtens, P., Raes, M., Hensen, K., Jongen, H., Rummens, J.-L., Boehm, G. & Vandenplas, Y. (2007) Effect of prebiotic galacto-oligosaccharide, long-chain fructo-oligosaccharide infant formula on serum cholesterol and triacylglycerol levels. *Nutrition*, **23**, 719–723.

Anonymous (2013) The science behind *Bifidobacterium* BB-12®. Technical bulletin. Chr. Hansen A/S, Bøge Allé 10-12, DK 2970 Hørsholm, Denmark.

Antal, M., Regoely-Merei, A., Biro, L., Arato, G., Schmidt, J., Nagy, K., Greiner, E., Lasztity, N., Szabo, C., Peter, S. & Martos, E. (2008) Effects of oligofructose containing diet in obese persons. *Orvosi Hetilap*, **149**, 1989–1995.

Archer, B.J., Johson, S.K., Devereux, H.M. & Baxter, A.L. (2004) Effect of fat replacement by inulin or lupin-kernel fibre on sausage patty acceptability, post-meal perceptions of satiety and food intake in men. *British Journal of Nutrition*, **91**, 591–599.

Arumugam, M., Raes, J., Pelletier, E., Le Paslier, D., Yamada, T., Mende, D.R., Fernandez, G.R., Tap, J., Bruls, T., Batto, J.M., Bertalan, M., Borruel, N., Casellas, F., Fernandez, L., Gautier, L., Hansen, T., Hattori, M., Hayashi, T., Kleerebezem, M., Kurokawa, K., Leclerc, M., Levenez, F., Manichanh, C., Nielsen, H.B., Nielsen, T., Pons, N., Poulain, J., Qin, J.J., Sicheritz-Ponten, T., Tims, S., Torrents, D., Ugarte, E., Zoetendal, E.G., Wang, J., Guarner, F., Pedersen, O., de Vos, W.M., Brunak, S., Dore, J., Weissenbach, J., Ehrlich, S.D., Bork, P. & MetaHIT Consortium (2011) Enterotypes of the human gut microbiome. *Nature*, **473**, 174–180.

Asp, N.G., Burvall, A., Dahlqvist, A., Hallgren, P. & Lundbald, A. (1980) Oligosaccharide formation during hydrolysis of lactose with *Saccharomyces lactis* lactase (maxilact). 2. Oligosaccharide structures. *Food Chemistry*, **5**, 147–153.

Bakker-Zierikzee, A.M., Alles, M.S., Knol, J., Kok, F.J., Tolboom, J.J. & Bindels, J.G. (2005) Effects of infant formula containing a mixture of galacto- and fructo-oligosaccharides or viable *Bifidobacterium animalis* on the intestinal microflora during the first 4 months of life. *British Journal of Nutrition*, **94**, 783–790.

Bakker-Zierikzee, A.M., van Tol, E.A.F., Kroes, H., Alles, M.S., Kok, F.J. & Bindels, J.G. (2006) Faecal SIgA secretion in infants fed on pre- or probiotic infant formula. *Pediatric Allergy and Immunology*, **17**, 134–140.

Ben, X.M., Li, J., Feng, Z.T., Shi, S.Y., Lu, Y.D., Chen, R. & Zhou, X.Y. (2008) Low level of galacto-oligosaccharide in infant formula stimulates growth of intestinal bifidobacteria and lactobacilli. *World Journal of Gastroenterology*, **14**, 6564–6568.

Benjamin, J.L., Hedin, C.R.H., Kousoumpas, A., Ng, S.C., McCarthy, N.E., Hart, A.L., Kamm, M.A., Sanderson J.D., Knight, S.C., Forbes, A., Stagg, A.J., Whelan, K. & Lindsay, J.O. (2011) Randomised, double-blind, placebo-controlled trial of fructo-oligosaccharides in active Crohn's disease. *Gut*, **60**, 923–929.

Benno, Y., Sawada, K. & Mitsuoka, T. (1984) The intestinal microflora of infants – composition of fecal flora in breast-fed and bottle-fed infants. *Microbiology and Immunology*, **28**, 975–986.

Bingham, S.A., Day, N.E., Luben, R., Ferrari, P., Slimani, N., Norat, T., Clavel-Chapelon, F., Kesse, E., Nieters, A., Boeing, H., Tjonneland, A., Overvad, K., Martinez, C., Dorronsoro, M., Gonzalez, C. A., Key, T.J., Trichopoulou, A., Naska, A., Vineis, P., Tumino, R., Krogh, V., Bueno-de-Mesquita, H.B., Peeters, P.H.M., Berglund, G., Hallmans, G., Lund, E., Skeie, G., Kaaks, R. & Riboli, E. (2003) Dietary fibre in food and protection against colorectal cancer in the European Prospective Investigation into Cancer and Nutrition (EPIC): an observational study. *Lancet*, **361**, 1496–1501.

Bloemen, J.G., Venema, K., van de Poll, M.C., Olde Damink, S.W., Buurman, W.A. & de Jong, C.H. (2009) Short chain fatty acids exchange across the gut and liver in humans measured at surgery. *Clinical Nutrition*, **28**, 657–661.

Bocquet, A., Lachambre, E., Kempe, C. & Beck, L. (2013) Effect of infant and follow-on formulas containing *B. lactis* and galacto- and fructo-oligosaccharides on infection in healthy term infants. *Journal of Pediatric Gastroenterology and Nutrition*, **57**, 180–187.

Boehm, G., Lidestri, M., Casetta, P., Jelinek, J., Negretti, F., Stahl, B. & Marini, A. (2002) Supplementation of a bovine milk formula with an oligosaccharide mixture increases counts of faecal bifidobacteria in preterm infants. *Archives of Disease in Childhood*, **86**, F178–F181.

Boffa, L.C., Vindali, G., Mann, R.S. & Allfrey, V.G. (1978) Suppression of histone deacetylation in vivo and in vitro by sodium butyrate. *Journal of Biological Chemistry*, **253**, 3364–3366.

Bordin, M., D'atri, F., Guillemot, L. & Citi, S. (2004) Histone deacetylase inhibitors up-regulate the expression of tight junction proteins. *Molecular Cancer Research*, **2**, 692–701.

Boudry, G., Jamin, A., Chatelais, L., Gras-Le Guen, C., Michel, C. & Le Huerou-Luron, I. (2013) Dietary protein excess during neonatal life alters colonic microbiota and mucosal response to inflammatory mediators later in life in female pigs. *Journal of Nutrition*, **143**, 1225–1232.

Bouhnik, Y., Flourie, B., Dagayabensour, L., Pochart, P., Gramet, G., Durand, M. & Rambaud, J.C. (1997) Administration of transgalacto-oligosaccharides increases fecal bifidobacteria and modifies colonic fermentation metabolism in healthy humans. *Journal of Nutrition*, **127**, 444–448.

Bouhnik, Y., Vahedi, K., Achour, L., Attar, A., Salfati, J., Pochart, P., Marteau, P., Flourie, B., Bornet, F. & Rambaud, J.C. (1999) Short-chain fructo-oligosaccharide administration dose-dependently increases fecal bifidobacteria in healthy humans. *Journal of Nutrition*, **129**, 113–116.

Bouhnik, Y., Raskine, L., Simoneau, G., Vicaut, E., Neut, C., Flourie, B., Brouns, F. & Bornet, F.R. (2004) The capacity of nondigestible carbohydrates to stimulate fecal bifidobacteria in healthy humans: a double-blind, randomized, placebo-controlled, parallel-group, dose-response relation study. *American Journal of Clinical Nutrition*, **80**, 1658–1664.

Boutron-Ruault, M.C., Marteau, P., Lavergne-Slove, A., Myara, A., Gerhardt, M.F., Franchisseur, C., Bornet, F. & Eripolyp Study Group (2005) Effects of a 3-mo consumption of short-chain fructo-oligosaccharides on parameters of colorectal carcinogenesis in patients with or without small or large colorectal adenomas. *Nutrition and Cancer – an International Journal*, **53**, 160–168.

Bozensky, J., Hill, M., Zelenka, R. & Skyba, T. (2015) Prebiotics do not influence the severity of atopic dermatitis in infants: a randomised controlled trial. *PLoS One*, **10**, e0142897.

Brighenti, F., Casiraghi, M.C., Canzi, E. & Ferrari, A. (1999) Effect of consumption of a ready-to-eat breakfast cereal containing inulin on the intestinal milieu and blood lipids in healthy male volunteers. *European Journal of Clinical Nutrition*, **53**, 726–733.

Brommage, R., Binacua, C., Antille, S. & Carrie, A.L. (1993) Intestinal calcium-absorption in rats is stimulated by dietary lactulose and other resistant sugars. *Journal of Nutrition*, **123**, 2186–2194.

Brown, A.J., Goldsworthy, S.M., Barnes, A.A., Eilert, M.M., Tcheang, L., Daniels, D., Muir, A.I., Wigglesworth, M.J., Kinghhorn, I., Fraser, N.J., Pike, N.B., Strum, J.C., Steplewski, K.M., Murdock, P.R., Holder, J.C., Marshall, F.H., Szekeres, P.G., Wilson, S., Ignar, D.M., Foord, S.M., Wise, A. & Dowell, S.J. (2003) The orphan G protein-coupled receptors GPR41 and GPR43 are activated by propionate and other short chain carboxylic acids. *Journal of Biological Chemistry*, **278**, 11312–11319.

Buddington, R.K., Williams, C.H., Chen, S.C. & Witherly, S.A. (1996) Dietary supplement of neosugar alters the fecal flora and decreases activities of some reductive enzymes in human subjects. *American Journal of Clinical Nutrition*, **63**, 709–716.

Buddington, K.K., Donahoo, J.B. & Buddington, R.K. (2002) Dietary oligofructose and inulin protect mice from enteric and systemic pathogens and tumor inducers. *Journal of Nutrition*, **132**, 472–477.

Burns, A.J. & Rowland, I.R. (2004) Antigenotoxicity of probiotics and prebiotics on faecal water-induced DNA damage in human colon adenocarcinoma cells. *Mutation Research – Fundamental and Molecular Mechanisms of Mutagenesis*, **551**, 233–243.

Cani, P.D., Joly, E., Hrsmans, Y. & Delzenne, N.M. (2006) Oligofructose promotes satiety in healthy human: a pilot study. *European Journal of Clinical Nutrition*, **60**, 567–572.

Canzi, E., Brighenti, F., Casiraghi, M.C., Del Puppo, E. & Ferrari, A. (1996) Prolonged consumption of inulin in ready-to-eat breakfast cereals: effects on intestinal ecosystem, bowel habits and lipid metabolism. *Proceedings of the COST Action 92 on Dietary Fiber and Fermentation in the Colon* (ed. Y. Mälkki and J.H. Cummings), 280–284, Espoo, Finland, 15–17 June 1995. European Commission, Directorate-General XII, Science, Research and Development Brussels.

Causey, J.L., Feirtag, J.M., Gallaher, D.D., Tungland, B.C. & Slavin, J.L. (2000) Effects of dietary inulin on serum lipids, blood glucose and the gastrointestinal, environment in hypercholesterolemic men. *Nutrition Research*, **20**, 191–201.

Chambers, E.S., Viardot, A., Psichas, A., Morrison, D.J., Murphy, K.G., Zac-Varghese, S.E., MacDougall, K., Preston T., Tedford, C., Finlayson, GS., Blundell, J.E., Bell, J.D., Thomas, E.L., Mt-Isa, S., Ashby. D., Gibson, G.R., Kolida, S., Dhillo, W.S., Bloom, S.R., Morley, W., Clegg, S. & Frost, G. (2015) Effects of targeted delivery of propionate to the human colon on appetite regulation, body weight maintenance and adiposity in overweight adults. *Gut*, **64**, 1744–1754.

Cherbut, C., Ferrier, L., Roze, C., Anini, Y., Blottiere, H., Lecannu, G. & Galimiche, J.P. (1998) Short-chain fatty acids modify colonic motility through nerves and polypeptide YY release in the rat. *American Journal of Physiology – Gastrointestinal and Liver Physiology*, **275**, G1415–G1422.

Chonan, O. & Watanuki, M. (1996) The effect of 6′-galactooligosaccharides on bone mineralization of rats adapted to different levels of dietary calcium. *International Journal for Vitamin and Nutrition Research*, **66**, 244–249.

Chonan, O., Matsumoto, K. & Watanuki, M. (1995) Effect of galactooligosaccharides on calcium-absorption and preventing bone loss in ovariectomized rats. *Bioscience Biotechnology and Biochemistry*, **59**, 236–239.

Chonan, O., Takahashi, R., Yasui, H. & Watanuki, M. (1996) Effects of beta 1→4 linked galactooligosaccharides on use of magnesium and calcification of the kidney and heart in rats fed excess dietary phosphorus and calcium. *Bioscience Biotechnology and Biochemistry*, **60**, 1735–1737.

Chouraqui, J.P., Grathwohl, D., Labaune, J.M., Hascoet, J.M., de Montgolfier, I., Leclaire, M. & Steenhout, P. (2008) Assessment of the safety, tolerance, and protective effect against diarrhea of infant formulas containing mixtures of probiotics or probiotics and prebiotics in a randomized controlled trial. *American Journal of Clinical Nutrition*, **87**, 1365–1373.

Clausen, M.R. & Mortensen, P.B. (1995) Kinetic-studies on colonocyte metabolism of short-chain fatty-acids and glucose in ulcerative-colitis. *Gut*, **37**, 684–689.

Commane, D.M., Shortt, C.T., Silvi, S., Cresci, A., Hughes, R.M. & Rowland, I.R. (2005) Effects of fermentation products of pro- and prebiotics on trans-epithelial electrical resistance in an in vitro model of the colon. *Nutrition and Cancer – an International Journal*, **51**, 102–109.

Costabile, A., Deaville, E.R., Morales, A.M. & Gibson, G.R. (2016) Prebiotic potential of a maize-based soluble fibre and impact of dose on the human gut microbiota. *PLoS One*, **11**, e0144457.

Coudray, C., Bellanger, J., Castiglia-Delavaud, C., Remesy, C., Vermorel, M. & Rayssignuier, Y. (1997) Effect of soluble or partly soluble dietary fibres supplementation on absorption and balance of calcium, magnesium, iron and zinc in healthy young men. *European Journal of Clinical Nutrition*, **51**, 375–380.

Cryan, J.F. & Dinan, T.G. (2012) Mind-altering microorganisms: the impact of the gut microbiota on brain and behaviour. *Nature Reviews Neuroscience*, **13**, 701–712.

Cuche, G., Cuber, J.C. & Malbert, C.H. (2000) Ileal short-chain fatty acids inhibit gastric motility by a humoral pathway. *American Journal of Physiology-Gastrointestinal and Liver Physiology*, **279**, G925–G930.

Culhane, C.T. & Tanugraha, L. (2013) *GRAS Notification Prepared for and Submitted* (No. 489). http://www.fda.gov/Food/IngredientsPackagingLabeling/GRAS/NoticeInventory/default.htm.

Cummings, J.H. (1981). Short chain fatty-acids in the human-colon. *Gut*, **22**, 763–779.

Cummings, D.E. & Overduin, J. (2007) Gastrointestinal regulation of food intake. *Journal of Clinical Investigation*, **117**, 13–23.

Cummings, J.H., Pomare, E.W., Branch, W.J., Naylor, C.P.E. & Macfarlane, G.T. (1987) Short chain fatty-acids in human large-intestine, portal, hepatic and venous-blood. *Gut*, **28**, 1221–1227.

Cummings, J.H., Christie, S. & Cole, T.J. (2001) A study of fructo oligosaccharides in the prevention of travellers' diarrhoea. *Alimentary Pharmacology & Therapeutics*, **15**, 1139–1145.

da Costa Ribeiro, H.J., Ribeiro, T.C., de Mattos, A.P., Pontes, M., Sarni, R.O., Cruz, M.L. & Steenhout, P. (2015) Normal growth of healthy infants born from HIV+ mothers fed a reduced protein infant formula containing the prebiotics galacto-oligosaccharides and fructo-oligosaccharides: a randomized controlled trial. *Clinical Medicine Insights: Pediatrics*, **9**, 37–47.

Dahl, W.J., Ford, A.L., Ukhanova, M., Radford, A., Christman, M.C., Waugh, S. & Mai, V. (2016) Resistant potato starches (type 4 RS) exhibit varying effects on laxation with and without phylum level changes in microbiota: a randomised trial in young adults. *Journal of Functional Foods*, **23**, 1–11.

Davidson, M.H., Maki, K.C., Synecki, C., Torri, S.A. & Drennan, K.B. (1998) Effects of dietary inulin on serum lipids in men and women with hypercholesterolemia. *Nutrition Research*, **18**, 503–517.

Davis, L.M.G., Martínez, I., Walter, J. & Hutkins, R. (2010) A dose dependent impact of prebiotic galactooligosaccharides on the intestinal microbiota of healthy adults. *International Journal of Food Microbiology*, **144**, 285–292.

Dehghan, P., Pourghassem Gargari, B. & Asgharijafarabadi, M. (2013) Effects of high performance inulin supplementation on glycemic status and lipid profile in women with type 2 diabetes: a randomized, placebo-controlled clinical trial. *Health Promotion Perspectives*, **3**, 55–63.

Delzenne, N., Aertssens, J., Verplaetse, H., Roccaro, M. & Roberfroid, M. (1995) Effect of fermentable fructo-oligosaccharides on mineral, nitrogen and energy digestive balance in the rat. *Life Sciences*, **57**, 1579–1587.

Depeint, F., Tzortzis, G., Vulevic, J., l'Anson, K. & Gibson, G.R. (2008) Prebiotic evaluation of a novel galactooligosaccharide mixture produced by the enzymatic activity of *Bifidobacterium bifidum* NCIMB 41171, in healthy humans: a randomized, double-blind, crossover, placebo-controlled intervention study. *American Journal of Clinical Nutrition*, **87**, 785–791.

Drakoularakou, A., Tzortzis, G., Rastall, R.A. & Gibson, G.R. (2010) A double-blind, placebo-controlled, randomized human study assessing the capacity of a novel galacto-oligosaccharide mixture in reducing travellers' diarrhoea. *European Journal of Clinical Nutrition*, **64**, 146–152.

Drucker, D.J. (2006) The biology of incretin hormones. *Cell Metabolism*, **3**, 153–165.

Duggan, C., Penny, M.E., Hibberd, P., Gil, A., Huapaya, A., Cooper, A. & Kleinman, R.E. (2003) Oligofructose-supplemented infant cereal: 2 randomized, blinded, community-based trials in Peruvian infants. *American Journal of Clinical Nutrition*, **77**, 937–942.

Eckburg, P.B., Bik, E.M., Bernstein, C.N., Purdom, E., Dethlefsen, L., Sargent, M. & Relman, D.A. (2005) Diversity of the human intestinal microbial flora. *Science*, **308**, 1635–1638.

Ellegard, L., Andersson, H. & Bosaeus, I. (1997) Inulin and oligofructose do not influence the absorption of cholesterol, or the excretion of cholesterol, Ca, Mg, Zn, Fe, or bile acids but increases energy excretion in ileostomy subjects. *European Journal of Clinical Nutrition*, **51**, 1–5.

Englyst, H.N., Kingman, S.M., Hudson, G.J. & Cummings, J.H. (1996) Measurement of resistant starch *in vitro* and *in vivo*. *British Journal of Nutrition*, **75**, 749–755.

Fanaro, S., Marten, B., Bagna, R., Vigi, V., Fabris, C., Peña-Quintana, L., Argüelles, F., Scholz-Ahrens, K.E., Sawatzki, G., Zelenka, R., Schrezenmeir, J., de Vrese, M. & Bertino, E. (2009) Galacto-oligosaccharides are bifidogenic and safe at weaning: a double-blind randomized multicenter study. *Journal of Pediatric Gastroenterology and Nutrition*, **48**, 82–88.

Femia, A.P., Luceri, C., Dolara, P., Giannini, A., Biggeri, A., Salvadori, M., Clune, Y., Collins, K.J., Paglierani, M. & Caderni, G. (2002) Antitumorigenic activity of the prebiotic inulin enriched with oligofructose in combination with the probiotics *Lactobacillus rhamnosus* and *Bifidobacterium lactis* on azoxymethane-induced colon carcinogenesis in rats. *Carcinogenesis*, **23**, 1953–1960.

Finegold, S.M., Li, Z., Summanen, P.H., Downes, J., Thames, G., Corbett, K., Dowd, S., Krak, M. & Heber D. (2014) Xylooligosaccharide increases bifidobacteria but not lactobacilli in human gut microbiota. *Food & Function*, **5**, 436–445.

Forcheron, F. & Beylot, M. (2007) Long-term administration of inulin-type fructans has no significant lipid-lowering effect in normolipidemic humans. *Metabolism*, **56**, 1093–1098.

Franck, A.A. (2002) Technological functionality of inulin and oligofructose. *British Journal of Nutrition*, **87**, 287–291.

Furrie, E., Macfarlane, S., Kennedy, A., Cummings, J.H., Walsh, S.V., O'Neil, D.A. & Macfarlane, G.T. (2005) Synbiotic therapy (*Bifidobacterium longum*/Synergy 1) initiates resolution of inflammation in patients with active ulcerative colitis: a randomised controlled pilot trial. *Gut*, **54**, 242–249.

Gibson, G.R. & Rastall, R.A. (2006) *Prebiotics Development & Application*. John Wiley & Sons, Chichester.

Gibson, G.R. & Roberfroid, M.B. (1995) Dietary modulation of the human colonic microbiota – introducing the concept of prebiotics. *Journal of Nutrition*, **125**, 1401–1412.

Gibson, G.R. & Wang, X. (1994) Enrichment of bifidobacteria from human gut contents by oligofructose using continuous-culture. *FEMS Microbiology Letters*, **118**, 121–127.

Gibson, G.R., Beatty, E.R., Wang, X. & Cummings, J.H. (1995) Selective stimulation of bifidobacteria in the human colon by oligofructose and inulin. *Gastroenterology*, **108**, 975–982.

Gibson, G.R., Probert, H.M., Loo, J.V., Rastall, R.A. & Roberfroid, M.B. (2004) Dietary modulation of the human colonic microbiota: updating the concept of prebiotics. *Nutrition Research Reviews*, **17**, 259–275.

Gibson, G.R., Scott, K.P., Rastall, R.A., Tuohy, K.M., Hotchkiss, A., Dubert-Ferrandon, A., Gareau, M., Murphy, E.F., Saulnier, D., Loh, G., Macfarlane, S., Delzenne, N., Ringel, Y., Kozianowski, G., Dickmann, R., Lenoir-Wijnkoop, I., Walker, C. & Buddington, R. (2010) Dietary prebiotics: current status and new definition. *Food Science and Technology Bulletin: Functional Foods*, **7**, 1–19.

Giovannini, M., Verduci, E., Gregori, D., Ballali, S., Soldi, S., Ghisleni, D. & Riva, E. (2014) Prebiotic effect of an infant formula supplemented with galacto-oligosaccharides: randomized multicenter trial. *Journal of the American College of Nutrition*, **33**, 385–393.

Greer, F.R., Krebs, N.F. & Comm, N. (2006) Optimizing bone health and calcium intakes of infants, children, and adolescents. *Pediatrics*, **117**, 578–585.

Griffin, I.J., Davila, P.M. & Abrams, S.A. (2002) Non-digestible oligosaccharides and calcium absorption in girls with adequate calcium intakes. *British Journal of Nutrition*, **87**, S187–S191.

Griffin, I.J., Hicks, P.M.D., Heaney, R.P. & Abrams, S.A. (2003) Enriched chicory inulin increases calcium absorption mainly in girls with lower calcium absorption. *Nutrition Research*, **23**, 901–909.

Gruber, C., van Stuijvenberg, M., Mosca, F., Moro, G., Chirico, G., Braegger, C.P. & Wahn, U. (2010) Reduced occurrence of early atopic dermatitis because of immunoactive prebiotics among low-atopy-risk infants. *Journal of Allergy and Clinical Immunology*, **126**, 791–797.

Guigoz, Y., Rochat, F., Perruisseau-Carrier, G., Rochat, I. & Schiffrin, E.J. (2002) Effects of oligosaccharide on the faecal flora and non-specific immune system in elderly people. *Nutrition Research*, **22**, 13–25.

Hafer, A., Kraemer, S., Duncker, S., Krueger, M., Manns, M.P. & Bischoff, S.C. (2007) Effect of oral lactulose on clinical and immunohistochemical parameters in patients with inflammatory bowel disease: a pilot study. *BMC Gastroenterology*, **7**, 36.

Hague, A., Elder, D.J.E., Hicks, D.J. & Paraskeva, C. (1995) Apoptosis in colorectal tumor-cells – induction by the short-chain fatty-acids butyrate, propionate and acetate and by the bile-salt deoxycholate. *International Journal of Cancer*, **60**, 400–406.

Harmsen, H.J.M., Wildeboer-Veloo, A.C.M., Raangs, G.C., Wagendorp, A.A., Klijn, N., Bindels, J.G. & Welling, G.W. (2000) Analysis of intestinal flora development in breast-fed and formula-fed infants by using molecular identification and detection methods. *Journal of Pediatric Gastroenterology and Nutrition*, **30**, 61–67.

Heavey, P.M., Savage, S.A.H., Parrett, A., Cecchini, C., Edwards, C.A. & Rowland, I.R. (2003) Protein-degradation products and bacterial enzyme activities in faeces of breast-fed and formula-fed infants. *British Journal of Nutrition*, **89**, 509–515.

Hidaka, H. (1991) Proliferation of bifidobacteria by oligosaccharides and their useful effect on human health. *Bifidobacteria and Microflora*, **10**, 65–79.

Hoebregs, H. (1997) Fructans in foods and food products, ion-exchange chromatographic method: collaborative study. *Journal of AOAC International*, **80**, 1029–1037.

Holloway, L., Moynihan, S., Abrams, S.A., Kent, K., Hsu, A.R. & Friedlander, A.L. (2007) Effects of oligofructose-enriched inulin on intestinal absorption of calcium and magnesium and bone turnover markers in postmenopausal women. *British Journal of Nutrition*, **97**, 365–372.

Holscher, H.D., Faust, K.L., Czerkies, L.A., Litov, R., Ziegler, E.E., Lessin, H. & Tappenden, K.A. (2012) Effects of prebiotic-containing infant formula on gastrointestinal tolerance and fecal microbiota in a randomized controlled trial. *Journal of Parenteral and Enteral Nutrition*, **36**, 95s–105s.

Hughes, R. & Rowland, I.R. (2001) Stimulation of apoptosis by two prebiotic chicory fructans in the rat colon. *Carcinogenesis*, **22**, 43–47.

Ito, M. (1993) Influence of galactooligosaccharides on the human fecal microflora. *Journal of Nutritional Science and Vitaminology*, **39**, 635–640.

Jackson, K.G., Taylor, G.R., Clohessy, A.M. & Williams, C.M. (1999) The effect of the daily intake of inulin on fasting lipid, insulin and glucose concentrations in middle-aged men and women. *British Journal of Nutrition*, **82**, 23–30.

Jackson, R.D., LaCroix, A.Z., Gass, M., Wallace, R.B., Robbins, J., Lewis, C.E., Bassford, T., Beresford, S.A.A., Black, H.R., Blanchette, P., Bonds, D.Ei, Brunner, R.L., Brzyski, R.G., Caan, B., Cauley, J.A., Chlebowski, R.T., Cummings, S.R., Granek, T., Hays, J., Heiss, G., Hendrix, S.L., Howard, B.V., Hsia, J., Hubbell, F.A., Johnson, K.C., Judd, H., Kotchen, J.M., Kuller, L.H., Langer, R.D., Norman L. Lasser, N.L., Limacher, M.C., Ludlam, S., Manson, J.E., Margolis, K.L., McGowan, J., Ockene, J.K., O'Sullivan, M.J., Phillips, L., Prentice, R.L., Sarto, G.E., Stefanick, M.L., Van Horn, L., Wactawski-Wende, J., Whitlock, E., Anderson, G.L., Assaf, A.R. & Barad, D. (2006) Calcium plus vitamin D supplementation and the risk of fractures. *New England Journal of Medicine*, **354**, 669–683.

Jose Gosalbes, M., Durbán, A., Pignatelli, M., Abellan, J.A., Jiménez-Hernández, N., Pérez-Cobas, A.E., Latorre, A. & Moya, A. (2011) Metatranscriptomic approach to analyze the functional human gut microbiota. *PLoS One*, **6**(3), e17447.

Kaplan, H. & Hutkins, R.W. (2000) Fermentation of fructooligosaccharides by lactic acid bacteria and bifidobacteria. *Applied and Environmental Microbiology*, **66**, 2682–2684.

Karaki, S.-I., Tazoe, H., Hayashi, H., Kashiwabara, H., Tooyama, K., Suzuki, Y. & Kuwahara, A. (2008) Expression of the short-chain fatty acid receptor, GPR43, in the human colon. *Journal of Molecular Histology*, **39**, 135–142.

Kelly, G. (2008) Inulin-type prebiotics – a review: part 1. *Alternative Medicine Review*, **13**, 315–329.

Kleessen, B., Sykura, B., Zunft, H.J. & Blaut, M. (1997) Effects of inulin and lactose on fecal microflora, microbial activity, and bowel habit in elderly constipated persons. *American Journal of Clinical Nutrition*, **65**, 1397–1402.

Klinder, A., Forster, A., Caderni, G., Femia, A.P. & Pool-Zobel, B.L. (2004) Fecal water genotoxicity is predictive of tumor-preventive activities by inulin-like oligofructoses, probiotics (*Lactobacillus rhamnosus* and *Bifidobacterium lactis*), and their synbiotic combination. *Nutrition and Cancer – an International Journal*, **49**, 144–155.

Knol, J., Scholtens, P., Kafka, C., Steenbakkers, J., Gro, S., Helm, K., Klarczyk, M., Schöpfer, H., Böckler, H.-M. & Wells, J. (2005) Colon microflora in infants fed formula with galacto- and fructo-oligosaccharides: more like breast-fed infants. *Journal of Pediatric Gastroenterology and Nutrition*, **40**, 36–42.

Kobata, A. & Ginsburg, V. (1969) Oligosaccharides of human milk: 2. Isolation and characterization of a new pentasaccharide, lacto-*N*-fucopentaose III. *Journal of Biological Chemistry*, **244**, 5496–5502.

Kobata, A. & Ginsburg, V. (1972) Oligosaccharides of human milk: 3. Isolation and characterization of a new hexasaccharide, lacto-*N*-hexaose. *Journal of Biological Chemistry*, **247**, 1525–1529.

Koeth, R.A., Wang, Z., Levison, B.S., Buffa, J.A., Org, E., Sheehy, B.T., Britt, E.B., Fu, X., Wu, Y., Li, L., Smith, J.D., Di Donato,J.A., Chen, J., Li, H., Wu, G.D., Lewis, J.D., Warrier, M., Brown, J.M., Krauss, R.M., Tang, W.H.W., Bushman, F.D., Lusis, A.J. & Hazen, S.L. (2013) Intestinal microbiota metabolism of L-carnitine, a nutrient in red meat, promotes atherosclerosis. *Nature Medicine*, **19**, 576–585.

Kok, N., Roberfroid, M., Robert, A. & Delzenne, N. (1996) Involvement of lipogenesis in the lower VLDL secretion induced by oligofructose in rats. *British Journal of Nutrition*, **76**, 881–890.

Kruger, M.C., Chan, Y.M., Kuhn-Sherlock, B., Lau, L.T., Lau, C., Chin, Y.S., Todd, J.M. & Schollum L.M. (2015) Differential effects of calcium- and vitamin D-fortified milk with FOS-inulin compared to regular milk, on bone biomarkers in Chinese pre- and postmenopausal women. *European Journal of Nutrition*, **55**, 1911–1921. doi:10.1007/s00394-015-1007-x

Kruse, H.P., Kleessen, B. & Blaut, M. (1999) Effects of inulin on faecal bifidobacteria in human subjects. *British Journal of Nutrition*, **82**, 375–382.

Le Poul, E., Loison, C., Struyf, S., Springael, J.-Y., Lannoy, V., Decobecq, M.-E., Brezillon, S., Dupriez, V., Vassart, G., Van Damme, J., Parmentier, M. & Detheux, M. (2003) Functional characterization of human receptors for short chain fatty acids and their role in polymorphonuclear cell activation. *Journal of Biological Chemistry*, **278**, 25481–25489.

Lecerf, J.-M., Depeint, F., Clerc, E., Dugenet, Y., Niamba, C.N., Rhazi, L. & Pouillart, P.R. (2012) Xylo-oligosaccharide (XOS) in combination with inulin modulates both the intestinal environment and immune status in healthy subjects, while XOS alone only shows prebiotic properties. *British Journal of Nutrition*, **108**, 1847–1858.

Lee, L.Y., Bharani, R., Biswas, A., Lee, J., Tran, L.A., Pecquet, S. & Steenhout, P. (2015) Normal growth of infants receiving an infant formula containing *Lactobacillus reuteri*, galacto-oligosaccharides, and fructo-oligosaccharide: a randomized controlled trial. *Maternal Health, Neonatology and Perinatology*, **1**, 9. doi:10.1186/s40748-015-0008-3

Letexier, D., Diraison, F. & Beylot, M. (2003) Addition of inulin to a moderately high-carbohydrate diet reduces hepatic lipogenesis and plasma triacylglycerol concentrations in humans. *American Journal of Clinical Nutrition*, **77**, 559–564.

Liber, A. & Szajewska, H. (2014) Effect of oligofructose supplementation on body weight in overweight and obese children: a randomised, double-blind, placebo-controlled trial. *British Journal of Nutrition*, **112**, 2068–2074.

Limburg, P.J., Mahoney, M.R., Ziegler, K.L., Sontag, S.J., Schoen, R.E., Benya, R., Lawson, M.J., Weinberg, D.S., Stoffel, E., Chiorean, M., Heigh, R., Levine, J., Della'Zanna, G., Rodriguez, L., Richmond, E., Gostout, C., Mandrekar, S.J. & Smyrk, T.C. (2011) Randomized phase II trial of sulindac, atorvastatin, and prebiotic dietary fiber for colorectal cancer chemoprevention. *Cancer Prevention Research (Philadelphia)*, **4**, 259–269.

Lindsay, J.O., Whelan, K., Stagg, A.J., Gobin, P., Al-Hassi, H.O., Rayment, N., Kamm, M.A., Knight, S.C. & Forbes, A. (2006) Clinical, microbiological, and immunological effects of fructo-oligosaccharide in patients with Crohn's disease. *Gut*, **55**, 348–355.

Lomax, A.R., Cheung, L.V., Noakes, P.S., Miles, E.A. & Calder, P.C. (2015) Inulin-type β2-1 fructans have some effect on the antibody response to seasonal influenza vaccination in healthy middle-aged humans. *Frontiers in Immunology*, **6**, 490. doi:10.3389/fimmu.2015.00490

Luo, J., Rizkalla, S.W., Alamowitch, C., Boussairi, A., Blayo, A., Barry, J.L., Laffitte, A., Guyon, F., Bornet, F.R. & Slama, G. (1996) Chronic consumption of short-chain fructooligosaccharides by healthy subjects decreased basal hepatic glucose production but had no effect on insulin-stimulated glucose metabolism. *American Journal of Clinical Nutrition*, **63**, 939–945.

Macfarlane, S. & Macfarlane, G.T. (2003) Regulation of short-chain fatty acid production. *Proceedings of the Nutrition Society*, **62**, 67–72.

Macfarlane, S., Cleary, S., Bahrami, B., Reynolds, N. & Macfarlane, G.T. (2013) Synbiotic consumption changes the metabolism and composition of the gut microbiota in older people and modifies inflammatory processes: a randomised, double-blind, placebo-controlled crossover study. *Alimentary Pharmacology & Therapeutics*, **38**, 804–816.

Mattarelli, P., Bonaparte, C., Pot, B. & Biavati, B. (2008) Proposal to reclassify the three biotypes of *Bifidobacterium longum* as three subspecies: *Bifidobacterium longum* subsp. *longum* subsp. nov., *Bifidobacterium longum* subsp. *infantis* comb. nov. and *Bifidobacterium longum* subsp. *suis* comb. nov. *International Journal of Systematic and Evolutionary Microbiology*, **58**, 767–772.

McCleary, B.V., Murphy, A. & Mugford, D.C. (2000) Measurement of total fructan in foods by enzymatic/spectrophotometric method: collaborative study. *Journal of AOAC International*, **83**, 356–364.

McKellar, R.C. & Modler, H.W. (1989) Metabolism of fructo-oligosaccharides by *Bifidobacterium* spp. *Applied Microbiology and Biotechnology*, **31**, 537–541.

McNeil, N.I., Cummings, J.H. & James, W.P.T. (1978) Short chain fatty-acid absorption by human large-intestine. *Gut*, **19**, 819–822.

Meijers, B.K.I., De Preter, V., Verbeke, K., Vanrenterghem, Y. & Evenepoel, P. (2010) p-Cresyl sulfate serum concentrations in haemodialysis patients are reduced by the prebiotic oligofructose-enriched inulin. *Nephrology Dialysis Transplantation*, **25**, 219–224.

Meijers, B.K.I. & Evenepoel, P. (2011) The gut-kidney axis: indoxyl sulfate, p-cresyl sulfate and CKD progression. *Nephrology Dialysis Transplantation*, **26**, 759–761.

Mevissenverhage, E.A.E., Marcelis, J.H., Devos, M.N., Harmsenvanamerongen, W.C.M. & Verhoef, J. (1987) *Bifidobacterium*, *Bacteroides*, and *Clostridium* spp. in fecal samples from breast-fed and bottle-fed infants with and without iron supplement. *Journal of Clinical Microbiology*, **25**, 285–289.

Moro, G., Arslanoglu, S., Stahl, B., Jelinek, J., Wahn, U. & Boehm, G. (2006) A mixture of prebiotic oligosaccharides reduces the incidence of atopic dermatitis during the first six months of age. *Archives of Disease in Childhood*, **91**, 814–819.

Moshfegh, A.J., Friday, J.E., Goldman, J.P. & Ahuja, J.K.C. (1999) Presence of inulin and oligofructose in the diets of Americans. *Journal of Nutrition*, **129**, 1407S–1411S.

Nagengast, F.M., Grubben, M.J. & van Munster, I.P. (1995) Role of bile acids in colorectal carcinogenesis. *European Journal of Cancer*, **31a**, 1067–1070.

Ohta, A., Ohtsuki, M., Baba, S., Takizawa, T., Adachi, T. & Kimura, S. (1995) Effects of fructooligosaccharides on the absorption of iron, calcium and magnesium in iron-deficient anemic rats. *Journal of Nutritional Science and Vitaminology*, **41**, 281–291.

Palframan, R.J., Gibson, G.R. & Rastall, R.A. (2003) Carbohydrate preferences of *Bifidobacterium* species isolated from the human gut. *Current Issues in Intestinal Microbiology*, **4**, 71–75.

Parnell, J.A. & Reimer, R.A. (2009) Weight loss during oligofructose supplementation is associated with decreased ghrelin and increased peptide YY in overweight and obese adults. *American Journal of Clinical Nutrition*, **89**, 1751–1759.

Pedersen, A., Sandstrom, B. & VanAmelsvoort, J.M.M. (1997) The effect of ingestion of inulin on blood lipids and gastrointestinal symptoms in healthy females. *British Journal of Nutrition*, **78**, 215–222.

Pereira, D.I.A. & Gibson, G.R. (2002) Effects of consumption of probiotics and prebiotics on serum lipid levels in humans. *Critical Reviews in Biochemistry and Molecular Biology*, **37**, 259–281.

Perrigue, M.M., Monsivais, P. & Drewnowski, A. (2009) Added soluble fiber enhances the satiating power of low-energy-density liquid yogurts. *Journal of the American Dietetic Association*, **109**, 1862–1868.

Peters, H.P.F., Boers, H.M., Haddeman, E., Melnikov, S.M. & Qvyjt, F. (2009) No effect of added β-glucan or of fructooligosaccharide on appetite or energy intake. *American Journal of Clinical Nutrition*, **89**, 58–63.

Piche, T., Bruley, S., Des Varannes, S.B., Sacher-Huvelin, S., Holst, J.J., Cuber, J.C. & Galmiche, J.P. (2003) Colonic fermentation influences lower esophageal sphincter function in gastroesophageal reflux disease. *Gastroenterology*, **124**, 894–902.

Pool-Zobel, B.L. (2005) Inulin-type fructans and reduction in colon cancer risk: review of experimental and human data. *British Journal of Nutrition*, **93**, S73–S90.

Rafter, J., Bennett, M., Caderni, G., Clune, Y., Hughes, R., Karlsson, P.C., Klinder, A., O'Riordan, M., O'Sullivan, G.C., Pool-Zobel, B., Rechkemmer, G., Roller, M., Rowland, I., Salvadori, M., Thijs, H., Van Loo, J., Watzl, B. & Collins, J.K. (2007) Dietary synbiotics reduce cancer risk factors in polypectomized and colon cancer patients. *American Journal of Clinical Nutrition*, **85**, 488–496.

Ribeiro, T.C.M., Hugo Jr.,C.-R., Almeida, P.S., Pontes, M.V., Leite, M.E.Q., Filadelfo, L.R., Khoury, J.C., Bean, Judy, A., Mitmesser, S.H., Vanderhoof, J.A. Scalabrin, & Deolinda, M.F. (2012) Stool pattern changes in toddlers consuming a follow-on formula supplemented with polydextrose and galactooligosaccharides. *Journal of Pediatric Gastroenterology and Nutrition*, **54**, 288–290.

Ridlon, J.M., Kang, D.J. & Hylemon, P.B. (2006) Bile salt biotransformations by human intestinal bacteria. *Journal of Lipid Research*, **47**, 241–259.

Roberfroid, M.B. (2005) Inulin-type fructans. *British Journal of Nutrition*, **93**(Suppl. 1), S13–S25

Roberfroid, M.B. (2007) Inulin-type fructans: functional food ingredients. *Journal of Nutrition*, **137**, 2493S–2502S.

Roller, M., Clune, Y., Collins, K., Rechkemmer, G. & Watzl, B. (2007) Consumption of prebiotic inulin enriched with oligofructose in combination with the probiotics *Lactobacillus rhamnosus* and *Bifidobacterium lactis* has minor effects on selected immune parameters in polypectomised and colon cancer patients. *British Journal of Nutrition*, **97**, 676–684.

Russell, W.R., Hoyles, L., Flint, H.J. & Dumas, M.-E. (2013) Colonic bacterial metabolites and human health. *Current Opinion in Microbiology*, **16**, 246–254.

Russo, F., Chimienti, G., Riezzo, G., Pepe, G., Petrosillo, G., Chiloiro, M. & Marconi, E. (2008) Inulin-enriched pasta affects lipid profile and Lp(a) concentrations in Italian young healthy male volunteers. *European Journal of Nutrition*, **47**, 453–459.

Russo, F., Clemente, C., Linsalata, M., Chiloiro, M., Orlando, A., Marconi, E., Chimienti, G. & Riezzo, G. (2011) Effects of a diet with inulin-enriched pasta on gut peptides and gastric emptying rates in healthy young volunteers. *European Journal of Nutrition*, **50**, 271–277.

Russo, F., Linsalata, M., Clemente, C., Chiloiro, M., Orlando, A., Marconi, E., Chimienti, G. & Riezzo, G. (2012) Inulin-enriched pasta improves intestinal permeability and modifies the circulating levels of zonulin and glucagon-like peptide 2 in healthy young volunteers. *Nutrition Research*, **32**, 940–946.

Rycroft, C.E., Jones, M.R., Gibson, G.R. & Rastall, R. A. (2001) A comparative in vitro evaluation of the fermentation properties of prebiotic oligosaccharides. *Journal of Applied Microbiology*, **91**, 878–887.

Sambucetti, L.C., Fischer, D.D., Zabludoff, S., Kwon, P.O. Chamberlin, H., Trogani, N., Hong Xu, H. & Cohen, D. (1999) Histone deacetylase inhibition selectively alters the activity and expression of cell cycle proteins leading to specific chromatin acetylation and antiproliferative effects. *Journal of Biological Chemistry*, **274**, 34940–34947.

Shadid, R., Haarman, M., Knol, J., Theis, W., Beermann, C., Rjosk-Dendorfer, D., Schendel, D.J., Koletzko, B.V. & Krauss-Etschmann, S. (2007) Effects of galactooligosaccharide and long-chain fructooligosaccharide supplementation during pregnancy on maternal and neonatal microbiota and immunity – a randomized, double-blind, placebo-controlled study. *American Journal of Clinical Nutrition*, **86**, 1426–1437.

Sierra, C., Bernal, M.-J., Blasco, J., Martınez, R., Dalmau, J., Ortuno, I., Espín, B., Vasallo, M.-I., Gil, D., Vidal, M.-L., Infante, D., Leis, R., Maldonado, J., Moreno, J.-M. & Roman, E. (2015) Prebiotic effect during the first year of life in healthy infants fed formula containing GOS as the only prebiotic: a multicentre, randomised, double-blind and placebo-controlled trial. *European Journal of Nutrition*, **54**, 89–99.

Slegte, J.D. (2002) Determination of trans-galactooligosaccharides in selected food products by ion-exchange chromatography: collaborative study. *Journal of AOAC International*, **85**, 417–423.

Slevin, M.M., Allsopp, P.J., Magee, P.J., Bonham, M.P., Naughton, V.R., Strain, J.J., Duffy, M.E., Wallace, J.M. & McSorley, E.M. (2014) Supplementation with calcium and short-chain fructo-oligosaccharides affects markers of one turnover but not bone mineral density in postmenopausal women. *Journal of Nutrition*, **144**, 297–304.

Stark, P.L. & Lee, A. (1982) The microbial ecology of the large bowel of breast-fed and formula-fed infants during the 1st year of life. *Journal of Medical Microbiology*, **15**, 189–203.

Stoddart, L.A., Smith, N.J. & Milligan, G. (2008) International Union of Pharmacology – LXXI – Free fatty acid receptors FFA-1, -2, and -3: pharmacology and pathophysiological functions. *Pharmacological Reviews*, **60**, 405–417.

Suzuki, T., Yoshida, S. & Hara, H. (2008) Physiological concentrations of short-chain fatty acids immediately suppress colonic epithelial permeability. *British Journal of Nutrition*, **100**, 297–305.

Tahiri, M., Tressol, J.C., Arnaud, J., Bornet, F., Bouteloup-Demange, C., Feillet-Coudray, C., Ducros, V., Pepin, D., Brouns, F., Roussel, A.M., Rayssiguier, Y. & Coudry, C. (2001) Five-week intake of short-chain fructo-oligosaccharides increases intestinal absorption and status of magnesium in postmenopausal women. *Journal of Bone and Mineral Research*, **16**, 2152–2160.

Tahiri, M., Tressol, J.C., Arnaud, J., Bornet, F., Bouteloup-Demange, C., Feillet-Coudray, C., Brandolini, M., Ducros, V., Pepin, D., Brouns, F., Roussel, A.M., Rayssiguier, Y. & Coudry, C. (2003) Effect of short-chain fructooligosaccharides on intestinal calcium absorption and calcium status in postmenopausal women: a stable-isotope study. *American Journal of Clinical Nutrition*, **77**, 449–457.

Takasugi, S., Ashida, K., Maruyama, S., Matsukiyo, Y., Kaneko, T. & Yamaji, T. (2013) A combination of a dairy product fermented by lactobacilli and galactooligosaccharides shows additive effects on mineral balances in growing rats with hypochlorhydria induced by a proton pump inhibitor. *Biological Trace Element Research*, **153**, 309–318.

Tanaka, R., Takayama, H., Morotomi, M., Kuroshima, T, Ueyama, S., Matsumoto, K., Kuroda, A. & Mutai, M. (1983) Effects of administration of TOS and *Bifidobacterium breve* 4006 on the human fecal flora. *Bifidobacteria and Microflora*, **2**, 17–24.

Tang, W.H.W., Wang, Z., Levison, B.S., Koeth, R.A., Britt, E.B., Fu, X., Wu, Y. & Hazen, S.L. (2013) Intestinal microbial metabolism of phosphatidylcholine and cardiovascular risk. *New England Journal of Medicine*, **368**, 1575–1584.

Tarini, J. & Wolever, T.M. (2010) The fermentable fibre inulin increases postprandial serum short-chain fatty acids and reduces free-fatty acids and ghrelin in healthy subjects. *Applied Physiology, Nutrition, and Metabolism*, **35**, 9–16.

Tateyama, I., Hashii, K., Johno, I., Iino, T., Hirai, K., Suwa, Y. & Kiso, Y. (2005) Effect of xylooligosaccharide intake on severe constipation in pregnant women. *Journal of Nutritional Science and Vitaminology*, **51**, 445–448.

Tazoe, H., Otomo, Y., Kaji, I., Tanaka, R., Karaki, S.I. & Kuwahara, A. (2008) Roles of short-chain fatty acids receptors, GPR41 and GPR43 on colonic functions. *Journal of Physiology and Pharmacology*, **59**, 251–262.

Tazoe, H., Otomo, Y., Karaki, S.-I., Kato, I., Fukami, Y., Terasaki, M. & Kuwahara, A. (2009) Expression of short-chain fatty acid receptor GPR41 in the human colon. *Biomedical Research – Tokyo*, **30**, 149–156.

Ten Bruggencate, S.J.M., Bovee-Oudenhoven, I.M.J., Lettink-Wissink, M.L.G., Katan, M.B. & van der Meer, R. (2006) Dietary fructooligosaccharides affect intestinal barrier function in healthy men. *Journal of Nutrition*, **136**, 70–74.

Upadhyaya, B., McCormack, L., Fardin-Kia, A.R., Juenemann, R., Nichenametla, S., Clapper, J. & Dey, M. (2016) Impact of dietary resistant starch type 4 on human gut microbiota and immunometabolic functions. *Scientific Reports*, **6**, 28797–28797.

van den Heuvel, E.G., Muijs, T., van Dokkum, W. & Schaafsma, G. (1999a) Lactulose stimulates calcium absorption in postmenopausal women. *Journal of Bone and Mineral Research*, **14**, 1211–1216.

van den Heuvel, E.G., Muys, T., van Dokkum, W. & Schaafsma, G. (1999b) Oligofructose stimulates calcium absorption in adolescents. *American Journal of Clinical Nutrition*, **69**, 544–548.

van den Heuvel, E.G., Muijs, T., Brouns, F. & Hendriks, H. F. (2009) Short-chain fructo-oligosaccharides improve magnesium absorption in adolescent girls with a low calcium intake. *Nutrition Research*, **29**, 229–237.

van den Heuvel, E.G., Schoterman, M.H. & Muijs, T. (2000) Transgalactooligosaccharides stimulate calcium absorption in postmenopausal women. *Journal of Nutrition*, **130**, 2938–2942.

van Dokkum, W., Wezendonk, B., Srikumar, T.S. & van den Heuvel, E.G. (1999) Effect of nondigestible oligosaccharides on large-bowel functions, blood lipid concentrations and glucose absorption in young healthy male subjects. *European Journal of Clinical Nutrition*, **53**, 1–7.

van Hoffen, E., Ruiter, B., Faber, J., M'Rabet, L., Knol, E.F., Stahl, B., Arslanoglu, S., Moro, G., Boehm, G. & Garssen, J. (2009) A specific mixture of short-chain galacto-oligosaccharides and long-chain fructo-oligosaccharides induces a beneficial immunoglobulin profile in infants at high risk for allergy. *Allergy*, **64**, 484–487.

van Loo, J., Coussement, P., Deleenheer, L., Hoebregs, H. & Smits, G. (1995) On the presence of inulin and oligofructose as natural ingredients in the western diet. *Critical Reviews in Food Science and Nutrition*, **35**, 525–552.

van Loo, J., Cummings, J., Delzenne, N., Englyst, H. Franck, A., Hopkins, M., Kok, N., Macfarlane, G., Newton, D., Quigley, M., Roberfroid, M., van Vliet, T. & van den Heuvel, E. (1999) Functional food properties of non-digestible oligosaccharides: a consensus report from the ENDO project (DGXII AIRII-CT94-1095). *British Journal of Nutrition*, **81**, 121–132.

Venkataraman, A., Sieber, J.R., Schmidt, A.W., Waldron, C., Theis, K.R. & Schmidt, T.M. (2016) Variable responses of human microbiomes to dietary supplementation with resistant starch. *Microbiome*, **4**, 33–33.

Virk, A., Mandrekar, J., Berbari, E.F., Boyce, T.G., Fischer, P.R., Kasten, M.J., Orenstein, R., Rosenblatt, J.E., Sampathkumar, P., Sia, I., Springer, D. & Witzig, T.E. (2013) A randomized, double blind, placebo-controlled trial of an oral synbiotic (AKSB) for prevention of travelers' diarrhea. *Journal of Travel Medicine*, **20**, 88–94.

Vulevic, J., Drakoularakou, A., Yaqoob, P., Tzortzis, G. & Gibson, G.R. (2008) Modulation of the fecal microflora profile and immune function by a novel trans-galactooligosaccharide mixture (B-GOS) in healthy elderly volunteers. *American Journal of Clinical Nutrition*, **88**, 1438–1446.

Vulevic, J., Juric, A., Tzortzis, G. & Gibson, G.R. (2013) A mixture of trans-galactooligosaccharides reduces markers of metabolic syndrome and modulates the fecal microbiota and immune function of overweight adults. *Journal of Nutrition*, **143**, 324–331.

Walton, G.E., van den Heuvel, E.G.H.M., Kosters, M.H.W., Rastall, R.A., Tuohy, K.M. & Gibson, G.R. (2012) A randomised crossover study investigating the effects of galacto-oligosaccharides on the faecal microbiota in men and women over 50 years of age. *British Journal of Nutrition*, **107**, 1466–1475.

Wang, X. & Gibson, G.R. (1993) Effects of the in-vitro fermentation of oligofructose and inulin by bacteria growing in the human large-intestine. *Journal of Applied Bacteriology*, **75**, 373–380.

Wang, Z., Klipfell, E., Bennett, B.J., Koeth, R., Levison, B.S., Dugar, B., Feldstein, A.E., Britt, E.B., Fu, X., Chung, Y.M., Wu, Y., Schauer, P., Smith, J.D., Allayee, H., Tang, W.H., DiDonato, J.A., Lusis, A.J. & Hazen, S.L. (2011) Gut flora metabolism of phosphatidylcholine promotes cardiovascular disease. *Nature*, **472**, 57–65.

Weaver, C.M., Martin, B.R., Nakatsu, C.H., Armstrong, A.P., Clavijo, A., McCabe, L.D., McCabe, G.P., Duignan, S., Schoterman. M.H. & van den Heuvel, E.G. (2011) Galactooligosaccharides improve mineral absorption and bone properties in growing rats through gut fermentation. *Journal of Agricultural and Food Chemistry*, **59**, 6501–6510.

Welters, C.F., Heineman, E., Thunnissen, F.B., van den Bogaard, A.E., Soeters, P.B. & Baeten, C.G. (2002) Effect of dietary inulin supplementation on inflammation of pouch mucosa in patients with an ileal pouch-anal anastomosis. *Diseases of the Colon and Rectum*, **45**, 621–627.

Westerbeek, E.A., Morch, E., Lafeber, H.N., Fetter, W.P., Twisk, J.W. & Van Elburg, R.M. (2011) Effect of neutral and acidic oligosaccharides on fecal IL-8 and fecal calprotectin in preterm infants. *Pediatric Research*, **69**, 255–258.

Whisner, C.M., Martin, B.R., Schoterman, M.H.C., Nakatsu, C.H., McCabe, L.D., McCabe, G.P., Wastney, M.E, van den Heuvel, E.G.H.M. & Weaver, C.M. (2013) Galacto-oligosaccharides increase calcium absorption and gut bifidobacteria in young girls: a double-blind cross-over trial. *British Journal of Nutrition*, **110**, 1292–1303.

Williams, C.H., Witherly, S.A. & Buddington, R.K. (1994) Influence of dietary neosugar on selected bacterial groups of the human fecal microbiota. *Microbial Ecology in Health and Disease*, **7**, 91–97.

Windey, K., De Preter, V. & Verbeke, K. (2012) Relevance of protein fermentation to gut health. *Molecular Nutrition & Food Research*, **56**, 184–196.

Xiong, Y.M., Miyamoto, N., Shibata, K., Valasek, M.A., Motoike, T., Kedzierski, R.M. & Yanagisawa, M. (2004) Short-chain fatty acids stimulate leptin production in adipocytes through the G protein-coupled receptor GPR41. *Proceedings of the National Academy of Sciences of the United States of America*, **101**, 1045–1050.

Yamashita, K. & Kobata, A. (1974) Oligosaccharides of human milk. 5. Isolation and characterization of a new trisaccharide, 6′-galactosyllactose. *Archives of Biochemistry and Biophysics*, **161**, 164–170.

Yamashita, K., Tachibana, Y. & Kobata, A. (1976a) Oligosaccharides of human milk – isolation and characterization of 2 new non-asaccharides, monofucosyllacto-*N*-octaose and monofucosyllacto-*N*-neooctaose. *Biochemistry*, **15**, 3950–3955.

Yamashita, K., Tachibana, Y. & Kobata, A. (1976b) Oligosaccharides of human milk – isolation and characterization of 3 new disialylfucosyl hexasaccharides. *Archives of Biochemistry and Biophysics*, **174**, 582–591.

Yamashita, K., Tachibana, Y. & Kobata, A. (1977a) Oligosaccharides of human milk – structures of 3 lacto-*N*-hexaose derivatives with H-haptenic structure. *Archives of Biochemistry and Biophysics*, **182**, 546–555.

Yamashita, K., Tachibana, Y. & Kobata, A. (1977b) Oligosaccharides of human milk. 10. Structural studies of 2 new octasaccharides, difucosyl derivatives of para-lacto-n-hexaose and para-lacto-*N*-neohexaose. *Journal of Biological Chemistry*, **252**, 5408–5411.

Yamashita, K., Kawai, K. & Itakura, M. (1984) Effects of fructo-oligosaccharides on blood-glucose and serum-lipids in diabetic subjects. *Nutrition Research*, **4**, 961–966.

Yano, J.M., Yu, K., Donaldson, G.P., Shastri, G.G., Ann, P., Ma, L., Nagler, C.R., Ismagilov, R.F., Mazmanian, S.K. & Hsiao, E.Y. (2015) Indigenous bacteria from the gut microbiota regulate host serotonin biosynthesis. *Cell*, **161**, 264–276.

8 An Overview of Probiotic Research: Human and Mechanistic Studies[1]

G. Zoumpopoulou, E. Tsakalidou and L.V. Thomas

8.1 Mechanisms underlying probiotic effects

The human gut microbiota, comprising 10^{14} microbial cells or more, has such a fundamental influence on health that experts now consider it to be an organ in the body (Marchesi *et al.*, 2016). Despite the fact that they contain much lower numbers of micro-organisms (usually in the order of 10^9 to 10^{10}), probiotics are associated with numerous health benefits (Bermudez-Brito *et al.*, 2012) as a result of several different mechanisms of activity, which can be categorised as follows (Hill *et al.*, 2014):

- Widespread mechanisms among commonly studied probiotic genera. Examples: colonisation resistance, competitive exclusion of pathogens, acid, short-chain fatty acid (SCFA) production, regulation of intestinal transit, normalisation of a perturbed microbiota and increased turnover of enterocytes.
- Frequently observed mechanisms amongst most strains of a probiotic species. Examples: vitamin synthesis, bile salt metabolism, direct antagonism, gut barrier reinforcement, enzymatic activity and neutralisation of carcinogens.
- Rare mechanisms, present in only a few strains of a given species. Examples: those responsible for neurological, immunological and endocrinological effects, and the production of specific bioactives.

Understanding how probiotics work is important for new product development, for exploring new health benefits and to substantiate claims. Mechanistic insights can be derived from *ex vivo*, animal model and *in vitro* studies as well as human intervention trials. The latter remain the cornerstone of evidence for probiotic effects, with randomised placebo-controlled trials considered the best design. Subjects in these trials can range from healthy people to those with sub-optimum health and patients with varying severities of illness. When novel areas of benefit are being explored, often smaller pilot trials are first conducted, to evaluate if further work is warranted as well as inform the design of subsequent confirmatory trials. The widespread probiotic consumption in several countries has meant that epidemiological studies can be

[1] This chapter is dedicated to the memory of Professor George MacFarlane, a co-author of the original chapter in 2005, in acknowledgement of the major contribution he made to gut microbiology research.

Probiotic Dairy Products, Second Edition. Edited by Adnan Y. Tamime and Linda V. Thomas.

conducted, even studies investigating how probiotic use could reduce healthcare costs (Toi *et al.*, 2013; Lenoir-Wijnkoop *et al.*, 2015).

8.1.1 *Probiotic effects on the gut microbiota and its metabolites*

Developments in molecular techniques for the analysis of microbial ecosystems have led to intense research into the human microbiome over the last decade (Koren *et al.*, 2013; Belizario & Napolitano, 2015). This has meant there is better understanding of how the gut microbiota changes over the course of a lifetime, and how it is affected by dietary and lifestyle factors, such as medication, infection, poor diet and stress (Jandhyala *et al.*, 2015; Jeffery *et al.*, 2015). Antibiotics in particular disrupt the gut microbiota, which can increase risk of diarrhoea in older people (Lopez *et al.*, 2014; Gillespie *et al.*, 2015). Antibiotic use in early life may also have long-term health consequences (Nobel *et al.*, 2015; Schulfer & Blaser, 2015). There have been frequent observations of compositional differences between the gut microbiota of healthy people compared to those with disease or increased risk of disease (Thomas *et al.*, 2014; Borges-Canha *et al.*, 2015), but it is not always clear whether this was the cause or result of disease. Probiotic trials may help clarify this. The increase of bifidobacteria or lactobacilli numbers in the gut is a beneficial and widespread effect associated with probiotics, and one reason why many products contain these bacteria (Tojo *et al.*, 2014; Di Cerbo *et al.*, 2015).

Colonisation resistance describes the ability of the commensal microbes to inhibit gut overgrowth or colonisation by pathogens or harmful bacteria. Several mechanisms may be involved in this. The commensal bacteria may competitively exclude invading microorganisms from nutrients and niches in the gut by producing SCFAs and/or bacteriocins, consuming available oxygen and enhancing immune and intestinal barrier functions (Lawley & Walker, 2013; Arques *et al.*, 2015). Disruption of the commensal microbiota reduces its protective capability, but it can be restored by probiotics. Probiotics can promote beneficial or non-harmful commensal species and inhibit harmful species, for example by producing bacteriocins (Dobson *et al.*, 2012) and immune modulation. Such mechanisms help the resilience of the gut microbiota and its ability to revert to a 'normal' profile following any disruption. Microbial metabolism in the gut is also affected by probiotics, in particular levels of SCFAs, such as butyrate, propionate, acetate and lactate (Flint *et al.*, 2012). These metabolites have many positive effects, such as lowering gut pH, downregulating inflammation, improving gut barrier function and regulating satiety and enterocyte growth (Kim *et al.*, 2014; Canfora *et al.*, 2015; Kelly *et al.*, 2015a). SCFAs also affect the release of glucagon-like peptide-1 (GLP-1), a hormone that regulates small intestinal transit (Wichmann *et al.*, 2013). Probiotics may produce SCFAs or promote the growth of commensal species producing these compounds. Other mechanisms attributed to probiotic effects on bowel function include neurological effects, bile deconjugation activity and reduction of methane-producing species (Choi & Chang, 2015).

Enzymes in the gut can also be influenced by probiotics. For example, the enzyme bile salt hydrolase, which is produced by certain commensals and probiotics, regulates lipid metabolism due to its ability to deconjugate bile salts, which are then excreted via the faeces (Joyce *et al.*, 2014). Cholesterol replaces bile acids lost in this way; this is the mechanism behind probiotic cholesterol-lowering effects (Shimada *et al.*, 1969; Begley *et al.*, 2006; Ishimwe *et al.*, 2015). Another enzyme that can be produced by probiotics is β-galactosidase, which helps with lactose intolerance (Almeida *et al.*, 2012, Savaiano, 2014). Probiotics have also been shown to improve levels of liver aminotransferases in non-alcoholic fatty liver disease (NAFLD) patients (Buss *et al.*, 2014).

Carbohydrate metabolism in the colon generally produces beneficial products, whereas putrefaction can produce toxins and carcinogens, such as ammonia, phenols, thiols and indoles (Smith & Macfarlane, 1996). Probiotic benefit for colorectal cancer, for instance, may be linked to their promotion of saccharolytic fermentation and reduction of bacterial enzymes linked to carcinogen production (De Preter *et al.*, 2008, 2011). Probiotic-associated increases in butyric acid may also be involved as this is an important regulator of the growth and apoptosis of intestinal epithelium cells (Goncalves & Martel, 2013; Kumar *et al.*, 2013; Zhong *et al.*, 2014). Other cancer-protective probiotic mechanisms include suppression of proteolytic fermentation (De Preter *et al.*, 2011) and carcinogenic secondary bile acids, the binding or degradation of carcinogens and mutagens, anti-genotoxic activity and enhancement of natural killer (NK) cell activity (Commane *et al.*, 2005; Chong, 2014). The latter is particularly important as low NK-cell activity has been linked to increased cancer risk (Fujiki *et al.*, 2000; Furue *et al.*, 2008). Altered NK-cell function has been demonstrated in healthy obese adults and smokers (Laue *et al.*, 2015). Protective effects are not exclusive to colorectal cancer (Raman *et al.*, 2013); encouraging results have been reported for other cancers, such as bladder and breast cancer (Ohashi *et al.*, 2002; Toi *et al.*, 2013).

Beneficial probiotic metabolites also include vitamins (B and K) (Bentley & Meganathan, 1982; Crittenden *et al.*, 2003; Resta, 2009; LeBlanc *et al.*, 2011) and conjugated linoleic acid (Fernandez *et al.*, 2015). Certain probiotics also produce or enhance levels of bioactive peptides with antimicrobial activity, such as defensins and bacteriocins (Schlee *et al.*, 2008; Dobson *et al.*, 2012). Gamma-aminobutyric acid (GABA), another bioactive peptide produced by probiotics, can benefit mild hypertension (Li & Cao, 2010; Khalesi *et al.*, 2014); GABA is also involved in regulation of depression, anxiety and hormone secretion. Other mechanisms relating to cardiovascular health include production of ACE-inhibitory peptides, modulation of the oral microbiota and effects on cholesterol metabolism (Ramchandran & Shah, 2008; Ettinger *et al.*, 2014).

8.1.2 *Probiotic immune modulation*

The majority of the body's lymph nodes are located in the gut-associated lymphoid tissue (GALT) with particular concentrations in the Peyer's patches of the small intestine (Forchielli & Walker, 2005; MacDonald & Bateman, 2007). After birth and throughout life, the gut microbiota is vital in establishing and programming a well-regulated immune response (Erturk-Hasdemir & Kasper, 2013; Weng & Walker, 2013; Peterson *et al.*, 2015). Commensal microbes (and probiotics) communicate with the immune system via pattern recognition receptors (e.g. Toll-like and nucleotide oligomerisation domain-like receptors) on the intestinal epithelial cells in the gut and dendritic cells that extend into the gut lumen to monitor antigens. Microfold cells also take up antigens from the gut lumen by endocytosis or phagocytosis, and transport these into the lamina propria for processing and antigen presentation (Janeway *et al.*, 2005; Tanoue *et al.*, 2010; Min & Rhee, 2015).

Immune modulation has been demonstrated for many probiotic strains (Ng *et al.*, 2009; Hardy *et al.*, 2013; Giorgetti *et al.*, 2015; Santiago-Lopez *et al.*, 2015; Wan *et al.*, 2015): human studies have shown effects on vaccine and pathogen antibody titres, NK-cell activity, salivary immunoglobulin A (IgA), T-cell activation and several cytokines. Lactobacilli have been particularly associated with induction of T helper 1 cytokines and maintenance of NK cells, and bifidobacteria with downregulation of inflammation (Dong *et al.*, 2012; Ashraf & Shah, 2014). The immune response induced by a probiotic, however, may depend on the gut environment and other bacteria present there (Shida *et al.*, 2011). Immune effects are complex and strain-specific, as was observed in a trial where biopsies taken from the proximal small intestine of healthy adults showed strain-specific responses of the mucosal immune system, involving many different gene-regulatory networks and pathways (Van Baarlen *et al.*, 2011).

8.1.3 *Probiotic effects on gut barrier function*

Any impairment to gut permeability results in temporary loss of intestinal homeostasis, and can lead to functional disorders and disease: infectious diarrhoea, inflammatory bowel disease (IBD), irritable bowel syndrome (IBS), allergies, obesity and metabolic diseases (Bischoff *et al.*, 2014). The many mechanisms whereby probiotics help protect the intestinal mucosa include regulation of cell division and apoptosis, synthesis of proteins, strengthening of the epithelial tight junctions, support of the immune system and protection of the mucus layer (Rao & Samak, 2013).

Lipopolysaccharide, a major component of Gram-negative bacterial cell walls, is an endotoxin. When gut permeability becomes impaired, the 'leaky' gut lets bacteria or their products to translocate into the body, which increases blood endotoxin level that triggers a low-grade chronic inflammation promoting metabolic diseases, such as type 2 diabetes mellitus (T2DM) and NAFLD (Cani *et al.*, 2007, 2009; Scarpellini *et al.*, 2014). Probiotic benefit for metabolic disease is linked to protection of gut barrier function (Rao & Samak, 2013; Bischoff *et al.*, 2014; Delzenne *et al.*, 2015).

8.1.4 *Probiotics and the gut–brain axis*

The two-way communication between the gut microbiota and the brain is regulated by the neural, endocrine and immune systems. The importance of the gut microbiota for brain development and behaviour is evident from germ-free animals, which display neurochemical differences and anxiety (Diaz Heijtz *et al.*, 2011; Neufeld *et al.*, 2011). Stress in early life has an impact on the gut microbiota, immune system and behaviour; disruption of microbial colonisation at birth also affects neurodevelopment. The vagus nerve, spinal cord and immune and neuroendocrine systems are all influenced by gut bacteria (Liu *et al.*, 2015).

The term 'psychobiotics' has been coined for probiotics that benefit mood, anxiety or cognition (Dinan *et al.*, 2013; Kelly *et al.*, 2015b). Animal studies have revealed how such strains affect the gut–brain axis, with the vagus nerve identified as a key route for this (Bravo *et al.*, 2011; Savignac *et al.*, 2015). Probiotics may also produce or influence neuroactive metabolites, such as GABA, serotonin, catecholamines and acetylcholine (Wall *et al.*, 2014). Gut permeability effects may also be involved, as well as effects on hormones produced in the gut including those involved in pain perception, mood and glucose metabolism (Bienenstock *et al.*, 2015; Chichlowski & Rudolph, 2015; Razmpoosh *et al.*, 2015).

8.1.5 *Probiotic mechanisms in the urogenital tract*

Lactobacillus, the dominant species in the vaginal microbiota of healthy women, is considered a biomarker for a healthy vaginal ecosystem. Lactobacilli are depleted in bacterial vaginosis (BV) (Petrova *et al.*, 2015), and probiotics applied orally or vaginally (sometimes with antibiotics) help restore numbers (Bisanz *et al.*, 2014; Macklaim *et al.*, 2015). This enhances the colonisation resistance of the vagina by increasing lactic acid production, by preventing pathogen adherence to vaginal cells, and possibly through bacteriocin production and stimulation of the host immune response. Similar mechanisms may underlie probiotic benefits for urinary tract infections (UTIs) (Grin *et al.*, 2013), human papillomavirus (Verhoeven *et al.*, 2013) and human immunodeficiency virus (HIV) (Hemsworth *et al.*, 2012; Wilson *et al.*, 2013).

8.1.6 *Survival of the gut microbiota through the gut*

Fermented dairy probiotics are normally taken orally, and it is important that the probiotic strains in these products are able to survive through the gut. Although *in vitro* or gut model studies are useful screening methods for new strains, gut survival of commercial strains should be proved in studies where volunteers ingest the product or an equivalent number of its probiotic strain, after which live cells of the probiotic strain should be detected and enumerated in the faeces (Tuohy *et al.*, 2007). Probiotics do not permanently colonise the gut; strains are usually detected for up to one week following cessation of the intervention.

8.2 Probiotic human studies: gastrointestinal conditions

8.2.1 *Inflammatory bowel disease (IBD)*

Inflammatory bowel disease is a heterogeneous group of chronic, relapsing, immune-mediated disorders of the gastrointestinal (GI) tract, which mainly includes Crohn's disease (CD) and ulcerative colitis (UC) (Mulder *et al.*, 2014). Although the aetiology of IBD remains unclear, it is thought to result from an inappropriate and continuing inflammatory response to commensal microbes in a genetically susceptible host (Khor *et al.*, 2011). This has prompted clinical trials of microbial-modulatory strategies, including probiotics (Hansen & Sartor, 2015).

Crohn's disease (CD)

Chron's disease is characterised by a discontinuous transmural inflammation that can affect any part of the GI tract (Mulder *et al.*, 2014). Early probiotic studies reported symptom benefit using the yeast '*Saccharomyces boulardii*' (presumed to be *Saccharomyces cerevisiae* var. *boulardii*; Rajkowska & Kunicka-Styczynska, 2009) (Plein & Hotz, 1993) and better maintenance of remission with *Escherichia coli* Nissle 1917 (Malchow, 1997), but there have been relatively few probiotic trials in CD patients, and those that have been conducted had significant limitations (Ghouri *et al.*, 2014). Most have investigated maintenance of remission rather than treatment of active episodes of disease.

As far as we know, in terms of CD trials, only two probiotics have been tested as single strains: *Lactobacillus rhamnosus* GG and *Lactobacillus johnsonii* LA-1. Administration of *Lb. rhamnosus* GG to 38 patients following surgical resection of diseased gut found no significant improvement in severity of recurrent CD lesions (Prantera *et al.*, 2002). *Lactobacillus rhamnosus* GG was also investigated in a double-blind, placebo-controlled, randomised trial (DBPCRT) involving 11 patients with moderate to active CD who were followed up for 6 months (Schultz *et al.*, 2004). Possibly because the patients had received antibiotics the week before the probiotic intervention as well as a tapering regime of corticosteroids in the first 12 weeks, *Lb. rhamnosus* GG had no effect in maintaining medically induced CD remission. In the largest trial to date, *Lb. rhamnosus* GG was tested as an adjunct to standard maintenance therapy in children with CD (Bousvaros *et al.*, 2005). Seventy-five children and adolescents (5 to 21 years) with CD but in remission were randomised to receive either probiotic or placebo; concomitant medications and low-dose alternate day corticosteroids were allowed. No significant effects of *Lb. rhamnosus* GG on relapse rate or median time to relapse were observed after 2 years. *Lactobacillus johnsonii* LA-1 was used in two trials investigating patients who had either undergone surgical resection of <1 m within the previous 21 d (Marteau *et al.*, 2006) or ileo-caecal resection 3–7 d before (Van Gossum *et al.*, 2007), but both studies showed no effect of *Lb. johnsonii* LA-1 in preventing early endoscopic recurrence of CD.

Trials with the yeast '*Sac. boulardii*' (presumed to be *Sac. cerevisiae* var. *boulardii*) have also been conducted. In an early study, 32 CD patients in clinical remission for at least 3 months were randomly treated for 6 months with mesalazine or mesalazine plus '*Sac. boulardii*' (presumed to be *Sac. cerevisiae* var. *boulardii*). The combination of the yeast with the standard treatment resulted in significantly fewer patients experiencing a clinical relapse (Guslandi *et al.*, 2000), but a later DBPCRT reported no benefit with '*Sac. boulardii*' (presumed to be *Sac. cerevisiae* var. *boulardii*) in a trial involving 165 CD patients in remission (due to steroid or salicylate therapy) (Bourreille *et al.*, 2013).

The multispecies bacterial cocktail VSL#3 was tested in a DBPCRT of CD patients following ileo-colonic surgical resection. The probiotic was associated with less severe endoscopic recurrence, reduced levels of pro-inflammatory cytokines [interleukin-1β (IL1β), tumour necrosis factor-α (TNFα) and interferon-γ (IFNγ)] and increased transforming growth factor-β (TGFβ) (Madsen *et al.*, 2008). Synbiotic preparations (mixtures of pro- and prebiotics) have also shown effectiveness in reducing symptoms of active CD (Fujimori *et al.*, 2007; Steed *et al.*, 2010), but

Synbiotic 2000 (a mix of four probiotic bacteria and four prebiotics) had no effect on post-operative CD recurrence (Chermesh *et al.*, 2007).

A recent systematic review of randomised clinical trials (RCTs) concluded synbiotics showed potential as therapies for active CD (Saez-Lara *et al.*, 2015), but Cochrane systematic reviews have concluded there is no evidence to support use of probiotics for the induction and the maintenance of remission of CD (Rolfe *et al.*, 2006; Butterworth *et al.*, 2008; Rahimi *et al.*, 2008). The lack of RCTs and the small patient numbers in the trials were noted. It was also concluded from another meta-analysis that probiotics were not effective in maintaining remission or preventing clinical and endoscopic relapse in CD. However, most trials investigating prevention of relapses in inactive CD include patients with both ileal and colonic predominant disease. This was not considered in the meta-analyses (Jonkers *et al.*, 2012); thus, the lack of positive results may be partly due to inter-individual differences of disease location (e.g. ileum *vs.* colon) (Jonkers *et al.*, 2012), as well as differences at the genetic level (such as polymorphisms in genes involved in microbial response) (Sokol, 2014).

Ulcerative colitis (UC)

In UC, the continuous superficial mucosal inflammation seen is restricted to the colon, which, together with particular features of its inflammation, differentiates it from CD (Mulder *et al.*, 2014). The various RCTs that have been conducted with probiotics have mainly focused on their ability to induce and/or maintain remission in active mild to moderate disease, and have compared them with either placebo or standard UC maintenance therapy. There have been encouraging, albeit conflicting, results (Orel & Kamhi Trop, 2014). In patients with active yet mild to moderate disease, for instance, a pilot study reported benefit from a 4-week intervention with a single strain of '*Sac. boulardii*' (presumed to be *Sac. cerevisiae* var. *boulardii*) combined with mesalazine maintenance treatment (Guslandi *et al.*, 2003). The most studied probiotic in UC has been the multispecies VSL#3 product that, when combined with balsalazide, proved to be significantly superior to balsalazide or mesalazine alone in achieving remission in patients (Tursi *et al.*, 2004). Over the last decade, three open-label studies with VSL#3 have shown it to help improve symptoms, which were assessed using the UC disease activity index (UCDAI), simple clinical colitis activity index (SCCAI) scores, endoscopic assessment and other inflammatory markers (Bibiloni *et al.*, 2005; Huynh *et al.*, 2009; Lee *et al.*, 2012). The VSL#3 cocktail has also now been tested in two DBPCRTs with adults (Sood *et al.*, 2009; Tursi *et al.*, 2010) and one with children (Miele *et al.*, 2009): it was effective in achieving clinical responses and remissions. This was also the case for another multispecies probiotic preparation, BIO-THREE [containing '*Streptococcus faecalis*' (presumed to be *Enterococcus faecalis*) (Devriese & Pot, 1995), *Clostridium butyricum* and *Bacillus mesentericus*], which was investigated for the treatment of mild to moderate distal UC refractory to conventional therapies (Tsuda *et al.*, 2007). In a DBPCRT, a *Bifidobacterium*-fermented milk (BFM) containing *Bifidobacterium breve* strain Yakult, *Bifidobacterium bifidum* and *Lactobacillus acidophilus* was administered to patients with active UC (Kato *et al.*, 2004), resulting in significant lowering of their clinical activity index (compared to placebo), as well as improved endoscopic activity index and histological scores. Patients treated with the probiotic had increases in total faecal SCFAs, especially for butyrate and propionate concentrations. When a '*Bifidobacterium longum*' (presumed to be *Bifidobacterium longum* subsp. *longum*) (Mattarelli *et al.*, 2008; Underwood *et al.*, 2015) strain and the prebiotic Synergy 1 were taken for one month in a DBPCRT of patients with active UC (Furrie *et al.*, 2005), this appeared to reduce inflammation. In another study, a year-long intervention with a synbiotic combination of *Bif. breve* strain Yakult plus prebiotic galactooligosaccharide (GOS) in active UC patients improved clinical symptoms (Ishikawa *et al.*, 2011). Recently, a combination of strains of species of *Lactobacillus delbrueckii* spp. and *Lactobacillus fermentum* showed promising effects on inflammatory mediators and nuclear factor (NF)-κB activation in active UC patients (Hegazy & El-Bedewy, 2010). Capsules of Bifid Triple Viable [a Chinese product containing '*Bacillus acidophilus*' (presumed to be *Lb. acidophilus*) (Rogosa, 1974a), *Bif. bifidum* and faecal enterococci] have also been investigated in active UC and were associated with the induction of remission, which was linked to immune effects (Li *et al.*, 2012a). Finally, rectal administration of probiotics has also been investigated with active distal UC, achieving promising results with *E. coli* strain Nissle 1917 (Matthes *et al.*, 2010), a *Lactobacillus casei* strain (D'Inca *et al.*, 2011) and a *Lactobacillus reuteri* strain (Oliva *et al.*, 2012).

A systematic review in 2007 identified only a limited number of probiotic studies investigating induction and maintenance of remission in active UC; these had many differences in methodology and results (Zigra *et al.*, 2007). A more recent meta-analysis concluded that, compared to placebo, overall probiotics achieved significantly higher remission rates in patients with active UC, although a subgroup-specific meta-analysis found only VSL#3 was effective (Mardini & Grigorian, 2014; Shen *et al.*, 2014). Interestingly, probiotics recommended for the induction of remission in UC by an expert group in 2014 included *E. coli* Nissle 1917 and VSL#3, but the evidence was rated as 'B' because some studies do not show positive effects (Floch, 2014).

Probiotic benefit for maintaining remission in UC has also been investigated, comparing effects to either placebo or standard medication, such as mesalazine (Cammarota *et al.*, 2015). *Escherichia coli* Nissle 1917 is the strain most widely investigated (Kruis *et al.*, 1997; Rembacken *et al.*, 1999; Kruis *et al.*, 2004; Henker *et al.*, 2008). Three trials with adults and one with children assessed clinical activity index scores, relapse rates and relapse-free periods in patients; in all cases, *E. coli* Nissle 1917 was shown to be effective, safe and equivalent to mesalazine in maintaining remission. In another study, *Lb. rhamnosus* GG alone was more effective than mesalazine alone in prolonging remission, but similar for relapse rates (Zocco *et al.*, 2006). The BFM supplementation mentioned in this chapter (Kato *et al.*, 2004) has also been investigated in UC patients in remission, and was effective in maintaining remission (Ishikawa *et al.*, 2003). Similarly, the Bifid Triple Viable capsules (now called BIFICO) were investigated in a DBPCRT of UC patients in remission: the efficacy of the probiotic in preventing relapse was determined by clinical, endoscopic and histological assessments plus faecal and immune analyses (Cui *et al.*, 2004). A DBPCRT using *Lactobacillus salivarius* and a '*Bifidobacterium infantis*' (presumed to be *Bifidobacterium longum* subsp. *longum*) strain with 157 patients, however, found no effect on UC relapse rates

(Shanahan *et al.*, 2006), and no significant benefit (compared to placebo) was reported for Probio-Tec AB-25, a combination of *Lb. acidophilus* LA-5 and *Bifidobacterium animalis* subsp. *lactis* BB-12, for patients in remission (Wildt *et al.*, 2011). Only one synbiotic trial on maintenance of UC remission has been reported: patients given the synbiotic had greater improvement of their quality of life compared to those on probiotic or prebiotic treatment alone, but the trial did not conduct any standard endoscopic or histological evaluation of disease activity (Fujimori *et al.*, 2009).

A Cochrane review on probiotic maintenance of UC remission found only four studies that met the inclusion criteria (Naidoo *et al.*, 2011), but a later review (Floch, 2014) gave a strong recommendation (grade 'A') for *E. coli* Nissle 1917 and VSL#3; thus, specific probiotics could be as efficient as standard maintenance therapy. Probiotics could also be useful for patients intolerant or allergic to medical treatment, or as an adjunct to standard therapy (Orel & Kamhi Trop, 2014).

Pouchitis

Pouchitis is the non-specific inflammation of the ileal reservoir that can occur in UC patients who have undergone restorative proctocolectomy with ileal pouch anal anastomosis. This problem can develop in more than half of such patients. Probiotics have been studied only in adults, and for prevention of initial post-operative onset of pouchitis, maintenance of pouchitis remission and treatment of mild to moderate pouchitis (Mack, 2011; Ritchie & Romanuk, 2012).

Positive results were reported in a DBPCRT of VSL#3 investigating prevention of post-operative pouchitis (Gionchetti *et al.*, 2003) and for *Lb. rhamnosus* GG in a retrospective open-label study (Gosselink *et al.*, 2004). In both trials, patients were followed up for one year. In a more recent trial, a 9-month intake period of a probiotic mix [cited by the authors as *Lb. acidophilus*, *Lb. delbrueckii* subsp. *bulgaricus* and '*Bifidobacterium bifidus*' (presumed to be *Bif. bifidum* as *Bif. bifidum*, *Bif. bifidus*, *Bacillus bifidus*, and *Actinomyces parabifidus* are considered as synonyms) (Rogosa, 1974b)] reduced the number of patients developing pouchitis and disease severity (Tomasz *et al.*, 2014). There are few trials on probiotic treatment of mild to moderate pouchitis, and they have only a small numbers of adult subjects (Fedorak & Demeria, 2012). Most research has been with VSL#3 (Gionchetti *et al.*, 2007; Pronio *et al.*, 2008), *Lb. rhamnosus* GG (Kuisma *et al.*, 2003) and a fermented milk product containing *Lb. acidophilus* LA-5 and '*Bif. lactis* BB-12' (Laake *et al.*, 2005); the latter is currently reclassified as *Bif. animalis* subsp. *lactis* BB-12 (Masco *et al.*, 2004; Anonymous, 2013). Efficacy was usually assessed by evaluation of the pouch disease activity index, which comprises clinical symptoms and endoscopic and histological findings. The VSL#3 and the probiotic fermented milk proved clinically effective, but not *Lb. rhamnosus* GG.

Pouchitis can also recur, so probiotic studies have investigated whether they can prevent relapses. The most studied product is VSL#3; a single daily high dose was effective in maintaining antibiotic-induced remission in two trials (Gionchetti *et al.*, 2000; Mimura *et al.*, 2004), but a subsequent uncontrolled trial in routine clinical practice gave disappointing results (Shen *et al.*, 2005), although it was later noted that the study had methodological weaknesses (Orel & Kamhi Trop, 2014). In a more recent study, positive effects of Ecologic 825 during antibiotic-induced remission correlated with changes in mucosal barrier function (Persborn *et al.*, 2013).

A meta-analysis highlighted VSL#3 as significantly reducing pouchitis relapse rates (Shen *et al.*, 2014), and a Cochrane review concluded this probiotic was effective for pouchitis treatment and preventing relapse (Holubar *et al.*, 2010). In 2014, a recommendation of VSL#3 as a probiotic for the maintenance of remission in pouchitis was based on evidence considered strong (Floch, 2014). Overall, the high efficacy of probiotics for treatment of pouchitis, as determined by systematic reviews and meta-analyses, may be partly because trials have only investigated a few probiotics (i.e. VSL#3 and *Lb. rhamnosus* GG) and subjects have all been adults (Ritchie & Romanuk, 2012).

8.2.2 *Irritable bowel syndrome (IBS)*

Irritable bowel syndrome is a common functional bowel disorder characterised by abdominal pain or discomfort in association with altered bowel function. The disease can be sub-divided into constipation-predominant IBS (IBS-C), diarrhoea-predominant IBS (IBS-D), IBS of mixed or alternating symptoms (IBS-M and IBS-A) and IBS of no subtype (IBS-U) (Longstreth *et al.*, 2006). Over the last decade, there have been numerous probiotic trials in different IBS patient groups evaluating effects on different symptoms, in particular abdominal pain, discomfort, bloating and distension, defecation frequency and flatulence, as well as overall IBS severity (Clarke *et al.*, 2012).

Studies with single strains indicate lactobacilli may not be as effective as bifidobacteria for IBS (Brandt *et al.*, 2009). Certainly, there is compelling evidence from trials with '*Bif. infantis* 35624' (presumed to be *Bif. longum* subsp. *infantis* 35624) (O'Mahony *et al.*, 2005; Whorwell *et al.*, 2006), a *Bif. animalis* subsp. *lactis* DN-173 010 yoghurt (Agrawal *et al.*, 2009) and *Bif. bifidum* MIMBb75 (Guglielmetti *et al.*, 2011); and, in general, not as good results with most probiotic lactobacilli studied (Sen *et al.*, 2002; Niv *et al.*, 2005; Ligaarden *et al.*, 2010; Thijssen *et al.*, 2016). There are some exceptions, however (Nobaek *et al.*, 2000; Niedzielin *et al.*, 2001), such as three DBPCRTs of *Lb. rhamnosus* GG in children (Bauserman & Michail, 2005; Gawronska *et al.*, 2007; Francavilla *et al.*, 2010) that resulted in a meta-analysis that concluded *Lb. rhamnosus* GG had a significant positive effect on intensity as well as frequency of pain with IBS (Horvath *et al.*, 2011). There have been promising results from trials with *E. coli* Nissle 1917 (Kruis *et al.*, 2012) and the yeast '*Sac. boulardii*' (presumed to be *Sac. cerevisiae* var. *boulardii*) (Choi *et al.*, 2011; Pineton de Chambrun *et al.*, 2015).

Among the multi-strain bacterial products tested, VSL#3 has proved effective in reducing bloating in both children (Guandalini *et al.*, 2010) and adults (Kim *et al.*, 2003), although this was not replicated in a subsequent adult study (Kim *et al.*, 2005). Mixtures of strains seem particularly effective for IBS symptoms relating to bowel habit (e.g. stool frequency and consistency) (Hosseini *et al.*, 2012; Ortiz-Lucas *et al.*, 2013). In general, mixtures have proved effective in alleviating different IBS symptoms (Saggioro, 2004; Drouault-Holowacz *et al.*, 2008; Enck *et al.*, 2008; Kajander *et al.*, 2008; Sinn *et al.*, 2008; Williams *et al.*, 2009; Simren *et al.*, 2010; Ringel-Kulka *et al.*,

2011; Ki Cha *et al.*, 2012; Yoon *et al.*, 2015), although this was not the case for two synbiotics (Min *et al.*, 2012; Cappello *et al.*, 2013).

There continues to be much interest in probiotics for alleviating and/or preventing IBS-associated symptoms, but the evidence is still not strong enough for general recommendation in clinical guidelines for IBS (Bixquert, 2013). This is because not only are probiotic effects strain-specific, but also it is difficult to evaluate the supportive evidence due to variations in trial design: the range of strains tested, as well as their dosage, duration and formulation, and the numbers and types of IBS patients in the trials (Rogers & Mousa, 2012; Mazurak *et al.*, 2015). Bearing this in mind, *Bifidobacterium* species and certain combinations that include *Bifidobacterium* strains seem more likely to be effective than single *Lactobacillus* probiotics (Ciorba, 2012; Simren *et al.*, 2013). It should also be noted that the most recent and largest meta-analysis to date, including 35 trials, concluded that probiotics are effective therapies for IBS in terms of improving symptoms overall, and improving abdominal pain, bloating and flatulence scores (Ford *et al.*, 2014). This is in line with previous positive meta-analyses (Nikfar *et al.*, 2008; Moayyedi *et al.*, 2010; Ortiz-Lucas *et al.*, 2013).

8.2.3 *Constipation*

Constipation is a well-characterised functional bowel disorder that can be classified into three broad categories: normal-transit constipation (or functional constipation), slow-transit constipation and disorders of defecation or rectal evacuation (Lembo & Camilleri, 2003). In general, probiotic trials have investigated functional constipation in adults, evaluating effects on intestinal transit time (ITT), stool frequency and consistency and defecation symptoms (Davis & Gamier, 2015).

The few DBPCRTs that have evaluated probiotics in adult patients with functional constipation have mostly been of short duration (maximum 4 weeks) (Cash, 2014). In four trials, dairy products such as fermented milk (Yang *et al.*, 2008; Takii *et al.*, 2012), a milk-like drink (Ishizuka *et al.*, 2012) and cheese (Favretto *et al.*, 2013), all containing *Bifidobacterium* strains, were shown to significantly improve constipation symptoms. Many other studies, however, have reported benefit after short-term administration of lactobacilli probiotics (Koebnick *et al.*, 2003; Krammer *et al.*, 2011; Sakai *et al.*, 2011, 2015; Riezzo *et al.*, 2012; Ojetti *et al.*, 2014; Tilley *et al.*, 2014; van den Nieuwboer *et al.*, 2015), except for one study that may have been of too short a duration to show any probiotic effect (Mazlyn *et al.*, 2013). Two recent trials with positive results have been conducted in infants with *Lb. reuteri* DSM 17938 (Coccorullo *et al.*, 2010; Indrio *et al.*, 2014b). There are also trials with synbiotic combinations, in most cases comparing this with either the probiotic alone or a placebo (De Paula *et al.*, 2008; Fateh *et al.*, 2011; Li *et al.*, 2012b; Liu *et al.*, 2012; Jayasimhan *et al.*, 2013; Waitzberg *et al.*, 2013; Magro *et al.*, 2014; Yeun & Lee, 2015).

A systematic review and meta-analysis of RCTs (Miller & Ouwehand, 2013) concluded that overall probiotics shortened ITT, particularly in constipated females; bifidobacteria were considered the most effective. Another systematic review, comparing probiotic interventions of 2–8 weeks with placebo, found probiotics significantly shortened whole and regional gut transit time, increased stool frequency and improved stool consistency (Dimidi *et al.*, 2014). A meta-analysis of just a few RCTs in adults with chronic idiopathic constipation showed probiotics were no more effective than placebo for symptom improvement but that probiotics were associated with significantly improved defecation frequency. There was also some evidence for synbiotic benefit (Ford *et al.*, 2014).

8.2.4 *Diarrhoeal diseases*

Diarrhoea as a consequence of antibiotic use is a major healthcare concern, accounting for significant morbidity and mortality, extended hospitalisation and greater healthcare costs, especially in patients who are elderly and/or have recurring episodes (Gillespie *et al.*, 2015). Antibiotic-associated diarrhoea (AAD) occurs in about 5–25% of adult patients and 11–40% of children upon antibiotic use, with a higher percentage in hospitalised patients (Guarino *et al.*, 2009; Lewis *et al.*, 2009). A common and severe form of AAD, which accounts for up to a quarter of cases, is caused by the spore-former *Clostridium difficile*; this is commonly referred to as *Cl. difficile* infection (CDI) or *Cl. difficile*-associated diarrhoea (CDAD) (Gao *et al.*, 2015). This pathogen can be carried asymptomatically in the gut (Furuya-Kanamori *et al.*, 2015). Although the pathophysiology of both AAD and CDI is not completely understood, disruption of the commensal gut microbiota and subsequent changes in the metabolism of carbohydrates, SCFAs and bile acids seem to play a key role (Antunes *et al.*, 2011; Pirker *et al.*, 2013; Khanna & Pardi, 2016).

Several studies have investigated the ability of probiotics to prevent AAD; some have also investigated AAD treatment (Hempel *et al.*, 2012; Xie *et al.*, 2015). Several *Lb. rhamnosus* GG trials have been conducted in both adults and children (Szajewska & Kolodziej, 2015). Two studies in children showed reduced incidence of AAD with *Lb. rhamnosus* GG (Arvola *et al.*, 1999; Vanderhoof *et al.*, 1999), but not one trial with adult patients (Thomas *et al.*, 2001). In addition, when *Lb. rhamnosus* GG was administered in a fermented milk also containing *Lb. acidophilus* LA-5 and *Bif. animalis* subsp. *lactis* BB-12, significantly fewer patients on the probiotic mixture developed AAD (compared to placebo) (Wenus *et al.*, 2008). Interestingly, when the combination was later tested without *Lb. rhamnosus* GG in a DBPCRT, it did not lower AAD incidence in adults (Chatterjee *et al.*, 2013). Several *Lactobacillus* spp. probiotics, as either single-strain (Lonnermark *et al.*, 2010; Cimperman *et al.*, 2011; Pirker *et al.*, 2013; Dietrich *et al.*, 2014; Wong *et al.*, 2014) or multi-strain preparations (Beausoleil *et al.*, 2007; Gao *et al.*, 2010; Sampalis *et al.*, 2010), have shown efficacy in ADD prevention. One of these trials (with *Lb. acidophilus* CL1285 and *Lb. casei* LBC80R) also investigated the dose response (Gao *et al.*, 2010). Remarkably, probiotic combinations of lactobacilli with other bacterial genera, such as *Bifidobacterium*, did not reduce AAD incidence in hospitalised patients (Stein *et al.*, 2007; Szymanski *et al.*, 2008; Allen *et al.*, 2013). Finally, in an RCT with children, daily intake of a commercial probiotic containing '*B. lactis* [*sic*]' (presumed *Bif. animalis* subsp. *lactis* [*sic*]) and *Str. thermophilus* strains was associated with a significant reduction in AAD (Correa *et al.*, 2005). The yeast '*Sac. boulardii*' (presumed to be *Sac. cerevisiae* var. *boulardii*) has also been extensively studied, but with mixed results (Kotowska *et al.*, 2005; Can *et al.*, 2006; Cindoruk *et al.*, 2007; Bravo *et al.*, 2008).

There are fewer trials investigating CDAD prevention, and in most cases this was a secondary endpoint for studies investigating AAD (Imhoff & Karpa, 2009). Another factor is the reduction in CDAD incidence in many hospitals due to improved infection control strategies. A probiotic combination of *Lb. acidophilus* and *Bif. bifidum* was associated with a reduction in *Cl. difficile* toxin detection in patients who developed diarrhoea (2.9% compared to 7.25% in the control group) (Plummer *et al.*, 2004). The beneficial effects have also been demonstrated of a daily probiotic drink containing strains of *Lb. casei*, *Lb. delbrueckii* subsp. *bulgaricus* and *Str. thermophilus* administered within 48 h of patients starting antibiotic therapy and continuing up to one week after antibiotics stopped (Hickson *et al.*, 2007). None of the 59 patients on probiotics developed CDI, compared to 9 of 53 on placebo.

Another line of probiotic research studies the prevention of recurrent episodes of CDI (RCDI). An initial report of benefit with *Lb. rhamnosus* GG (Pochapin, 2000) was not confirmed later (Lawrence *et al.*, 2005). There has been evidence of RCDI benefit with *Lactobacillus plantarum* 299V (Wullt *et al.*, 2003), *Lb. casei* Shirota (Lee *et al.*, 2013) and '*Sac. boulardii*' (presumed to be *Sac. cerevisiae* var. *boulardii*) (McFarland *et al.*, 1994; Surawicz *et al.*, 2000). It should be noted that in all cases, the patients received standard antibiotic therapy to treat CDI.

The first meta-analyses on probiotics and AAD were Cremonini *et al.* (2002) and D'Souza *et al.* (2002): both concluded that *Lb. rhamnosus* GG and '*Sac. boulardii*' (presumed to be *Sac. cerevisiae* var. *boulardii*) in particular were beneficial in preventing AAD. A later meta-analysis (Videlock & Cremonini, 2012) examined 34 studies with 4138 patients. The pooled relative risk (RR) for AAD for probiotics *vs.* placebo was 0.53 (95% CI: 0.44–0.63), corresponding to a number needed to treat (NNT) of 8 (95% CI: 7–11). The probiotic preventive effect remained significant even when grouped by probiotic species, population age group, relative duration of antibiotics and probiotics, study risk of bias and probiotic administered. Another meta-analysis published in the same year, which included 82 RCTs, also concluded there was sufficient evidence of probiotic prevention of AAD: a pooled RR of 0.58 (95% CI: 0.50–0.68; $P<0.001$) (Hempel *et al.*, 2012). Finally, a large Cochrane review of 16 RCTs of children (3432 patients from 2 weeks to 17 years of age) also found decreased incidence of AAD associated with probiotics, a dose–response effect and no report of serious adverse events reported (Johnston *et al.*, 2011).

A Cochrane review in 2008 concluded that there were insufficient data to support probiotic use as sole or adjunct treatment for CDI (Pillai & Nelson, 2008). Other reviews have also examined the effectiveness of probiotics in preventing CDI, and concluded there was only moderate-quality evidence to support this (Segarra-Newnham, 2007; Johnston *et al.*, 2012). The most recent systematic review and meta-analysis, however, which involved 26 RCTs and 7957 patients, concluded probiotics did significantly reduce risk of CDAD (RR: 0.395; 95% CI: 0.294–0.531; $P<0.001$) (Lau & Chamberlain, 2016). Probiotic use for CDI prevention remains an area of clinical interest.

8.2.5 *Paediatric conditions*

There is continued interest in probiotic use for infants and children; the strongest indications for probiotics are for GI-related disorders (Thomas *et al.*, 2015). As well as the conditions discussed above, research in children has also focused on the treatment and prevention of acute infectious diarrhoea, necrotising enterocolitis and infantile colic (Vandenplas *et al.*, 2015).

Acute diarrhoea

Acute infectious gastroenteritis is generally defined as a decrease in stool consistency (loose or liquid) and/or an increase in frequency of evacuation (typically ≥3 in 24 h), with or without fever or vomiting (Szajewska & Karas, 2014). It remains a major cause of childhood morbidity and mortality worldwide, especially in neonates and infants under 5 years of age (Black *et al.*, 2010); rotavirus is the most common cause, followed by adenovirus and norovirus. Bacterial and parasitic infections appear to be decreasing (Wiegering *et al.*, 2011).

A few community studies with probiotics have investigated the prevention of acute infectious diarrhoea (Caffarelli *et al.*, 2015). In trials of healthy infants attending day care centres, 3-month daily regimes of *Bif. animalis* subsp. *lactis* BB-12, *Lb. reuteri* ATCC 55730 (Weizman *et al.*, 2005) and *Lb. reuteri* DSM 17938 (Gutierrez-Castrellon *et al.*, 2014) resulted in fewer and shorter episodes of diarrhoea compared to placebo intervention. In a large study of 3758 children aged 1–5 years in India, a 12-week intake of a fermented milk drink containing *Lb. casei* Shirota reduced episodes of acute diarrhoea by 14% (Sur *et al.*, 2011). In another double-blind RCT, a fermented milk drink containing *Lb. casei* DN-114 001 reduced overall infectious GI disease episodes in children aged 3–6 years in a city in the United States of America (USA) (Merenstein *et al.*, 2010). Other community studies, however, have shown only modest effects. Synbiotic combinations containing different probiotic strains together with GOS and short-chain fructooligosaccharides (FOS) were tested in one trial. No effects on diarrhoea were noted during the intervention period, but at the one-year follow up, infants who had received the synbiotic containing '*Bif. longum* BL999' (presumed to be *Bif. longum* subsp. *longum* BL999) and *Lb. rhamnosus* LPR had lower rates of diarrhoea (Chouraqui *et al.*, 2008). In another study, a 3-month daily intervention of milk fortified with *Bif. animalis* subsp. *lactis* HN019 and prebiotic oligosaccharide resulted in significant reduction of dysentery, respiratory morbidity and febrile illness but overall had no significant effect on diarrhoea (Sazawal *et al.*, 2010). A later DBPCRT showed only *Lb. reuteri* DSM 17938, and not *Lb. casei* CRL431, to be effective in reducing diarrhoea, especially in children with lower nutritional status (Agustina *et al.*, 2012).

Research has mainly focused on using probiotics to treat paediatric diarrhoea, usually in parallel with rehydration therapy (Cruchet *et al.*, 2015). Often, trials investigate the same strains tested in adults. The efficacy of both *Lb. rhamnosus* GG and '*Sac. boulardii*' (presumed to be *Sac. cerevisiae* var. *boulardii*) in reducing the duration of acute diarrhoea in children has been repeatedly demonstrated (Allen *et al.*, 2010). For example, a multicentre European trial of *Lb. rhamnosus* GG in combination with oral rehydration solution in 287 children with acute diarrhoea showed this treatment shortened the duration of rotavirus diarrhoea (Guandalini *et al.*, 2000). Positive effects of *Lb. rhamnosus* GG have been shown in several other studies (Szajewska & Mrukowicz, 2001; Canani *et al.*, 2007; Basu *et al.*, 2009), but not all (Misra *et al.*, 2009; Ritchie *et al.*, 2010). Numerous studies with '*Sac. boulardii*' (presumed to be *Sac. cerevisiae* var. *boulardii*) in children also indicate a protective effect of this probiotic in reducing duration of, and/or protecting against, acute diarrhoea (Kurugol & Koturoglu, 2005; Billoo *et al.*, 2006; Villarruel *et al.*, 2007; Htwe *et al.*, 2008). Recently, *Lb. reuteri* DSM 17938 was used in three clinical trials investigating reducing the duration of diarrhoea in hospitalised children (Francavilla *et al.*, 2012; Wanke & Szajewska, 2012; Dinleyici *et al.*, 2014), but a meta-analysis advised caution in interpreting the evidence (Szajewska *et al.*, 2014). Various synbiotic combinations have also shown promise in reducing diarrhoeal duration and severity (Shamir *et al.*, 2005; Vandenplas *et al.*, 2011; Passariello *et al.*, 2012).

The results of one of the first meta-analyses performed for the efficacy of probiotics in acute diarrhoea in children suggested that *Lactobacillus* strains are safe and effective as a treatment by reducing diarrhoea duration by approximately two-thirds of a day and also reducing the frequency of diarrhoea on the second day of treatment (Van Niel *et al.*, 2002). Although encouraging results, a later review concluded more research was needed to identify which particular probiotics should be used and for which patients (Allen *et al.*, 2010). Meta-analyses have also been conducted for specific strains. For example, a meta-analysis of 15 RCTs including 2963 participants showed that *Lb. rhamnosus* GG decreased diarrhoea duration by 1.1 days in a dose-dependent manner (Szajewska *et al.*, 2013). Similar conclusions resulted from a meta-analysis of three studies of *Lb. reuteri* ATCC 55730 and its variant, *Lb. reuteri* DSM 17938 (Szajewska *et al.*, 2014). A meta-analysis of nine RCTs of 1117 participants (2 months to 12 years old) (Szajewska & Skorka, 2009), and a later systematic review and meta-analysis of 17 studies with 2012 participants, concluded '*Sac. boulardii*' (presumed to be *Sac. cerevisiae* var. *boulardii*) was beneficial in reducing duration of acute diarrhoea in children (Feizizadeh *et al.*, 2014).

Finally, a Working Group from the European Society for Pediatric Gastroenterology, Hepatology, and Nutrition reviewed published RCTs on the use of probiotics for the prevention of AAD in children and recommended *Lb. rhamnosus* GG or '*Sac. boulardii*' (presumed to be *Sac. cerevisiae* var. *boulardii*) (both with moderate quality of evidence and a strong recommendation). A conditional recommendation was given for '*Sac. boulardii*' (presumed to be *Sac. cerevisiae* var. *boulardii*) to prevent CDAD in children, based on a low quality of evidence (Szajewska *et al.*, 2016). Further evidence is needed for other strains or combinations of strains that have been tested.

Necrotising enterocolitis

Necrotising enterocolitis (NEC), an acute inflammatory necrosis of the intestinal tract, is the most common GI emergency in neonatal intensive care units and a major cause of morbidity in preterm infants (Wu *et al.*, 2012). The strong evidence that the initial bacterial colonisation process after birth plays a pivotal role in NEC development points to the potential of probiotics as a way to reduce NEC incidence (Vongbhavit & Underwood, 2016).

Most probiotics in NEC trials have been bifidobacteria, used either as single strains (Mohan *et al.*, 2008; Underwood *et al.*, 2013; Dilli *et al.*, 2015) or in combination with lactobacilli (Lin *et al.*, 2008; Braga *et al.*, 2011). In most (but not all) cases, their administration to preterm and low-birthweight infants has shown clinical benefit. There are fewer trials with purely *Lactobacillus* strains (Awad *et al.*, 2010; Manzoni *et al.*, 2011; Oncel *et al.*, 2014) and in most cases, no significant reduction in overall rates of NEC and/or death was observed although reductions in sepsis frequency, feeding intolerance and duration of hospital stay have been reported. More recent RCTs have shown a mixture of four bifidobacterial strains (*Bif. breve*, *Bif. bifidum*, *Bif. longum* subsp. *infantis* and *Bif. longum* subsp. *longum*), and *Lb. rhamnosus* GG (Janvier *et al.*, 2014), '*Bif. lactis*' (presumed to be *Bif. animalis* subsp. *lactis*) (Dilli *et al.*, 2015) and *Bif. breve* M-16V (Patole *et al.*, 2016) can significantly reduce NEC rates in neonatal intensive care units.

Recent meta-analyses have concluded there is benefit for probiotics (particularly bifidobacteria strains) in reducing risk of NEC in preterm and very low-birthweight babies (Robinson, 2014; Aceti *et al.*, 2015; Baucells *et al.*, 2015; Cruchet *et al.*, 2015; Lau & Chamberlain, 2015; Olsen *et al.*, 2016).

Infantile colic

This is a common disorder occurring mainly in the first 3 months of life. The benefit of probiotics for its treatment and prevention has been a focus of research over the past 10 years (Barnes & Yeh, 2015). One of the earliest trials found that a milk-based formula containing *Bif. animalis* subsp. *lactis* BB-12 and a *Str. thermophilus* strain reduced episodes of colic or irritability (Saavedra *et al.*, 2004). In a later trial, *Lb. reuteri* ATCC 55730 was shown to be as effective as simethicone (a medical treatment) in reducing crying times in a cohort of breastfed colicky infants (Savino *et al.*, 2007). Similarly, positive effects on colic symptoms and crying times, as well as in modulating the intestinal microbiota, were confirmed in later trials with *Lb. reuteri* DSM 17938 (Roos *et al.*, 2013; Chau *et al.*, 2015). In contrast, however, a recent large clinical trial with this strain did not find any benefit for breastfed or formula-fed neonates (Sung *et al.*, 2014), but despite this result, later systematic reviews have concluded that *Lb. reuteri* DSM 17938 is effective for the management of infantile colic (Harb *et al.*, 2015; Schreck Bird *et al.*, 2016). There are also indications that this strain could have prophylactic benefit (Indrio *et al.*, 2014a; Savino *et al.*, 2015; Xu *et al.*, 2015). Other lactobacilli and bifidobacteria strains have been screened for their potential for colic treatment (Savino *et al.*, 2011; Aloisio *et al.*, 2012).

8.3 Probiotic research: human studies investigating extra-intestinal conditions

Initially, most probiotic studies in humans focused on GI tract diseases, but the last decade has seen a growing research interest in disorders associated with other parts of the human body.

8.3.1 *Common infectious diseases*

Common infectious diseases remain a predominant cause of morbidity and mortality worldwide, particularly in the ageing population (Yoshikawa, 2000). A panel of established probiotics has been used against respiratory tract infections, with primary endpoints being the number of acute episodes and average duration of episodes (Alexandre *et al.*, 2014). Most trials have investigated infants and children, but geriatric patients as well as athletes have also been studied (Hao *et al.*, 2015).

Trials in infants and children have examined a range of respiratory tract infections, including the common cold, influenza, sinusitis, pharyngitis and acute otitis (Araujo *et al.*, 2015). For example, *Lb. acidophilus* NCFM alone or in combination with *Bif. animalis* subsp. *lactis* Bi-07 was administered for 6 months to children (3 to 5 years of age); both strains were associated with reduced incidence of fever, rhinorrhoea and cough (Leyer *et al.*, 2009). Various RCTs have also shown reduced incidence of upper respiratory tract infections (URTIs) with *Lb. rhamnosus* GG (Hatakka *et al.*, 2001; Hojsak *et al.*, 2010; Kumpu *et al.*, 2012), *Bif. animalis* subsp. *lactis* BB-12 (Taipale *et al.*, 2012) and *Bif. animalis* subsp. *lactis* Lab4 combined with vitamin C (Garaiova *et al.*, 2015), while no effect was reported for another probiotic formula containing *Bif. animalis* subsp. *lactis* BB-12 and *Lb. reuteri* ATCC 55730 (Weizman *et al.*, 2005).

Data regarding the use of oral and topical probiotics for otitis media (OM) have emerged recently but indicate variable efficacy (Marom *et al.*, 2016). No reduction of OM incidence and duration of acute episodes was observed with the following oral probiotics: a combination of *Lb.* rhamnosus GG, *Lb. rhamnosus* LC 705, *Bif. breve* 99 and *Propionibacterium freudenreichii* subsp. *shermanii* JS (Hatakka *et al.*, 2007); *Lb. rhamnosus* GG and *Bif. animalis* subsp. *lactis* BB-12 (Rautava *et al.*, 2009); *Lb. rhamnosus* GG alone (Tapiovaara *et al.*, 2014); or a synbiotic (*Str. thermophilus* NCC 2496, *Streptococcus salivarius* DSM 13084, *Lb. rhamnosus* LPR CGMCC 1.3724 and Raftilose/Raftiline) (Cohen *et al.*, 2013). In contrast, topical administration of probiotics by nasal spray has been considered promising for moderately recurrent OM in children (Roos *et al.*, 2001; Skovbjerg *et al.*, 2009; Marchisio *et al.*, 2015).

Three clinical trials in elderly people have tested fermented dairy products containing *Lb. casei* Shirota (Fujita *et al.*, 2013), *Lb. casei* DN-114 001 (Guillemard *et al.*, 2010) or *Lb. delbrueckii* subsp. *bulgaricus* OLL1073R-1 (Makino *et al.*, 2010). In the first two studies, the probiotics were associated with shorter average URTI episode duration; in the third study, the probiotic was associated with reduction of the risk of common cold incidence. A 3-month daily regimen of *Lb. casei* Shirota fermented milk also reduced the incidence and duration of URTIs in healthy middle-aged office workers (Shida *et al.*, 2017). Moreover, a 3-month administration of *Lb. rhamnosus* GG and *Bif. animalis* subsp. *lactis* BB-12 to students reduced URTI median duration by 2 d, and gave a 34% significant reduction of the median severity score (compared to placebo) (Smith *et al.*, 2013). Reduced URTI incidence in healthy physically active adults was reported with *Bif. animalis* subsp. *lactis* BI-04 (West *et al.*, 2014); reduced URTI incidence was also reported in athletes taking a daily *Lb. casei* Shirota fermented milk drink (Gleeson *et al.*, 2011). Probiotics have since been recommended as a nutritional supplement for athletes (Gleeson, 2016), but effects may be strain specific as certain strains have not shown benefit (West *et al.*, 2011; Gleeson *et al.*, 2012).

A recent Cochrane review reported a meta-analysis of 14 RCTs (the majority cited above) and concluded that probiotics were efficient in reducing URTI episodes and their mean duration, although the quality of the evidence was considered low or very low (Hao et al., 2015). Another recent review identified 20 probiotic RCTs in adults and children; meta-analysis found significant effects of probiotics in reducing days of illness per person, episode duration and days' sick leave (King *et al.*, 2014). These two reviews formed the basis of a modelling study that concluded general probiotic use in France would reduce common respiratory tract infections, resulting in significant reductions in sick leave and antibiotic courses, particularly with children, active smokers and people most exposed to infection (Lenoir-Wijnkoop *et al.*, 2015).

8.3.2 *Allergic diseases*

The global rise in children and adults with allergic disorders (e.g. atopic dermatitis, asthma, allergic rhinitis and food-related allergy) is now a major public health concern. The rationale for probiotic research in this area has been supported by different things: for example, the 'hygiene hypothesis' (reduced exposure to the microbial stimulus early in childhood promotes disease), differences in the gut microbiota before and after development of allergy, and the realisation of the key role of the commensal gut microbiota in the maturation of the early immune system (Szajewska, 2013).

Atopic dermatitis

Atopic dermatitis, a type of eczema, is a common inflammatory skin condition that is a considerable social and economic burden because its prevalence has significantly increased over the last two decades in many parts of the world (Deckers *et al.*, 2012). Human studies into the prevention of atopic dermatitis testing various probiotic bacteria, as either single or mixed strains, have been reported but results have varied (Szajewska, 2013).

Most studies have investigated probiotics given to pregnant women, usually in the last 2 months of pregnancy, as well as breastfeeding women (mainly up to 6–24 months following the birth). Fewer studies have evaluated just post-natal intervention and only one trial has investigated just prenatal intervention (Ismail *et al.*, 2013). An early study investigating the effects of pre- and post-natal administration of *Lb. rhamnosus* GG reported a reduction of eczema risk (Kalliomaki *et al.*, 2001). When the children were followed up for a further 7 years, this effect was still evident (Kalliomaki *et al.*, 2003, 2007). However, no significant effects were reported in other *Lb. rhamnosus* GG studies (Kopp *et al.*, 2008; Ou *et al.*, 2012). Among other trials investigating combined pre- and post-natal treatment, the effects of *Lb. rhamnosus* HN001 and *Bif. animalis* subsp. *lactis* HN019 were separately evaluated in one study (Wickens *et al.*, 2008). Only the *Lb. rhamnosus* strain substantially reduced cumulative prevalence of eczema, and this benefit persisted up to the age of

4 years (Wickens *et al.*, 2012). No benefits were observed with pre- and post-natal intervention of *Lb. reuteri* ATCC 55730 for immunoglobulin E (IgE)-associated eczema (Abrahamsson *et al.*, 2007). Other trials have reported benefit with probiotics (mainly bifidobacteria and lactobacilli) administered to mothers, both antenatally and postnatally (Niers *et al.*, 2009; Dotterud *et al.*, 2010; Kim *et al.*, 2010; Rautava *et al.*, 2012), although not all trials have been positive (Huurre *et al.*, 2008). Only two of the above trials (Huurre *et al.*, 2008; Dotterud *et al.*, 2010) were conducted in birth cohorts not selected for high allergy risk. Probiotic feeding of infants was investigated in three trials with *Lb. rhamnosus* LPR and '*Bif. longum* BL999' (presumed to be *Bif. longum* subsp. *longum* BL999) (Soh *et al.*, 2009), *Lb. acidophilus* LAVRI-A1 (Taylor *et al.*, 2007) or *Lb. paracasei* subsp. *paracasei* F19 (West *et al.*, 2009), with only the latter reporting beneficial effects for the infants. To our knowledge, there is only one study investigating pre-natal - only probiotic administration: *Lb. rhamnosus* GG administered from 36 weeks' gestation until delivery to high-risk pregnant women was not effective in reducing risk of eczema or IgE-associated eczema (Boyle *et al.*, 2011).

Since the first small-scale *Lb. rhamnosus* GG study (Majamaa & Isolauri, 1997), many trials have assessed probiotic benefit for atopic dermatitis, usually using the Scoring Atopic Dermatitis (SCORAD) index (Yao *et al.*, 2010). Conflicting results from infant studies may reflect the strain-specific nature of probiotic effects. For example, administration of *Lb. fermentum* VRI-003 PCC for 8 weeks significantly reduced SCORAD scores in infants with moderate to severe atopic dermatitis (Weston *et al.*, 2005), but there was no benefit in infant studies with *Lb. rhamnosus* GG (Folster-Holst *et al.*, 2006; Gruber *et al.*, 2007), or hydrolysed whey-based formula supplemented either with *Lb. rhamnosus* GG or another *Lb. rhamnosus* strain (Brouwer *et al.*, 2006). No significant benefits were reported in a large study with 230 infants given *Lb. rhamnosus* GG or a probiotic mixture of four strains for 4 weeks, although *Lb. rhamnosus* GG alleviated symptoms in IgE-sensitised infants (Viljanen *et al.*, 2005). In two studies of older children, administration of *Lb. rhamnosus* HN001 with *Bif. animalis* subsp. *lactis* HN019 (Sistek *et al.*, 2006) and *Lb. rhamnosus* 19070-2 with *Lb. reuteri* DSM 122460 (Rosenfeldt *et al.*, 2003) was associated with reduced SCORAD scores in children with food sensitisation or atopic dermatitis, respectively. Promising results were obtained in two more recent studies into symptom alleviation in children with moderate to severe eczema receiving either a synbiotic (*Lb. acidophilus* DDS-1, *Bif. animalis* subsp. *lactis* UABLA-12, and FOS) (Gerasimov *et al.*, 2010) or an '*Lb. sakei*' strain (Woo *et al.*, 2010), presumed to be *Lb. sakei* subsp. *sakei* strain (https://www.atcc.org/products/all/15521.aspx). Only three small-scale studies have investigated probiotic effects on eczema in adults. Two crossover studies found little efficacy for either a *Bif. lactis* subsp. *animalis* LKM512-containing yoghurt for one month (Matsumoto *et al.*, 2007), or a *Lb. paracasei* subsp. *paracasei* Lpc-37, *Lb. acidophilus* 74-2 and *Bif. animalis* subsp. *lactis* DGCC 420 containing probiotic drink for 2 months (Roessler *et al.*, 2008). In contrast, adults treated with heat-killed *Lb. paracasei* K71 (presumed to be *Lb. paracasei* subsp. *paracasei* K71) for 3 months showed significant reduction in skin severity scores compared to those treated with placebo (Moroi *et al.*, 2011). It should be noted, however, that this cannot be considered a probiotic because the administered strains were not live.

A number of meta-analyses have now been published, with most concluding that probiotics are effective for prevention of atopic dermatitis (Lee *et al.*, 2008; Pelucchi *et al.*, 2012; Kuitunen, 2013; Zuccotti *et al.*, 2015). A later Cochrane review had concluded that there was insufficient evidence to recommend adding probiotics to infant foods to prevent allergic disease (Osborn & Sinn, 2007). This conclusions was, however, in line with the last report of the World Allergy Organisation (WAO), which pointed out the very low-quality evidence for probiotic use in allergy, although it did give a conditional recommendation for both pre- and post-natal treatments (Fiocchi *et al.*, 2015) while at the same time highlighting the range of methodological variations in the trials, such as selected or unselected populations for allergy risk, single or multi-strain intervention, as well as timing and duration of administration (Szajewska, 2013). Finally, there is no clear evidence to substantiate probiotics for the treatment of established eczema (Boyle *et al.*, 2008; Michail *et al.*, 2008): trials in older children or adults have generally shown minor or no benefit (Ismail *et al.*, 2013).

Allergic rhinitis and asthma

Human studies evaluating the benefits of probiotics in terms of treatment or prevention of allergic rhinitis and asthma have also given conflicting results (Yao *et al.*, 2010). No clinical benefit was reported in trials of adults with allergic rhinitis given fermented dairy products containing *Lb. acidophilus* L-92 (Ishida *et al.*, 2005b), '*Bif. longum* BB536' (presumed to be *Bif. longum* subsp. *longum* BB536) (Xiao *et al.*, 2006) and *Lb. casei* Shirota (Tamura *et al.*, 2007; Ivory *et al.*, 2013), although modification of an allergen-induced immune response was observed with *Lb. casei* Shirota in a previous trial not assessing clinical outcome (Ivory *et al.*, 2008). Significant improvement in nasal symptom–medication scores in adults was reported for certain single *Lactobacillus* strains (Ishida *et al.*, 2005a; Nagata *et al.*, 2010; Wassenberg *et al.*, 2011). Several studies have included mixed populations of children and adults with allergic rhinitis and/or asthma (Ismail *et al.*, 2013). *Lactobacillus rhamnosus* GG administration, for example, did not alleviate allergic symptoms in both young adults and teenagers (Helin *et al.*, 2002), but there was benefit with live (Wang *et al.*, 2004) and heat-killed '*Lb. paracasei*' 33 (presumed to be *Lb. paracasei* subsp. *paracasei* 33) (Peng & Hsu, 2005) for children and adults with allergic rhinitis. A more recent trial of 105 schoolchildren with asthma and allergic rhinitis showed significantly reduced clinical symptoms after *Lb. gasseri* intervention (Chen *et al.*, 2010). In another study with children given a fermented milk drink containing *Lb. casei* DN-114 001, the probiotic showed benefit for rhinitis but not asthma (Giovannini *et al.*, 2007). A few studies have evaluated probiotics for the treatment of asthma but with poor results in children (Stockert *et al.*, 2007) and adults (Wheeler *et al.*, 1997; van de Pol *et al.*, 2011). Meta-analysis of the efficacy of probiotics for treatment of allergic rhinitis and asthma is difficult due to the considerable heterogeneity of the relevant studies (Yao *et al.*, 2010), which have yielded inconsistent results; thus, currently there is judged to be insufficient evidence to support a role for probiotics in the treatment of allergic rhinitis and asthma (Ismail *et al.*, 2013).

8.3.3 *Urogenital conditions*

Urogenital conditions suffered by women are commonly either UTI or BV. Women can also suffer from yeast vaginitis, which is mostly caused by *Candida albicans* and also referred as vulvovaginal candidiasis (VVC). The dominance of *Lactobacillus* strains in the healthy vaginal microbiota together with observations of vaginal dysbiosis and depletion of lactobacilli during urogenital disorders supported the theory that oral or vaginal administration of probiotic *Lactobacillus* strains could promote a lactobacilli-dominant vaginal microbiota that would be more protective and resilient (Macklaim *et al.*, 2015). Several trials have investigated probiotics administered vaginally or orally, as treatment or preventive measures (MacPhee *et al.*, 2010; Homayouni *et al.*, 2014), and also as adjuncts to standard antibiotic therapy (Heczko *et al.*, 2015; Recine *et al.*, 2016).

Extensive research into BV has been conducted on a combination of *Lb. rhamnosus* GR-1 and *Lb. reuteri* RC-14. Administration of this formulation, both orally (Anukam *et al.*, 2006a; Martinez *et al.*, 2009) and directly to the vagina (Anukam *et al.*, 2006b), resulted in better cure rates for women diagnosed with BV compared to those treated with placebo or antibiotic. An earlier DBPCRT had shown that a 2-month daily oral intake of this combination restored a lactobacilli-dominant microbiota in women with BV (Reid *et al.*, 2004). The same probiotic mixture (taken orally) also improved the vaginal microbiota of post-menopausal women with intermediate vaginal microbiota (Nugent score 4–6) (Petricevic *et al.*, 2008). Various lactobacilli have also been tested as single or multiple strains in the form of vaginal tablets or capsules, with improvements in the vaginal environment being reported (i.e. lower pH, lower Nugent score and higher counts of lactobacilli) (Tomusiak *et al.*, 2015). Restoration of the vaginal microbiota has often been demonstrated in studies investigating the efficacy of probiotics (Bradshaw & Brotman, 2015; Macklaim *et al.*, 2015). A small-scale trial of a vaginal tablet containing three lactobacilli improved cure rates and symptoms of BV (Mastromarino *et al.*, 2009), and vaginal tablets containing a *Lb. rhamnosus* strain decreased itching, vaginal discharge and burning in BV patients, even after 24 months (Rossi *et al.*, 2010). In an open-label pilot study, treatment of 40 women with BV for 6 d with a douche containing a *Lb. acidophilus* strain helped restore the vaginal environment (Drago *et al.*, 2007); this was also observed with vaginal administration of another *Lb. acidophilus* strain in combination with a low dose of an oestrogen in a trial involving 240 women (Ozkinay *et al.*, 2005). Lactobacilli-impregnated tampons administered during the first menstruation period after vaginal clindamycin treatment for BV, however, failed to improve cure rates (Eriksson *et al.*, 2005). Vaginal application of a *Lb. rhamnosus* strain following conventional metronidazole therapy, and also vaginal capsules containing two *Lactobacillus* strains with a *Str. thermophilus* strain, have both been shown to reduce BV recurrence (Marcone *et al.*, 2008; Ya *et al.*, 2010). This was also observed with vaginal capsules of *Lb. gasseri* (Lba EB01-DSM 14869) and *Lb. rhamnosus* (Lbp PB01-DSM 14870) administered once daily for 10 days during three consecutive menstrual cycles following clindamycin cream therapy (Larsson *et al.*, 2008). A more recent study found that concomitant treatment of metronidazole and an oral probiotic containing three different lactobacilli (prOVag) increased the duration of remission periods between episodes in women with a history of recurrent BV and aerobic vaginitis (Heczko *et al.*, 2015).

A Cochrane review in 2009 concluded that there was some evidence of benefit for BV with probiotics combined with metronidazole or oestriol, but more evidence was needed (Senok *et al.*, 2009). A later meta-analysis in 2014, which identified 1304 patients from 12 RCTs, found that probiotics could significantly improve the adult BV cure rate but, as before, noted the limited nature of the evidence and the heterogeneity of studies (Huang *et al.*, 2014). Other recent systematic reviews concluded there was evidence for probiotic benefit for prevention or treatment of BV recurrence (Homayouni *et al.*, 2014) and for prophylactic use following antibiotic treatment (Parma *et al.*, 2014).

There are fewer studies on the effects of probiotics on VVC, and the data are more conflicting. For example, a large-scale RCT in Australia of women taking short courses of antibiotics found no effect of *Lactobacillus* preparations (taken orally, vaginally or both for up to 4 d after finishing the antibiotics) for prevention of post-antibiotic VVC (Pirotta *et al.*, 2004), but slow-release vaginal tablets of *Lb. fermentum* LF10 (DSM 19187) and *Lb. acidophilus* LA02 (DSM 21717) strains proved successful as treatment for VVC. This also helped prevent infection recurrence (Vicariotto *et al.*, 2012). Other research showing positive effects of vaginally applied probiotics in preventing VVC recurrence include a study with *Lb. plantarum* P17630 capsules (De Seta *et al.*, 2014), and another with a mixture of *Lb. acidophilus*, *Lb. rhamnosus*, *Str. thermophilus* and *Lb. delbrueckii* subsp. *bulgaricus* strains, which also improved the outcome of antibiotic treatment for *Candida albicans* (Kovachev & Vatcheva-Dobrevska, 2015).

With regard to prevention of recurrent UTIs, data from 294 patients across five studies was included in a meta-analysis that concluded there was no statistically significant effect of *Lactobacillus* probiotics in reducing risk of recurrent UTIs (Grin *et al.*, 2013), although, after excluding ineffective strains and safety studies, the authors suggested that probiotics might be effective. A more recent Cochrane review examined a total of 735 participants in nine studies, although these varied. Most included patients and compared probiotic effects with placebo, no treatment or antibiotics; one other placebo study studied healthy women (Schwenger *et al.*, 2015). The variation in study design and lack of data may explain why it was concluded that there was no evidence of probiotic benefit for prevention of UTI (compared to placebo), but that further research is warranted. For example, in a DBPCRT investigating a *Lactobacillus crispatus* intravaginal intervention, the probiotic was effective in reducing risk of recurrent UTI in premenopausal women (Stapleton *et al.*, 2011). Interestingly, comparison of the efficacy of the *Lb. rhamnosus* GR-1 and *Lb. reuteri* RC-14 combination (discussed above) with standard antibiotic treatment for preventing recurrent UTIs found the probiotic was not as good as antibiotics in preventing UTIs, but it did have one advantage: unlike the antibiotic, probiotic use was not associated with increased detection of antibiotic-resistant *E. coli* from the women (Beerepoot *et al.*, 2012).

8.3.4 *Obesity-related disease*

Although it is widely known that there are health risks from being overweight or obese, these conditions are increasing in all ages, to the extent that obesity-related disease is now considered a major global health challenge (Ng *et al.*, 2014). This, together with evidence of the gut microbiota's influence on energy homeostasis and weight management, has prompted probiotic research (Delzenne & Cani, 2011).

There are, as yet, relatively few RCTs examining the effects of probiotics on weight management. One DBPCRT, investigating a 12-week intervention with a fermented milk containing *Lb. gasseri* SBT2055, showed this was associated with significant reductions in abdominal visceral and subcutaneous fat areas, body weight and body mass index (BMI) (Kadooka *et al.*, 2010). More recent studies with the same strain also showed significant reductions in abdominal adiposity, as well as postprandial and fasting serum non-esterified fatty acid levels (Kadooka *et al.*, 2013), which suggest that this probiotic might help reduce risk of obesity and obesity-related disease such as T2DM (Ogawa *et al.*, 2014). A weight-reducing effect in obese adults has also been reported for a combination of *Lb. fermentum* and *Lactobacillus amylovorus* strains (Omar *et al.*, 2013). *Lactobacillus* strains, given as capsules, have also improved weight loss and vitamin B_{12} levels in morbidly obese patients undergoing gastric bypass surgery (Woodard *et al.*, 2009). An 8-week combination of a low-calorie diet and yoghurt containing *Lb. acidophilus* LA-5, *Bif. animalis* subsp. *lactis* BB-12 and *Lb. casei* DN001 significantly reduced BMI and body fat percentage in overweight and obese individuals, when compared to either a low-calorie diet combined with non-probiotic yoghurt, or just the probiotic yoghurt with no diet restriction (Zarrati *et al.*, 2014). In a DBPCRT, a synbiotic formulation of a *Lb. rhamnosus* strain combined with oligo-fructose and inulin achieved significantly better weight loss in obese women but not men (Sanchez *et al.*, 2014). A study in children showed that an 8-week daily intervention of a synbiotic capsule containing seven probiotic strains with FOS significantly reduced BMI, waist circumference, serum triglycerides and total- and low-density lipoprotein cholesterol levels compared to the placebo treatment (Safavi *et al.*, 2013). In another tactic, a DBPCRT of a fermented dairy beverage containing *Lb. acidophilus* and a strain of *Pro. freudenreichii* spp. promoted satiety in normal-weight women (Ruijschop *et al.*, 2008). In addition, there has been interest in seeing how gut modulation in early life may help prevent subsequent weight gain, for example in a 10-year follow-up study on perinatal use of *Lb. rhamnosus* GG (Luoto *et al.*, 2010). Only one meta-analysis has examined evidence for probiotic effects on weight management; this included 17 RCTs, 51 studies with farm animals and 14 experimental models. It highlighted the strain-dependent effects of probiotic lactobacilli and concluded that there was an association for *Lb. gasseri* and weight loss in obese humans (Million *et al.*, 2012).

The last decade has also seen increasing research interest in probiotic potential for metabolic disorders such as T2DM, with several promising studies. For example, a DBPCRT of a 6-week intervention with yoghurt containing *Lb. acidophilus* LA-5 and *Bif. animalis* subsp. *lactis* BB-12 in people with T2DM significantly decreased fasting blood glucose and haemoglobin A1c (HbA1c), and improved ratios of total cholesterol:high-density lipoprotein cholesterol (HDL-C) and of low-density lipoprotein cholesterol (LDL-C):HDL-C (Ejtahed *et al.*, 2011, 2012). A synbiotic study in older people with T2DM showed that a 30-day daily intake of a drink containing strains of *Lb. acidophilus* and *Bif. bifidum* with FOS resulted in significant increase in HDL-C and a decline of fasting glycaemia levels (Moroti *et al.*, 2012). The effects of oral synbiotics comprising seven bacterial strains and FOS on metabolic profiles, high-sensitivity C-reactive protein (hs-CRP) and oxidative stress have also been evaluated in

T2DM patients - with positive outcomes (Asemi *et al.*, 2013). In another study with T2DM subjects, a 6-week daily regimen of a fermented milk containing *Lb. acidophilus* LA-5 and *Bif. animalis* subsp. *lactis* BB-12 showed significant improvement in HbA1c, total cholesterol and LDL-C values (Tonucci *et al.*, 2017). A multi-strain preparation containing three *Lactobacillus* and three *Bifidobacterium* strains has been associated with decreased HbA1c and fasting insulin in T2DM patients (Firouzi *et al.*, 2017), and the protective effects of *Lb. casei* Shirota have been demonstrated in healthy adults consuming a short-term, high-fat and overfeeding diet. The unhealthy diet reduced insulin sensitivity by ~27% in subjects not taking probiotic, but glycaemic control and insulin sensitivity were preserved in those subjects who had consumed the probiotic prior to, and during, the overfeeding period (Hulston *et al.*, 2015). In another recent study, intervention with a mix of four strains had positive effects on glucose metabolism and weight gain in pregnant women with gestational diabetes mellitus (Dolatkhah *et al.*, 2015). A study with *Lb. acidophilus* NCFM showed efficacy in preserving insulin sensitivity but no effect on systemic inflammatory response (Andreasen *et al.*, 2010).

Not all studies have given promising results, perhaps because of the strain-specific nature of these effects. For example, no effect on lipid profile, glycaemic control, insulin level, oxidative stress and inflammatory markers was observed for T2DM patients after taking a 6-week course of capsules containing four probiotic *Lactobacillus* strains (Mazloom *et al.*, 2013). More recently, no effects on glycaemic control were observed in overweight adults taking *Lb. acidophilus* LA-5 and *Bif. animalis* subsp. *lactis* BB-12, administered either in yoghurt or as capsules (Ivey *et al.*, 2014), or in another study investigating cardiovascular risk factors (Ivey *et al.*, 2015).

It is clear, however, that obesity-related disease remains a current focus of probiotic research (Alokail *et al.*, 2013; Yan *et al.*, 2015), and there are now sufficient studies to warrant meta-analysis. One that identified eight trials with 438 people with T2DM concluded probiotics had significant effects in reducing HbA1c levels and insulin resistance but had no effect on fasting plasma glucose, insulin, CRP levels and lipid profiles (Kasinska & Drzewoski, 2015). A meta-analysis of 17 RCTs concluded that probiotics might help glycaemic control (Ruan *et al.*, 2015); and a more recent meta-analysis of T2DM trials, comprising seven studies, concluded probiotics could improvement glucose metabolism to a modest degree, particularly if multispecies were taken for more than 8 weeks (Sun *et al.*, 2016). Another recent analysis of 11 trials found significant effects for probiotics in reducing glucose, HbA1c and insulin resistance in people with diabetes (Yang *et al.*, 2016b).

The blood pressure-lowering effect of probiotics has been investigated in only few trials. A significant decrease in the systolic blood pressure (BP), for example, was observed in heavy smokers after taking a drink containing *Lb. plantarum* 299v for 6 weeks (Naruszewicz *et al.*, 2002). In a pilot study of obese hypertensive patients, a hypocaloric diet supplemented with a probiotic cheese containing *Lb. plantarum* TENSIA was associated with lower BMI and arterial BP values (Sharafedtinov *et al.*, 2013). In another trial, 3 weeks' intake of the same strain (in yoghurt or cheese) lowered diastolic and systolic BP (Hutt *et al.*, 2015). The same effects were observed in people with T2DM drinking soy milk containing *Lb. plantarum* A7 (Hariri *et al.*, 2015).

A recent systematic review and meta-analysis of RCTs found that probiotics (particularly multispecies) might moderately improve BP, particularly in people with hypertension and if taken for at least 8 weeks (Khalesi *et al.*, 2014).

8.3.5 *Liver disease*

Liver disease has also been a field of probiotic research, prompted partly by observations of gut microbiota changes associated with the pathogenesis of disease (Minemura & Shimizu, 2015).

Non-alcoholic fatty liver disease (NAFLD) is, as its name suggests, a condition when fat accumulates in the liver (hepatic steatosis), but not because of excessive alcohol intake. It causes a spectrum of disorders, ranging from steatosis, to steatohepatitis (when inflammation develops), to advanced fibrosis and cirrhosis (Adams & Angulo, 2005). Worldwide, NAFLD is the most common liver disease in both adults and children, and it is often associated with obesity, insulin resistance and diabetes (Younossi *et al.*, 2016). A limited number of RCTs have evaluated the efficacy of probiotics for NAFLD treatment (Gao *et al.*, 2016). VSL#3 have shown positive effects in both adults (Loguercio *et al.*, 2005) and children (Alisi *et al.*, 2014). Significant reductions in serum levels of liver aminotransferase (a biomarker of hepatocellular injury) levels were observed in NAFLD patients after a 3-month treatment with tablets containing strains of species '*Lb. bulgaricus*' (presumed to be *Lb. delbrueckii* subsp. *bulgaricus*) and *Str. thermophilus* (Aller *et al.*, 2011), and in obese children after a 2-month treatment with *Lb. rhamnosus* GG (Vajro *et al.*, 2011). Similar effects were observed in non-alcoholic steatohepatitis (NASH) patients given a multispecies probiotic formula (Lepicol) (Wong *et al.*, 2013), as well as NAFLD patients consuming yoghurt containing *Lb. acidophilus* LA-5 and *Bif. animalis* subsp. *lactis* BB-12 (Nabavi *et al.*, 2014). Lifestyle improvements proved to be more successful for NASH patients if they were combined with intake of a synbiotic preparation of a '*Bif. longum*' (presumed to be *Bif. longum* subsp. *longum*) strain and FOS (Malaguarnera *et al.*, 2012). The most recent meta-analysis of probiotics and NAFLD identified nine RCTs (535 cases) measuring a range of endpoints. It concluded that probiotics could improve outcomes of homeostasis model assessment, total cholesterol, high-density lipoprotein and inflammatory cytokine TNFα levels (Gao *et al.*, 2016).

Urease-producing species, such as *Klebsiella* and *Proteus*, have been associated with increased serum levels of ammonia and lipopolysaccharide, which can cause hepatic encephalopathy (HE), a complication of liver cirrhosis (Bajaj *et al.*, 2012). A synbiotic combination of four freeze-dried, non-urease-producing strains (*Pediococcus pentosaceus*, a *Leuconostoc mesenteroides* species, *Lb. paracasei* subsp. *paracasei* and *Lb. plantarum*) with fermentable fibres were shown to modulate the gut microbiota, reduce blood ammonia levels and reverse minimal HE (Liu *et al.*, 2004). A synbiotic combination of '*Bif. longum*' (presumed to be *Bif. longum* subsp. *longum*) with FOS in cirrhotic patients also improved both minimal (Malaguarnera *et al.*, 2007) and overt HE (Malaguarnera *et al.*, 2010). Promising results in minimal HE have also been reported for a probiotic yoghurt (Bajaj *et al.*, 2008), *Lb. rhamnosus* GG (in reducing endotoxaemia) (Bajaj *et al.*, 2014) and VSL#3 (Lunia *et al.*, 2014). Probiotics may also help prevent HE in cirrhotic patients, as was shown in studies with VSL#3 where significantly reduced episodes and patient hospitalisation were observed (Agrawal *et al.*, 2012; Dhiman *et al.*, 2014). While it is important to note that studies have not always given positive results (Pereg *et al.*, 2011; Saji *et al.*, 2011), two different meta-analyses acknowledge the potential of probiotics for HE (Shukla *et al.*, 2011; Holte *et al.*, 2012). A Cochrane review highlighted that, although probiotics may reduce plasma ammonia concentration, further research is needed before clinical recommendation (McGee *et al.*, 2011).

Alcoholic liver disease, a major cause of morbidity and mortality worldwide, has also been researched (Li *et al.*, 2016; Marchesi *et al.*, 2016). For example, *E. coli* Nissle 1917 given to cirrhotic patients for 42 days resulted in reduced endotoxaemia and improved liver function (Lata *et al.*, 2007). Restoration of the normal gut microbiota and improved liver function were also reported for cirrhotic patients after a shorter (5-day) regimen of *Bif. bifidum* and *Lb. plantarum* 8PA3 (Kirpich *et al.*, 2008). A proof-of-concept study with a fermented milk drink containing *Lb. casei* Shirota showed this restored neutrophil function (Stadlbauer *et al.*, 2008). A more recent study reported evidence of improved gut microbiota and endotoxaemia in patients hospitalised with alcoholic hepatitis given a 7-day course of *Lactobacillus subtilis* and '*Streptococcus faecium*' (presumed to be *Ent. faecium*) (Han *et al.*, 2015).

8.3.6 *Cancer*

The potential of probiotics for preventing or slowing disease has been investigated with several types of cancer but, not surprisingly, most research has focused on colorectal cancer (CRC) because of the influence of colonic bacteria on gut metabolism, the immune systems and colonic cell division (Commane *et al.*, 2005; Bultman, 2014; Marchesi *et al.*, 2016). There are still relatively few trials, however, but in one large RCT, patients with a clinical history of CRC (and previous removal of tumours) were given *Lb. casei* Shirota for 4 years. The probiotic significantly reduced the rate of development of tumours of moderate or severe atypia (Ishikawa *et al.*, 2005). A European Union (EU)-sponsored DBPCRT of CRC patients, including those with previous removal of colonic polyps, investigated 12 weeks' supplementation with a synbiotic comprising oligo-fructose-enriched inulin and the *Lb. rhamnosus* GG and *Bif. animalis* subsp. *lactis* BB-12 strains (Rafter *et al.*, 2007). Significant reductions in several biomarkers of CRC risk were observed. Other aspects of probiotic effects on patients have been investigated: patients with previous CRC diagnosis had less gut problems and better quality of life after taking a combination of a *Lb. rhamnosus* and a *Lb. acidophilus* strain (Lee *et al.*, 2014). In another synbiotic study, a 4-week intervention with a *Bif. animalis* subsp. *lactis* strain and resistant starch improved the gut microbiota but not biomarkers of CRC (Worthley *et al.*, 2009). Finally, two research groups have evaluated the effects of peri-operative probiotics for CRC patients. A DBPCRT conducted with 31 subjects undergoing elective colorectal resection for cancer found that a mixture of *Bif. longum* BB536 and *Lb. johnsonii* LA-1 given 3 d both before and after surgery resulted in beneficial changes to the gut microbiota and immune markers (Gianotti *et al.*, 2010). The other research group conducted two RCTs in CRC patients given an encapsulated mixture of three strains (*Lb. plantarum* CGMCC 1258, *Lb. acidophilus* 11 and '*Bif. longum* 88' – presumed to be *Bif. longum* subsp. *longum* 88) for 6 and 10 d pre- and post-operatively, respectively (Liu *et al.*, 2011, 2013). There was evidence of benefit for gut barrier function, for the gut microbiota and in reducing post-operative infectious complications.

As yet, *Lb. casei* Shirota is possibly the only probiotic investigated for bladder cancer benefit. An early trial demonstrated a powder preparation of this strain significantly reduced recurrence of superficial bladder cancer after transurethral resection (Aso *et al.*, 1995). Further positive evidence came from a case–control study investigating a fermented milk product containing the same strain (Ohashi *et al.*, 2002). Following transurethral resection, regular intake of *Lb. casei* Shirota in combination with standard epirubicin treatment has also shown efficacy with regard to preventing bladder cancer recurrence (Naito *et al.*, 2008).

There are relatively few publications on probiotic studies and breast cancer. In 1989, a case–control study in the Netherlands correlated high consumption of fermented milk products with reduced risk of breast cancer (Van't Veer *et al.*, 1989), and more recently, a case–control population study of Japanese women showed that regular consumption of fermented milk drinks containing *Lb. casei* Shirota since adolescence was inversely associated with reduced breast cancer incidence (Toi *et al.*, 2013).

8.3.7 *Immune disorders: HIV*

Based on their potential to influence gut barrier function and mucosal immunity, probiotics have also been suggested for HIV (Sinha & Rubens, 2014). A RCT of HIV-infected children showed that a 2-month intervention with *Bif. bifidum* and *Str. thermophilus* strains increased their CD4+ T-cell counts (Trois *et al.*, 2008). Similarly, positive effects were shown in a trial where a yoghurt containing *Lb. rhamnosus* GR-1 and *Lb. reuteri* RC-14 was given for 4 weeks to HIV-infected women in sub-Saharan Africa (Anukam *et al.*, 2008), although a later trial in Canada with these two strains showed no benefit from a 25-week intake by HIV-positive women (Hummelen *et al.*, 2011a). A trial of micronutrient-fortified yoghurt containing *Lb. rhamnosus* GR-1 showed no effect on CD4+ T-cell counts in antiretroviral therapy (ART)-naïve HIV-positive subjects (Hummelen *et al.*, 2011b), but a later trial with this yoghurt showed some positive effects in subjects on highly active ART, including improving their energy and ability to perform daily activities (Hemsworth *et al.*, 2012). A small trial in men on ART given a fermented milk drink containing *Lb. casei* Shirota also reported beneficial immune changes (increases in T lymphocytes and CD56+ cells) and some evidence of reduced inflammation and cardiovascular risk (Falasca *et al.*, 2015). Other markers of benefit have been investigated: a dietary supplement containing a multi-species probiotic was associated with significant reduction of inflammation and markers of microbial translocation in HIV patients on ART (D'Ettorre *et al.*, 2015). Similar outcomes for HIV-positive individuals on ART have been reported with other probiotics: for example, *Bacillus coagulans* GBI-30 (Yang *et al.*, 2014) and '*Sac. boulardii*' (presumed to be *Sac. cerevisiae* var. *boulardii*) (Villar-Garcia *et al.*, 2015).

There is limited research with synbiotics, but one trial in HIV-positive ART-naïve subjects showed that a 16-week intervention with a combination of *Lb. rhamnosus* HN001, *Bif. animalis* subsp. *lactis* Bi-07 and FOS improved immune function, significantly increasing CD4+ T cells and downregulating inflammatory cytokines (Gonzalez-Hernandez *et al.*, 2012). A trial investigating a 4-week intervention with Synbiotic 2000, however, showed no effect on microbial translocation or immune function (Schunter *et al.*, 2012).

8.3.8 *Trials investigating aspects of the gut–brain axis*

Although most evidence is still from animal models, research in this field clearly shows the crucial role of the gut microbiota as a signalling component of the gut–brain axis (Cryan & O'Mahony, 2011). There are a limited number of clinical trials with probiotics in this field, mostly investigating IBS, mood and/or psychological distress. For example, a trial in healthy adults given a *Lb. casei* Shirota fermented milk drink showed that daily consumption for 3 weeks improved mood, although only in subjects who had low mood at the start of the trial (Benton *et al.*, 2007). A DBPCRT in chronic fatigue syndrome patients, given the same strain as a powder for 2 months, lowered their anxiety levels (Rao *et al.*, 2009). In another group, for cancer patients scheduled for laryngectomy, a course of tablets of a *Cl. butyricum* probiotic strain helped relieve the patient's anxiety as they awaited surgery (Yang *et al.*, 2016a). A 30-day intervention with *Lactobacillus helveticus* R0052 and '*Bif. longum* R0175' (presumed to be *Bif. longum* subsp. *longum* R0175) significantly lessened psychological distress (assessed by questionnaires) and lowered urinary cortisol levels in healthy volunteers (Messaoudi *et al.*, 2011). A trial investigating 4-week consumption by healthy women of a fermented milk product containing probiotic strains *Bif. animalis* subsp. *lactis*, *Str. thermophilus*, *Lb. delbrueckii* subsp. *bulgaricus* species and *Lactococcus lactis* subsp. *lactis* led to the conclusion that probiotics could affect the activity of brain regions controlling central processing of emotion and sensation (Tillisch *et al.*, 2013).

Observations of GI problems and gut microbiota disturbances in people with autistic spectrum disorder (ASD), including abnormally high levels of certain clostridia in the gut (Parracho *et al.*, 2005), as well as new research into the microbiota–gut–brain axis (Li & Zhou, 2016; Mangiola *et al.*, 2016), particularly animal studies (Gilbert *et al.*, 2013), has led to speculation that probiotics might benefit people with ASDs (Rosenfeld, 2015). It has even been suggested that the degree of gut dysbiosis correlates with disease severity (De Angelis *et al.*, 2015). There have been, to our knowledge, only two trials with probiotics given for ASD. Administration of *Lb. plantarum* WCFS1 to children with ASD resulted in beneficial changes to the gut microbiota, which was associated with changes in the stool consistency and the children's behaviour. The study did highlight the importance of using a study design appropriate for ASD subjects (who are likely to drop out) and the high inter-individual variability of responses to the probiotic (Parracho *et al.*, 2010). A more recent study in children with ASD showed that a 4-month intake of multispecies probiotic application beneficially modulated the intestinal microbiota, but there was no assessment of whether this affected the children's behaviour (Tomova *et al.*, 2015). In a related study, researchers in Finland took a retrospective look at 13-year-old children who had received *Lb. rhamnosus* GG (or placebo) in the first 6 months of their life: attention-deficit hyperactivity disorder and Asperger syndrome were diagnosed in 6/35 (17.1%) of the placebo group but in none of the probiotic group (Partty *et al.*, 2015). These few, but promising findings, suggest that further probiotic research should be conducted in ASD (Frye *et al.*, 2015).

8.4 Conclusions

It has been a decade since the Society of Dairy Technology published the first edition of this book, and over this time an enormous amount of probiotic research has been published. The studies have explored new areas of probiotic benefit, established mechanisms of activity to explain their effects and collected evidence to support approval of probiotic health claims and in clinical guidelines. Sufficient data have accumulated to enable systematic reviews and meta-analyses in many different health areas to be conducted – and even analyses of individual probiotic strains or products. It is important to bear in mind, however, that probiotic effects are considered strain-specific, and most reviews conclude that more research is needed to identify which strains are the most effective and at what dosage.

References

Abrahamsson, T.R., Jakobsson, T., Bottcher, M.F., Fredrikson, M., Jenmalm, M.C., Bjorksten, B. & Oldaeus, G. (2007) Probiotics in prevention of IgE-associated eczema: a double-blind, randomized, placebo-controlled trial. *Journal of Allergy and Clinical Immunology*, **119**, 1174–1180.

Aceti, A., Gori, D., Barone, G., Callegari, M.L., Di Mauro, A., Fantini, M.P., Indrio, F., Maggio, L., Meneghin, F., Morelli, L., Zuccotti, G., Corvaglia, L. & Italian Society of Neonatology (2015) Probiotics for prevention of necrotizing enterocolitis in preterm infants: systematic review and meta-analysis. *Italian Journal of Pediatrics*, **41**, 89.

Adams, L.A. & Angulo, P. (2005) Recent concepts in non-alcoholic fatty liver disease. *Diabetic Medicine*, **22**, 1129–1133.

Agrawal, A., Houghton, L.A., Morris, J., Reilly, B., Guyonnet, D., Goupil Feuillerat, N., Schlumberger, A., Jakob, S. & Whorwell, P.J. (2009) Clinical trial: the effects of a fermented milk product containing *Bifidobacterium lactis* DN-173 010 on abdominal distension and gastrointestinal transit in irritable bowel syndrome with constipation. *Alimentary Pharmacology and Therapeutics*, **29**, 104–114.

Agrawal, A., Sharma, B.C., Sharma, P. & Sarin, S.K. (2012) Secondary prophylaxis of hepatic encephalopathy in cirrhosis: an open-label, randomized controlled trial of lactulose, probiotics, and no therapy. *American Journal of Gastroenterology*, **107**, 1043–1050.

Agustina, R., Kok, F.J., Van De Rest, O., Fahmida, U., Firmansyah, A., Lukito, W., Feskens, E.J., Van Den Heuvel, E.G., Albers, R. & Bovee-Oudenhoven, I.M. (2012) Randomized trial of probiotics and calcium on diarrhea and respiratory tract infections in Indonesian children. *Pediatrics*, **129**, e1155–1164.

Alexandre, Y., Le Blay, G., Boisrame-Gastrin, S., Le Gall, F., Hery-Arnaud, G., Gouriou, S., Vallet, S. & Le Berre, R. (2014) Probiotics: a new way to fight bacterial pulmonary infections? *Médecine et Maladies Infectieuses*, **44**, 9–17.

Alisi, A., Bedogni, G., Baviera, G., Giorgio, V., Porro, E., Paris, C., Giammaria, P., Reali, L., Anania, F. & Nobili, V. (2014) Randomised clinical trial: the beneficial effects of VSL#3 in obese children with non-alcoholic steatohepatitis. *Alimentary Pharmacology and Therapeutics*, **39**, 1276–1285.

Allen, S.J., Martinez, E.G., Gregorio, G.V. & Dans, L.F. (2010) Probiotics for treating acute infectious diarrhoea. *Cochrane Database of Systematic Reviews*, CD003048.

Allen, S.J., Wareham, K., Wang, D., Bradley, C., Hutchings, H., Harris, W., Dhar, A., Brown, H., Foden, A., Gravenor, M.B. & Mack, D. (2013) Lactobacilli and bifidobacteria in the prevention of antibiotic-associated diarrhoea and *Clostridium difficile* diarrhoea in older inpatients (PLACIDE): a randomised, double-blind, placebo-controlled, multicentre trial. *Lancet*, **382**, 1249–1257.

Aller, R., De Luis, D.A., Izaola, O., Conde, R., Gonzalez Sagrado, M., Primo, D., De La Fuente, B. & Gonzalez, J. (2011) Effect of a probiotic on liver aminotransferases in nonalcoholic fatty liver disease patients: a double blind randomized clinical trial. *European Review for Medical and Pharmacological Sciences*, **15**, 1090–1095.

Almeida, C.C., Lorena, S.L., Pavan, C.R., Akasaka, H.M. & Mesquita, M.A. (2012) Beneficial effects of long-term consumption of a probiotic combination of *Lactobacillus casei* Shirota and *Bifidobacterium breve* Yakult may persist after suspension of therapy in lactose-intolerant patients. *Nutrition in Clinical Practice*, **27**, 247–251.

Aloisio, I., Santini, C., Biavati, B., Dinelli, G., Cencic, A., Chingwaru, W., Mogna, L. & Di Gioia, D. (2012) Characterization of *Bifidobacterium* spp. strains for the treatment of enteric disorders in newborns. *Applied Microbiology and Biotechnology*, **96**, 1561–1576.

Alokail, M.S., Sabico, S., Al-Saleh, Y., Al-Daghri, N.M., Alkharfy, K.M., Vanhoutte, P.M. & Mcternan, P.G. (2013) Effects of probiotics in patients with diabetes mellitus type 2: study protocol for a randomized, double-blind, placebo-controlled trial. *Trials*, **14**, 195.

Andreasen, A.S., Larsen, N., Pedersen-Skovsgaard, T., Berg, R.M., Moller, K., Svendsen, K.D., Jakobsen, M. & Pedersen, B.K. (2010) Effects of *Lactobacillus acidophilus* NCFM on insulin sensitivity and the systemic inflammatory response in human subjects. *British Journal of Nutrition*, **104**, 1831–1838.

Anonymous (2013) The science behind *Bifidobacterium* BB-12®. Technical Bulletin. Ch. Hansen A/S, Bøge Allé 10-12, DK 2970 Hørsholm, Denmark.

Antunes, L.C., Han, J., Ferreira, R.B., Lolic, P., Borchers, C.H. & Finlay, B.B. (2011) Effect of antibiotic treatment on the intestinal metabolome. *Antimicrobial Agents and Chemotherapy*, **55**, 1494–1503.

Anukam, K., Osazuwa, E., Ahonkhai, I., Ngwu, M., Osemene, G., Bruce, A.W. & Reid, G. (2006a) Augmentation of antimicrobial metronidazole therapy of bacterial vaginosis with oral probiotic *Lactobacillus rhamnosus* GR-1 and *Lactobacillus reuteri* RC-14: randomized, double-blind, placebo controlled trial. *Microbes and Infection*, **8**, 1450–1454.

Anukam, K.C., Osazuwa, E., Osemene, G.I., Ehigiagbe, F., Bruce, A.W. & Reid, G. (2006b) Clinical study comparing probiotic *Lactobacillus* GR-1 and RC-14 with metronidazole vaginal gel to treat symptomatic bacterial vaginosis. *Microbes and Infection*, **8**, 2772–2776.

Anukam, K.C., Osazuwa, E.O., Osadolor, H.B., Bruce, A.W. & Reid, G. (2008) Yogurt containing probiotic *Lactobacillus rhamnosus* GR-1 and *L. reuteri* RC-14 helps resolve moderate diarrhea and increases CD4 count in HIV/AIDS patients. *Journal of Clinical Gastroenterology*, **42**, 239–243.

Araujo, G.V., Oliveira Junior, M.H., Peixoto, D.M. & Sarinho, E.S. (2015) Probiotics for the treatment of upper and lower respiratory-tract infections in children: systematic review based on randomized clinical trials. *Jornal de Pediatría*, **91**, 413–427.

Arques, J.L., Rodriguez, E., Langa, S., Landete, J.M. & Medina, M. (2015) Antimicrobial activity of lactic acid bacteria in dairy products and gut: effect on pathogens. *BioMed Research International*, **2015**, 584183.

Arvola, T., Laiho, K., Torkkeli, S., Mykkanen, H., Salminen, S., Maunula, L. & Isolauri, E. (1999) Prophylactic *Lactobacillus* GG reduces antibiotic-associated diarrhea in children with respiratory infections: a randomized study. *Pediatrics*, **104**, e64.

Asemi, Z., Zare, Z., Shakeri, H., Sabihi, S.S. & Esmaillzadeh, A. (2013) Effect of multispecies probiotic supplements on metabolic profiles, hs-CRP, and oxidative stress in patients with type 2 diabetes. *Annals of Nutrition and Metabolism*, **63**, 1–9.

Ashraf, R. & Shah, N.P. (2014) Immune system stimulation by probiotic microorganisms. *Critical Reviews in Food Science and Nutrition*, **54**, 938–956.

Aso, Y., Akaza, H., Kotake, T., Tsukamoto, T., Imai, K. & Naito, S. (1995) Preventive effect of a *Lactobacillus casei* preparation on the recurrence of superficial bladder cancer in a double-blind trial – The BLP Study Group. *European Urology*, **27**, 104–109.

Awad, H., Mokhtar, H., Imam, S.S., Gad, G.I., Hafez, H. & Aboushady, N. (2010) Comparison between killed and living probiotic usage versus placebo for the prevention of necrotizing enterocolitis and sepsis in neonates. *Pakistan Journal of Biological Sciences*, **13**, 253–262.

Bajaj, J.S., Saeian, K., Christensen, K.M., Hafeezullah, M., Varma, R.R., Franco, J., Pleuss, J.A., Krakower, G., Hoffmann, R.G. & Binion, D.G. (2008) Probiotic yogurt for the treatment of minimal hepatic encephalopathy. *American Journal of Gastroenterology*, **103**, 1707–1715.

Bajaj, J.S., Hylemon, P.B., Ridlon, J.M., Heuman, D.M., Daita, K., White, M.B., Monteith, P., Noble, N.A., Sikaroodi, M. & Gillevet, P.M. (2012) Colonic mucosal microbiome differs from stool microbiome in cirrhosis and hepatic encephalopathy and is linked to cognition and inflammation. *American Journal of Physiology: Gastrointestinal and Liver Physiology*, **303**, G675–685.

Bajaj, J.S., Heuman, D.M., Hylemon, P.B., Sanyal, A.J., Puri, P., Sterling, R.K., Luketic, V., Stravitz, R.T., Siddiqui, M.S., Fuchs, M., Thacker, L.R., Wade, J.B., Daita, K., Sistrun, S., White, M.B., Noble, N.A., Thorpe, C., Kakiyama, G., Pandak, W.M., Sikaroodi, M. & Gillevet, P.M. (2014) Randomised clinical trial: *Lactobacillus* GG modulates gut microbiome, metabolome and endotoxemia in patients with cirrhosis. *Alimentary Pharmacology and Therapeutics*, **39**, 1113–1125.

Barnes, D. & Yeh, A.M. (2015) Bugs and guts: practical applications of probiotics for gastrointestinal disorders in children. *Nutrition in Clinical Practice*, **30**, 747–759.

Basu, S., Paul, D.K., Ganguly, S., Chatterjee, M. & Chandra, P.K. (2009) Efficacy of high-dose *Lactobacillus rhamnosus* GG in controlling acute watery diarrhea in Indian children: a randomized controlled trial. *Journal of Clinical Gastroenterology*, **43**, 208–213.

Baucells, B.J., Mercadal Hally, M., Alvarez Sanchez, A.T. & Figueras Aloy, J. (2015) Probiotic associations in the prevention of necrotising enterocolitis and the reduction of late-onset sepsis and neonatal mortality in preterm infants under 1500 g: a systematic review. *Anales de Pediatría*, **85**(5), 247–255.

Bauserman, M. & Michail, S. (2005) The use of *Lactobacillus* GG in irritable bowel syndrome in children: a double-blind randomized control trial. *Journal of Pediatrics*, **147**, 197–201.

Beausoleil, M., Fortier, N., Guenette, S., L'Ecuyer, A., Savoie, M., Franco, M., Lachaine, J. & Weiss, K. (2007) Effect of a fermented milk combining *Lactobacillus acidophilus* Cl1285 and *Lactobacillus casei* in the prevention of antibiotic-associated diarrhea: a randomized, double-blind, placebo-controlled trial. *Canadian Journal of Gastroenterology*, **21**, 732–736.

Beerepoot, M.A., Ter Riet, G., Nys, S., Van Der Wal, W.M., De Borgie, C.A., De Reijke, T.M., Prins, J.M., Koeijers, J., Verbon, A., Stobberingh, E. & Geerlings, S.E. (2012) Lactobacilli vs antibiotics to prevent urinary tract infections: a randomized, double-blind, noninferiority trial in postmenopausal women. *Archives of Internal Medicine*, **172**, 704–712.

Begley, M., Hill, C. & Gahan, C.G. (2006) Bile salt hydrolase activity in probiotics. *Applied and Environmental Microbiology*, **72**, 1729–1738.

Belizario, J.E. & Napolitano, M. (2015) Human microbiomes and their roles in dysbiosis, common diseases, and novel therapeutic approaches. *Frontiers in Microbiology*, **6**, 1050.

Bentley, R. & Meganathan, R. (1982) Biosynthesis of vitamin K (menaquinone) in bacteria. *Microbiological Reviews*, **46**, 241–280.

Benton, D., Williams, C. & Brown, A. (2007) Impact of consuming a milk drink containing a probiotic on mood and cognition. *European Journal of Clinical Nutrition*, **61**, 355–361.

Bermudez-Brito, M., Plaza-Diaz, J., Munoz-Quezada, S., Gomez-Llorente, C. & Gil, A. (2012) Probiotic mechanisms of action. *Annals of Nutrition and Metabolism*, **61**, 160–174.

Bibiloni, R., Fedorak, R.N., Tannock, G.W., Madsen, K.L., Gionchetti, P., Campieri, M., De Simone, C. & Sartor, R.B. (2005) VSL#3 probiotic-mixture induces remission in patients with active ulcerative colitis. *American Journal of Gastroenterology*, **100**, 1539–1546.

Bienenstock, J., Kunze, W. & Forsythe, P. (2015) Microbiota and the gut-brain axis. *Nutrition Reviews*, **73**, 28–31.

Billoo, A.G., Memon, M.A., Khaskheli, S.A., Murtaza, G., Iqbal, K., Saeed Shekhani, M. & Siddiqi, A.Q. (2006) Role of a probiotic (*Saccharomyces boulardii*) in management and prevention of diarrhoea. *World Journal of Gastroenterology*, **12**, 4557–4560.

Bisanz, J.E., Seney, S., Mcmillan, A., Vongsa, R., Koenig, D., Wong, L., Dvoracek, B., Gloor, G.B., Sumarah, M., Ford, B., Herman, D., Burton, J.P. & Reid, G. (2014) A systems biology approach investigating the effect of probiotics on the vaginal microbiome and host responses in a double blind, placebo-controlled clinical trial of post-menopausal women. *PLoS One*, **9**, e104511.

Bischoff, S.C., Barbara, G., Buurman, W., Ockhuizen, T., Schulzke, J.D., Serino, M., Tilg, H., Watson, A. & Wells, J.M. (2014) Intestinal permeability: a new target for disease prevention and therapy. *BMC Gastroenterology*, **14**, 189.

Bixquert, M. (2013) Treatment of irritable bowel syndrome with probiotics: growing evidence. *Indian Journal of Medical Research*, **138**, 175–177.

Black, R.E., Cousens, S., Johnson, H.L., Lawn, J.E., Rudan, I., Bassani, D.G., Jha, P., Campbell, H., Walker, C.F., Cibulskis, R., Eisele, T., Liu, L., Mathers, C., Child Health Epidemiology Reference Group Of, W.H.O. & Unicef (2010) Global, regional, and national causes of child mortality in 2008: a systematic analysis. *Lancet*, **375**, 1969–1987.

Borges-Canha, M., Portela-Cidade, J.P., Dinis-Ribeiro, M., Leite-Moreira, A.F. & Pimentel-Nunes, P. (2015) Role of colonic microbiota in colorectal carcinogenesis: a systematic review. *Revista Española de Enfermedades Digestivas*, **107**, 659–671.

Bourreille, A., Cadiot, G., Le Dreau, G., Laharie, D., Beaugerie, L., Dupas, J.L., Marteau, P., Rampal, P., Moyse, D., Saleh, A., Le Guern, M.E., Galmiche, J.P. & Group, F.S. (2013) *Saccharomyces boulardii* does not prevent relapse of Crohn's disease. *Clinical Gastroenterology and Hepatology*, **11**, 982–987.

Bousvaros, A., Guandalini, S., Baldassano, R.N., Botelho, C., Evans, J., Ferry, G.D., Goldin, B., Hartigan, L., Kugathasan, S., Levy, J., Murray, K.F., Oliva-Hemker, M., Rosh, J.R., Tolia, V., Zholudev, A., Vanderhoof, J.A. & Hibberd, P.L. (2005) A randomized, double-blind trial of *Lactobacillus* GG versus placebo in addition to standard maintenance therapy for children with Crohn's disease. *Inflammatory Bowel Diseases*, **11**, 833–839.

Boyle, R.J., Bath-Hextall, F.J., Leonardi-Bee, J., Murrell, D.F. & Tang, M.L. (2008) Probiotics for treating eczema. *Cochrane Database of Systematic Reviews*, CD006135.

Boyle, R.J., Ismail, I.H., Kivivuori, S., Licciardi, P.V., Robins-Browne, R.M., Mah, L.J., Axelrad, C., Moore, S., Donath, S., Carlin, J.B., Lahtinen, S.J. & Tang, M.L. (2011) *Lactobacillus* GG treatment during pregnancy for the prevention of eczema: a randomized controlled trial. *Allergy*, **66**, 509–516.

Bradshaw, C.S. & Brotman, R.M. (2015) Making inroads into improving treatment of bacterial vaginosis – striving for long-term cure. *BMC Infectious Diseases*, **15**, 292.

Braga, T.D., Da Silva, G.A., De Lira, P.I. & De Carvalho Lima, M. (2011) Efficacy of *Bifidobacterium breve* and *Lactobacillus casei* oral supplementation on necrotizing enterocolitis in very-low-birth-weight preterm infants: a double-blind, randomized, controlled trial. *American Journal of Clinical Nutrition*, **93**, 81–86.

Brandt, L.J., Chey, W.D., Fox-Orenstein, A.E., Schiller, L.R., Schoenfeld, P.S., Spiegel, B.M., Talley, N.J. & Quigley, E.M. (2009) An evidence-based position statement on the management of irritable bowel syndrome. *American Journal of Gastroenterology*, **104**, S1–S35.

Bravo, M.V., Bunout, D., Leiva, L., De La Maza, M.P., Barrera, G., De La Maza, J. & Hirsch, S. (2008) Effect of probiotic *Saccharomyces boulardii* on prevention of antibiotic-associated diarrhea in adult outpatients with amoxicillin treatment (in Spanish). *Revista Medica de Chile*, **136**, 981–988.

Bravo, J.A., Forsythe, P., Chew, M.V., Escaravage, E., Savignac, H.M., Dinan, T.G., Bienenstock, J. & Cryan, J.F. (2011) Ingestion of *Lactobacillus* strain regulates emotional behavior and central GABA receptor expression in a mouse via the vagus nerve. *Proceedings of the National Academy of Sciences of the United States of America*, **108**, 16050–16055.

Brouwer, M.L., Wolt-Plompen, S.A., Dubois, A.E., Van Der Heide, S., Jansen, D.F., Hoijer, M.A., Kauffman, H.F. & Duiverman, E.J. (2006) No effects of probiotics on atopic dermatitis in infancy: a randomized placebo-controlled trial. *Clinical and Experimental Allergy*, **36**, 899–906.

Bultman, S.J. (2014) Emerging roles of the microbiome in cancer. *Carcinogenesis*, **35**, 249–255.

Buss, C., Valle-Tovo, C., Miozzo, S. & Alves De Mattos, A. (2014) Probiotics and synbiotics may improve liver aminotransferases levels in non-alcoholic fatty liver disease patients. *Annals of Hepatology*, **13**, 482–488.

Butterworth, A.D., Thomas, A.G. & Akobeng, A.K. (2008) Probiotics for induction of remission in Crohn's disease. *Cochrane Database of Systematic Reviews*, CD006634.

Caffarelli, C., Cardinale, F., Povesi-Dascola, C., Dodi, I., Mastrorilli, V. & Ricci, G. (2015) Use of probiotics in pediatric infectious diseases. *Expert Review of Anti-Infective Therapy*, **13**, 1517–1535.

Cammarota, G., Ianiro, G., Cianci, R., Bibbo, S., Gasbarrini, A. & Curro, D. (2015) The involvement of gut microbiota in inflammatory bowel disease pathogenesis: potential for therapy. *Pharmacology and Therapeutics*, **149**, 191–212.

Can, M., Besirbellioglu, B.A., Avci, I.Y., Beker, C.M. & Pahsa, A. (2006) Prophylactic *Saccharomyces boulardii* in the prevention of antibiotic-associated diarrhea: a prospective study. *Medical Science Monitor*, **12**, PI19–PI22.

Canani, R.B., Cirillo, P., Terrin, G., Cesarano, L., Spagnuolo, M.I., De Vincenzo, A., Albano, F., Passariello, A., De Marco, G., Manguso, F. & Guarino, A. (2007) Probiotics for treatment of acute diarrhoea in children: randomised clinical trial of five different preparations. *BMJ*, **335**, 340.

Canfora, E.E., Jocken, J.W. & Blaak, E.E. (2015) Short-chain fatty acids in control of body weight and insulin sensitivity. *Nature Reviews: Endocrinology*, **11**(10), 577–591.

Cani, P.D., Amar, J., Iglesias, M.A., Poggi, M., Knauf, C., Bastelica, D., Neyrinck, A.M., Fava, F., Tuohy, K.M., Chabo, C., Waget, A., Delmee, E., Cousin, B., Sulpice, T., Chamontin, B., Ferrieres, J., Tanti, J.F., Gibson, G.R., Casteilla, L., Delzenne, N.M., Alessi, M.C. & Burcelin, R. (2007) Metabolic endotoxemia initiates obesity and insulin resistance. *Diabetes*, **56**, 1761–1772.

Cani, P.D., Possemiers, S., Van De Wiele, T., Guiot, Y., Everard, A., Rottier, O., Geurts, L., Naslain, D., Neyrinck, A., Lambert, D.M., Muccioli, G.G. & Delzenne, N.M. (2009) Changes in gut microbiota control inflammation in obese mice through a mechanism involving GLP-2-driven improvement of gut permeability. *Gut*, **58**, 1091–1103.

Cappello, C., Tremolaterra, F., Pascariello, A., Ciacci, C. & Iovino, P. (2013) A randomised clinical trial (RCT) of a symbiotic mixture in patients with irritable bowel syndrome (IBS): effects on symptoms, colonic transit and quality of life. *International Journal of Colorectal Disease*, **28**, 349–358.

Cash, B.D. (2014) Emerging role of probiotics and antimicrobials in the management of irritable bowel syndrome. *Current Medical Research and Opinion*, **30**, 1405–1415.

Chatterjee, S., Kar, P., Das, T., Ray, S., Gangulyt, S., Rajendiran, C. & Mitra, M. (2013) Randomised placebo-controlled double blind multicentric trial on efficacy and safety of *Lactobacillus acidophilus* LA-5 and *Bifidobacterium* BB-12 for prevention of antibiotic-associated diarrhoea. *Journal of the Association of Physicians of India*, **61**, 708–712.

Chau, K., Lau, E., Greenberg, S., Jacobson, S., Yazdani-Brojeni, P., Verma, N. & Koren, G. (2015) Probiotics for infantile colic: a randomized, double-blind, placebo-controlled trial investigating *Lactobacillus reuteri* DSM 17938. *Journal of Pediatrics*, **166**, 74–78.

Chen, Y.S., Jan, R.L., Lin, Y.L., Chen, H.H. & Wang, J.Y. (2010) Randomized placebo-controlled trial of *Lactobacillus* on asthmatic children with allergic rhinitis. *Pediatric Pulmonology*, **45**, 1111–1120.

Chermesh, I., Tamir, A., Reshef, R., Chowers, Y., Suissa, A., Katz, D., Gelber, M., Halpern, Z., Bengmark, S. & Eliakim, R. (2007) Failure of Synbiotic 2000 to prevent postoperative recurrence of Crohn's disease. *Digestive Diseases and Sciences*, **52**, 385–389.

Chichlowski, M. & Rudolph, C. (2015) Visceral pain and gastrointestinal microbiome. *Journal of Neurogastroenterol and Motility*, **21**, 172–181.

Choi, C.H. & Chang, S.K. (2015) Alteration of gut microbiota and efficacy of probiotics in functional constipation. *Journal of Neurogastroenterology and Motility*, **21**, 4–7.

Choi, C.H., Jo, S.Y., Park, H.J., Chang, S.K., Byeon, J.S. & Myung, S.J. (2011) A randomized, double-blind, placebo-controlled multicenter trial of *Saccharomyces boulardii* in irritable bowel syndrome: effect on quality of life. *Journal of Clinical Gastroenterology*, **45**, 679–683.

Chong, E.S. (2014) A potential role of probiotics in colorectal cancer prevention: review of possible mechanisms of action. *World Journal of Microbiology & Biotechnology*, **30**, 351–374.

Chouraqui, J.P., Grathwohl, D., Labaune, J.M., Hascoet, J.M., De Montgolfier, I., Leclaire, M., Giarre, M. & Steenhout, P. (2008) Assessment of the safety, tolerance, and protective effect against diarrhea of infant formulas containing mixtures of probiotics or probiotics and prebiotics in a randomized controlled trial. *American Journal of Clinical Nutrition*, **87**, 1365–1373.

Cimperman, L., Bayless, G., Best, K., Diligente, A., Mordarski, B., Oster, M., Smith, M., Vatakis, F., Wiese, D., Steiber, A. & Katz, J. (2011) A randomized, double-blind, placebo-controlled pilot study of *Lactobacillus reuteri* ATCC 55730 for the prevention of antibiotic-associated diarrhea in hospitalized adults. *Journal of Clinical Gastroenterology*, **45**, 785–789.

Cindoruk, M., Erkan, G., Karakan, T., Dursun, A. & Unal, S. (2007) Efficacy and safety of *Saccharomyces boulardii* in the 14-day triple anti-*Helicobacter pylori* therapy: a prospective randomized placebo-controlled double-blind study. *Helicobacter*, **12**, 309–316.

Ciorba, M.A. (2012) A gastroenterologist's guide to probiotics. *Clinical Gastroenterology and Hepatology*, **10**, 960–968.

Clarke, G., Cryan, J.F., Dinan, T.G. & Quigley, E.M. (2012) Review article: probiotics for the treatment of irritable bowel syndrome: focus on lactic acid bacteria. *Alimentary Pharmacology and Therapeutics*, **35**, 403–413.

Coccorullo, P., Strisciuglio, C., Martinelli, M., Miele, E., Greco, L. & Staiano, A. (2010) *Lactobacillus reuteri* (DSM 17938) in infants with functional chronic constipation: a double-blind, randomized, placebo-controlled study. *Journal of Pediatrics*, **157**, 598–602.

Cohen, R., Martin, E. de la Rocque, F., Thollot, F., Pecquet, S., Werner, A. Boucherat, M. Varon, E., Bingen, E. & Levy, C. (2013) Probiotics and prebiotics in preventing episodes of acute otitis media in high-risk children: a randomized, double-blind, placebo-controlled study. *Pediatric Infectious Disease Journal*, **32**, 810–814.

Commane, D., Hughes, R., Shortt, C. & Rowland, I. (2005) The potential mechanisms involved in the anti-carcinogenic action of probiotics. *Mutation Research*, **591**, 276–289.

Correa, N.B., Peret Filho, L.A., Penna, F.J., Lima, F.M. & Nicoli, J.R. (2005) A randomized formula controlled trial of *Bifidobacterium lactis* and *Streptococcus thermophilus* for prevention of antibiotic-associated diarrhea in infants. *Journal of Clinical Gastroenterology*, **39**, 385–389.

Cremonini, F., Di Caro, S., Nista, E.C., Bartolozzi, F., Capelli, G., Gasbarrini, G. & Gasbarrini, A. (2002) Meta-analysis: the effect of probiotic administration on antibiotic-associated diarrhoea. *Alimentary Pharmacology and Therapeutics*, **16**, 1461–1467.

Crittenden, R.G., Martinez, N.R. & Playne, M.J. (2003) Synthesis and utilisation of folate by yoghurt starter cultures and probiotic bacteria. *International Journal of Food Microbiology*, **80**, 217–222.

Cruchet, S., Furnes, R., Maruy, A., Hebel, E., Palacios, J., Medina, F., Ramirez, N., Orsi, M., Rondon, L., Sdepanian, V., Xochihua, L., Ybarra, M. & Zablah, R.A. (2015) The use of probiotics in pediatric gastroenterology: a review of the literature and recommendations by Latin-American experts. *Paediatric Drugs*, **17**, 199–216.

Cryan, J.F. & O'Mahony, S.M. (2011) The microbiome-gut-brain axis: from bowel to behavior. *Neurogastroenterology and Motility*, **23**, 187–192.

Cui, H.H., Chen, C.L., Wang, J.D., Yang, Y.J., Cun, Y., Wu, J.B., Liu, Y.H., Dan, H.L., Jian, Y.T. & Chen, X.Q. (2004) Effects of probiotic on intestinal mucosa of patients with ulcerative colitis. *World Journal of Gastroenterology*, **10**, 1521–1525.

D'Ettorre, G., Ceccarelli, G., Giustini, N., Serafino, S., Calantone, N., De Girolamo, G., Bianchi, L., Bellelli, V., Ascoli-Bartoli, T., Marcellini, S., Turriziani, O., Brenchley, J.M. & Vullo, V. (2015) Probiotics reduce inflammation in antiretroviral treated, HIV-infected individuals: results of the "Probio-HIV" clinical trial. *PLoS One*, **10**, e0137200.

D'Inca, R., Barollo, M., Scarpa, M., Grillo, A.R., Brun, P., Vettorato, M.G., Castagliuolo, I. & Sturniolo, G.C. (2011) Rectal administration of *Lactobacillus casei* DG modifies flora composition and Toll-like receptor expression in colonic mucosa of patients with mild ulcerative colitis. *Digestive Diseases and Sciences*, **56**, 1178–1187.

D'Souza, A.L., Rajkumar, C., Cooke, J. & Bulpitt, C.J. (2002) Probiotics in prevention of antibiotic associated diarrhoea: meta-analysis. *BMJ*, **324**, 1361.

Davis, M. & Gamier, P. (2015) New options in constipation management. *Current Oncology Reports*, **17**, 55.

De Angelis, M., Francavilla, R., Piccolo, M., De Giacomo, A. & Gobbetti, M. (2015) Autism spectrum disorders and intestinal microbiota. *Gut Microbes*, **6**, 207–213.

Devriese, L.A. & Pot, B. (1995) The genus *Enterococcus*. In *The Genera of Lactic Acid Bacteria* (ed. B.J.B Wood & W.H. Holzapfel), Vol. **2**, 327–367. Blackie Academic & Professional, London.

De Paula, J.A., Carmuega, E. & Weill, R. (2008) Effect of the ingestion of a symbiotic yogurt on the bowel habits of women with functional constipation. *Acta Gastroenterologica Latinoamericana*, **38**, 16–25.

De Preter, V., Raemen, H., Cloetens, L., Houben, E., Rutgeerts, P. & Verbeke, K. (2008) Effect of dietary intervention with different pre- and probiotics on intestinal bacterial enzyme activities. *European Journal of Clinical Nutrition*, **62**, 225–231.

De Preter, V., Hamer, H.M., Windey, K. & Verbeke, K. (2011) The impact of pre- and/or probiotics on human colonic metabolism: does it affect human health? *Molecular Nutrition & Food Research*, **55**, 46–57.

De Seta, F., Parazzini, F., De Leo, R., Banco, R., Maso, G.P., De Santo, D., Sartore, A., Stabile, G., Inglese, S., Tonon, M. & Restaino, S. (2014) *Lactobacillus plantarum* P17630 for preventing *Candida* vaginitis recurrence: a retrospective comparative study. *European Journal of Obstetrics, Gynecology, and Reproductive Biology*, **182**, 136–139.

Deckers, I.A., McLean, S., Linssen, S., Mommers, M., Van Schayck, C.P. & Sheikh, A. (2012) Investigating international time trends in the incidence and prevalence of atopic eczema 1990-2010: a systematic review of epidemiological studies. *PLoS One*, **7**, e39803.

Delzenne, N.M. & Cani, P.D. (2011) Interaction between obesity and the gut microbiota: relevance in nutrition. *Annual Review of Nutrition*, **31**, 15–31.

Delzenne, N.M., Cani, P.D., Everard, A., Neyrinck, A.M. & Bindels, L.B. (2015) Gut microorganisms as promising targets for the management of type 2 diabetes. *Diabetologia*, **58**, 2206–2217.

Dhiman, R.K., Rana, B., Agrawal, S., Garg, A., Chopra, M., Thumburu, K.K., Khattri, A., Malhotra, S., Duseja, A. & Chawla, Y.K. (2014) Probiotic VSL#3 reduces liver disease severity and hospitalization in patients with cirrhosis: a randomized, controlled trial. *Gastroenterology*, **147**, 1327–1337.

Di Cerbo, A., Palmieri, B., Aponte, M., Morales-Medina, J.C. & Iannitti, T. (2015) Mechanisms and therapeutic effectiveness of lactobacilli. *Journal of Clinical Pathology*, **69**, 187–203.

Diaz Heijtz, R., Wang, S., Anuar, F., Qian, Y., Bjorkholm, B., Samuelsson, A., Hibberd, M.L., Forssberg, H. & Pettersson, S. (2011) Normal gut microbiota modulates brain development and behavior. *Proceedings of the National Academy of Sciences of the United States of America*, **108**, 3047–3052.

Dietrich, C.G., Kottmann, T. & Alavi, M. (2014) Commercially available probiotic drinks containing *Lactobacillus casei* DN-114001 reduce antibiotic-associated diarrhea. *World Journal of Gastroenterology*, **20**, 15837–15844.

Dilli, D., Aydin, B., Fettah, N.D., Ozyazici, E., Beken, S., Zenciroglu, A., Okumus, N., Ozyurt, B.M., Ipek, M.S., Akdag, A., Turan, O. & Bozdag, S. (2015) The propre-save study: effects of probiotics and prebiotics alone or combined on necrotizing enterocolitis in very low birth weight infants. *Journal of Pediatrics*, **166**, 545–551.

Dimidi, E., Christodoulides, S., Fragkos, K.C., Scott, S.M. & Whelan, K. (2014) The effect of probiotics on functional constipation in adults: a systematic review and meta-analysis of randomized controlled trials. *American Journal of Clinical Nutrition*, **100**, 1075–1084.

Dinan, T.G., Stanton, C. & Cryan, J.F. (2013) Psychobiotics: a novel class of psychotropic. *Biological Psychiatry*, **74**, 720–726.

Dinleyici, E.C., Group, P.S. & Vandenplas, Y. (2014) *Lactobacillus reuteri* DSM 17938 effectively reduces the duration of acute diarrhoea in hospitalised children. *Acta Paediatrica*, **103**, e300–305.

Dobson, A., Cotter, P.D., Ross, R.P. & Hill, C. (2012) Bacteriocin production: a probiotic trait? *Applied and Environmental Microbiology*, **78**, 1–6.

Dolatkhah, N., Hajifaraji, M., Abbasalizadeh, F., Aghamohammadzadeh, N., Mehrabi, Y. & Mesgari Abbasi, M. (2015) Is there a value for probiotic supplements in gestational diabetes mellitus? A randomized clinical trial. *Journal of Health, Population, and Nutrition*, **33**, 25.

Dong, H., Rowland, I. & Yaqoob, P. (2012) Comparative effects of six probiotic strains on immune function *in vitro*. *British Journal of Nutrition*, **108**, 459–470.

Dotterud, C.K., Storro, O., Johnsen, R. & Oien, T. (2010) Probiotics in pregnant women to prevent allergic disease: a randomized, double-blind trial. *British Journal of Dermatology*, **163**, 616–623.

Drago, L., De Vecchi, E., Nicola, L., Zucchetti, E., Gismondo, M.R. & Vicariotto, F. (2007) Activity of a *Lactobacillus acidophilus*-based douche for the treatment of bacterial vaginosis. *Journal of Alternative and Complementary Medicine*, **13**, 435–438.

Drouault-Holowacz, S., Bieuvelet, S., Burckel, A., Cazaubiel, M., Dray, X. & Marteau, P. (2008) A double blind randomized controlled trial of a probiotic combination in 100 patients with irritable bowel syndrome. *Gastroenterologie Clinique et Biologique*, **32**, 147–152.

Ejtahed, H.S., Mohtadi-Nia, J., Homayouni-Rad, A., Niafar, M., Asghari-Jafarabadi, M., Mofid, V. & Akbarian-Moghari, A. (2011) Effect of probiotic yogurt containing *Lactobacillus acidophilus* and *Bifidobacterium lactis* on lipid profile in individuals with type 2 diabetes mellitus. *Journal of Dairy Science*, **94**, 3288–3294.

Ejtahed, H.S., Mohtadi-Nia, J., Homayouni-Rad, A., Niafar, M., Asghari-Jafarabadi, M. & Mofid, V. (2012) Probiotic yogurt improves antioxidant status in type 2 diabetic patients. *Nutrition*, **28**, 539–543.

Enck, P., Zimmermann, K., Menke, G., Muller-Lissner, S., Martens, U. & Klosterhalfen, S. (2008) A mixture of *Escherichia coli* (DSM 17252) and *Enterococcus faecalis* (DSM 16440) for treatment of the irritable bowel syndrome: a randomized controlled trial with primary care physicians. *Neurogastroenterology and Motility*, **20**, 1103–1109.

Eriksson, K., Carlsson, B., Forsum, U. & Larsson, P.G. (2005) A double-blind treatment study of bacterial vaginosis with normal vaginal lactobacilli after an open treatment with vaginal clindamycin ovules. *Acta Dermato-Venereologica*, **85**, 42–46.

Erturk-Hasdemir, D. & Kasper, D.L. (2013) Resident commensals shaping immunity. *Current Opinion in Immunology*, **25**, 450–455.

Ettinger, G., MacDonald, K., Reid, G. & Burton, J.P. (2014) The influence of the human microbiome and probiotics on cardiovascular health. *Gut Microbes*, **5**, 719–728.

Falasca, K., Vecchiet, J., Ucciferri, C., Di Nicola, M., D'Angelo, C. & Reale, M. (2015) Effect of probiotic supplement on cytokine levels in HIV-infected individuals: a preliminary study. *Nutrients*, **7**, 8335–8347.

Fateh, R., Iravani, S., Frootan, M., Rasouli, M.R. & Saadat, S. (2011) Synbiotic preparation in men suffering from functional constipation: a randomised controlled trial. *Swiss Medical Weekly*, **141**, w13239.

Favretto, D.C., Pontin, B. & Moreira, T.R. (2013) Effect of the consumption of a cheese enriched with probiotic organisms (*Bifidobacterium lactis* bi-07) in improving symptoms of constipation. *Arquivos de Gastroenterologia*, **50**, 196–201.

Fedorak, R. & Demeria, D. (2012) Probiotic bacteria in the prevention and the treatment of inflammatory bowel disease. *Gastroenterology Clinics of North America*, **41**, 821–842.

Feizizadeh, S., Salehi-Abargouei, A. & Akbari, V. (2014) Efficacy and safety of *Saccharomyces boulardii* for acute diarrhea. *Pediatrics*, **134**, e176–e191.

Fernandez, M., Hudson, J.A., Korpela, R. & De Los Reyes-Gavilan, C.G. (2015) Impact on human health of microorganisms present in fermented dairy products: an overview. *Biomed Research International*, **2015**, 412714.

Fiocchi, A., Pawankar, R., Cuello-Garcia, C., Ahn, K., Al-Hammadi, S., Agarwal, A., Beyer, K., Burks, W., Canonica, G.W., Ebisawa, M., Gandhi, S., Kamenwa, R., Lee, B.W., Li, H., Prescott, S., Riva, J.J., Rosenwasser, L., Sampson, H., Spigler, M., Terracciano, L., Vereda-Ortiz, A., Waserman, S., Yepes-Nunez, J.J., Brozek, J.L. & Schunemann, H.J. (2015) World Allergy Organization–McMaster University guidelines for allergic disease prevention (GLAD-P): probiotics. *World Allergy Organization Journal*, **8**, 4.

Firouzi, S., Majid, H.A., Ismail, A., Kamaruddin, N.A. & Barakatun-Nisak, M.Y. (2017) Effect of multi-strain probiotics (multi-strain microbial cell preparation) on glycemic control and other diabetes-related outcomes in people with type 2 diabetes: a randomized controlled trial. *European Journal of Nutrition*, **56**, 1535. doi:10.1007/s00394-016-1199-8

Flint, H.J., Scott, K.P., Louis, P. & Duncan, S.H. (2012) The role of the gut microbiota in nutrition and health. *Nature Reviews: Gastroenterology & Hepatology*, **9**, 577–589.

Floch, M.H. (2014) Recommendations for probiotic use in humans: a 2014 update. *Pharmaceuticals (Basel, Switzerland)*, **7**, 999–1007.

Folster-Holst, R., Muller, F., Schnopp, N., Abeck, D., Kreiselmaier, I., Lenz, T., Von Ruden, U., Schrezenmeir, J., Christophers, E. & Weichenthal, M. (2006) Prospective, randomized controlled trial on *Lactobacillus rhamnosus* in infants with moderate to severe atopic dermatitis. *British Journal of Dermatology*, **155**, 1256–1261.

Forchielli, M.L. & Walker, W.A. (2005) The role of gut-associated lymphoid tissues and mucosal defence. *British Journal of Nutrition*, **93**, S41–S48.

Ford, A.C., Quigley, E.M., Lacy, B.E., Lembo, A.J., Saito, Y.A., Schiller, L.R., Soffer, E.E., Spiegel, B.M. & Moayyedi, P. (2014) Efficacy of prebiotics, probiotics, and synbiotics in irritable bowel syndrome and chronic idiopathic constipation: systematic review and meta-analysis. *American Journal of Gastroenterology*, **109**, 1547–1561.

Francavilla, R., Miniello, V., Magista, A.M., De Canio, A., Bucci, N., Gagliardi, F., Lionetti, E., Castellaneta, S., Polimeno, L., Peccarisi, L., Indrio, F. & Cavallo, L. (2010) A randomized controlled trial of *Lactobacillus* GG in children with functional abdominal pain. *Pediatrics*, **126**, e1445–e1452.

Francavilla, R., Lionetti, E., Castellaneta, S., Ciruzzi, F., Indrio, F., Masciale, A., Fontana, C., La Rosa, M.M., Cavallo, L. & Francavilla, A. (2012) Randomised clinical trial: *Lactobacillus reuteri* DSM 17938 vs. placebo in children with acute diarrhoea – a double-blind study. *Alimentary Pharmacology and Therapeutics*, **36**, 363–369.

Frye, R.E., Slattery, J., MacFabe, D.F., Allen-Vercoe, E., Parker, W., Rodakis, J., Adams, J.B., Krajmalnik-Brown, R., Bolte, E., Kahler, S., Jennings, J., James, J., Cerniglia, C.E. & Midtvedt, T. (2015) Approaches to studying and manipulating the enteric microbiome to improve autism symptoms. *Microbial Ecology in Health and Disease*, **26**, 26878.

Fujiki, H., Suganuma, M., Okabe, S., Sueoka, E., Suga, K., Imai, K. & Nakachi, K. (2000) A new concept of tumor promotion by tumor necrosis factor-alpha, and cancer preventive agents (-)-epigallocatechin gallate and green tea: a review. *Cancer Detection and Prevention*, **24**, 91–99.

Fujimori, S., Tatsuguchi, A., Gudis, K., Kishida, T., Mitsui, K., Ehara, A., Kobayashi, T., Sekita, Y., Seo, T. & Sakamoto, C. (2007) High dose probiotic and prebiotic cotherapy for remission induction of active Crohn's disease. *Journal of Gastroenterology and Hepatology*, **22**, 1199–1204.

Fujimori, S., Gudis, K., Mitsui, K., Seo, T., Yonezawa, M., Tanaka, S., Tatsuguchi, A. & Sakamoto, C. (2009) A randomized controlled trial on the efficacy of synbiotic versus probiotic or prebiotic treatment to improve the quality of life in patients with ulcerative colitis. *Nutrition*, **25**, 520–525.

Fujita, R., Iimuro, S., Shinozaki, T., Sakamaki, K., Uemura, Y., Takeuchi, A., Matsuyama, Y. & Ohashi, Y. (2013) Decreased duration of acute upper respiratory tract infections with daily intake of fermented milk: a multicenter, double-blinded, randomized comparative study in users of day care facilities for the elderly population. *American Journal of Infection Control*, **41**, 1231–1235.

Furrie, E., MacFarlane, S., Kennedy, A., Cummings, J.H., Walsh, S.V., O'Neil D, A. & MacFarlane, G.T. (2005) Synbiotic therapy (*Bifidobacterium longum*/Synergy 1) initiates resolution of inflammation in patients with active ulcerative colitis: a randomised controlled pilot trial. *Gut*, **54**, 242–249.

Furue, H., Matsuo, K., Kumimoto, H., Hiraki, A., Suzuki, T., Yatabe, Y., Komori, K., Kanemitsu, Y., Hirai, T., Kato, T., Ueda, M., Ishizaki, K. & Tajima, K. (2008) Decreased risk of colorectal cancer with the high natural killer cell activity NKG2D genotype in Japanese. *Carcinogenesis*, **29**, 316–320.

Furuya-Kanamori, L., Marquess, J., Yakob, L., Riley, T.V., Paterson, D.L., Foster, N.F., Huber, C.A. & Clements, A.C. (2015) Asymptomatic *Clostridium difficile* colonization: epidemiology and clinical implications. *BMC Infectious Diseases*, **15**, 516.

Gao, X.W., Mubasher, M., Fang, C.Y., Reifer, C. & Miller, L.E. (2010) Dose-response efficacy of a proprietary probiotic formula of *Lactobacillus acidophilus* CL1285 and *Lactobacillus casei* LBC80R for antibiotic-associated diarrhea and *Clostridium difficile*-associated diarrhea prophylaxis in adult patients. *American Journal of Gastroenterology*, **105**, 1636–1641.

Gao, T., He, B., Pan, Y., Deng, Q., Sun, H., Liu, X., Chen, J., Wang, S. & Xia, Y. (2015) Association of *Clostridium difficile* infection in hospital mortality: a systematic review and meta-analysis. *American Journal of Infection Control*, **43**, 1316–1320.

Gao, X., Zhu, Y., Wen, Y., Liu, G. & Wan, C. (2016) Efficacy of probiotics in nonalcoholic fatty liver disease in adult and children: a meta-analysis of randomized controlled trials. *Hepatology Research*, **46**, 1226–1233.

Garaiova, I., Muchova, J., Nagyova, Z., Wang, D., Li, J.V., Orszaghova, Z., Michael, D.R., Plummer, S.F. & Durackova, Z. (2015) Probiotics and vitamin C for the prevention of respiratory tract infections in children attending preschool: a randomised controlled pilot study. *European Journal of Clinical Nutrition*, **69**, 373–379.

Gawronska, A., Dziechciarz, P., Horvath, A. & Szajewska, H. (2007) A randomized double-blind placebo-controlled trial of *Lactobacillus* GG for abdominal pain disorders in children. *Alimentary Pharmacology and Therapeutics*, **25**, 177–184.

Gerasimov, S.V., Vasjuta, V.V., Myhovych, O.O. & Bondarchuk, L.I. (2010) Probiotic supplement reduces atopic dermatitis in preschool children: a randomized, double-blind, placebo-controlled, clinical trial. *American Journal of Clinical Dermatology*, **11**, 351–361.

Ghouri, Y.A., Richards, D.M., Rahimi, E.F., Krill, J.T., Jelinek, K.A. & Dupont, A.W. (2014) Systematic review of randomized controlled trials of probiotics, prebiotics, and synbiotics in inflammatory bowel disease. *Clinical and Experimental Gastroenterology*, **7**, 473–487.

Gianotti, L., Morelli, L., Galbiati, F., Rocchetti, S., Coppola, S., Beneduce, A., Gilardini, C., Zonenschain, D., Nespoli, A. & Braga, M. (2010) A randomized double-blind trial on perioperative administration of probiotics in colorectal cancer patients. *World Journal of Gastroenterology*, **16**, 167–175.

Gilbert, J.A., Krajmalnik-Brown, R., Porazinska, D.L., Weiss, S.J. & Knight, R. (2013) Toward effective probiotics for autism and other neurodevelopmental disorders. *Cell*, **155**, 1446–1448.

Gillespie, D., Hood, K., Bayer, A., Carter, B., Duncan, D., Espinasse, A., Evans, M., Nuttall, J., Stanton, H., Acharjya, A., Allen, S., Cohen, D., Groves, S., Francis, N., Howe, R., Johansen, A., Mantzourani, E., Thomas-Jones, E., Toghill, A., Wood, F., Wigglesworth, N., Wootton, M. & Butler, C.C. (2015) Antibiotic prescribing and associated diarrhoea: a prospective cohort study of care home residents. *Age and Ageing*, **44**, 853–860.

Gionchetti, P., Rizzello, F., Venturi, A., Brigidi, P., Matteuzzi, D., Bazzocchi, G., Poggioli, G., Miglioli, M. & Campieri, M. (2000) Oral bacteriotherapy as maintenance treatment in patients with chronic pouchitis: a double-blind, placebo-controlled trial. *Gastroenterology*, **119**, 305–309.

Gionchetti, P., Rizzello, F., Helwig, U., Venturi, A., Lammers, K.M., Brigidi, P., Vitali, B., Poggioli, G., Miglioli, M. & Campieri, M. (2003) Prophylaxis of pouchitis onset with probiotic therapy: a double-blind, placebo-controlled trial. *Gastroenterology*, **124**, 1202–1209.

Gionchetti, P., Rizzello, F., Morselli, C., Poggioli, G., Tambasco, R., Calabrese, C., Brigidi, P., Vitali, B., Straforini, G. & Campieri, M. (2007) High-dose probiotics for the treatment of active pouchitis. *Diseases of the Colon and Rectum*, **50**, 2075–2082.

Giorgetti, G., Brandimarte, G., Fabiocchi, F., Ricci, S., Flamini, P., Sandri, G., Trotta, M.C., Elisei, W., Penna, A., Lecca, P.G., Picchio, M. & Tursi, A. (2015) Interactions between innate immunity, microbiota, and probiotics. *Journal of Immunology Research*, **2015**, 501361.

Giovannini, M., Agostoni, C., Riva, E., Salvini, F., Ruscitto, A., Zuccotti, G.V., Radaelli, G. & Felicita Study, G. (2007) A randomized prospective double blind controlled trial on effects of long-term consumption of fermented milk containing *Lactobacillus casei* in pre-school children with allergic asthma and/or rhinitis. *Pediatric Research*, **62**, 215–220.

Gleeson, M. (2016) Immunological aspects of sport nutrition. *Immunology and Cell Biology*, **94**, 117–123.

Gleeson, M., Bishop, N.C., Oliveira, M. & Tauler, P. (2011) Daily probiotic's (*Lactobacillus casei* Shirota) reduction of infection incidence in athletes. *International Journal of Sport Nutrition and Exercise Metabolism*, **21**, 55–64.

Gleeson, M., Bishop, N.C., Oliveira, M., McCauley, T., Tauler, P. & Lawrence, C. (2012) Effects of a *Lactobacillus salivarius* probiotic intervention on infection, cold symptom duration and severity, and mucosal immunity in endurance athletes. *International Journal of Sport Nutrition and Exercise Metabolism*, **22**, 235–242.

Goncalves, P. & Martel, F. (2013) Butyrate and colorectal cancer: the role of butyrate transport. *Current Drug Metabolism*, **14**, 994–1008.

Gonzalez-Hernandez, L.A., Jave-Suarez, L.F., Fafutis-Morris, M., Montes-Salcedo, K.E., Valle-Gutierrez, L.G., Campos-Loza, A.E., Enciso-Gomez, L.F. & Andrade-Villanueva, J.F. (2012) Synbiotic therapy decreases microbial translocation and inflammation and improves immunological status in HIV-infected patients: a double-blind randomized controlled pilot trial. *Nutrition Journal*, **11**, 90.

Gosselink, M.P., Schouten, W.R., Van Lieshout, L.M., Hop, W.C., Laman, J.D. & Ruseler-Van Embden, J.G. (2004) Delay of the first onset of pouchitis by oral intake of the probiotic strain *Lactobacillus rhamnosus* GG. *Diseases of the Colon and Rectum*, **47**, 876–884.

Grin, P.M., Kowalewska, P.M., Alhazzan, W. & Fox-Robichaud, A.E. (2013) *Lactobacillus* for preventing recurrent urinary tract infections in women: meta-analysis. *Canadian Journal of Urology*, **20**, 6607–6614.

Gruber, C., Wendt, M., Sulser, C., Lau, S., Kulig, M., Wahn, U., Werfel, T. & Niggemann, B. (2007) Randomized, placebo-controlled trial of *Lactobacillus rhamnosus* GG as treatment of atopic dermatitis in infancy. *Allergy*, **62**, 1270–1276.

Guandalini, S., Pensabene, L., Zikri, M.A., Dias, J.A., Casali, L.G., Hoekstra, H., Kolacek, S., Massar, K., Micetic-Turk, D., Papadopoulou, A., De Sousa, J.S., Sandhu, B., Szajewska, H. & Weizman, Z. (2000) *Lactobacillus* GG administered in oral rehydration solution to children with acute diarrhea: a multicenter European trial. *Journal of Pediatric Gastroenterology and Nutrition*, **30**, 54–60.

Guandalini, S., Magazzu, G., Chiaro, A., La Balestra, V., Di Nardo, G., Gopalan, S., Sibal, A., Romano, C., Canani, R.B., Lionetti, P. & Setty, M. (2010) VSL#3 improves symptoms in children with irritable bowel syndrome: a multicenter, randomized, placebo-controlled, double-blind, crossover study. *Journal of Pediatric Gastroenterology and Nutrition*, **51**, 24–30.

Guarino, A., Lo Vecchio, A. & Canani, R.B. (2009) Probiotics as prevention and treatment for diarrhea. *Current Opinion in Gastroenterology*, **25**, 18–23.

Guglielmetti, S., Mora, D., Gschwender, M. & Popp, K. (2011) Randomised clinical trial: *Bifidobacterium bifidum* MIMBb75 significantly alleviates irritable bowel syndrome and improves quality of life: a double-blind, placebo-controlled study. *Alimentary Pharmacology and Therapeutics*, **33**, 1123–1132.

Guillemard, E., Tondu, F., Lacoin, F. & Schrezenmeir, J. (2010) Consumption of a fermented dairy product containing the probiotic *Lactobacillus casei* DN-114001 reduces the duration of respiratory infections in the elderly in a randomised controlled trial. *British Journal of Nutrition*, **103**, 58–68.

Guslandi, M., Mezzi, G., Sorghi, M. & Testoni, P.A. (2000) *Saccharomyces boulardii* in maintenance treatment of Crohn's disease. *Digestive Diseases and Sciences*, **45**, 1462–1464.

Guslandi, M., Giollo, P. & Testoni, P.A. (2003) A pilot trial of *Saccharomyces boulardii* in ulcerative colitis. *European Journal of Gastroenterology and Hepatology*, **15**, 697–698.

Gutierrez-Castrellon, P., Lopez-Velazquez, G., Diaz-Garcia, L., Jimenez-Gutierrez, C., Mancilla-Ramirez, J., Estevez-Jimenez, J. & Parra, M. (2014) Diarrhea in preschool children and *Lactobacillus reuteri*: a randomized controlled trial. *Pediatrics*, **133**, e904–e909.

Han, S.H., Suk, K.T., Kim, D.J., Kim, M.Y., Baik, S.K., Kim, Y.D., Cheon, G.J., Choi, D.H., Ham, Y.L., Shin, D.H. & Kim, E.J. (2015) Effects of probiotics (cultured *Lactobacillus subtilis/Streptococcus faecium*) in the treatment of alcoholic hepatitis: randomized-controlled multicenter study. *European Journal of Gastroenterology and Hepatology*, **27**, 1300–1306.

Hansen, J.J. & Sartor, R.B. (2015) Therapeutic manipulation of the microbiome in IBD: current results and future approaches. *Current Treatment Options in Gastroenterology*, **13**, 105–120.

Hao, Q., Dong, B.R. & Wu, T. (2015) Probiotics for preventing acute upper respiratory tract infections. *Cochrane Database of Systematic Reviews*, **2**, CD006895.

Harb, T., Matsuyama, M., David, M. & Hill, R. (2015) Infant colic – what works: a systematic review of interventions for breastfed infants. *Journal of Pediatric Gastroenterology and Nutrition*, **62**, 668–686.

Hardy, H., Harris, J., Lyon, E., Beal, J. & Foey, A.D. (2013) Probiotics, prebiotics and immunomodulation of gut mucosal defences: homeostasis and immunopathology. *Nutrients*, **5**, 1869–1912.

Hariri, M., Salehi, R., Feizi, A., Mirlohi, M., Kamali, S. & Ghiasvand, R. (2015) The effect of probiotic soy milk and soy milk on anthropometric measures and blood pressure in patients with type II diabetes mellitus: a randomized double-blind clinical trial. *ARYA Atherosclerosis*, **11**, 74–80.

Hatakka, K., Savilahti, E., Ponka, A., Meurman, J.H., Poussa, T., Nase, L., Saxelin, M. & Korpela, R. (2001) Effect of long term consumption of probiotic milk on infections in children attending day care centres: double blind, randomised trial. *BMJ*, **322**, 1327.

Hatakka, K., Blomgren, K., Pohjavuori, S., Kaijalainen, T., Poussa, T., Leinonen, M., Korpela, R. & Pitkaranta, A. (2007) Treatment of acute otitis media with probiotics in otitis-prone children: a double-blind, placebo-controlled randomised study. *Clinical Nutrition*, **26**, 314–321.

Heczko, P.B., Tomusiak, A., Adamski, P., Jakimiuk, A.J., Stefanski, G., Mikolajczyk-Cichonska, A., Suda-Szczurek, M. & Strus, M. (2015) Supplementation of standard antibiotic therapy with oral probiotics for bacterial vaginosis and aerobic vaginitis: a randomised, double-blind, placebo-controlled trial. *BMC Women's Health*, **15**, 115.

Hegazy, S.K. & El-Bedewy, M.M. (2010) Effect of probiotics on pro-inflammatory cytokines and NF-kappaB activation in ulcerative colitis. *World Journal of Gastroenterology*, **16**, 4145–4151.

Helin, T., Haahtela, S. & Haahtela, T. (2002) No effect of oral treatment with an intestinal bacterial strain, *Lactobacillus rhamnosus* (ATCC 53103), on birch-pollen allergy: a placebo-controlled double-blind study. *Allergy*, **57**, 243–246.

Hempel, S., Newberry, S.J., Maher, A.R., Wang, Z., Miles, J.N., Shanman, R., Johnsen, B. & Shekelle, P.G. (2012) Probiotics for the prevention and treatment of antibiotic-associated diarrhea: a systematic review and meta-analysis. *Journal of the American Medical Association*, **307**, 1959–1969.

Hemsworth, J.C., Hekmat, S. & Reid, G. (2012) Micronutrient supplemented probiotic yogurt for HIV-infected adults taking HAART in London, Canada. *Gut Microbes*, **3**, 414–419.

Henker, J., Muller, S., Laass, M.W., Schreiner, A. & Schulze, J. (2008) Probiotic *Escherichia coli* Nissle 1917 (EcN) for successful remission maintenance of ulcerative colitis in children and adolescents: an open-label pilot study. *Zeitschrift für Gastroenterologie*, **46**, 874–875.

Hickson, M., D'Souza, A.L., Muthu, N., Rogers, T.R., Want, S., Rajkumar, C. & Bulpitt, C.J. (2007) Use of probiotic *Lactobacillus* preparation to prevent diarrhoea associated with antibiotics: randomised double blind placebo controlled trial. *BMJ*, **335**, 80.

Hill, C., Guarner, F., Reid, G., Gibson, G.R., Merenstein, D.J., Pot, B., Morelli, L., Canani, R.B., Flint, H.J., Salminen, S., Calder, P.C. & Sanders, M.E. (2014) Expert consensus document. The International Scientific Association for Probiotics and Prebiotics consensus statement on the scope and appropriate use of the term probiotic. *Nature Reviews: Gastroenterology & Hepatology*, **11**, 506–514.

Hojsak, I., Abdovic, S., Szajewska, H., Milosevic, M., Krznaric, Z. & Kolacek, S. (2010) *Lactobacillus* GG in the prevention of nosocomial gastrointestinal and respiratory tract infections. *Pediatrics*, **125**, e1171–e1177.

Holte, K., Krag, A. & Gluud, L.L. (2012) Systematic review and meta-analysis of randomized trials on probiotics for hepatic encephalopathy. *Hepatology Research*, **42**, 1008–1015.

Holubar, S.D., Cima, R.R., Sandborn, W.J. & Pardi, D.S. (2010) Treatment and prevention of pouchitis after ileal pouch-anal anastomosis for chronic ulcerative colitis. *Cochrane Database of Systematic Reviews*, CD001176.

Homayouni, A., Bastani, P., Ziyadi, S., Mohammad-Alizadeh-Charandabi, S., Ghalibaf, M., Mortazavian, A.M. & Mehrabany, E.V. (2014) Effects of probiotics on the recurrence of bacterial vaginosis: a review. *Journal of Lower Genital Tract Disease*, **18**, 79–86.

Horvath, A., Dziechciarz, P. & Szajewska, H. (2011) Meta-analysis: *Lactobacillus rhamnosus* GG for abdominal pain-related functional gastrointestinal disorders in childhood. *Alimentary Pharmacology and Therapeutics*, **33**, 1302–1310.

Hosseini, A., Nikfar, S. & Abdollahi, M. (2012) Probiotics use to treat irritable bowel syndrome. *Expert Opinion on Biological Therapy*, **12**, 1323–1334.

Htwe, K., Yee, K.S., Tin, M. & Vandenplas, Y. (2008) Effect of *Saccharomyces boulardii* in the treatment of acute watery diarrhea in Myanmar children: a randomized controlled study. *American Journal of Tropical Medicine and Hygiene*, **78**, 214–216.

Huang, H., Song, L. & Zhao, W. (2014) Effects of probiotics for the treatment of bacterial vaginosis in adult women: a meta-analysis of randomized clinical trials. *Archives of Gynecology and Obstetrics*, **289**, 1225–1234.

Hulston, C.J., Churnside, A.A. & Venables, M.C. (2015) Probiotic supplementation prevents high-fat, overfeeding-induced insulin resistance in human subjects. *British Journal of Nutrition*, **113**, 596–602.

Hummelen, R., Changalucha, J., Butamanya, N.L., Koyama, T.E., Cook, A., Habbema, J.D. & Reid, G. (2011a) Effect of 25 weeks probiotic supplementation on immune function of HIV patients. *Gut Microbes*, **2**, 80–85.

Hummelen, R., Hemsworth, J., Changalucha, J., Butamanya, N.L., Hekmat, S., Habbema, J.D. & Reid, G. (2011b) Effect of micronutrient and probiotic fortified yogurt on immune-function of anti-retroviral therapy naive HIV patients. *Nutrients*, **3**, 897–909.

Hutt, P., Songisepp, E., Ratsep, M., Mahlapuu, R., Kilk, K. & Mikelsaar, M. (2015) Impact of probiotic *Lactobacillus plantarum* TENSIA in different dairy products on anthropometric and blood biochemical indices of healthy adults. *Beneficial Microbes*, **6**, 233–243.

Huurre, A., Laitinen, K., Rautava, S., Korkeamaki, M. & Isolauri, E. (2008) Impact of maternal atopy and probiotic supplementation during pregnancy on infant sensitization: a double-blind placebo-controlled study. *Clinical and Experimental Allergy*, **38**, 1342–1348.

Huynh, H.Q., Debruyn, J., Guan, L., Diaz, H., Li, M., Girgis, S., Turner, J., Fedorak, R. & Madsen, K. (2009) Probiotic preparation VSL#3 induces remission in children with mild to moderate acute ulcerative colitis: a pilot study. *Inflammatory Bowel Diseases*, **15**, 760–768.

Imhoff, A. & Karpa, K. (2009) Is there a future for probiotics in preventing *Clostridium difficile*-associated disease and treatment of recurrent episodes? *Nutrition in Clinical Practice*, **24**, 15–32.

Indrio, F., Di Mauro, A. & Riezzo, G. (2014a) Prophylactic use of a probiotic in the prevention of colic, regurgitation, and functional constipation: reply. *Journal of the American Medical Association Pediatrics*, **168**, 777–778.

Indrio, F., Di Mauro, A., Riezzo, G., Civardi, E., Intini, C., Corvaglia, L., Ballardini, E., Bisceglia, M., Cinquetti, M., Brazzoduro, E., Del Vecchio, A., Tafuri, S. & Francavilla, R. (2014b) Prophylactic use of a probiotic in the prevention of colic, regurgitation, and functional constipation: a randomized clinical trial. *Journal of the American Medical Association Pediatrics*, **168**, 228–233.

Ishida, Y., Nakamura, F., Kanzato, H., Sawada, D., Hirata, H., Nishimura, A., Kajimoto, O. & Fujiwara, S. (2005a) Clinical effects of *Lactobacillus acidophilus* strain L-92 on perennial allergic rhinitis: a double-blind, placebo-controlled study. *Journal of Dairy Science*, **88**, 527–533.

Ishida, Y., Nakamura, F., Kanzato, H., Sawada, D., Yamamoto, N., Kagata, H., Oh-Ida, M., Takeuchi, H. & Fujiwara, S. (2005b) Effect of milk fermented with *Lactobacillus acidophilus* strain L-92 on symptoms of Japanese cedar pollen allergy: a randomized placebo-controlled trial. *Bioscience, Biotechnology, and Biochemistry*, **69**, 1652–1660.

Ishikawa, H., Akedo, I., Umesaki, Y., Tanaka, R., Imaoka, A. & Otani, T. (2003) Randomized controlled trial of the effect of bifidobacteria-fermented milk on ulcerative colitis. *Journal of the American College of Nutrition*, **22**, 56–63.

Ishikawa, H., Akedo, I., Otani, T., Suzuki, T., Nakamura, T., Takeyama, I., Ishiguro, S., Miyaoka, E., Sobue, T. & Kakizoe, T. (2005) Randomized trial of dietary fiber and *Lactobacillus casei* administration for prevention of colorectal tumors. *International Journal of Cancer*, **116**, 762–767.

Ishikawa, H., Matsumoto, S., Ohashi, Y., Imaoka, A., Setoyama, H., Umesaki, Y., Tanaka, R. & Otani, T. (2011) Beneficial effects of probiotic bifidobacterium and galacto-oligosaccharide in patients with ulcerative colitis: a randomized controlled study. *Digestion*, **84**, 128–133.
Ishimwe, N., Daliri, E.B., Lee, B.H., Fang, F. & Du, G. (2015) The perspective on cholesterol-lowering mechanisms of probiotics. *Molecular Nutrition and Food Research*, **59**, 94–105.
Ishizuka, A., Tomizuka, K., Aoki, R., Nishijima, T., Saito, Y., Inoue, R., Ushida, K., Mawatari, T. & Ikeda, T. (2012) Effects of administration of *Bifidobacterium animalis* subsp. *lactis* GCL2505 on defecation frequency and bifidobacterial microbiota composition in humans. *Journal of Bioscience and Bioengineering*, **113**, 587–591.
Ismail, I.H., Licciardi, P.V. & Tang, M.L. (2013) Probiotic effects in allergic disease. *Journal of Paediatrics and Child Health*, **49**, 709–715.
Ivey, K.L., Hodgson, J.M., Kerr, D.A., Lewis, J.R., Thompson, P.L. & Prince, R.L. (2014) The effects of probiotic bacteria on glycaemic control in overweight men and women: a randomised controlled trial. *European Journal of Clinical Nutrition*, **68**, 447–452.
Ivey, K.L., Hodgson, J.M., Kerr, D.A., Thompson, P.L., Stojceski, B. & Prince, R.L. (2015) The effect of yoghurt and its probiotics on blood pressure and serum lipid profile: a randomised controlled trial. *Nutrition, Metabolism, and Cardiovascular Diseases*, **25**, 46–51.
Ivory, K., Chambers, S.J., Pin, C., Prieto, E., Arques, J.L. & Nicoletti, C. (2008) Oral delivery of *Lactobacillus casei* Shirota modifies allergen-induced immune responses in allergic rhinitis. *Clinical and Experimental Allergy*, **38**, 1282–1289.
Ivory, K., Wilson, A.M., Sankaran, P., Westwood, M., Mccarville, J., Brockwell, C., Clark, A., Dainty, J.R., Zuidmeer-Jongejan, L. & Nicoletti, C. (2013) Oral delivery of a probiotic induced changes at the nasal mucosa of seasonal allergic rhinitis subjects after local allergen challenge: a randomised clinical trial. *PLoS One*, **8**, e78650.
Jandhyala, S.M., Talukdar, R., Subramanyam, C., Vuyyuru, H., Sasikala, M. & Nageshwar Reddy, D. (2015) Role of the normal gut microbiota. *World Journal of Gastroenterology*, **21**, 8787–8803.
Janeway, C., Travers, P., Walport, M. & Shlomchik, M. (eds.) (2005) *Immunobiology: The Immune System in Health and Disease*. Garland Science Publishing, New York and London.
Janvier, A., Malo, J. & Barrington, K.J. (2014) Cohort study of probiotics in a North American neonatal intensive care unit. *Journal of Pediatrics*, **164**, 980–985.
Jayasimhan, S., Yap, N.Y., Roest, Y., Rajandram, R. & Chin, K.F. (2013) Efficacy of microbial cell preparation in improving chronic constipation: a randomized, double-blind, placebo-controlled trial. *Clinical Nutrition*, **32**, 928–934.
Jeffery, I.B., Lynch, D.B. & O'Toole, P.W. (2015) Composition and temporal stability of the gut microbiota in older persons. *ISME Journal*, **10**, 170–182.
Johnston, B.C., Goldenberg, J.Z., Vandvik, P.O., Sun, X. & Guyatt, G.H. (2011) Probiotics for the prevention of pediatric antibiotic-associated diarrhea. *Cochrane Database of Systematic Reviews*, CD004827.
Johnston, B.C., Ma, S.S., Goldenberg, J.Z., Thorlund, K., Vandvik, P.O., Loeb, M. & Guyatt, G.H. (2012) Probiotics for the prevention of *Clostridium difficile*-associated diarrhea: a systematic review and meta-analysis. *Annals of Internal Medicine*, **157**, 878–888.
Jonkers, D., Penders, J., Masclee, A. & Pierik, M. (2012) Probiotics in the management of inflammatory bowel disease: a systematic review of intervention studies in adult patients. *Drugs*, **72**, 803–823.
Joyce, S.A., Shanahan, F., Hill, C. & Gahan, C.G. (2014) Bacterial bile salt hydrolase in host metabolism: potential for influencing gastrointestinal microbe-host crosstalk. *Gut Microbes*, **5**, 669–674.
Kadooka, Y., Sato, M., Imaizumi, K., Ogawa, A., Ikuyama, K., Akai, Y., Okano, M., Kagoshima, M. & Tsuchida, T. (2010) Regulation of abdominal adiposity by probiotics (*Lactobacillus gasseri* SBT2055) in adults with obese tendencies in a randomized controlled trial. *European Journal of Clinical Nutrition*, **64**, 636–643.

Kadooka, Y., Sato, M., Ogawa, A., Miyoshi, M., Uenishi, H., Ogawa, H., Ikuyama, K., Kagoshima, M. & Tsuchida, T. (2013) Effect of *Lactobacillus gasseri* SBT2055 in fermented milk on abdominal adiposity in adults in a randomised controlled trial. *British Journal of Nutrition*, **110**, 1696–1703.

Kajander, K., Myllyluoma, E., Rajilic-Stojanovic, M., Kyronpalo, S., Rasmussen, M., Jarvenpaa, S., Zoetendal, E.G., De Vos, W.M., Vapaatalo, H. & Korpela, R. (2008) Clinical trial: multispecies probiotic supplementation alleviates the symptoms of irritable bowel syndrome and stabilizes intestinal microbiota. *Alimentary Pharmacology and Therapeutics*, **27**, 48–57.

Kalliomaki, M., Salminen, S., Arvilommi, H., Kero, P., Koskinen, P. & Isolauri, E. (2001) Probiotics in primary prevention of atopic disease: a randomised placebo-controlled trial. *Lancet*, **357**, 1076–1079.

Kalliomaki, M., Salminen, S., Poussa, T., Arvilommi, H. & Isolauri, E. (2003) Probiotics and prevention of atopic disease: 4-year follow-up of a randomised placebo-controlled trial. *Lancet*, **361**, 1869–1871.

Kalliomaki, M., Salminen, S., Poussa, T. & Isolauri, E. (2007) Probiotics during the first 7 years of life: a cumulative risk reduction of eczema in a randomized, placebo-controlled trial. *Journal of Allergy and Clinical Immunology*, **119**, 1019–1021.

Kasinska, M.A. & Drzewoski, J. (2015) Effectiveness of probiotics in type 2 diabetes: a meta-analysis. *Polskie Archiwum Medycyny Wewnetrznej*, **125**, 803–813.

Kato, K., Mizuno, S., Umesaki, Y., Ishii, Y., Sugitani, M., Imaoka, A., Otsuka, M., Hasunuma, O., Kurihara, R., Iwasaki, A. & Arakawa, Y. (2004) Randomized placebo-controlled trial assessing the effect of bifidobacteria-fermented milk on active ulcerative colitis. *Alimentary Pharmacology and Therapeutics*, **20**, 1133–1141.

Kelly, C.J., Zheng, L., Campbell, E.L., Saeedi, B., Scholz, C.C., Bayless, A.J., Wilson, K.E., Glover, L.E., Kominsky, D.J., Magnuson, A., Weir, T.L., Ehrentraut, S.F., Pickel, C., Kuhn, K.A., Lanis, J.M., Nguyen, V., Taylor, C.T. & Colgan, S.P. (2015a) Crosstalk between microbiota-derived short-chain fatty acids and intestinal epithelial HIF augments tissue barrier function. *Cell Host Microbe*, **17**, 662–671.

Kelly, J.R., Kennedy, P.J., Cryan, J.F., Dinan, T.G., Clarke, G. & Hyland, N.P. (2015b) Breaking down the barriers: the gut microbiome, intestinal permeability and stress-related psychiatric disorders. *Frontiers in Cellular Neuroscience*, **9**, 392.

Khalesi, S., Sun, J., Buys, N. & Jayasinghe, R. (2014) Effect of probiotics on blood pressure: a systematic review and meta-analysis of randomized, controlled trials. *Hypertension*, **64**, 897–903.

Khanna, S. & Pardi, D.S. (2016) Clinical implications of antibiotic impact on gastrointestinal microbiota and *Clostridium difficile* infection. *Expert Review of Gastroenterology & Hepatology*, March 16. (Epub ahead of print)

Khor, B., Gardet, A. & Xavier, R.J. (2011) Genetics and pathogenesis of inflammatory bowel disease. *Nature*, **474**, 307–317.

Ki Cha, B., Mun Jung, S., Hwan Choi, C., Song, I.D., Woong Lee, H., Joon Kim, H., Hyuk, J., Kyung Chang, S., Kim, K., Chung, W.S. & Seo, J.G. (2012) The effect of a multispecies probiotic mixture on the symptoms and fecal microbiota in diarrhea-dominant irritable bowel syndrome: a randomized, double-blind, placebo-controlled trial. *Journal of Clinical Gastroenterology*, **46**, 220–227.

Kim, H.J., Camilleri, M., McKinzie, S., Lempke, M.B., Burton, D.D., Thomforde, G.M. & Zinsmeister, A.R. (2003) A randomized controlled trial of a probiotic, VSL#3, on gut transit and symptoms in diarrhoea-predominant irritable bowel syndrome. *Alimentary Pharmacology and Therapeutics*, **17**, 895–904.

Kim, H.J., Vazquez Roque, M.I., Camilleri, M., Stephens, D., Burton, D.D., Baxter, K., Thomforde, G. & Zinsmeister, A.R. (2005) A randomized controlled trial of a probiotic combination VSL#3 and placebo in irritable bowel syndrome with bloating. *Neurogastroenterology and Motility*, **17**, 687–696.

Kim, J.Y., Kwon, J.H., Ahn, S.H., Lee, S.I., Han, Y.S., Choi, Y.O., Lee, S.Y., Ahn, K.M. & Ji, G.E. (2010) Effect of probiotic mix (*Bifidobacterium bifidum*, *Bifidobacterium lactis*, *Lactobacillus acidophilus*) in the primary prevention of eczema: a double-blind, randomized, placebo-controlled trial. *Pediatric Allergy and Immunology*, **21**, e386–e393.

Kim, C.H., Park, J. & Kim, M. (2014) Gut microbiota-derived short-chain fatty acids, T cells, and inflammation. *Immune Network*, **14**, 277–288.

King, S., Glanville, J., Sanders, M.E., Fitzgerald, A. & Varley, D. (2014) Effectiveness of probiotics on the duration of illness in healthy children and adults who develop common acute respiratory infectious conditions: a systematic review and meta-analysis. *British Journal of Nutrition*, **112**, 41–54.

Kirpich, I.A., Solovieva, N.V., Leikhter, S.N., Shidakova, N.A., Lebedeva, O.V., Sidorov, P.I., Bazhukova, T.A., Soloviev, A.G., Barve, S.S., Mcclain, C.J. & Cave, M. (2008) Probiotics restore bowel flora and improve liver enzymes in human alcohol-induced liver injury: a pilot study. *Alcohol*, **42**, 675–682.

Koebnick, C., Wagner, I., Leitzmann, P., Stern, U. & Zunft, H.J. (2003) Probiotic beverage containing *Lactobacillus casei* Shirota improves gastrointestinal symptoms in patients with chronic constipation. *Canadian Journal of Gastroenterology*, **17**, 655–659.

Kopp, M.V., Hennemuth, I., Heinzmann, A. & Urbanek, R. (2008) Randomized, double-blind, placebo-controlled trial of probiotics for primary prevention: no clinical effects of *Lactobacillus* GG supplementation. *Pediatrics*, **121**, e850–e856.

Koren, O., Knights, D., Gonzalez, A., Waldron, L., Segata, N., Knight, R., Huttenhower, C. & Ley, R.E. (2013) A guide to enterotypes across the human body: meta-analysis of microbial community structures in human microbiome datasets. *PLoS Computational Biology*, **9**, e1002863.

Kotowska, M., Albrecht, P. & Szajewska, H. (2005) *Saccharomyces boulardii* in the prevention of antibiotic-associated diarrhoea in children: a randomized double-blind placebo-controlled trial. *Alimentary Pharmacology and Therapeutics*, **21**, 583–590.

Kovachev, S.M. & Vatcheva-Dobrevska, R.S. (2015) Local probiotic therapy for vaginal *Candida albicans* infections. *Probiotics and Antimicrobial Proteins*, **7**, 38–44.

Krammer, H.J., Von Seggem, H., Schaumberg, J. & Neumer, F. (2011) Effect of *Lactobacillus casei* Shirota on colonic transit time in patients with chronic constipation. *Coloproctology*, **33**, 109–113.

Kruis, W., Schutz, E., Fric, P., Fixa, B., Judmaier, G. & Stolte, M. (1997) Double-blind comparison of an oral *Escherichia coli* preparation and mesalazine in maintaining remission of ulcerative colitis. *Alimentary Pharmacology and Therapeutics*, **11**, 853–858.

Kruis, W., Fric, P., Pokrotnieks, J., Lukas, M., Fixa, B., Kascak, M., Kamm, M.A., Weismueller, J., Beglinger, C., Stolte, M., Wolff, C. & Schulze, J. (2004) Maintaining remission of ulcerative colitis with the probiotic *Escherichia coli* Nissle 1917 is as effective as with standard mesalazine. *Gut*, **53**, 1617–1623.

Kruis, W., Chrubasik, S., Boehm, S., Stange, C. & Schulze, J. (2012) A double-blind placebo-controlled trial to study therapeutic effects of probiotic *Escherichia coli* Nissle 1917 in subgroups of patients with irritable bowel syndrome. *International Journal of Colorectal Disease*, **27**, 467–474.

Kuisma, J., Mentula, S., Jarvinen, H., Kahri, A., Saxelin, M. & Farkkila, M. (2003) Effect of *Lactobacillus rhamnosus* GG on ileal pouch inflammation and microbial flora. *Alimentary Pharmacology and Therapeutics*, **17**, 509–515.

Kuitunen, M. (2013) Probiotics and prebiotics in preventing food allergy and eczema. *Current Opinion in Allergy and Clinical Immunology*, **13**, 280–286.

Kumar, M., Nagpal, R., Verma, V., Kumar, A., Kaur, N., Hemalatha, R., Gautam, S.K. & Singh, B. (2013) Probiotic metabolites as epigenetic targets in the prevention of colon cancer. *Nutrition Reviews*, **71**, 23–34.

Kumpu, M., Kekkonen, R.A., Kautiainen, H., Jarvenpaa, S., Kristo, A., Huovinen, P., Pitkaranta, A., Korpela, R. & Hatakka, K. (2012) Milk containing probiotic *Lactobacillus rhamnosus* GG and respiratory illness in children: a randomized, double-blind, placebo-controlled trial. *European Journal of Clinical Nutrition*, **66**, 1020–1023.

Kurugol, Z. & Koturoglu, G. (2005) Effects of *Saccharomyces boulardii* in children with acute diarrhoea. *Acta Paediatrica*, **94**, 44–47.

Laake, K.O., Bjorneklett, A., Aamodt, G., Aabakken, L., Jacobsen, M., Bakka, A. & Vatn, M.H. (2005) Outcome of four weeks' intervention with probiotics on symptoms and endoscopic appearance after surgical reconstruction with a J-configurated ileal-pouch-anal-anastomosis in ulcerative colitis. *Scandinavian Journal of Gastroenterology*, **40**, 43–51.

Larsson, P.G., Stray-Pedersen, B., Ryttig, K.R. & Larsen, S. (2008) Human lactobacilli as supplementation of clindamycin to patients with bacterial vaginosis reduce the recurrence rate: a 6-month, double-blind, randomized, placebo-controlled study. *BMC Women's Health*, **8**, 3.

Lata, J., Novotny, I., Pribramska, V., Jurankova, J., Fric, P., Kroupa, R. & Stiburek, O. (2007) The effect of probiotics on gut flora, level of endotoxin and Child-Pugh score in cirrhotic patients: results of a double-blind randomized study. *European Journal of Gastroenterology and Hepatology*, **19**, 1111–1113.

Lau, C.S. & Chamberlain, R.S. (2015) Probiotic administration can prevent necrotizing enterocolitis in preterm infants: a meta-analysis. *Journal of Pediatric Surgery*, **50**, 1405–1412.

Lau, C.S. & Chamberlain, R.S. (2016) Probiotics are effective at preventing *Clostridium difficile*-associated diarrhea: a systematic review and meta-analysis. *International Journal of General Medicine*, **9**, 27–37.

Laue, T., Wrann, C.D., Hoffmann-Castendiek, B., Pietsch, D., Hubner, L. & Kielstein, H. (2015) Altered NK cell function in obese healthy humans. *BMC Obesity*, **2**, 1.

Lawley, T.D. & Walker, A.W. (2013) Intestinal colonization resistance. *Immunology*, **138**, 1–11.

Lawrence, S.J., Korzenik, J.R. & Mundy, L.M. (2005) Probiotics for recurrent *Clostridium difficile* disease. *Journal of Medical Microbiology*, **54**, 905–906.

Leblanc, J.G., Laino, J.E., Del Valle, M.J., Vannini, V., Van Sinderen, D., Taranto, M.P., De Valdez, G.F., De Giori, G.S. & Sesma, F. (2011) B-group vitamin production by lactic acid bacteria: current knowledge and potential applications. *Journal of Applied Microbiology*, **111**, 1297–1309.

Lee, J., Seto, D. & Bielory, L. (2008) Meta-analysis of clinical trials of probiotics for prevention and treatment of pediatric atopic dermatitis. *Journal of Allergy and Clinical Immunology*, **121**, 116–121.e11.

Lee, J.H., Moon, G., Kwon, H.J., Jung, W.J., Seo, P.J., Baec, T.Y., Lee, J.H. & Kim, H.S. (2012) Effect of a probiotic preparation (VSL#3) in patients with mild to moderate ulcerative colitis (in Korean). *Korean Journal of Gastroenterology*, **60**, 94–101.

Lee, L.Y.W., Golmohamad, R. & MacFaul, G. (2013) Prevention of relapse following *Clostridium difficile* infection using probiotic *Lactobacillus casei* Shirota. *International Journal of Probiotics and Prebiotics*, **8**, 145–148.

Lee, J.Y., Chu, S.H., Jeon, J.Y., Lee, M.K., Park, J.H., Lee, D.C., Lee, J.W. & Kim, N.K. (2014) Effects of 12 weeks of probiotic supplementation on quality of life in colorectal cancer survivors: a double-blind, randomized, placebo-controlled trial. *Digestive and Liver Disease*, **46**, 1126–1132.

Lembo, A. & Camilleri, M. (2003) Chronic constipation. *New England Journal of Medicine*, **349**, 1360–1368.

Lenoir-Wijnkoop, I., Gerlier, L., Bresson, J.L., Le Pen, C. & Berdeaux, G. (2015) Public health and budget impact of probiotics on common respiratory tract infections: a modelling study. *PLoS One*, **10**, e0122765.

Lewis, J.N., Thomas, L.V. & Weir, W. (2009) The potential of probiotic fermented milk products in reducing risk of antibiotic-associated diarrhoea and *Clostridium difficile* disease. *International Journal of Dairy Technology* **62**, 461–471.

Leyer, G.J., Li, S., Mubasher, M.E., Reifer, C. & Ouwehand, A.C. (2009) Probiotic effects on cold and influenza-like symptom incidence and duration in children. *Pediatrics*, **124**, e172–e179.

Li, H. & Cao, Y. (2010) Lactic acid bacterial cell factories for gamma-aminobutyric acid. *Amino Acids*, **39**, 1107–1116.

Li, Q. & Zhou, J.M. (2016) The microbiota-gut-brain axis and its potential therapeutic role in autism spectrum disorder. *Neuroscience*, **324**, 131–139.

Li, G., Zeng, S., Liao, W. & Lv, N. (2012a) The effect of bifid triple viable on immune function of patients with ulcerative colitis. *Gastroenterology Research and Practice*, **2012**, 404752.

Li, S.R., Wang, H.H., Wu, Z.Y., Liu, R.H., Tong, M.H., Wang, C.H., Wang, R.L., Zhao, H.C. & Wei, W. (2012b) Efficacies of lactulose plus live combined *Bacillus subtilis* and *Enterococcus faecium* capsules in the treatment of functional constipation: a multicenter, randomized, double blind, controlled trial (in Chinese). *Zhonghua Yi Xue Za Zhi*, **92**, 2955–2960.

Li, F., Duan, K., Wang, C., Mcclain, C. & Feng, W. (2016) Probiotics and alcoholic liver disease: treatment and potential mechanisms. *Gastroenterology Research and Practice*, **2016**, 5491465.

Ligaarden, S.C., Axelsson, L., Naterstad, K., Lydersen, S. & Farup, P.G. (2010) A candidate probiotic with unfavourable effects in subjects with irritable bowel syndrome: a randomised controlled trial. *BMC Gastroenterology*, **10**, 16.

Lin, H.C., Hsu, C.H., Chen, H.L., Chung, M.Y., Hsu, J.F., Lien, R.I., Tsao, L.Y., Chen, C.H. & Su, B.H. (2008) Oral probiotics prevent necrotizing enterocolitis in very low birth weight preterm infants: a multicenter, randomized, controlled trial. *Pediatrics*, **122**, 693–700.

Liu, Q., Duan, Z.P., Ha, D.K., Bengmark, S., Kurtovic, J. & Riordan, S.M. (2004) Synbiotic modulation of gut flora: effect on minimal hepatic encephalopathy in patients with cirrhosis. *Hepatology*, **39**, 1441–1449.

Liu, Z., Qin, H., Yang, Z., Xia, Y., Liu, W., Yang, J., Jiang, Y., Zhang, H., Yang, Z., Wang, Y. & Zheng, Q. (2011) Randomised clinical trial: the effects of perioperative probiotic treatment on barrier function and post-operative infectious complications in colorectal cancer surgery: a double-blind study. *Alimentary Pharmacology and Therapeutics*, **33**, 50–63.

Liu, Y.P., Liu, X. & Dong, L. (2012) Lactulose plus live binary *Bacillus subtilis* in the treatment of elders with functional constipation (in Chinese). *Zhonghua Yi Xue Za Zhi*, **92**, 2961–2964.

Liu, Z.H., Huang, M.J., Zhang, X.W., Wang, L., Huang, N.Q., Peng, H., Lan, P., Peng, J.S., Yang, Z., Xia, Y., Liu, W.J., Yang, J., Qin, H.L. & Wang, J.P. (2013) The effects of perioperative probiotic treatment on serum zonulin concentration and subsequent postoperative infectious complications after colorectal cancer surgery: a double-center and double-blind randomized clinical trial. *American Journal of Clinical Nutrition*, **97**, 117–126.

Liu, X., Cao, S. & Zhang, X. (2015) Modulation of gut microbiota-brain axis by probiotics, prebiotics and diets. *Journal of Agricultural and Food Chemistry* **63**, 7885–7895.

Loguercio, C., Federico, A., Tuccillo, C., Terracciano, F., D'Auria, M.V., De Simone, C. & Del Vecchio Blanco, C. (2005) Beneficial effects of a probiotic VSL#3 on parameters of liver dysfunction in chronic liver diseases. *Journal of Clinical Gastroenterology*, **39**, 540–543.

Longstreth, G.F., Thompson, W.G., Chey, W.D., Houghton, L.A., Mearin, F. & Spiller, R.C. (2006) Functional bowel disorders. *Gastroenterology*, **130**, 1480–1491.

Lonnermark, E., Friman, V., Lappas, G., Sandberg, T., Berggren, A. & Adlerberth, I. (2010) Intake of *Lactobacillus plantarum* reduces certain gastrointestinal symptoms during treatment with antibiotics. *Journal of Clinical Gastroenterology*, **44**, 106–112.

Lopez, C.A., Kingsbury, D.D., Velazquez, E.M. & Baumler, A.J. (2014) Collateral damage: microbiota-derived metabolites and immune function in the antibiotic era. *Cell Host Microbe*, **16**, 156–163.

Lunia, M.K., Sharma, B.C., Sharma, P., Sachdeva, S. & Srivastava, S. (2014) Probiotics prevent hepatic encephalopathy in patients with cirrhosis: a randomized controlled trial. *Clinical Gastroenterology and Hepatology*, **12**, 1003–1008.

Luoto, R., Kalliomaki, M., Laitinen, K. & Isolauri, E. (2010) The impact of perinatal probiotic intervention on the development of overweight and obesity: follow-up study from birth to 10 years. *International Journal of Obesity*, **34**, 1531–1537.

MacDonald, T. & Bateman, A. (2007) *Immunology and Diseases of the Gut*. Remedica, London.

Mack, D.R. (2011) Probiotics in inflammatory bowel diseases and associated conditions. *Nutrients*, **3**, 245–264.

Macklaim, J.M., Clemente, J.C., Knight, R., Gloor, G.B. & Reid, G. (2015) Changes in vaginal microbiota following antimicrobial and probiotic therapy. *Microbial Ecology in Health and Disease*, **26**, 27799.

MacPhee, R.A., Hummelen, R., Bisanz, J.E., Miller, W.L. & Reid, G. (2010) Probiotic strategies for the treatment and prevention of bacterial vaginosis. *Expert Opinion on Pharmacotherapy*, **11**, 2985–2995.

Madsen, K., Backer, J., Leddin, D., Dieleman, L., Bitton, A., Feagan, B., Petrunia, D.M., Chiba, N., Enns, R.A. & Fedotak, R. (2008) A randomized controlled trial of VSL#3 for the prevention of endoscopic recurrence following surgery for Crohn's disease. *Gastroenterology*, **134**, A-361.

Magro, D.O., De Oliveira, L.M., Bernasconi, I., Ruela Mde, S., Credidio, L., Barcelos, I.K., Leal, R.F., Ayrizono Mde, L., Fagundes, J.J., Teixeira Lde, B., Ouwehand, A.C. & Coy, C.S. (2014) Effect of yogurt containing polydextrose, *Lactobacillus acidophilus* NCFM and *Bifidobacterium lactis* HN019: a randomized, double-blind, controlled study in chronic constipation. *Nutrition Journal*, **13**, 75.

Majamaa, H. & Isolauri, E. (1997) Probiotics: a novel approach in the management of food allergy. *Journal of Allergy and Clinical Immunology*, **99**, 179–185.

Makino, S., Ikegami, S., Kume, A., Horiuchi, H., Sasaki, H. & Orii, N. (2010) Reducing the risk of infection in the elderly by dietary intake of yoghurt fermented with *Lactobacillus delbrueckii* ssp. *bulgaricus* OLL1073R-1. *British Journal of Nutrition*, **104**, 998–1006.

Malaguarnera, M., Greco, F., Barone, G., Gargante, M.P., Malaguarnera, M. & Toscano, M.A. (2007) *Bifidobacterium longum* with fructo-oligosaccharide (FOS) treatment in minimal hepatic encephalopathy: a randomized, double-blind, placebo-controlled study. *Digestive Diseases and Sciences*, **52**, 3259–3265.

Malaguarnera, M., Gargante, M.P., Malaguarnera, G., Salmeri, M., Mastrojeni, S., Rampello, L., Pennisi, G., Li Volti, G. & Galvano, F. (2010) *Bifidobacterium* combined with fructo-oligosaccharide versus lactulose in the treatment of patients with hepatic encephalopathy. *European Journal of Gastroenterology and Hepatology*, **22**, 199–206.

Malaguarnera, M., Vacante, M., Antic, T., Giordano, M., Chisari, G., Acquaviva, R., Mastrojeni, S., Malaguarnera, G., Mistretta, A., Li Volti, G. & Galvano, F. (2012) *Bifidobacterium longum* with fructo-oligosaccharides in patients with non alcoholic steatohepatitis. *Digestive Diseases and Sciences*, **57**, 545–553.

Malchow, H.A. (1997) Crohn's disease and *Escherichia coli*. A new approach in therapy to maintain remission of colonic Crohn's disease? *Journal of Clinical Gastroenterology*, **25**, 653–658.

Mangiola, F., Ianiro, G., Franceschi, F., Fagiuoli, S., Gasbarrini, G. & Gasbarrini, A. (2016) Gut microbiota in autism and mood disorders. *World Journal of Gastroenterology*, **22**, 361–368.

Manzoni, P., Lista, G., Gallo, E., Marangione, P., Priolo, C., Fontana, P., Guardione, R. & Farina, D. (2011) Routine *Lactobacillus rhamnosus* GG administration in VLBW infants: a retrospective, 6-year cohort study. *Early Human Development*, **87**, S35–S38.

Marchesi, J.R., Adams, D.H., Fava, F., Hermes, G.D., Hirschfield, G.M., Hold, G., Quraishi, M.N., Kinross, J., Smidt, H., Tuohy, K.M., Thomas, L.V., Zoetendal, E.G. & Hart, A. (2016) The gut microbiota and host health: a new clinical frontier. *Gut*, **65**, 330–339.

Marchisio, P., Santagati, M., Scillato, M., Baggi, E., Fattizzo, M., Rosazza, C., Stefani, S., Esposito, S. & Principi, N. (2015) *Streptococcus salivarius* 24SMB administered by nasal spray for the prevention of acute otitis media in otitis-prone children. *European Journal of Clinical Microbiology and Infectious Diseases*, **34**, 2377–2383.

Marcone, V., Calzolari, E. & Bertini, M. (2008) Effectiveness of vaginal administration of *Lactobacillus rhamnosus* following conventional metronidazole therapy: how to lower the rate of bacterial vaginosis recurrences. *New Microbiologica*, **31**, 429–433.

Mardini, H.E. & Grigorian, A.Y. (2014) Probiotic mix VSL#3 is effective adjunctive therapy for mild to moderately active ulcerative colitis: a meta-analysis. *Inflammatory Bowel Diseases*, **20**, 1562–1567.

Marom, T., Marchisio, P., Tamir, S.O., Torretta, S., Gavriel, H. & Esposito, S. (2016) Complementary and alternative medicine treatment options for otitis media: a systematic review. *Medicine (Baltimore)*, **95**, e2695.

Marteau, P., Lemann, M., Seksik, P., Laharie, D., Colombel, J.F., Bouhnik, Y., Cadiot, G., Soule, J.C., Bourreille, A., Metman, E., Lerebours, E., Carbonnel, F., Dupas, J.L., Veyrac, M., Coffin, B., Moreau, J., Abitbol, V., Blum-Sperisen, S. & Mary, J.Y. (2006) Ineffectiveness of *Lactobacillus johnsonii* LA1 for prophylaxis of postoperative recurrence in Crohn's disease: a randomised, double blind, placebo controlled GETAID trial. *Gut*, **55**, 842–847.

Martinez, R.C., Franceschini, S.A., Patta, M.C., Quintana, S.M., Gomes, B.C., De Martinis, E.C. & Reid, G. (2009) Improved cure of bacterial vaginosis with single dose of tinidazole (2 g), *Lactobacillus rhamnosus* GR-1, and *Lactobacillus reuteri* RC-14: a randomized, double-blind, placebo-controlled trial. *Canadian Journal of Microbiology*, **55**, 133–138.

Masco, L., Ventura, M., Zink, R., Huys, G. & Swings, J. (2004) Polyphasic taxonomic analysis of *Bifidobacterium animalis* and *Bifidobacterium lactis* reveals relatedness at the subspecies level: reclassification of *Bifidobacterium animalis* as *Bifidobacterium animalis* subsp. *animalis* subsp. nov. and *Bifidobacterium lactis* as *Bifidobacterium animalis* subsp. *lactis* subsp. *nov*. *International Journal of Systematic and Evolutionary Microbiology*, **54**, 1137–1143.

Mastromarino, P., Macchia, S., Meggiorini, L., Trinchieri, V., Mosca, L., Perluigi, M. & Midulla, C. (2009) Effectiveness of *Lactobacillus*-containing vaginal tablets in the treatment of symptomatic bacterial vaginosis. *Clinical Microbiology and Infection*, **15**, 67–74.

Matsumoto, M., Aranami, A., Ishige, A., Watanabe, K. & Benno, Y. (2007) LKM512 yogurt consumption improves the intestinal environment and induces the T-helper type 1 cytokine in adult patients with intractable atopic dermatitis. *Clinical and Experimental Allergy*, **37**, 358–370.

Mattarelli, P., Bonaparte, C., Pot, B. & Biavati, B. (2008) Proposal to reclassify the three biotypes of *Bifidobacterium longum* as three subspecies: *Bifidobacterium longum* subsp. *longum* subsp. nov., *Bifidobacterium longum* subsp. *infantis* comb. nov. and *Bifidobacterium longum* subsp. *suis* comb. nov. *International Journal of Systematic and Evolutionary Microbiology*, **58**, 767–772.

Matthes, H., Krummenerl, T., Giensch, M., Wolff, C. & Schulze, J. (2010) Clinical trial: probiotic treatment of acute distal ulcerative colitis with rectally administered *Escherichia coli* Nissle 1917 (EcN). *BMC Complementary and Alternative Medicine*, **10**, 13.

Mazloom, Z., Yousefinejad, A. & Dabbaghmanesh, M.H. (2013) Effect of probiotics on lipid profile, glycemic control, insulin action, oxidative stress, and inflammatory markers in patients with type 2 diabetes: a clinical trial. *Iranian Journal of Medical Sciences*, **38**, 38–43.

Mazlyn, M.M., Nagarajah, L.H., Fatimah, A., Norimah, A.K. & Goh, K.L. (2013) Effects of a probiotic fermented milk on functional constipation: a randomized, double-blind, placebo-controlled study. *Journal of Gastroenterology and Hepatology*, **28**, 1141–1147.

Mazurak, N., Broelz, E., Storr, M. & Enck, P. (2015) Probiotic therapy of the irritable bowel syndrome: why is the evidence still poor and what can be done about it? *Journal of Neurogastroenterology and Motility*, **21**, 471–485.

McFarland, L.V., Surawicz, C.M., Greenberg, R.N., Fekety, R., Elmer, G.W., Moyer, K.A., Melcher, S.A., Bowen, K.E., Cox, J.L., Noorani, Z. & *et al*. (1994) A randomized placebo-controlled trial of *Saccharomyces boulardii* in combination with standard antibiotics for *Clostridium difficile* disease. *Journal of the American Medical Association*, **271**, 1913–1918.

McGee, R.G., Bakens, A., Wiley, K., Riordan, S.M. & Webster, A.C. (2011) Probiotics for patients with hepatic encephalopathy. *Cochrane Database of Systematic Reviews*, CD008716.

Merenstein, D., Murphy, M., Fokar, A., Hernandez, R.K., Park, H., Nsouli, H., Sanders, M.E., Davis, B.A., Niborski, V., Tondu, F. & Shara, N.M. (2010) Use of a fermented dairy probiotic drink containing *Lactobacillus casei* (DN-114 001) to decrease the rate of illness in kids: the DRINK study: a patient-oriented, double-blind, cluster-randomized, placebo-controlled, clinical trial. *European Journal of Clinical Nutrition*, **64**, 669–677.

Messaoudi, M., Lalonde, R., Violle, N., Javelot, H., Desor, D., Nejdi, A., Bisson, J.F., Rougeot, C., Pichelin, M., Cazaubiel, M. & Cazaubiel, J.M. (2011) Assessment of psychotropic-like properties of a probiotic formulation (*Lactobacillus helveticus* R0052 and *Bifidobacterium longum* R0175) in rats and human subjects. *British Journal of Nutrition*, **105**, 755–764.

Michail, S.K., Stolfi, A., Johnson, T. & Onady, G.M. (2008) Efficacy of probiotics in the treatment of pediatric atopic dermatitis: a meta-analysis of randomized controlled trials. *Annals of Allergy, Asthma, and Immunology*, **101**, 508–516.

Miele, E., Pascarella, F., Giannetti, E., Quaglietta, L., Baldassano, R.N. & Staiano, A. (2009) Effect of a probiotic preparation (VSL#3) on induction and maintenance of remission in children with ulcerative colitis. *American Journal of Gastroenterology*, **104**, 437–443.

Miller, L.E. & Ouwehand, A.C. (2013) Probiotic supplementation decreases intestinal transit time: meta-analysis of randomized controlled trials. *World Journal of Gastroenterology*, **19**, 4718–4725.

Million, M., Angelakis, E., Paul, M., Armougom, F., Leibovici, L. & Raoult, D. (2012) Comparative meta-analysis of the effect of *Lactobacillus* species on weight gain in humans and animals. *Microbial Pathogenesis*, **53**, 100–108.

Mimura, T., Rizzello, F., Helwig, U., Poggioli, G., Schreiber, S., Talbot, I.C., Nicholls, R.J., Gionchetti, P., Campieri, M. & Kamm, M.A. (2004) Once daily high dose probiotic therapy (VSL#3) for maintaining remission in recurrent or refractory pouchitis. *Gut*, **53**, 108–114.

Min, Y.W. & Rhee, P.L. (2015) The role of microbiota on the gut immunology. *Clinical Therapeutics*, **37**, 968–975.

Min, Y.W., Park, S.U., Jang, Y.S., Kim, Y.H., Rhee, P.L., Ko, S.H., Joo, N., Kim, S.I., Kim, C.H. & Chang, D.K. (2012) Effect of composite yogurt enriched with acacia fiber and *Bifidobacterium lactis*. *World Journal of Gastroenterology*, **18**, 4563–4569.

Minemura, M. & Shimizu, Y. (2015) Gut microbiota and liver diseases. *World Journal of Gastroenterology*, **21**, 1691–1702.

Misra, S., Sabui, T.K. & Pal, N.K. (2009) A randomized controlled trial to evaluate the efficacy of *Lactobacillus* GG in infantile diarrhea. *Journal of Pediatrics*, **155**, 129–132.

Moayyedi, P., Ford, A.C., Talley, N.J., Cremonini, F., Foxx-Orenstein, A.E., Brandt, L.J. & Quigley, E.M. (2010) The efficacy of probiotics in the treatment of irritable bowel syndrome: a systematic review. *Gut*, **59**, 325–332.

Mohan, R., Koebnick, C., Schildt, J., Mueller, M., Radke, M. & Blaut, M. (2008) Effects of *Bifidobacterium lactis* Bb12 supplementation on body weight, fecal pH, acetate, lactate, calprotectin, and IgA in preterm infants. *Pediatric Research*, **64**, 418–422.

Moroi, M., Uchi, S., Nakamura, K., Sato, S., Shimizu, N., Fujii, M., Kumagai, T., Saito, M., Uchiyama, K., Watanabe, T., Yamaguchi, H., Yamamoto, T., Takeuchi, S. & Furue, M. (2011) Beneficial effect of a diet containing heat-killed *Lactobacillus paracasei* K71 on adult type atopic dermatitis. *Journal of Dermatology*, **38**, 131–139.

Moroti, C., Souza Magri, L.F., De Rezende Costa, M., Cavallini, D.C. & Sivieri, K. (2012) Effect of the consumption of a new symbiotic shake on glycemia and cholesterol levels in elderly people with type 2 diabetes mellitus. *Lipids in Health and Disease*, **11**, 29.

Mulder, D.J., Noble, A.J., Justinich, C.J. & Duffin, J.M. (2014) A tale of two diseases: the history of inflammatory bowel disease. *Journal of Crohn's and Colitis*, **8**, 341–348.

Nabavi, S., Rafraf, M., Somi, M.H., Homayouni-Rad, A. & Asghari-Jafarabadi, M. (2014) Effects of probiotic yogurt consumption on metabolic factors in individuals with nonalcoholic fatty liver disease. *Journal of Dairy Science*, **97**, 7386–7393.

Nagata, Y., Yoshida, M., Kitazawa, H., Araki, E. & Gomyo, T. (2010) Improvements in seasonal allergic disease with *Lactobacillus plantarum* No. 14. *Bioscience, Biotechnology, and Biochemistry*, **74**, 1869–1877.

Naidoo, K., Gordon, M., Fagbemi, A.O., Thomas, A.G. & Akobeng, A.K. (2011) Probiotics for maintenance of remission in ulcerative colitis. *Cochrane Database of Systematic Reviews*, CD007443.

Naito, S., Koga, H., Yamaguchi, A., Fujimoto, N., Hasui, Y., Kuramoto, H., Iguchi, A., Kinukawa, N. & Kyushu University Urological Oncology, G. (2008) Prevention of recurrence with epirubicin and *Lactobacillus casei* after transurethral resection of bladder cancer. *Journal of Urology*, **179**, 485–490.

Naruszewicz, M., Johansson, M.L., Zapolska-Downar, D. & Bukowska, H. (2002) Effect of *Lactobacillus plantarum* 299v on cardiovascular disease risk factors in smokers. *American Journal of Clinical Nutrition*, **76**, 1249–1255.

Neufeld, K.M., Kang, N., Bienenstock, J. & Foster, J.A. (2011) Reduced anxiety-like behavior and central neurochemical change in germ-free mice. *Neurogastroenterology and Motility*, **23**, 255–264.e119.

Ng, S.C., Hart, A.L., Kamm, M.A., Stagg, A.J. & Knight, S.C. (2009) Mechanisms of action of probiotics: recent advances. *Inflammatory Bowel Diseases*, **15**, 300–310.

Ng, M., Fleming, T., Robinson, M., Thomson, B., Graetz, N., Margono, C., Mullany, E.C., Biryukov, S., Abbafati, C., Abera, S.F., Abraham, J.P., Abu-Rmeileh, N.M., Achoki, T., Albuhairan, F.S., Alemu, Z.A., Alfonso, R., Ali, M.K., Ali, R., Guzman, N.A., Ammar, W., Anwari, P., Banerjee, A., Barquera, S., Basu, S., Bennett, D.A., Bhutta, Z., Blore, J., Cabral, N., Nonato, I.C., Chang, J.C., Chowdhury, R., Courville, K.J., Criqui, M.H., Cundiff, D.K., Dabhadkar, K.C., Dandona, L., Davis, A., Dayama, A., Dharmaratne, S.D., Ding, E.L., Durrani, A.M., Esteghamati, A., Farzadfar, F., Fay, D.F., Feigin, V.L., Flaxman, A., Forouzanfar, M.H., Goto, A., Green, M.A., Gupta, R., Hafezi-Nejad, N., Hankey, G.J., Harewood, H.C., Havmoeller, R., Hay, S., Hernandez, L., Husseini, A., Idrisov, B.T., Ikeda, N., Islami, F., Jahangir, E., Jassal, S.K., Jee, S.H., Jeffreys, M., Jonas, J.B., Kabagambe, E.K., Khalifa, S.E., Kengne, A.P., Khader, Y.S., Khang, Y.H., Kim, D., Kimokoti, R.W., Kinge, J.M., Kokubo, Y., Kosen, S., Kwan, G., Lai, T., Leinsalu, M., Li, Y., Liang, X., Liu, S., Logroscino, G., Lotufo, P.A., Lu, Y., Ma, J., Mainoo, N.K., Mensah, G.A., Merriman, T.R., Mokdad, A.H., Moschandreas, J., Naghavi, M., Naheed, A., Nand, D., Narayan, K.M., Nelson, E.L., Neuhouser, M.L., Nisar, M.I., Ohkubo, T., Oti, S.O., Pedroza, A. & *et al.* (2014) Global, regional, and national prevalence of overweight and obesity in children and adults during 1980–2013: a systematic analysis for the global burden of disease study 2013. *Lancet*, **384**, 766–781.

Niedzielin, K., Kordecki, H. & Birkenfeld, B. (2001) A controlled, double-blind, randomized study on the efficacy of *Lactobacillus plantarum* 299V in patients with irritable bowel syndrome. *European Journal of Gastroenterology and Hepatology*, **13**, 1143–1147.

Niers, L., Martin, R., Rijkers, G., Sengers, F., Timmerman, H., Van Uden, N., Smidt, H., Kimpen, J. & Hoekstra, M. (2009) The effects of selected probiotic strains on the development of eczema (the PandA study). *Allergy*, **64**, 1349–1358.

Nikfar, S., Rahimi, R., Rahimi, F., Derakhshani, S. & Abdollahi, M. (2008) Efficacy of probiotics in irritable bowel syndrome: a meta-analysis of randomized, controlled trials. *Diseases of the Colon and Rectum*, **51**, 1775–1780.

Niv, E., Naftali, T., Hallak, R. & Vaisman, N. (2005) The efficacy of *Lactobacillus reuteri* ATCC 55730 in the treatment of patients with irritable bowel syndrome – a double blind, placebo-controlled, randomized study. *Clinical Nutrition*, **24**, 925–931.

Nobaek, S., Johansson, M.L., Molin, G., Ahrne, S. & Jeppsson, B. (2000) Alteration of intestinal microflora is associated with reduction in abdominal bloating and pain in patients with irritable bowel syndrome. *American Journal of Gastroenterology*, **95**, 1231–1238.

Nobel, Y.R., Cox, L.M., Kirigin, F.F., Bokulich, N.A., Yamanishi, S., Teitler, I., Chung, J., Sohn, J., Barber, C.M., Goldfarb, D.S., Raju, K., Abubucker, S., Zhou, Y., Ruiz, V.E., Li, H., Mitreva, M., Alekseyenko, A.V., Weinstock, G.M., Sodergren, E. & Blaser, M.J. (2015) Metabolic and metagenomic outcomes from early-life pulsed antibiotic treatment. *Nature Communications*, **6**, 7486.

O'Mahony, L., McCarthy, J., Kelly, P., Hurley, G., Luo, F., Chen, K., O'Sullivan, G.C., Kiely, B., Collins, J.K., Shanahan, F. & Quigley, E.M. (2005) *Lactobacillus* and *Bifidobacterium* in irritable bowel syndrome: symptom responses and relationship to cytokine profiles. *Gastroenterology*, **128**, 541–551.

Ogawa, A., Kadooka, Y., Kato, K., Shirouchi, B. & Sato, M. (2014) *Lactobacillus gasseri* SBT2055 reduces postprandial and fasting serum non-esterified fatty acid levels in Japanese hypertriacylglycerolemic subjects. *Lipids in Health and Disease*, **13**, 36.

Ohashi, Y., Nakai, S., Tsukamoto, T., Masumori, N., Akaza, H., Miyanaga, N., Kitamura, T., Kawabe, K., Kotake, T., Kuroda, M., Naito, S., Koga, H., Saito, Y., Nomata, K., Kitagawa, M. & Aso, Y. (2002) Habitual intake of lactic acid bacteria and risk reduction of bladder cancer. *Urologia Internationalis*, **68**, 273–280.

Ojetti, V., Ianiro, G., Tortora, A., D'angelo, G., Di Rienzo, T.A., Bibbo, S., Migneco, A. & Gasbarrini, A. (2014) The effect of *Lactobacillus reuteri* supplementation in adults with chronic functional constipation: a randomized, double-blind, placebo-controlled trial. *Journal of Gastrointestinal and Liver Diseases*, **23**, 387–391.

Oliva, S., Di Nardo, G., Ferrari, F., Mallardo, S., Rossi, P., Patrizi, G., Cucchiara, S. & Stronati, L. (2012) Randomised clinical trial: the effectiveness of *Lactobacillus reuteri* ATCC 55730 rectal enema in children with active distal ulcerative colitis. *Alimentary Pharmacology and Therapeutics*, **35**, 327–334.

Olsen, R., Greisen, G., Schroder, M. & Brok, J. (2016) Prophylactic probiotics for preterm infants: a systematic review and meta-analysis of observational studies. *Neonatology*, **109**, 105–112.

Omar, J.M., Chan, Y.-M., Jones, M.L., Pakrash, S. & Jones, P.J.H. (2013) *Lactobacillus fermentum* and *Lactobacillus amylovorus* as probiotics alter body adiposity and gut microflora in healthy person. *Journal of Functional Foods*, **5**, 116–123.

Oncel, M.Y., Sari, F.N., Arayici, S., Guzoglu, N., Erdeve, O., Uras, N., Oguz, S.S. & Dilmen, U. (2014) *Lactobacillus reuteri* for the prevention of necrotising enterocolitis in very low birthweight infants: a randomised controlled trial. *Archives of Disease in Childhood: Fetal and Neonatal Edition*, **99**, F110–F115.

Orel, R. & Kamhi Trop, T. (2014) Intestinal microbiota, probiotics and prebiotics in inflammatory bowel disease. *World Journal of Gastroenterology*, **20**, 11505–11524.

Ortiz-Lucas, M., Tobias, A., Saz, P. & Sebastian, J.J. (2013) Effect of probiotic species on irritable bowel syndrome symptoms: a bring up to date meta-analysis. *Revista Española de Enfermedades Digestivas*, **105**, 19–36.

Osborn, D.A. & Sinn, J.K. (2007) Probiotics in infants for prevention of allergic disease and food hypersensitivity. *Cochrane Database of Systematic Reviews*, CD006475.

Ou, C.Y., Kuo, H.C., Wang, L., Hsu, T.Y., Chuang, H., Liu, C.A., Chang, J.C., Yu, H.R. & Yang, K.D. (2012) Prenatal and postnatal probiotics reduces maternal but not childhood allergic diseases: a randomized, double-blind, placebo-controlled trial. *Clinical and Experimental Allergy*, **42**, 1386–1396.

Ozkinay, E., Terek, M.C., Yayci, M., Kaiser, R., Grob, P. & Tuncay, G. (2005) The effectiveness of live lactobacilli in combination with low dose oestriol (Gynoflor) to restore the vaginal flora after treatment of vaginal infections. *BJOG: An International Journal of Obstetrics and Gynaecology*, **112**, 234–240.

Parma, M., Stella Vanni, V., Bertini, M. & Candiani, M. (2014) Probiotics in the prevention of recurrences of bacterial vaginosis. *Alternative Therapies in Health and Medicine*, **20**, 52–57.

Parracho, H.M., Bingham, M.O., Gibson, G.R. & McCartney, A.L. (2005) Differences between the gut microflora of children with autistic spectrum disorders and that of healthy children. *Journal of Medical Microbiology*, **54**, 987–991.

Parracho, H.M.R.T., Gibson, G.R., Knott, F., Bosscher, D., Lkleerebezem, M. & McCartney, A.L. (2010) A double-blind, placebo-controlled, crossover-designed probiotic feeding study in children diagnosed with autistic spectum disorders. *International Journal of Probiotics & Prebiotics*, **5**, 69–74.

Partty, A., Kalliomaki, M., Wacklin, P., Salminen, S. & Isolauri, E. (2015) A possible link between early probiotic intervention and the risk of neuropsychiatric disorders later in childhood: a randomized trial. *Pediatric Research*, **77**, 823–828.

Passariello, A., Terrin, G., Cecere, G., Micillo, M., De Marco, G., Di Costanzo, M., Cosenza, L., Leone, L., Nocerino, R. & Canani, R.B. (2012) Randomised clinical trial: efficacy of a new synbiotic formulation containing *Lactobacillus paracasei* B21060 plus arabinogalactan and xilooligosaccharides in children with acute diarrhoea. *Alimentary Pharmacology and Therapeutics*, **35**, 782–788.

Patole, S.K., Rao, S.C., Keil, A.D., Nathan, E.A., Doherty, D.A. & Simmer, K.N. (2016) Benefits of *Bifidobacterium breve* M-16V supplementation in preterm neonates – A retrospective cohort study. *PLoS One*, **11**, e0150775.

Pelucchi, C., Chatenoud, L., Turati, F., Galeone, C., Moja, L., Bach, J.F. & La Vecchia, C. (2012) Probiotics supplementation during pregnancy or infancy for the prevention of atopic dermatitis: a meta-analysis. *Epidemiology*, **23**, 402–414.

Peng, G.C. & Hsu, C.H. (2005) The efficacy and safety of heat-killed *Lactobacillus paracasei* for treatment of perennial allergic rhinitis induced by house-dust mite. *Pediatric Allergy and Immunology*, **16**, 433–438.

Pereg, D., Kotliroff, A., Gadoth, N., Hadary, R., Lishner, M. & Kitay-Cohen, Y. (2011) Probiotics for patients with compensated liver cirrhosis: a double-blind placebo-controlled study. *Nutrition*, **27**, 177–181.

Persborn, M., Gerritsen, J., Wallon, C., Carlsson, A., Akkermans, L.M. & Soderholm, J.D. (2013) The effects of probiotics on barrier function and mucosal pouch microbiota during maintenance treatment for severe pouchitis in patients with ulcerative colitis. *Alimentary Pharmacology and Therapeutics*, **38**, 772–783.

Peterson, C.T., Sharma, V., Elmen, L. & Peterson, S.N. (2015) Immune homeostasis, dysbiosis and therapeutic modulation of the gut microbiota. *Clinical and Experimental Immunology*, **179**, 363–377.

Petricevic, L., Unger, F.M., Viernstein, H. & Kiss, H. (2008) Randomized, double-blind, placebo-controlled study of oral lactobacilli to improve the vaginal flora of postmenopausal women. *European Journal of Obstetrics, Gynecology, and Reproductive Biology*, **141**, 54–57.

Petrova, M.I., Lievens, E., Malik, S., Imholz, N. & Lebeer, S. (2015) *Lactobacillus* species as biomarkers and agents that can promote various aspects of vaginal health. *Frontiers in Physiology*, **6**, 81.

Pillai, A. & Nelson, R. (2008) Probiotics for treatment of *Clostridium difficile*-associated colitis in adults. *Cochrane Database of Systematic Reviews*, CD004611.

Pineton De Chambrun, G., Neut, C., Chau, A., Cazaubiel, M., Pelerin, F., Justen, P. & Desreumaux, P. (2015) A randomized clinical trial of *Saccharomyces cerevisiae* versus placebo in the irritable bowel syndrome. *Digestive and Liver Disease*, **47**, 119–124.

Pirker, A., Stockenhuber, A., Remely, M., Harrant, A., Hippe, B., Kamhuber, C., Adelmann, K., Stockenhuber, F. & Haslberger, A. (2013) Effects of antibiotic therapy on the gastrointestinal microbiota and the influence of *Lactobacillus casei* Shirota. *Food and Agricultural Immunology*, **24**, 315–330.

Pirotta, M., Gunn, J., Chondros, P., Grover, S., O'malley, P., Hurley, S. & Garland, S. (2004) Effect of lactobacillus in preventing post-antibiotic vulvovaginal candidiasis: a randomised controlled trial. *BMJ*, **329**, 548.

Plein, K. & Hotz, J. (1993) Therapeutic effects of *Saccharomyces boulardii* on mild residual symptoms in a stable phase of Crohn's disease with special respect to chronic diarrhea – a pilot study. *Zeitschrift für Gastroenterologie*, **31**, 129–134.

Plummer, S., Weaver, M.A., Harris, J.C., Dee, P. & Hunter, J. (2004) *Clostridium difficile* pilot study: effects of probiotic supplementation on the incidence of *C. difficile* diarrhoea. *International Microbiology*, **7**, 59–62.

Pochapin, M. (2000) The effect of probiotics on *Clostridium difficile* diarrhea. *American Journal of Gastroenterology*, **95**, S11–S13.

Prantera, C., Scribano, M.L., Falasco, G., Andreoli, A. & Luzi, C. (2002) Ineffectiveness of probiotics in preventing recurrence after curative resection for Crohn's disease: a randomised controlled trial with *Lactobacillus* GG. *Gut*, **51**, 405–409.

Pronio, A., Montesani, C., Butteroni, C., Vecchione, S., Mumolo, G., Vestri, A., Vitolo, D. & Boirivant, M. (2008) Probiotic administration in patients with ileal pouch-anal anastomosis for ulcerative colitis is associated with expansion of mucosal regulatory cells. *Inflammatory Bowel Diseases*, **14**, 662–668.

Rafter, J., Bennett, M., Caderni, G., Clune, Y., Hughes, R., Karlsson, P.C., Klinder, A., O'Riordan, M., O'Sullivan, G.C., Pool-Zobel, B., Rechkemmer, G., Roller, M., Rowland, I., Salvadori, M., Thijs, H., Van Loo, J., Watzl, B. & Collins, J.K. (2007) Dietary synbiotics reduce cancer risk factors in polypectomized and colon cancer patients. *American Journal of Clinical Nutrition*, **85**, 488–496.

Rahimi, R., Nikfar, S., Rahimi, F., Elahi, B., Derakhshani, S., Vafaie, M. & Abdollahi, M. (2008) A meta-analysis on the efficacy of probiotics for maintenance of remission and prevention of clinical and endoscopic relapse in Crohn's disease. *Digestive Diseases and Sciences*, **53**, 2524–2531.

Rajkowska, K. & Kunicka-Styczynska, A. (2009) Phenotypic and genotypic characterization of probiotic yeasts. *Biotechnology & Biotechnological Equipment*, **23**, 662–665.

Raman, M., Ambalam, P., Kondepudi, K.K., Pithva, S., Kothari, C., Patel, A.T., Purama, R.K., Dave, J.M. & Vyas, B.R. (2013) Potential of probiotics, prebiotics and synbiotics for management of colorectal cancer. *Gut Microbes*, **4**, 181–192.

Ramchandran, L. & Shah, N.P. (2008) Proteolytic profiles and angiotensin-I converting enzyme and alpha-glucosidase inhibitory activities of selected lactic acid bacteria. *Journal of Food Science*, **73**, M75–M81.

Rao, R.K. & Samak, G. (2013) Protection and restitution of gut barrier by probiotics: nutritional and clinical implications. *Current Nutrition and Food Science*, **9**, 99–107.

Rao, A.V., Bested, A.C., Beaulne, T.M., Katzman, M.A., Iorio, C., Berardi, J.M. & Logan, A.C. (2009) A randomized, double-blind, placebo-controlled pilot study of a probiotic in emotional symptoms of chronic fatigue syndrome. *Gut Pathogens*, **1**, 6.

Rautava, S., Salminen, S. & Isolauri, E. (2009) Specific probiotics in reducing the risk of acute infections in infancy: a randomised, double-blind, placebo-controlled study. *British Journal of Nutrition*, **101**, 1722–1726.

Rautava, S., Kainonen, E., Salminen, S. & Isolauri, E. (2012) Maternal probiotic supplementation during pregnancy and breast-feeding reduces the risk of eczema in the infant. *Journal of Allergy and Clinical Immunology*, **130**, 1355–1360.

Razmpoosh, E., Javadi, M., Ejtahed, H.S. & Mirmiran, P. (2015) Probiotics as beneficial agents in the management of diabetes mellitus: a systematic review. *Diabetes/Metabolism Research and Reviews*, **32**, 143–168.

Recine, N., Palma, E., Domenici, L., Giorgini, M., Imperiale, L., Sassu, C., Musella, A., Marchetti, C., Muzii, L. & Benedetti Panici, P. (2016) Restoring vaginal microbiota: biological control of bacterial vaginosis: a prospective case-control study using *Lactobacillus rhamnosus* BMX 54 as adjuvant treatment against bacterial vaginosis. *Archives of Gynecology and Obstetrics*, **293**, 101–107.

Reid, G., Burton, J., Hammond, J.A. & Bruce, A.W. (2004) Nucleic acid-based diagnosis of bacterial vaginosis and improved management using probiotic lactobacilli. *Journal of Medicinal Food*, **7**, 223–228.

Rembacken, B.J., Snelling, A.M., Hawkey, P.M., Chalmers, D.M. & Axon, A.T. (1999) Non-pathogenic *Escherichia coli* versus mesalazine for the treatment of ulcerative colitis: a randomised trial. *Lancet*, **354**, 635–639.

Resta, S.C. (2009) Effects of probiotics and commensals on intestinal epithelial physiology: implications for nutrient handling. *Journal of Physiology*, **587**, 4169–4174.

Riezzo, G., Orlando, A., D'attoma, B., Guerra, V., Valerio, F., Lavermicocca, P., De Candia, S. & Russo, F. (2012) Randomised clinical trial: efficacy of *Lactobacillus paracasei*-enriched artichokes in the treatment of patients with functional constipation: a double-blind, controlled, crossover study. *Alimentary Pharmacology and Therapeutics*, **35**, 441–450.

Ringel-Kulka, T., Palsson, O.S., Maier, D., Carroll, I., Galanko, J.A., Leyer, G. & Ringel, Y. (2011) Probiotic bacteria *Lactobacillus acidophilus* NCFM and *Bifidobacterium lactis* Bi-07 versus placebo for the symptoms of bloating in patients with functional bowel disorders: a double-blind study. *Journal of Clinical Gastroenterology*, **45**, 518–525.

Ritchie, M.L. & Romanuk, T.N. (2012) A meta-analysis of probiotic efficacy for gastrointestinal diseases. *PLoS One*, **7**, e34938.

Ritchie, B.K., Brewster, D.R., Tran, C.D., Davidson, G.P., Mcneil, Y. & Butler, R.N. (2010) Efficacy of *Lactobacillus* GG in aboriginal children with acute diarrhoeal disease: a randomised clinical trial. *Journal of Pediatric Gastroenterology and Nutrition*, **50**, 619–624.

Robinson, J. (2014) Cochrane in context: probiotics for prevention of necrotizing enterocolitis in preterm infants. *Evidence-Based Child Health*, **9**, 672–674.

Roessler, A., Friedrich, U., Vogelsang, H., Bauer, A., Kaatz, M., Hipler, U.C., Schmidt, I. & Jahreis, G. (2008) The immune system in healthy adults and patients with atopic dermatitis seems to be affected differently by a probiotic intervention. *Clinical and Experimental Allergy*, **38**, 93–102.

Rogers, N.J. & Mousa, S.A. (2012) The shortcomings of clinical trials assessing the efficacy of probiotics in irritable bowel syndrome. *Journal of Alternative and Complementary Medicine*, **18**, 112–119.

Rogosa, M. (1974a) Genus *Lactobacillus*. In *Bergey's Manual of Determinative Bacteriology* (ed. R.E. Buchanan & N.E. Gibson), 8th ed., 576–593. Williams & Wilkins Company, Baltimore.

Rogosa, M. (1974b) Genus *Bifidobacterium*. In *Bergey's Manual of Determinative Bacteriology* (ed. R.E. Buchanan & N.E. Gibson), 8th ed., 669–676. Williams & Wilkins Company, Baltimore.

Rolfe, V.E., Fortun, P.J., Hawkey, C.J. & Bath-Hextall, F. (2006) Probiotics for maintenance of remission in Crohn's disease. *Cochrane Database of Systematic Reviews*, CD004826.

Roos, K., Hakansson, E.G. & Holm, S. (2001) Effect of recolonisation with "interfering" alpha streptococci on recurrences of acute and secretory otitis media in children: randomised placebo controlled trial. *BMJ*, **322**, 210–212.

Roos, S., Dicksved, J., Tarasco, V., Locatelli, E., Ricceri, F., Grandin, U. & Savino, F. (2013) 454 pyrosequencing analysis on faecal samples from a randomized DBPC trial of colicky infants treated with *Lactobacillus reuteri* DSM 17938. *PLoS One*, **8**, e56710.

Rosenfeld, C.S. (2015) Microbiome disturbances and autism spectrum disorders. *Drug Metabolism and Disposition: The Biological Fate of Chemicals*, **43**, 1557–1571.

Rosenfeldt, V., Benfeldt, E., Nielsen, S.D., Michaelsen, K.F., Jeppesen, D.L., Valerius, N.H. & Paerregaard, A. (2003) Effect of probiotic *Lactobacillus* strains in children with atopic dermatitis. *Journal of Allergy and Clinical Immunology*, **111**, 389–395.

Rossi, A., Rossi, T., Bertini, M. & Caccia, G. (2010) The use of *Lactobacillus rhamnosus* in the therapy of bacterial vaginosis: evaluation of clinical efficacy in a population of 40 women treated for 24 months. *Archives of Gynecology and Obstetrics*, **281**, 1065–1069.

Ruan, Y., Sun, J., He, J., Chen, F., Chen, R. & Chen, H. (2015) Effect of probiotics on glycemic control: a systematic review and meta-analysis of randomized, controlled trials. *PLoS One*, **10**, e0132121.

Ruijschop, R.M.J., Boelfijk, A.E.M. & Te Giffel, M.C. (2008) Satiety effects of a dairy beverage fermented with propionic acid bacteria. *International Dairy Journal*, **18**, 945–950.

Saavedra, J.M., Abi-Hanna, A., Moore, N. & Yolken, R.H. (2004) Long-term consumption of infant formulas containing live probiotic bacteria: tolerance and safety. *American Journal of Clinical Nutrition*, **79**, 261–267.

Saez-Lara, M.J., Gomez-Llorente, C., Plaza-Diaz, J. & Gil, A. (2015) The role of probiotic lactic acid bacteria and bifidobacteria in the prevention and treatment of inflammatory bowel disease and other related diseases: a systematic review of randomized human clinical trials. *Biomed Research International*, **2015**, 505878.

Safavi, M., Farajian, S., Kelishadi, R., Mirlohi, M. & Hashemipour, M. (2013) The effects of synbiotic supplementation on some cardio-metabolic risk factors in overweight and obese children: a randomized triple-masked controlled trial. *International Journal of Food Sciences and Nutrition*, **64**, 687–693.

Saggioro, A. (2004) Probiotics in the treatment of irritable bowel syndrome. *Journal of Clinical Gastroenterology*, **38**, S104–S106.

Saji, S., Kumar, S. & Thomas, V. (2011) A randomized double blind placebo controlled trial of probiotics in minimal hepatic encephalopathy. *Tropical Gastroenterology*, **32**, 128–132.

Sakai, T., Makino, H., Ishikawa, E., Oishi, K. & Kushiro, A. (2011) Fermented milk containing *Lactobacillus casei* strain Shirota reduces incidence of hard or lumpy stools in healthy population. *International Journal of Food Sciences and Nutrition*, **62**, 423–430.

Sakai, T., Kubota, H., Gawad, A., Gheyle, L., Ramael, S. & Oishi, K. (2015) Effect of fermented milk containing *Lactobacillus casei* strain Shirota on constipation-related symptoms and haemorrhoids in women during puerperium. *Beneficial Microbes*, **6**, 253–262.

Sampalis, J., Psaradellis, E. & Rampakakis, E. (2010) Efficacy of BIO K+ CL1285 in the reduction of antibiotic-associated diarrhea: a placebo controlled double-blind randomized, multicenter study. *Archives of Medical Science*, **6**, 56–64.

Sanchez, M., Darimont, C., Drapeau, V., Emady-Azar, S., Lepage, M., Rezzonico, E., Ngom-Bru, C., Berger, B., Philippe, L., Ammon-Zuffrey, C., Leone, P., Chevrier, G., St-Amand, E., Marette, A., Dore, J. & Tremblay, A. (2014) Effect of *Lactobacillus rhamnosus* CGMCC1.3724 supplementation on weight loss and maintenance in obese men and women. *British Journal of Nutrition*, **111**, 1507–1519.

Santiago-Lopez, L., Hernandez-Mendoza, A., Garcia, H., Mata-Haro, V., Vallejo-Cordoba, B. & Gonzalez-Cordova, A. (2015) The effects of consuming probiotic-fermented milk on the immune system: a review of scientific evidence. *International Journal of Dairy Technology*, **68**, 153–165.

Savaiano, D.A. (2014) Lactose digestion from yogurt: mechanism and relevance. *American Journal of Clinical Nutrition*, **99**, 1251S–1255S.

Savignac, H.M., Tramullas, M., Kiely, B., Dinan, T.G. & Cryan, J.F. (2015) Bifidobacteria modulate cognitive processes in an anxious mouse strain. *Behavioural Brain Research*, **287**, 59–72.

Savino, F., Pelle, E., Palumeri, E., Oggero, R. & Miniero, R. (2007) *Lactobacillus reuteri* (American Type Culture Collection Strain 55730) versus simethicone in the treatment of infantile colic: a prospective randomized study. *Pediatrics*, **119**, e124–e130.

Savino, F., Cordisco, L., Tarasco, V., Locatelli, E., Di Gioia, D., Oggero, R. & Matteuzzi, D. (2011) Antagonistic effect of *Lactobacillus* strains against gas-producing coliforms isolated from colicky infants. *BMC Microbiology*, **11**, 157.

Savino, F., Ceratto, S., Poggi, E., Cartosio, M.E., Cordero Di Montezemolo, L. & Giannattasio, A. (2015) Preventive effects of oral probiotic on infantile colic: a prospective, randomised, blinded, controlled trial using *Lactobacillus reuteri* DSM 17938. *Beneficial Microbes*, **6**, 245–251.

Sazawal, S., Dhingra, U., Hiremath, G., Sarkar, A., Dhingra, P., Dutta, A., Verma, P., Menon, V.P. & Black, R.E. (2010) Prebiotic and probiotic fortified milk in prevention of morbidities among children: community-based, randomized, double-blind, controlled trial. *PLoS One*, **5**, e12164.

Scarpellini, E., Lupo, M., Iegri, C., Gasbarrini, A., De Santis, A. & Tack, J. (2014) Intestinal permeability in non-alcoholic fatty liver disease: the gut-liver axis. *Reviews on Recent Clinical Trials*, **9**, 141–147.

Schlee, M., Harder, J., Koten, B., Stange, E.F., Wehkamp, J. & Fellermann, K. (2008) Probiotic lactobacilli and VSL#3 induce enterocyte beta-defensin 2. *Clinical and Experimental Immunology*, **151**, 528–535.

Schreck Bird, A., Gregory, P.J., Jalloh, M.A., Risoldi Cochrane, Z. & Hein, D.J. (2017) Probiotics for the treatment of infantile colic: a systematic review. *Journal of Pharmacy Practice*, **30**, 366–374.

Schulfer, A. & Blaser, M.J. (2015) Risks of antibiotic exposures early in life on the developing microbiome. *PLoS Pathogens*, **11**, e1004903.

Schultz, M., Timmer, A., Herfarth, H.H., Sartor, R.B., Vanderhoof, J.A. & Rath, H.C. (2004) *Lactobacillus* GG in inducing and maintaining remission of Crohn's disease. *BMC Gastroenterology*, **4**, 5.

Schunter, M., Chu, H., Hayes, T.L., Mcconnell, D., Crawford, S.S., Luciw, P.A., Bengmark, S., Asmuth, D.M., Brown, J., Bevins, C.L., Shacklett, B.L. & Critchfield, J.W. (2012) Randomized pilot trial of a synbiotic dietary supplement in chronic HIV-1 infection. *BMC Complementary and Alternative Medicine*, **12**, 84.

Schwenger, E.M., Tejani, A.M. & Loewen, P.S. (2015) Probiotics for preventing urinary tract infections in adults and children. *Cochrane Database of Systematic Reviews*, **12**, CD008772.

Segarra-Newnham, M. (2007) Probiotics for *Clostridium difficile*-associated diarrhea: focus on *Lactobacillus rhamnosus* GG and *Saccharomyces boulardii*. *Annals of Pharmacotherapy*, **41**, 1212–1221.

Sen, S., Mullan, M.M., Parker, T.J., Woolner, J.T., Tarry, S.A. & Hunter, J.O. (2002) Effect of *Lactobacillus plantarum* 299v on colonic fermentation and symptoms of irritable bowel syndrome. *Digestive Diseases and Sciences*, **47**, 2615–2620.

Senok, A.C., Verstraelen, H., Temmerman, M. & Botta, G.A. (2009) Probiotics for the treatment of bacterial vaginosis. *Cochrane Database of Systematic Reviews*, CD006289.

Shamir, R., Makhoul, I.R., Etzioni, A. & Shehadeh, N. (2005) Evaluation of a diet containing probiotics and zinc for the treatment of mild diarrheal illness in children younger than one year of age. *Journal of the American College of Nutrition*, **24**, 370–375.

Shanahan, F., Guarner, F., Von Wright, A., Vilipponene-Salmela, T., O'Donoghue, D. & Kiely, B.A. (2006) A one year, randomised, double blind, placebo controlled trial of a *Lactobacillus* or *Bifidobacterium* probiotic for maintenance of steroid-induced remission of ulcerative colitis. *Gastroenterology*, **130**, A44.

Sharafedtinov, K.K., Plotnikova, O.A., Alexeeva, R.I., Sentsova, T.B., Songisepp, E., Stsepetova, J., Smidt, I. & Mikelsaar, M. (2013) Hypocaloric diet supplemented with probiotic cheese improves body mass index and blood pressure indices of obese hypertensive patients: a randomized double-blind placebo-controlled pilot study. *Nutrition Journal*, **12**, 138.

Shen, B., Brzezinski, A., Fazio, V.W., Remzi, F.H., Achkar, J.P., Bennett, A.E., Sherman, K. & Lashner, B.A. (2005) Maintenance therapy with a probiotic in antibiotic-dependent pouchitis: experience in clinical practice. *Alimentary Pharmacology and Therapeutics*, **22**, 721–728.

Shen, J., Zuo, Z.X. & Mao, A.P. (2014) Effect of probiotics on inducing remission and maintaining therapy in ulcerative colitis, Crohn's disease, and pouchitis: meta-analysis of randomized controlled trials. *Inflammatory Bowel Diseases*, **20**, 21–35.

Shida, K., Nanno, M. & Nagata, S. (2011) Flexible cytokine production by macrophages and T cells in response to probiotic bacteria: a possible mechanism by which probiotics exert multifunctional immune regulatory activities. *Gut Microbes*, **2**, 109–114.

Shida, K., Sato, T., Iizuka, R., Hoshi, R., Watanabe, O., Igarashi, T., Miyazaki, K., Nanno, M. & Ishikawa, F. (2017) Daily intake of fermented milk with *Lactobacillus casei* strain Shirota reduces the incidence and duration of upper respiratory tract infections in healthy middle-aged office workers. *European Journal of Nutrition*, **56**, 45–53.

Shimada, K., Bricknell, K.S. & Finegold, S.M. (1969) Deconjugation of bile acids by intestinal bacteria: review of literature and additional studies. *Journal of Infectious Diseases*, **119**, 273–281.

Shukla, S., Shukla, A., Mehboob, S. & Guha, S. (2011) Meta-analysis: the effects of gut flora modulation using prebiotics, probiotics and synbiotics on minimal hepatic encephalopathy. *Alimentary Pharmacology and Therapeutics*, **33**, 662–671.

Simren, M., Ohman, L., Olsson, J., Svensson, U., Ohlson, K., Posserud, I. & Strid, H. (2010) Clinical trial: the effects of a fermented milk containing three probiotic bacteria in patients with irritable bowel syndrome: a randomized, double-blind, controlled study. *Alimentary Pharmacology and Therapeutics*, **31**, 218–227.

Simren, M., Barbara, G., Flint, H.J., Spiegel, B.M., Spiller, R.C., Vanner, S., Verdu, E.F., Whorwell, P.J., Zoetendal, E.G. & Rome Foundation, C. (2013) Intestinal microbiota in functional bowel disorders: a Rome Foundation report. *Gut*, **62**, 159–176.

Sinha, B. & Rubens, M. (2014) Systemic immune activation in HIV and potential therapeutic options. *Immunopharmacology and Immunotoxicology*, **36**, 89–95.

Sinn, D.H., Song, J.H., Kim, H.J., Lee, J.H., Son, H.J., Chang, D.K., Kim, Y.H., Kim, J.J., Rhee, J.C. & Rhee, P.L. (2008) Therapeutic effect of *Lactobacillus acidophilus* SDC 2012, 2013 in patients with irritable bowel syndrome. *Digestive Diseases and Sciences*, **53**, 2714–2718.

Sistek, D., Kelly, R., Wickens, K., Stanley, T., Fitzharris, P. & Crane, J. (2006) Is the effect of probiotics on atopic dermatitis confined to food sensitized children? *Clinical and Experimental Allergy*, **36**, 629–633.

Skovbjerg, S., Roos, K., Holm, S.E., Grahn Hakansson, E., Nowrouzian, F., Ivarsson, M., Adlerberth, I. & Wold, A.E. (2009) Spray bacteriotherapy decreases middle ear fluid in children with secretory otitis media. *Archives of Disease in Childhood*, **94**, 92–98.

Smith, E.A. & MacFarlane, G.T. (1996) Enumeration of human colonic bacteria producing phenolic and indolic compounds: effects of pH, carbohydrate availability and retention time on dissimilatory aromatic amino acid metabolism. *Journal of Applied Bacteriology*, **81**, 288–302.

Smith, T.J., Rigassio-Radler, D., Denmark, R., Haley, T. & Touger-Decker, R. (2013) Effect of *Lactobacillus rhamnosus* LGG(R) and *Bifidobacterium animalis* ssp. *lactis* BB-12(R) on health-related quality of life in college students affected by upper respiratory infections. *British Journal of Nutrition*, **109**, 1999–2007.

Soh, S.E., Aw, M., Gerez, I., Chong, Y.S., Rauff, M., Ng, Y.P., Wong, H.B., Pai, N., Lee, B.W. & Shek, L.P. (2009) Probiotic supplementation in the first 6 months of life in at risk Asian infants: effects on eczema and atopic sensitization at the age of 1 year. *Clinical and Experimental Allergy*, **39**, 571–578.

Sokol, H. (2014) Probiotics and antibiotics in IBD. *Digestive Diseases*, **32**(Suppl. 1), 10–17.

Sood, A., Midha, V., Makharia, G.K., Ahuja, V., Singal, D., Goswami, P. & Tandon, R.K. (2009) The probiotic preparation, VSL#3 induces remission in patients with mild-to-moderately active ulcerative colitis. *Clinical Gastroenterology and Hepatology*, **7**, 1202–1209.

Stadlbauer, V., Mookerjee, R.P., Hodges, S., Wright, G.A., Davies, N.A. & Jalan, R. (2008) Effect of probiotic treatment on deranged neutrophil function and cytokine responses in patients with compensated alcoholic cirrhosis. *Journal of Hepatology*, **48**, 945–951.

Stapleton, A.E., Au-Yeung, M., Hooton, T.M., Fredricks, D.N., Roberts, P.L., Czaja, C.A., Yarova-Yarovaya, Y., Fiedler, T., Cox, M. & Stamm, W.E. (2011) Randomized, placebo-controlled phase 2 trial of a *Lactobacillus crispatus* probiotic given intravaginally for prevention of recurrent urinary tract infection. *Clinical Infectious Diseases*, **52**, 1212–1217.

Steed, H., MacFarlane, G.T., Blackett, K.L., Bahrami, B., Reynolds, N., Walsh, S.V., Cummings, J.H. & MacFarlane, S. (2010) Clinical trial: the microbiological and immunological effects of synbiotic consumption: a randomized double-blind placebo-controlled study in active Crohn's disease. *Alimentary Pharmacology and Therapeutics*, **32**, 872–883.

Stein, G.Y., Nanim, R., Karniel, E., Moskowitz, I. & Zeidman, A. (2007) Probiotics as prophylactic agents against antibiotic-associated diarrhea in hospitalized patients (in Hebrew). *Harefuah*, **146**, 520–522, 575.

Stockert, K., Schneider, B., Porenta, G., Rath, R., Nissel, H. & Eichler, I. (2007) Laser acupuncture and probiotics in school age children with asthma: a randomized, placebo-controlled pilot study of therapy guided by principles of Traditional Chinese Medicine. *Pediatric Allergy and Immunology*, **18**, 160–166.

Sun, W., Shi, L., Ye, Z., Mu, Y., Liu, C., Zhao, J., Chen, L., Li, Q., Yang, T., Yan, L., Wan, Q., Wu, S., Liu, Y., Wang, G., Luo, Z., Tang, X., Chen, G., Huo, Y., Gao, Z., Su, Q., Wang, Y., Qin, G., Deng, H., Yu, X., Shen, F., Chen, L., Zhao, L., Sun, J., Ding, L., Xu, Y., Xu, M., Dai, M., Wang, T., Zhang, D., Lu, J., Bi, Y., Lai, S., Li, D., Wang, W., Ning, G. & Group, R.S. (2016) Association between the change in body mass index from early adulthood to midlife and subsequent type 2 diabetes mellitus. *Obesity (Silver Spring)*, **24**, 703–709.

Sung, V., Hiscock, H., Tang, M.L., Mensah, F.K., Nation, M.L., Satzke, C., Heine, R.G., Stock, A., Barr, R.G. & Wake, M. (2014) Treating infant colic with the probiotic *Lactobacillus reuteri*: double blind, placebo controlled randomised trial. *BMJ*, **348**, g2107.

Sur, D., Manna, B., Niyogi, S.K., Ramamurthy, T., Palit, A., Nomoto, K., Takahashi, T., Shima, T., Tsuji, H., Kurakawa, T., Takeda, Y., Nair, G.B. & Bhattacharya, S.K. (2011) Role of probiotic in preventing acute diarrhoea in children: a community-based, randomized, double-blind placebo-controlled field trial in an urban slum. *Epidemiology and Infection*, **139**, 919–926.

Surawicz, C.M., Mcfarland, L.V., Greenberg, R.N., Rubin, M., Fekety, R., Mulligan, M.E., Garcia, R.J., Brandmarker, S., Bowen, K., Borjal, D. & Elmer, G.W. (2000) The search for a better treatment for recurrent *Clostridium difficile* disease: use of high-dose vancomycin combined with *Saccharomyces boulardii*. *Clinical Infectious Diseases*, **31**, 1012–1017.

Szajewska, H. (2013) Understanding the role of probiotics and prebiotics in preventing allergic disease: evidence and methodological issues. *Immunotherapy*, **5**, 869–878.

Szajewska, H. & Karas, J. (2014) Acute gastroenteritis – the COMMENT working group on acute diarrhea: where are we now and where are we going? *Journal of Clinical Gastroenterology*, **48**, S32–S33.

Szajewska, H., Canani, R.B., Guarino, A., Hojsak, I., Indrio, F., Kolacek, S., Orel, R., Shamir, R., Vandenplas, Y., Van Goudoever, J.B., Weizman, Z. & ESPGHAN Working Group for Probiotics and Prebiotics (2016) Probiotics for the prevention of antibiotic-associated diarrhea in children. *Journal of Pediatric Gastroenterology and Nutrition*, **62**, 495–506.

Szajewska, H. & Kolodziej, M. (2015) Systematic review with meta-analysis: *Lactobacillus rhamnosus* GG in the prevention of antibiotic-associated diarrhoea in children and adults. *Alimentary Pharmacology and Therapeutics*, **42**, 1149–1157.

Szajewska, H. & Mrukowicz, J.Z. (2001) Probiotics in the treatment and prevention of acute infectious diarrhea in infants and children: a systematic review of published randomized, double-blind, placebo-controlled trials. *Journal of Pediatric Gastroenterology and Nutrition*, **33**, S17–S25.

Szajewska, H. & Skorka, A. (2009) *Saccharomyces boulardii* for treating acute gastroenteritis in children: updated meta-analysis of randomized controlled trials. *Alimentary Pharmacology and Therapeutics*, **30**, 960–961.

Szajewska, H., Skorka, A., Ruszczynski, M. & Gieruszczak-Bialek, D. (2013) Meta-analysis: *Lactobacillus* GG for treating acute gastroenteritis in children: updated analysis of randomised controlled trials. *Alimentary Pharmacology and Therapeutics*, **38**, 467–476.

Szajewska, H., Urbanska, M., Chmielewska, A., Weizman, Z. & Shamir, R. (2014) Meta-analysis: *Lactobacillus reuteri* strain DSM 17938 (and the original strain ATCC 55730) for treating acute gastroenteritis in children. *Beneficial Microbes*, **5**, 285–293.

Szymanski, H., Armanska, M., Kowalska-Duplaga, K. & Szajewska, H. (2008) *Bifidobacterium longum* PL03, *Lactobacillus rhamnosus* KL53A, and *Lactobacillus plantarum* PL02 in the prevention of antibiotic-associated diarrhea in children: a randomized controlled pilot trial. *Digestion*, **78**, 13–17.

Taipale, T., Pienihakkinen, K., Salminen, S., Jokela, J. & Soderling, E. (2012) *Bifidobacterium animalis* subsp. *lactis* BB-12 administration in early childhood: a randomized clinical trial of effects on oral colonization by mutans streptococci and the probiotic. *Caries Research*, **46**, 69–77.

Takii, H., Nishijima, T., Takami, K., Tanaka, Y., Uinugami, M., Mawatari, T., Sugimura, H. & Aoki, R. (2012) Effects of fermented milk containing *Bifidobacterium animalis* subsp. *lactis*

GCL2505 on improvement of defecation, fecal properties, and intestinal microflora. *Japan Pharmacology & Therapeutics*, **40**, 657–665.

Tamura, M., Shikina, T., Morihana, T., Hayama, M., Kajimoto, O., Sakamoto, A., Kajimoto, Y., Watanabe, O., Nonaka, C., Shida, K. & Nanno, M. (2007) Effects of probiotics on allergic rhinitis induced by Japanese cedar pollen: randomized double-blind, placebo-controlled clinical trial. *International Archives of Allergy and Immunology*, **143**, 75–82.

Tanoue, T., Umesaki, Y. & Honda, K. (2010) Immune responses to gut microbiota-commensals and pathogens. *Gut Microbes*, **1**, 224–233.

Tapiovaara, L., Lehtoranta, L., Swanljung, E., Makivuokko, H., Laakso, S., Roivainen, M., Korpela, R. & Pitkaranta, A. (2014) *Lactobacillus rhamnosus* GG in the middle ear after randomized, double-blind, placebo-controlled oral administration. *International Journal of Pediatric Otorhinolaryngology*, **78**, 1637–1641.

Taylor, A.L., Dunstan, J.A. & Prescott, S.L. (2007) Probiotic supplementation for the first 6 months of life fails to reduce the risk of atopic dermatitis and increases the risk of allergen sensitization in high-risk children: a randomized controlled trial. *Journal of Allergy and Clinical Immunology*, **119**, 184–191.

Thijssen, A.Y., Clemens, C.H., Vankerckhoven, V., Goossens, H., Jonkers, D.M. & Masclee, A.A. (2016) Efficacy of *Lactobacillus casei* Shirota for patients with irritable bowel syndrome. *European Journal of Gastroenterology and Hepatology*, **28**, 8–14.

Thomas, M.R., Litin, S.C., Osmon, D.R., Corr, A.P., Weaver, A.L. & Lohse, C.M. (2001) Lack of effect of *Lactobacillus* GG on antibiotic-associated diarrhea: a randomized, placebo-controlled trial. *Mayo Clinic Proceedings*, **76**, 883–889.

Thomas, L.V., Ockhuizen, T. & Suzuki, K. (2014) Exploring the influence of the gut microbiota and probiotics on health: a symposium report. *British Journal of Nutrition*, **112**, S1–S18.

Thomas, L.V., Suzuki, K. & Zhao, J. (2015) Probiotics: a proactive approach to health: a symposium report. *British Journal of Nutrition*, **114**, S1–S15.

Tilley, L., Keppens, K., Kushiro, A., Takada, T., Sakai, T., Vaneechoutte, M. & Degeest, B. (2014) A probiotic fermented milk drink containing *Lactobacillus casei* strain Shirota improves stool consistency of subjects with hard stools. *International Journal of Probiotics and Prebiotics*, **9**, 23–30.

Tillisch, K., Labus, J., Kilpatrick, L., Jiang, Z., Stains, J., Ebrat, B., Guyonnet, D., Legrain-Raspaud, S., Trotin, B., Naliboff, B. & Mayer, E.A. (2013) Consumption of fermented milk product with probiotic modulates brain activity. *Gastroenterology*, **144**, 1394–1401.

Toi, M., Hirota, S., Tomotaki, A., Sato, N., Hozumi, Y., Anan, K., Nagashima, T., Tokuda, Y., Masuda, N., Ohsumi, S., Ohno, S., Takahashi, M., Hayashi, H., Yamamoto, S. & Ohashi, Y. (2013) Probiotic beverage with soy isoflavone consumption for breast cancer prevention: a case-control study. *Current Nutrition and Food Science*, **9**, 194–200.

Tojo, R., Suarez, A., Clemente, M.G., De Los Reyes-Gavilan, C.G., Margolles, A., Gueimonde, M. & Ruas-Madiedo, P. (2014) Intestinal microbiota in health and disease: role of bifidobacteria in gut homeostasis. *World Journal of Gastroenterology*, **20**, 15163–15176.

Tomasz, B., Zoran, S., Jaroslaw, W., Ryszard, M., Marcin, G., Robert, B., Piotr, K., Lukasz, K., Jacek, P., Piotr, G., Przemyslaw, P. & Michal, D. (2014) Long-term use of probiotics *Lactobacillus* and *Bifidobacterium* has a prophylactic effect on the occurrence and severity of pouchitis: a randomized prospective study. *Biomedical Research International*, **2014**, 208064.

Tomova, A., Husarova, V., Lakatosova, S., Bakos, J., Vlkova, B., Babinska, K. & Ostatnikova, D. (2015) Gastrointestinal microbiota in children with autism in Slovakia. *Physiology and Behavior*, **138**, 179–187.

Tomusiak, A., Strus, M., Heczko, P.B., Adamski, P., Stefanski, G., Mikolajczyk-Cichonska, A. & Suda-Szczurek, M. (2015) Efficacy and safety of a vaginal medicinal product containing three strains of probiotic bacteria: a multicenter, randomized, double-blind, and placebo-controlled trial. *Drug Design, Development and Therapy*, **9**, 5345–5354.

Tonucci, L.B., Olbrich Dos Santos, K.M., Licursi De Oliveira, L., Rocha Ribeiro, S.M. & Duarte Martino, H.S. (2017) Clinical application of probiotics in type 2 diabetes mellitus: a randomized, double-blind, placebo-controlled study. *Clinical Nutrition*, **36**, 85–92.

Trois, L., Cardoso, E.M. & Miura, E. (2008) Use of probiotics in HIV-infected children: a randomized double-blind controlled study. *Journal of Tropical Pediatrics*, **54**, 19–24.

Tsuda, Y., Yoshimatsu, Y., Aoki, H., Nakamura, K., Irie, M., Fukuda, K., Hosoe, N., Takada, N., Shirai, K. & Suzuki, Y. (2007) Clinical effectiveness of probiotics therapy (BIO-THREE) in patients with ulcerative colitis refractory to conventional therapy. *Scandinavian Journal of Gastroenterology*, **42**, 1306–1311.

Tuohy, K.M., Pinart-Gilberga, M., Jones, M., Hoyles, L., McCartney, A.L. & Gibson, G.R. (2007) Survivability of a probiotic *Lactobacillus casei* in the gastrointestinal tract of healthy human volunteers and its impact on the faecal microflora. *Journal of Applied Microbiology*, **102**, 1026–1032.

Tursi, A., Brandimarte, G., Giorgetti, G.M., Forti, G., Modeo, M.E. & Gigliobianco, A. (2004) Low-dose balsalazide plus a high-potency probiotic preparation is more effective than balsalazide alone or mesalazine in the treatment of acute mild-to-moderate ulcerative colitis. *Medical Science Monitor*, **10**, PI126–PI131.

Tursi, A., Brandimarte, G., Papa, A., Giglio, A., Elisei, W., Giorgetti, G.M., Forti, G., Morini, S., Hassan, C., Pistoia, M.A., Modeo, M.E., Rodino, S., D'Amico, T., Sebkova, L., Sacca, N., Di Giulio, E., Luzza, F., Imeneo, M., Larussa, T., Di Rosa, S., Annese, V., Danese, S. & Gasbarrini, A. (2010) Treatment of relapsing mild-to-moderate ulcerative colitis with the probiotic VSL#3 as adjunctive to a standard pharmaceutical treatment: a double-blind, randomized, placebo-controlled study. *American Journal of Gastroenterology*, **105**, 2218–2227.

Underwood, M.A., Kalanetra, K.M., Bokulich, N.A., Lewis, Z.T., Mirmiran, M., Tancredi, D.J. & Mills, D.A. (2013) A comparison of two probiotic strains of bifidobacteria in premature infants. *Journal of Pediatrics*, **163**, 1585–1591.

Underwood, M.A., Bruce German, J., Lebrilla, C.B. & David A. Mills, D.A. (2015) *Bifidobacterium longum* subspecies *infantis*: champion colonizer of the infant gut. *Pediatr Resarch*, **77**, 229–235. doi:10.1038/pr.2014.156

Vajro, P., Mandato, C., Licenziati, M.R., Franzese, A., Vitale, D.F., Lenta, S., Caropreso, M., Vallone, G. & Meli, R. (2011) Effects of *Lactobacillus rhamnosus* strain GG in pediatric obesity-related liver disease. *Journal of Pediatric Gastroenterology and Nutrition*, **52**, 740–743.

Van't Veer, P., Dekker, J.M., Lamers, J.W., Kok, F.J., Schouten, E.G., Brants, H.A., Sturmans, F. & Hermus, R.J. (1989) Consumption of fermented milk products and breast cancer: a case-control study in The Netherlands. *Cancer Research*, **49**, 4020–4023.

Van Baarlen, P., Troost, F., Van Der Meer, C., Hooiveld, G., Boekschoten, M., Brummer, R.J. & Kleerebezem, M. (2011) Human mucosal *in vivo* transcriptome responses to three lactobacilli indicate how probiotics may modulate human cellular pathways. *Proceedings of the National Academy of Sciences of the United States of America*, **108**, 4562–4569.

Van De Pol, M.A., Lutter, R., Smids, B.S., Weersink, E.J. & Van Der Zee, J.S. (2011) Synbiotics reduce allergen-induced T-helper 2 response and improve peak expiratory flow in allergic asthmatics. *Allergy*, **66**, 39–47.

Van Den Nieuwboer, M., Klomp-Hogeterp, A., Verdoorn, S., Metsemakers-Brameijer, L., Vriend, T.M., Claassen, E. & Larsen, O.F. (2015) Improving the bowel habits of elderly residents in a nursing home using probiotic fermented milk. *Beneficial Microbes*, **6**, 397–403.

Van Gossum, A., Dewit, O., Louis, E., De Hertogh, G., Baert, F., Fontaine, F., Devos, M., Enslen, M., Paintin, M. & Franchimont, D. (2007) Multicenter randomized-controlled clinical trial of probiotics (*Lactobacillus johnsonii*, LA1) on early endoscopic recurrence of Crohn's disease after lleo-caecal resection. *Inflammatory Bowel Diseases*, **13**, 135–142.

Van Niel, C.W., Feudtner, C., Garrison, M.M. & Christakis, D.A. (2002) *Lactobacillus* therapy for acute infectious diarrhea in children: a meta-analysis. *Pediatrics*, **109**, 678–684.

Vandenplas, Y., De Hert, S.G. & Group, P.R.-S. (2011) Randomised clinical trial: the synbiotic food supplement Probiotical vs. placebo for acute gastroenteritis in children. *Alimentary Pharmacology and Therapeutics*, **34**, 862–867.

Vandenplas, Y., Huys, G. & Daube, G. (2015) Probiotics: an update. *Jornal de Pediatría*, **91**, 6–21.

Vanderhoof, J.A., Whitney, D.B., Antonson, D.L., Hanner, T.L., Lupo, J.V. & Young, R.J. (1999) *Lactobacillus* GG in the prevention of antibiotic-associated diarrhea in children. *Journal of Pediatrics*, **135**, 564–568.

Verhoeven, V., Renard, N., Makar, A., Van Royen, P., Bogers, J.P., Lardon, F., Peeters, M. & Baay, M. (2013) Probiotics enhance the clearance of human papillomavirus-related cervical lesions: a prospective controlled pilot study. *European Journal of Cancer Prevention*, **22**, 46–51.

Vicariotto, F., Del Piano, M., Mogna, L. & Mogna, G. (2012) Effectiveness of the association of 2 probiotic strains formulated in a slow release vaginal product, in women affected by vulvovaginal candidiasis: a pilot study. *Journal of Clinical Gastroenterology*, **46**, S73–S80.

Videlock, E.J. & Cremonini, F. (2012) Meta-analysis: probiotics in antibiotic-associated diarrhoea. *Alimentary Pharmacology and Therapeutics*, **35**, 1355–1369.

Viljanen, M., Savilahti, E., Haahtela, T., Juntunen-Backman, K., Korpela, R., Poussa, T., Tuure, T. & Kuitunen, M. (2005) Probiotics in the treatment of atopic eczema/dermatitis syndrome in infants: a double-blind placebo-controlled trial. *Allergy*, **60**, 494–500.

Villar-Garcia, J., Hernandez, J.J., Guerri-Fernandez, R., Gonzalez, A., Lerma, E., Guelar, A., Saenz, D., Sorli, L., Montero, M., Horcajada, J.P. & Knobel Freud, H. (2015) Effect of probiotics (*Saccharomyces boulardii*) on microbial translocation and inflammation in HIV-treated patients: a double-blind, randomized, placebo-controlled trial. *Journal of Acquired Immune Deficiency Syndromes*, **68**, 256–263.

Villarruel, G., Rubio, D.M., Lopez, F., Cintioni, J., Gurevech, R., Romero, G. & Vandenplas, Y. (2007) *Saccharomyces boulardii* in acute childhood diarrhoea: a randomized, placebo-controlled study. *Acta Paediatrica*, **96**, 538–541.

Vongbhavit, K. & Underwood, M.A. (2016) Prevention of necrotizing enterocolitis through manipulation of the intestinal microbiota of the premature infant. *Clinical Therapeutics*, **38**, 716–732.

Waitzberg, D.L., Logullo, L.C., Bittencourt, A.F., Torrinhas, R.S., Shiroma, G.M., Paulino, N.P. & Teixeira-Da-Silva, M.L. (2013) Effect of synbiotic in constipated adult women - a randomized, double-blind, placebo-controlled study of clinical response. *Clinical Nutrition*, **32**, 27–33.

Wall, R., Cryan, J.F., Ross, R.P., Fitzgerald, G.F., Dinan, T.G. & Stanton, C. (2014) Bacterial neuroactive compounds produced by psychobiotics. *Advances in Experimental Medicine and Biology*, **817**, 221–239.

Wan, L.Y., Chen, Z.J., Shah, N.P. & El-Nezami, H. (2016) Modulation of intestinal epithelial defense responses by probiotic bacteria. *Critical Reviews in Food Science and Nutrition*, **56**, 2628–2641.

Wang, M.F., Lin, H.C., Wang, Y.Y. & Hsu, C.H. (2004) Treatment of perennial allergic rhinitis with lactic acid bacteria. *Pediatric Allergy and Immunology*, **15**, 152–158.

Wanke, M. & Szajewska, H. (2012) Lack of an effect of *Lactobacillus reuteri* DSM 17938 in preventing nosocomial diarrhea in children: a randomized, double-blind, placebo-controlled trial. *Journal of Pediatrics*, **161**, 40–43.

Wassenberg, J., Nutten, S., Audran, R., Barbier, N., Aubert, V., Moulin, J., Mercenier, A. & Spertini, F. (2011) Effect of *Lactobacillus paracasei* ST11 on a nasal provocation test with grass pollen in allergic rhinitis. *Clinical and Experimental Allergy*, **41**, 565–573.

Weizman, Z., Asli, G. & Alsheikh, A. (2005) Effect of a probiotic infant formula on infections in child care centers: comparison of two probiotic agents. *Pediatrics*, **115**, 5–9.

Weng, M. & Walker, W.A. (2013) The role of gut microbiota in programming the immune phenotype. *Journal of Developmental Origins of Health and Disease*, **4**, 203–214.

Wenus, C., Goll, R., Loken, E.B., Biong, A.S., Halvorsen, D.S. & Florholmen, J. (2008) Prevention of antibiotic-associated diarrhoea by a fermented probiotic milk drink. *European Journal of Clinical Nutrition*, **62**, 299–301.

West, C.E., Hammarstrom, M.L. & Hernell, O. (2009) Probiotics during weaning reduce the incidence of eczema. *Pediatric Allergy and Immunology*, **20**, 430–437.

West, N.P., Pyne, D.B., Cripps, A.W., Hopkins, W.G., Eskesen, D.C., Jairath, A., Christophersen, C.T., Conlon, M.A. & Fricker, P.A. (2011) *Lactobacillus fermentum* (PCC-R) supplementation and gastrointestinal and respiratory-tract illness symptoms: a randomised control trial in athletes. *Nutrition Journal*, **10**, 30.

West, N.P., Horn, P.L., Pyne, D.B., Gebski, V.J., Lahtinen, S.J., Fricker, P.A. & Cripps, A.W. (2014) Probiotic supplementation for respiratory and gastrointestinal illness symptoms in healthy physically active individuals. *Clinical Nutrition*, **33**, 581–587.

Weston, S., Halbert, A., Richmond, P. & Prescott, S.L. (2005) Effects of probiotics on atopic dermatitis: a randomised controlled trial. *Archives of Disease in Childhood*, **90**, 892–897.

Wheeler, J.G., Shema, S.J., Bogle, M.L., Shirrell, M.A., Burks, A.W., Pittler, A. & Helm, R.M. (1997) Immune and clinical impact of *Lactobacillus acidophilus* on asthma. *Annals of Allergy, Asthma, and Immunology*, **79**, 229–233.

Whorwell, P.J., Altringer, L., Morel, J., Bond, Y., Charbonneau, D., O'Mahony, L., Kiely, B., Shanahan, F. & Quigley, E.M. (2006) Efficacy of an encapsulated probiotic *Bifidobacterium infantis* 35624 in women with irritable bowel syndrome. *American Journal of Gastroenterology*, **101**, 1581–1590.

Wichmann, A., Allahyar, A., Greiner, T.U., Plovier, H., Lunden, G.O., Larsson, T., Drucker, D.J., Delzenne, N.M., Cani, P.D. & Backhed, F. (2013) Microbial modulation of energy availability in the colon regulates intestinal transit. *Cell Host Microbe*, **14**, 582–590.

Wickens, K., Black, P.N., Stanley, T.V., Mitchell, E., Fitzharris, P., Tannock, G.W., Purdie, G., Crane, J. & Probiotic Study Group. (2008) A differential effect of 2 probiotics in the prevention of eczema and atopy: a double-blind, randomized, placebo-controlled trial. *Journal of Allergy and Clinical Immunology*, **122**, 788–794.

Wickens, K., Black, P., Stanley, T.V., Mitchell, E., Barthow, C., Fitzharris, P., Purdie, G. & Crane, J. (2012) A protective effect of *Lactobacillus rhamnosus* HN001 against eczema in the first 2 years of life persists to age 4 years. *Clinical and Experimental Allergy*, **42**, 1071–1079.

Wiegering, V., Kaiser, J., Tappe, D., Weissbrich, B., Morbach, H. & Girschick, H.J. (2011) Gastroenteritis in childhood: a retrospective study of 650 hospitalized pediatric patients. *International Journal of Infectious Diseases*, **15**, e401–e407.

Wildt, S., Nordgaard, I., Hansen, U., Brockmann, E. & Rumessen, J.J. (2011) A randomised double-blind placebo-controlled trial with *Lactobacillus acidophilus* LA-5 and *Bifidobacterium animalis* subsp. *lactis* BB-12 for maintenance of remission in ulcerative colitis. *Journal of Crohn's and Colitis*, **5**, 115–121.

Williams, E.A., Stimpson, J., Wang, D., Plummer, S., Garaiova, I., Barker, M.E. & Corfe, B.M. (2009) Clinical trial: a multistrain probiotic preparation significantly reduces symptoms of irritable bowel syndrome in a double-blind placebo-controlled study. *Alimentary Pharmacology and Therapeutics*, **29**, 97–103.

Wilson, N.L., Moneyham, L.D. & Alexandrov, A.W. (2013) A systematic review of probiotics as a potential intervention to restore gut health in HIV infection. *Journal of the Association of Nurses in AIDS Care*, **24**, 98–111.

Wong, V.W., Won, G.L., Chim, A.M., Chu, W.C., Yeung, D.K., Li, K.C. & Chan, H.L. (2013) Treatment of nonalcoholic steatohepatitis with probiotics: a proof-of-concept study. *Annals of Hepatology*, **12**, 256–262.

Wong, S., Jamous, A., O'Driscoll, J., Sekhar, R., Weldon, M., Yau, C.Y., Hirani, S.P., Grimble, G. & Forbes, A. (2014) A *Lactobacillus casei* Shirota probiotic drink reduces antibiotic-associated diarrhoea in patients with spinal cord injuries: a randomised controlled trial. *British Journal of Nutrition*, **111**, 672–678.

Woo, S.I., Kim, J.Y., Lee, Y.J., Kim, N.S. & Hahn, Y.S. (2010) Effect of *Lactobacillus sakei* supplementation in children with atopic eczema-dermatitis syndrome. *Annals of Allergy, Asthma, and Immunology*, **104**, 343–348.

Woodard, G.A., Encarnacion, B., Downey, J.R., Peraza, J., Chong, K., Hernandez-Boussard, T. & Morton, J.M. (2009) Probiotics improve outcomes after Roux-en-Y gastric bypass surgery: a prospective randomized trial. *Journal of Gastrointestinal Surgery*, **13**, 1198–1204.

Worthley, D.L., Le Leu, R.K., Whitehall, V.L., Conlon, M., Christophersen, C., Belobrajdic, D., Mallitt, K.A., Hu, Y., Irahara, N., Ogino, S., Leggett, B.A. & Young, G.P. (2009) A human, double-blind, placebo-controlled, crossover trial of prebiotic, probiotic, and synbiotic supplementation: effects on luminal, inflammatory, epigenetic, and epithelial biomarkers of colorectal cancer. *American Journal of Clinical Nutrition*, **90**, 578–586.

Wu, S.F., Caplan, M. & Lin, H.C. (2012) Necrotizing enterocolitis: old problem with new hope. *Pediatrics and Neonatology*, **53**, 158–163.

Wullt, M., Hagslatt, M.L. & Odenholt, I. (2003) *Lactobacillus plantarum* 299v for the treatment of recurrent *Clostridium difficile*-associated diarrhoea: a double-blind, placebo-controlled trial. *Scandinavian Journal of Infectious Diseases*, **35**, 365–367.

Xiao, J.Z., Kondo, S., Yanagisawa, N., Takahashi, N., Odamaki, T., Iwabuchi, N., Miyaji, K., Iwatsuki, K., Togashi, H., Enomoto, K. & Enomoto, T. (2006) Probiotics in the treatment of Japanese cedar pollinosis: a double-blind placebo-controlled trial. *Clinical and Experimental Allergy*, **36**, 1425–1435.

Xie, C., Li, J., Wang, K., Li, Q. & Chen, D. (2015) Probiotics for the prevention of antibiotic-associated diarrhoea in older patients: a systematic review. *Travel Medicine and Infectious Disease*, **13**, 128–134.

Xu, M., Wang, J., Wang, N., Sun, F., Wang, L. & Liu, X.H. (2015) The efficacy and safety of the probiotic bacterium *Lactobacillus reuteri* DSM 17938 for infantile colic: a meta-analysis of randomized controlled trials. *PLoS One*, **10**, e0141445.

Ya, W., Reifer, C. & Miller, L.E. (2010) Efficacy of vaginal probiotic capsules for recurrent bacterial vaginosis: a double-blind, randomized, placebo-controlled study. *American Journal of Obstetrics and Gynecology*, **203**, 120.e1–120.e6.

Yan, Q., Li, X. & Feng, B. (2015) The efficacy and safety of probiotics intervention in preventing conversion of impaired glucose tolerance to diabetes: study protocol for a randomized, double-blinded, placebo controlled trial of the Probiotics Prevention Diabetes Programme (PPDP). *BMC Endocrine Disorders*, **15**, 74.

Yang, Y.X., He, M., Hu, G., Wei, J., Pages, P., Yang, X.H. & Bourdu-Naturel, S. (2008) Effect of a fermented milk containing *Bifidobacterium lactis* DN-173010 on Chinese constipated women. *World Journal of Gastroenterology*, **14**, 6237–6243.

Yang, O.O., Kelesidis, T., Cordova, R. & Khanlou, H. (2014) Immunomodulation of antiretroviral drug-suppressed chronic HIV-1 infection in an oral probiotic double-blind placebo-controlled trial. *AIDS Research and Human Retroviruses*, **30**, 988–995.

Yang, H., Zhao, X., Tang, S., Huang, H., Zhao, X., Ning, Z., Fu, X. & Zhang, C. (2016a) Probiotics reduce psychological stress in patients before laryngeal cancer surgery. *Asia-Pacific Journal of Clinical Oncology*, **12**, e92–e96.

Yang, S.H., Li, S., Zhang, Y., Xu, R.X., Guo, Y.L., Zhu, C.G., Wu, N.Q., Cui, C.J., Sun, J. & Li, J.J. (2016b) Positive correlation of plasma PCSK9 levels with HbA1c in patients with type 2 diabetes. *Diabetes/Metabolism Research and Reviews*, **32**, 193–199.

Yao, T.C., Chang, C.J., Hsu, Y.H. & Huang, J.L. (2010) Probiotics for allergic diseases: realities and myths. *Pediatric Allergy and Immunology*, **21**, 900–919.

Yeun, Y. & Lee, J. (2015) Effect of a double-coated probiotic formulation on functional constipation in the elderly: a randomized, double blind, controlled study. *Archives of Pharmacal Research*, **38**, 1345–1350.

Yoon, H., Park, Y.S., Lee, D.H., Seo, J.G., Shin, C.M. & Kim, N. (2015) Effect of administering a multi-species probiotic mixture on the changes in fecal microbiota and symptoms of irritable bowel syndrome: a randomized, double-blind, placebo-controlled trial. *Journal of Clinical Biochemistry and Nutrition*, **57**, 129–134.

Yoshikawa, T.T. (2000) Epidemiology and unique aspects of aging and infectious diseases. *Clinical Infectious Diseases*, **30**, 931–933.

Younossi, Z.M., Koenig, A.B., Abdelatif, D., Fazel, Y., Henry, L. & Wymer, M. (2016) Global epidemiology of non-alcoholic fatty liver disease – Meta-analytic assessment of prevalence, incidence and outcomes. *Hepatology*, **64**, 73–84.

Zarrati, M., Salehi, E., Nourijelyani, K., Mofid, V., Zadeh, M.J., Najafi, F., Ghaflati, Z., Bidad, K., Chamari, M., Karimi, M. & Shidfar, F. (2014) Effects of probiotic yogurt on fat distribution and gene expression of proinflammatory factors in peripheral blood mononuclear cells in overweight and obese people with or without weight-loss diet. *Journal of the American College of Nutrition*, **33**, 417–425.

Zhong, L., Zhang, X. & Covasa, M. (2014) Emerging roles of lactic acid bacteria in protection against colorectal cancer. *World Journal of Gastroenterology*, **20**, 7878–7886.

Zigra, P.I., Maipa, V.E. & Alamanos, Y.P. (2007) Probiotics and remission of ulcerative colitis: a systematic review. *Netherlands Journal of Medicine*, **65**, 411–418.

Zocco, M.A., Dal Verme, L.Z., Cremonini, F., Piscaglia, A.C., Nista, E.C., Candelli, M., Novi, M., Rigante, D., Cazzato, I.A., Ojetti, V., Armuzzi, A., Gasbarrini, G. & Gasbarrini, A. (2006) Efficacy of *Lactobacillus* GG in maintaining remission of ulcerative colitis. *Alimentary Pharmacology and Therapeutics*, **23**, 1567–1574.

Zuccotti, G., Meneghin, F., Aceti, A., Barone, G., Callegari, M.L., Di Mauro, A., Fantini, M.P., Gori, D., Indrio, F., Maggio, L., Morelli, L., Corvaglia, L. & Italian Society of Neonatology. (2015) Probiotics for prevention of atopic diseases in infants: systematic review and meta-analysis. *Allergy*, **70**, 1356–1371.

9 Production of Vitamins, Exopolysaccharides and Bacteriocins by Probiotic Bacteria

D.M. Linares, G. Fitzgerald, C. Hill, C. Stanton and P. Ross

9.1 Introduction

Probiotics are 'live micro-organisms, which when consumed in adequate amounts confer a health benefit on the host' (FAO/WHO, 2001; Hill *et al.*, 2014). In this respect, for a culture to be termed a probiotic it needs to have associated clinical evidence of health promotion in humans or animals. While this is the case for a limited number of probiotic bacterial strains, the precise mechanism by which micro-organisms exert a health effect *in vivo* is often not understood. One aspect that is clear, however, is that some intestinal strains produce certain health-promoting metabolites (the so-called pharmabiotics), which are desirable from a nutritional and/or health promotion perspective. Production of such compounds in fermented dairy products means that many are often produced in the food prior to consumption. However, it should be emphasised that ingestion of probiotic fermented foods opens up the possibility that these health-promoting metabolites may be produced *in vivo* as well. This chapter will detail the production of three types of complex biomolecules by probiotic bacteria, namely vitamins, exopolysaccharides (EPS) and bacteriocins, and will discuss their potential for health promotion in humans.

9.2 Vitamin production by probiotic bacteria

9.2.1 *Background*

Vitamins are involved in important biochemical reactions in all living cells, and such deficiency is being linked to neural tube defects, anaemia, certain forms of cancer, poor cognitive performance and coronary heart diseases, among others (Divya & Nampoothiri, 2015). Most vitamins (particularly folate, riboflavin, thiamine and cobalamin) must be obtained exogenously due to the inability of humans to synthesise them (LeBlanc *et al.*, 2011; De Angelis *et al.*, 2014), although it is well known that some intestinal bacteria do produce certain vitamins. In this respect, it has been suggested that vitamin production is one of a number of functional characteristics associated with probiotic bacteria and gastrointestinal (GI) microbiota (Holzapfel & Schillinger, 2002; Linares *et al.*, 2016).

Probiotic Dairy Products, Second Edition. Edited by Adnan Y. Tamime and Linda V. Thomas.

Furthermore, a large number of lactic acid bacteria (LAB), including *Bifidobacterium* spp., have been reported to produce vitamins, such as folate (vitamin B_9), cobalamin (vitamin B_{12}), menaquinone (vitamin K_2), riboflavin (vitamin B_2) and thiamine (vitamin B_1). As such, the use of these cultures in food fermentation potentially provides a route to not only enhance the nutritional profile of the food, but also deliver micro-organisms to the gut where they can synthesise such vitamins *in vivo*.

9.2.2 *Folate*

Folate, an essential component of the human diet involved in cell metabolism, cell proliferation and DNA replication, is a generic term used to describe the salts of folic acid (pteroylmonoglutamic acid). In this chapter, the generic term 'folate' will refer to natural folate derivatives, such as 5-methyltetrahydrofolate and folylglutamates, which are naturally produced, but not the synthetic form of folic acid commonly used for food fortification and nutritional supplements (LeBlanc *et al.*, 2011).

Folates are receiving increasing attention due to the link between folate deficiency and neural tube defects in developing embryos during pregnancy and the ability of folates to protect against some forms of cancer (LeBlanc *et al.*, 2011; Divya & Nampoothiri, 2015). Previous studies established that the endogenous folate levels in plasma are insufficient for biological functions and could be a risk factor in the development of coronary heart disease (Morrison *et al.*, 1996; Divya & Nampoothiri, 2015). Thus, dietary folate is essential for humans, since it cannot be synthesised by mammalian cells and is required at high levels by tissues with high growth rates, such as leucocytes, erythrocytes and the intestinal mucosa (Crittenden *et al.*, 2003; Rossi *et al.*, 2011).

It has been reported that milk contains between 20 and 50 $\mu g\,L^{-1}$ of folate (Crittenden *et al.*, 2003). Considering this, an average adult person would need to consume 6–12 $L\,d^{-1}$ of milk to meet their daily folate requirement; however, the levels of folate in fermented milk have been shown to be higher (200 $\mu g\,L^{-1}$), due to folate production by *Streptococcus thermophilus* (LeBlanc *et al.*, 2011). This increased level of folate is due to the metabolic activity of LAB during the fermentation process. Indeed, a large number of LAB and *Bifidobacterium* spp. have been reported to produce folate (Smid *et al.*, 2001; Crittenden *et al.*, 2003), including two '*Bifidobacterium longum*' (presumed to be *Bifidobacterium longum* subsp. *longum*; Mattarelli *et al.*, 2008), *Lactobacillus acidophilus* and *Lactobacillus delbrueckii* subsp. *bulgaricus* strains, which were found to produce folate levels in reconstituted skimmed milk far in excess of the levels produced in synthetic media (M17, MRS) (Lin & Young, 2000). The levels of folate produced after 6 h of incubation varied from 53.9 $\mu g\,L^{-1}$ by *Lb. acidophilus* 4356 to 99.2 $\mu g\,L^{-1}$ by '*B. longum* B6' (presumed to be *B. longum* subsp. *longum* B6).

Bifidobacteria have been reported to produce folate in a number of studies. Strains of *Bifidobacterium bifidum* and '*Bifidobacterium infantis*' (presumed to be *Bifidobacterium longum* subsp. *infantis*; Mattarelli *et al.*, 2008) were classified as high folate accumulators. In addition to this, *Bifidobacterium adolescentis* and *Bifodobacterium pseudocatenulatum* strains significantly increased folate concentration in the faeces of healthy subjects (Rossi *et al.*, 2011). For instance, animal trials have reported that rats fed human milk solids had increased levels of both plasma folate and total caecal material folate, which coincided with a seven- and onefold increase in caecal and colonic *Bifidobacterium* spp. densities, respectively (Krause *et al.*, 1996). In addition, a study has reported that a number of other probiotic species of bifidobacteria, which included *Bifidobacterium animalis* subsp. *lactis*, *Bifidobacterium animalis* subsp. *animalis*, '*Bif infantis*' (presumed *Bif. longum* subsp. *infantis*) and *Bifidobacterium breve*, produce folate in reconstituted skimmed milk, with values ranging from 20 to 45 ng g^{-1} (Crittenden *et al.*, 2003). The use of mixed cultures, which include bifidobacteria, can also give further increases in the levels of folate. For example, the folate levels in probiotic yoghurt, containing *Bif. animalis* subsp. *lactis*, were over 33 ng g^{-1}, compared with levels under 25 ng g^{-1} in conventional yoghurt, while a mixed culture of *Bif. animalis* subsp. *animalis* and *Str. thermophilus* generated over 70 ng g^{-1} folate (Crittenden *et al.*, 2003). Other bifidobacteria strains, such as *Bif. breve*, *Bif. bifidum*, the *Bifidobacterium catenulatum* group (which includes *Bif. catenulatum* and *Bifidobacterium pseudocatenulatum*; Alegría *et al.*, 2014), *Bifidobacterium dentium* and '*Bif. longum*' (presumed to be *Bif. longum* subsp. *longum*), were able to produce folate in synthetic media (Rossi *et al.*, 2011).

In contrast to bifidobacteria, it has been reported that strains of *Lactobacillus* spp., used as both starter cultures and probiotic bacteria, generally utilise more folate than they produce. There are, however, exceptions with a number of *Lactobacillus* strains able to generate excess folate in the fermented dairy products, including *Lactobacillus plantarum*, *Lb. acidophilus* and *Lb. delbrueckii* subsp. *bulgaricus* (Crittenden *et al.*, 2003; Sybesma *et al.*, 2003a; LeBlanc *et al.*, 2011). Strains belonging to other *Lactobacillus* species, such as *Lactobacillus fermentum* and *Lb. reuteri*, are also able to produce significant amounts of folate *in vitro* (Cárdenas *et al.*, 2015; Presti *et al.*, 2015). The genetic determinants encoding the folate biosynthesis pathway were also identified in *Lactobacillus rossiae* (De Angelis *et al.*, 2014). Since many potential probiotic *Lactobacillus* strains utilise folate, many researchers have employed metabolic-engineering strategies to develop a folate-overproducing *Lactobacillus* spp. A recent study has described the transformation of a folate-utilising probiotic strain, *Lactobacillus gasseri*, into a folate producer through metabolic engineering (Wegkamp *et al.*, 2004).

The genetic determinants for folate biosynthesis by *Lactococcus lactis* subsp. *lactis* were identified (Sybesma *et al.*, 2003a) and, subsequently, introduced into *Lb. gasseri* on a broad-spectrum host range vector. The resulting strain was capable of producing up to 75 ng mL^{-1} total folate (Wegkamp *et al.*, 2004). This and other examples using *Lac. lactis* subsp. *lactis* (Sybesma *et al.*, 2003b, 2003c) show the potential which exists to develop genetically modified probiotic strains to overproduce folate, thereby possibly providing 100% of the required folate intake in the human diet instead of the current 15–20% (Hugenholtz *et al.*, 2002b). In addition, two strains of *Lac. lactis* spp. (CM22 and CM28) were reported to produce folate in skimmed milk and have a high GI survival rate (Divya & Nampoothiri, 2015).

In addition to the aforementioned probiotic strains, two *Leuconostoc* strains, namely '*Leuconostoc lactis*' (presumed to be *Leuconostoc mesenteroides* subsp. *lactis*) and '*Leuconostoc paramesenteroides*' (presumed to be *Leuconostoc mesenteroides* subsp. *paramesenteroides*; see https://www.ncbi.nlm.nih.gov/Taxonomy/Browser/wwwtax.

chttps://www.ncbi.nlm.nih.gov/Taxonomy/Browser/wwwtax.cgi?mode=Undef&id=1243&lvl=3&keep=1&srchmode=1&unlock), also produce folate (Sybesma *et al.*, 2003b). Even though *Leuconostoc* spp. may not be considered as being probiotic, they are, however, used in probiotic products (Goldin, 1998; Holzapfel *et al.*, 1998). Furthermore, there is increasing evidence to support the potential of propionibacteria as probiotic cultures (Zarate *et al.*, 2002a, 2002b, 2004), as this genus has been reported to produce bioactive fatty acids, such as conjugated linoleic acid (CLA) (Jiang *et al.*, 1998), vitamin B_{12} and folate (Hugenholtz *et al.*, 2002a). It has been reported that the levels of folate produced by different propionibacteria were greater than or equal to the levels produced by the well-known producer *Str. thermophilus* (Hugenholtz *et al.*, 2002a). Fourteen strains of propionibacteria were reported to produce folate with total levels varying from 9 to 78 ng mL^{-1}, with strains possessing varying abilities to excrete the vitamin (Hugenholtz *et al.*, 2002a).

9.2.3 *Vitamin B_{12}*

Vitamin B_{12}, also known as cobalamin, is an important cofactor for the metabolism of amino acids, carbohydrates, fatty acids and nucleic acids (Basavanna & Prapulla, 2013). Although very few food-grade micro-organisms possess the ability to produce this vitamin (Hugenholtz *et al.*, 2002a), there are some members of the genera *Propionibacterium*, *Bifidobacterium* and *Lactobacillus* that have been reported to have this rare activity (Deguchi *et al.*, 1985; Hugenholtz *et al.*, 2002a; Taranto *et al.*, 2003). As chemical synthesis of vitamin B_{12} is too expensive, industrial production takes place mostly via bacterial fermentation using strains such as *Pseudomonas denitrificans*, *Bacillus megaterium* and '*Propionibacterium freudenreichii*' (presumed to be *Propionibacterium freudenreichii* subsp. *freudenreichii*) (Burgess *et al.*, 2009). Therefore, since dairy propionibacteria are the only food-grade commercial producers of vitamin B_{12}, the pathway used for its biosynthesis has been well characterised and is reported to involve at least 25 steps (Burgess *et al.*, 2009). The fermentation process for the production of cobalamin can be optimised by controlling the aerobic and anaerobic phases of the process to obtain yields on glucose of up to 200 mg vitamin B_{12} kg^{-1} fermentation mesh (Hunik, 2002).

A number of *Bifidobacterium* species have also been reported to produce cobalamin, such as *Bif. adolescentis* (0.35 ng mL^{-1}), *Bif. bifidum* (0.65 ng mL^{-1}), *Bif. breve* (0.49 ng mL^{-1}), '*Bif. infantis*' (presumed to be *Bif. longum* subsp. *infantis*) (0.39 ng mL^{-1}) and '*Bif. longum*' (presumed to be *Bif. longum* subsp. *longum*) (0.46 ng mL^{-1}) (Deguchi *et al.*, 1985). In addition to propionibacteria and bifidobacteria, *Lb. reuteri* CRL1098, a probiotic strain that exhibits a hypocholesterolaemic effect in animal trials (Taranto *et al.*, 2000), has also been shown to produce cobalamin and reverse its deficiency in a murine model (Taranto *et al.*, 2003; LeBlanc *et al.*, 2011). One drawback associated with the use of this strain in the supplementation of foods with cobalamin is that the vitamin is not excreted from the cells; however, the genes encoding the pathways involved in cobalamin biosynthesis were identified, which may allow metabolic-engineering strategies to be exploited to overexpress the genes in a strain that would release the vitamin during gastrointestinal transit (Taranto *et al.*, 2003). Other strains, such as *Lb. reuteri* DSM 20016, JCM 1112 and ATCC 55730 and *Lb. rossiae* DSM 15814T, also contain the genes *cbi*, *cob* and *hem*, required for cobalamin biosynthesis (LeBlanc *et al.*, 2011; De Angelis *et al.*, 2014; Cárdenas *et al.*, 2015). Other LAB strains, such as *Lb. plantarum* PBS067, *Lb. fermentum* CFR 2195 (isolated from healthy infant faeces) and *Lb. rhamnosus* PBS070, have been reported to produce vitamin B_{12} in culture media (Basavanna & Prapulla, 2013; Presti *et al.*, 2015).

9.2.4 *Riboflavin and thiamine*

Deficiencies in riboflavin (vitamin B_2), a co-enzyme involved in numerous redox reactions, can lead to both liver and skin disorders (Russo *et al.*, 2014); while deficiencies in thiamine (vitamin B_1), also a cofactor of key metabolic enzymes, can cause changes in brain glucose metabolism (Hakim & Pappius, 1981). It has been reported that bifidobacteria can produce riboflavin and thiamine. '*Bifidobacterim longum*' and '*Bif. infantis*' (presumed to be *Bif. longum* subsp. *longum* and *Bif. longum* subsp. *infantis*) strains have been reported to increase the levels of both B vitamins during a 48 h fermentation in soymilk (Hou *et al.*, 2000; LeBlanc *et al.*, 2011). In these cases, the riboflavin content increased from 73.6 to 83.4 mg L^{-1} and 88.8 mg L^{-1} during the fermentations with '*Bif. infantis*' (presumed to be *Bif. longum* subsp. *infantis*) CCRC 14633 and '*Bif. longum* B6' (presumed to be *Bif. longum* subsp. *longum* B6), respectively, while the levels of thiamine increased from 3.3 to 3.8 mg L^{-1} and 3.7 mg L^{-1} during the fermentations, respectively.

Among lactobacilli, *Lb. fermentum* MTCC 8711 was found to produce 2.29 mg L^{-1} riboflavin after 24 h of growth in a defined medium (LeBlanc *et al.*, 2011). Other strains, such as *Lb. fermentum* CECT 5716 (isolated from human milk), were also reported to produce 0.62 mg L^{-1} of this vitamin (Cárdenas *et al.*, 2015). Strains *Lb. plantarum* LZ 227 and *Lb. rossiae* DSM 15814T are known to have the genetic pathway required for riboflavin biosynthesis (De Angelis *et al.*, 2014; Li *et al.*, 2016). In addition, strains *Lb. plantarum* CECT 8328 and *Lb. fermentum* CECT 8448 may be able to produce riboflavin in the human intestinal environment, to thus make this vitamin available to the host (Arena *et al.*, 2014).

As with folate, there is huge potential to develop genetically modified probiotic strains with the ability to overproduce riboflavin. Indeed, strategies have been employed to overproduce riboflavin in association with folate in a riboflavin-utilising strain of *Lac. lactis* subsp. *lactis* by overexpressing the riboflavin biosynthesis genes *ribG*, *ribH*, *ribB* and *ribA* (Sybesma *et al.*, 2004; LeBlanc *et al.*, 2011). Since some other LAB strains possess all four genes required for riboflavin biosynthesis (Burgess *et al.*, 2004), perhaps similar strategies could be applied to generate a genetically modified riboflavin-overproducing probiotic strain. For example, food-grade strains of *Lb. plantarum*, '*P. freudenreichii* B2336' (presumed to be *Propionibacterium freudenreichii* subsp. *freudenreichii* B2336) and *Leu. mesenteroides* spp. were reported to overproduce riboflavin in the presence of roseoflavin, an analogue of riboflavin (Burgess *et al.*, 2009; LeBlanc *et al.*, 2011).

9.2.5 *Vitamin K*

Vitamin K is an important cofactor, involved in the posttranslational carboxylation of glutamate residues to produce γ-carboxyglutamic acid (Gla), found in proteins involved in blood clotting, tissue calcification, and atherosclerotic plaque and tissues including bones and kidneys (LeBlanc *et al.*, 2011). This vitamin occurs in two forms: firstly, phylloquinone (vitamin K_1), which is present in green plants; and, secondly, menaquinone (MK) (vitamin K_2), which is produced by some intestinal bacteria (LeBlanc *et al.*, 2011). A range of LAB from a number of genera have been screened for the ability to produce MK. These included strains from the genera *Lactococcus*, *Lactobacillus*, *Enterococcus*, *Bifidobacterium*, *Leuconostoc* and *Streptococcus* (Morishita *et al.*, 1999). Four lactococcal strains and *Lac. lactis* subsp. *lactis* YIT 3001, which produced 648 nmol quinones g^{-1} lyophilised cells, yielded in excess of 250 nmol quinones g^{-1} lyophilised cells. While lactococci are not intestinal microbes *per se*, they are applied in probiotic products (Goldin, 1998; Holzapfel *et al.*, 1998), and can be incorporated into 'multi-species' products, which have been defined as 'containing strains of different probiotic species that belong to one or preferentially more genera', such as *Lb. acidophilus*, '*Bif. longum*' (presumed to be *Bif. longum* subsp. *longum*), *Enterococcus faecium* and *Lac. lactis* subsp. *lactis* (Timmerman *et al.*, 2004).

9.3 Exopolysaccharides (EPS) production by probiotic bacteria

9.3.1 *Introduction*

A number of LAB, propionibacteria and bifidobacteria can synthesise EPS, which are excreted from the bacterial cells and which may or may not be loosely attached to the cell wall (Laws *et al.*, 2001). In cases where they are actually bound to the cell surface, they are referred to as capsular polysaccharides (Harutoshi, 2013). EPS can contribute to the improved stability, rheology and texture of fermented dairy products, and may also offer protection to bacterial cells against bacteriophage attack, desiccation and osmotic stress (Ruas-Madiedo *et al.*, 2002). In addition to the technological characteristics they confer on dairy products, EPS have been reported to exert a number of beneficial health effects, including prebiotic, immunostimulatory, antiviral, antioxidant, anti-tumoural and blood cholesterol-lowering activities (Nakajima *et al.*, 1992; Hidalgo-Cantabrana *et al.*, 2012; Patel *et al.*, 2012; Harutoshi, 2013; Li *et al.*, 2014). A large number of strains from the genus *Lactobacillus* have been reported to produce EPS (Ricciardi & Clementi, 2000; Ruas-Madiedo *et al.*, 2002), while it is a less common phenomenon associated with bifidobacteria (Hidalgo-Cantabrana *et al.*, 2012). Some very well-studied examples are *Bif. breve* UCC 2003 (Fanning *et al.*, 2012), some *Bif. animalis* subsp. *lactis* strains (Hidalgo-Cantabrana *et al.*, 2014), '*Bif. infantis* ATCC 15697' (presumed to be *Bif. longum* subsp. *infantis* ATCC 15697), *Bif. catenulatum* YIT4016, '*Bif. longum* YIT 4028' (presumed to be *Bif. longum* subsp. *longum* YIT 4028) (Ruas-Madiedo *et al.*, 2009) and *Bif. bifidum* WBIN03 (Li *et al.*, 2014).

9.3.2 *Classification of exopolysaccharides*

EPS can be categorised into two types: (a) homopolysaccharides (HoPS), which consist of a single type of monosaccharide; and (b) heteropolysaccharides (HePS), which consist of repeating units of polysaccharides and non-carbohydrate units, including phosphate, acetyl and glycerol (Ruas-Madiedo *et al.*, 2002; Salazar *et al.*, 2016). The HoPS from microbial groups are categorised into fructans (e.g. levan and inulin types) and glucans (e.g. dextran, mutan, alternan and reuteran), whereas HePS comprise gellan, xanthan and kefiran (Ruas-Madiedo *et al.*, 2009; Patel *et al.*, 2012). HoPS are synthesised by excreted or anchored enzymes (glycansucrases) following the assembly of monosaccharides outside the cell, and they require the substrate sucrose, which provides the energy for elongation (Duboc & Mollet, 2001; Jolly *et al.*, 2002; Ruas-Madiedo *et al.*, 2002). In contrast, the precursors for HePS are synthesised within the cell involving isoprenoid glycosyl carrier lipids and, subsequently, transferred across the membrane by glycosyltransferases and assembled extracellularly (De Vuyst *et al.*, 2001; Ruas-Madiedo *et al.*, 2002; Salazar *et al.*, 2016).

HePS generally contain D-glucose, D-galactose and L-rhamnose and in some cases *N*-acetylglucosamine, *N*-acetylgalactosamine or glucuronic acid (Ruas-Madiedo *et al.*, 2002). Their synthesis is closely related to carbon metabolism and the production of nucleotide sugars in the cell (Hidalgo-Cantabrana *et al.*, 2014; Salazar *et al.*, 2016).

9.3.3 *Health benefits of exopolysaccharides*

The majority of studies on EPS tend to focus on their technological aspects rather than their physiological benefits, with much of the work focusing on EPS structure, genetics, yield improvement and industrial applications. This section will focus on the potential health benefits associated with the consumption of EPS produced by LAB and bifidobacteria, and also the potential of EPS-producing bacteria as probiotic cultures and their incorporation into functional foods.

Prebiotic effect of exopolysaccharides

One of the suggested health benefits of EPS consumption is its prebiotic effect. A prebiotic is described as 'a non-digestible food ingredient that beneficially affects the host by selectively stimulating the growth and/or activity of one or a limited number of bacteria in the colon, and thus improves host health' (Kumar *et al.*, 2015; Linares *et al.*, 2016); recently updated to a substrate that is selectively utilised by host micro-organisms conferring a health benefit' (Gibson *et al.*, 2017)'. Thus, it is essential that prebiotics are not degraded in the stomach and small intestine (Crociani *et al.*, 1994). Indeed, EPS may meet this criterion as they have been reported to withstand *in vivo* passage through the GI tract using a model simulating the human digestive process. Feed porridge containing 102 mg of EPS produced by *Lactococcus lactis* subsp. *cremoris* NZ 4010 given to rats resulted in the recovery of 96% (98 mg EPS 4.44 g^{-1} of faeces) of the EPS in their faeces, demonstrating that it was not degraded in the GI tract (Looijesteijn *et al.*, 2001).

It was also reported that EPS produced by strains from different genera were biodegraded to varying extents, with EPS produced by *Lactobacillus sakei* 0-1 and *Lactobacillus helveticus* Lh59 being among the most durable (Ruijssenaars *et al.*, 2000). Moreover, it has been reported that the levan-type EPS produced by *Lactobacillus sanfranciscensis* TMW 1.392 supported the growth of *Bif. bifidum*, *Bif. breve*, '*Bif. infantis*' (presumed to be *Bif. longum* subsp. *infantis*) and *Bif. adolescentis* (Korakli *et al.*, 2002), strains that are commonly employed as probiotic cultures (Goldin, 1998; Holzapfel *et al.*, 1998). During pH-controlled fermentation, all of the *Bifidobacterium* spp. exhibited diauxic growth, with fructose being metabolised first, and the EPS being utilised after a second lag phase (Korakli *et al.*, 2002). These results would suggest that there is potential for EPS to be used as prebiotics; however, further *in vivo* studies regarding the degradability of EPS in the stomach and small intestine and its utilisation by gut microbiota are required.

Exopolysaccharides and intestinal health

Microbially biosynthesised EPS have been associated with the promotion of intestinal health and prevention of intestinal disease. Recently, Hidalgo-Cantabrana *et al.* (2016) reported that the EPS-producing bacterial species *Bif. animalis* subsp. *lactis* has the capability to survive murine GI tract transit and improve clinical outcomes in a chemically induced colitis model. By a similar mechanism, fermented milk made with EPS-producing *Str. thermophilus* strains prevented chronic gastritis in an *in vivo* model of chronic gastritis (BALB/c mice) (Rodríguez *et al.*, 2009).

EPS can also exert their function of preventing adhesion of potential pathogens to the intestinal mucosa through epithelial barrier maintenance and competitive exclusion. For example, the EPS produced by the natural dairy isolate *Lactobacillus paracasei* subsp. *paracasei* BGSJ 2-8 decreased the association of *Escherichia coli* with Caco-2 cells (Živković *et al.*, 2016). Oral administration of yoghurt fermented with *Lb. delbrueckii* ssp. *bulgaricus* OLL 1073R-1 and its EPS protected mice against influenza virus infection (Nagai *et al.*, 2011). A pathogen protection effect was also described for the EPS produced by *Bif. breve* UCC 2003 (Fanning *et al.*, 2012).

Immunostimulatory activity of exopolysaccharides

A number of reports have suggested that LAB can induce immunological responses in macrophages and T cells (Marin *et al.*, 1998), and that the EPS they produce can play a role in eliciting these immunopotentiating activities. The EPS of strains from a number of food-grade genera, including *Bifidobacterium*, *Lactococcus* and *Lactobacillus*, have been reported to be immunostimulatory. Some recent examples include the EPS isolated from '*Lb. paracasei* DG' (presumed to be *Lb. paracasei* subsp. *paracasei* DG), which has immunostimulatory properties and activates THP-1 human monocytic cells; these properties may therefore contribute to the ability of this probiotic strain to interact with the immune system (Balzaretti *et al.*, 2016). EPS-overproducing '*Lb. paracasei* KB28' (presumed to be *Lb. paracasei* subsp. *paracasei* KB28), which was isolated from kimchi (a Korean fermented product), induced cytokines in mouse peritoneal macrophages via modulation of NF-κβ (nuclear factor kappa activated B cells) and mitogen-activated protein kinases (MAPKs) (Kang *et al.*, 2011). Similarly, the EPS fraction from *Pediococcus pentosaceus* KFT18 induced immunostimulatory activity in macrophages and immunosuppressed mice (Shin *et al.*, 2016). A similar mechanism was described for the EPS isolated from *Lactobacillus rhamnosus* KF5, which exerted splenocyte proliferation *in vitro*, indicating a potential immunomodulatory activity (Shao *et al.*, 2014). Other EPS isolated from *Lb. rhamnosus* KL37 was shown to have immunoregulatory potential against the production of inflammatory mediators by mouse macrophages (Ciszek-Lenda *et al.*, 2011). The EPS derived from yoghurt fermented with *Lb. delbrueckii* ssp. *bulgaricus* OLL 1073R-1 enhanced natural killer cell activation (Makino *et al.*, 2016). Furthermore, the EPS produced by *Lb. fermentum* Lf2 acted as a moderate immunomodulator, modifying immunoglobulin A (s-IgA, or IgA) and interleukin-6 (IL6) levels in the small intestine when added to yoghurt and milk, respectively (Ale *et al.*, 2016). *Leuconostoc mesenteroides* subsp. *mesenteroides* was found to produce large amounts of EPS with IgA-inducing activity; however, dietary supplementation with strain NTM048 induced a significant increase in the faecal IgA content and plasma IgA levels of BALB/cA mice (Matsuzaki *et al.*, 2014).

Among those produced by bifidobacteria, some EPS have also been reported to have the capacity to modulate the immune system. For example, EPS polymers produced by *Bif. animalis* subsp. *lactis* strains can elicit different responses from immune cells from blood and gut-associated lymphoid tissue (Hidalgo-Cantabrana *et al.*, 2014), and surface EPS produced by *Bif. breve* UCC 2003 was shown to be a beneficial trait mediating commensal–host interaction through immune modulation and pathogen protection (Fanning *et al.*, 2012). The water-soluble polysaccharide fraction of *Bif. adolescentis* M101-4 cells was reported to increase the [^{3}H] thymidine uptake of murine splenocytes and Peyer's patches, an indication of cell proliferation, relative to whole cells (Hosono *et al.*, 1997).

Anti-tumoural activity of exopolysaccharides

It has been suggested that yoghurt demonstrates anti-tumour activity (Perdigon *et al.*, 1998), and that one of the mechanisms may involve the EPS produced by the cultures during yoghurt production (Kitazawa *et al.*, 2000). Indeed, the extracts from milk fermented by *Lb. delbrueckii* subsp. *bulgaricus* and *Str. thermophilus* were shown to have anti-mutagenic activity (Bodana & Rao, 1990). In a separate study, EPS isolated from the supernatant of '*Bif. longum* PS$^+$' (presumed to be *Bif. longum* subsp. *longum* PS$^+$) has been reported to exhibit an anti-mutagenic effect against a known mutagen (Sreekumar & Hosono, 1998) in a study where 60 μL of crude polysaccharide solution (1.323 g L^{-1}) caused 89.4% inhibition of the mutagenicity of the mutagen Trp-P-1. Furthermore, *Lb. acidophilus* was reported to produce an EPS able to inhibit the expression of genes involved in tumour angiogenesis and survival in colon cancer cell lines *in vitro* (Deepak *et al.*, 2016). Novel cell-bound EPS isolated from *Lb. helveticus* MB2-1 and *Lb. plantarum* 70810 significantly inhibited the proliferation of HepG-2, BGC-823 and, in particular, HT-29 cancer cells (Wang *et al.*, 2014; Li *et al.*, 2015).

Exopolysaccharides and cholesterol-lowering effects

The EPS from *Lactococcus lactis* subsp. *cremoris* have been reported to have a cholesterol-lowering effect in rats (Nakajima *et al.*, 1992). In this study, the serum cholesterol concentration of rats fed ropy fermented milk (84.0 mg d^{-1}) was lower than that of rats fed non-ropy fermented milk (95.7 mg d^{-1}) or milk acidified with lactic acid (102.0 mg d^{-1}). In addition, the ratio of high-density lipoprotein (HDL) cholesterol to total cholesterol was significantly higher in rats fed ropy fermented milk compared with the other two products. *Lactobacillus mucosae* DPC 6426 has been reported to synthesise EPS with cholesterol-lowering properties in an animal model of lipid-driven atherosclerosis (London *et al.*, 2014; Ryan *et al.*, 2015). Similarly, EPS from *Lb. rhamnosus* GG has an anti-obesity effect. Fat pads of mice injected with EPS (50 mg kg^{-1}) every 2 days for 2 weeks became significantly reduced in size, with much smaller adipocytes. The levels of triacylglycerol and cholesterol ester in liver and serum were decreased in EPS-injected mice (Zhang *et al.*, 2016).

9.4 Production of bacteriocins by probiotic cultures

9.4.1 *Background*

Bacteriocins are ribosomally synthesised antimicrobial peptides produced by one bacterium that are active against other bacteria. It has been suggested that one of the desirable properties of a probiotic strain is the ability to produce antimicrobial substances, such as bacteriocins, which potentially offers a competitive advantage in colonisation and competition in the GI tract (Hegarty *et al.*, 2016). The best-known example is nisin, a broad host-range bacteriocin produced by many strains of *Lac. lactis* subsp. *lactis* and considered to be the prototype LAB bacteriocin. It was first described in 1928, when it was observed to have inhibitory effect to other LAB (Rogers & Whittier, 1928). Nisin has since been studied extensively and, in 1988, was awarded US Food and Drug Administration (FDA) approval for its use as a biopreservative in many processed foods. This major breakthrough, with nisin being the first bacteriocin to gain widespread commercial application, led to considerable interest in bacteriocins produced by generally regarded as safe (GRAS) organisms. As a result, the field has grown dramatically, and has led to the discovery and detailed characterisation of a great number of bacteriocins from LAB.

Concomitant with the discovery of these new bacteriocins, several intestinal and/or probiotic strains have been shown to display the ability to produce these inhibitory peptides. In this respect, most of the probiotic bacteriocins characterised to date are of *Lactobacillus* origin. Knowledge on bacteriocin producers *in situ* and their function in the gut of healthy animals is still limited due to a scarcity of *in vivo* studies (Umu *et al.*, 2016).

However, looking at their influence on the intestinal gut microbiota, animal studies have shown reduction in *Enterobacteriaceae* due to these antimicrobial metabolites (Gardiner *et al.*, 2004), and clinical trials reported that humans fed *Lactobacillus johnsonii* fermented milk had significantly reduced density of *Helicobacter pylori*, a pathogen associated with stomach ulcers (Felley *et al.*, 2001). Since the most widely used cultures for probiotic applications belong to the genera *Lactobacillus* and *Bifidobacterium*, the antimicrobial activities of species belonging to these genera are the topic of this section.

9.4.2 *Production of antimicrobials as a probiotic trait*

There are many examples where bacteriocins have been shown to have a dominant influence on complex microbial populations. A very good example of this is in the oral cavity, where production of the bacteriocin mutacin by *Streptococcus mutans* gives these bacteria a selective advantage in persisting, colonising and aggressively displacing the indigenous *Str. mutans* population (Hillman, 2002). It was found that 14 years post treatment of a concentrated cell suspension of a recombinant non-acidogenic *Str. mutans* mutacin-producing strain, all of the available subjects remained colonised by this strain and no other mutans streptococci were observed. This example clearly illustrates the potential of bacteriocins within complex ecosystems. Furthermore, bacteriocins play a role in achieving predominance over other micro-organisms in food fermentation systems. When the bacteriocin producer *Lb. plantarum* was used to ferment Spanish-style green olives, the bacteriocin producer outnumbered all the other individual naturally occurring *Lactobacillus* populations over the course of the 12-week fermentation (Ruiz-Barba *et al.*, 1994; Ruiz-Barba & Jimenz-Diaz, 1994). In contrast, the bacteriocin-negative isolate of the strain was not detected after 7 weeks. Similarly, lacticin 3147 has shown evidence of control of food pathogens in a variety of food fermentations. Ryan *et al.* (1996) demonstrated that lacticin 3147-producing starter cultures provided a means to control the microbiota developing in ripened fermented products. There is ample evidence to show that bacteriocins play a fundamental role in influencing complex microbial populations. There is a constant search for novel probiotic strains with bacteriocin traits (e.g. a selective advantage to persist and colonise), with strains being actively screened every day. Therefore, mining the gut microbiota metagenome for novel bacteriocins and antimicrobial compounds presents a rational approach for selection of new probiotic strains.

9.4.3 *Classification of bacteriocins*

Generally, bacteriocins are ribosomally synthesised polypeptides, which are normally posttranslationally modified to some degree, with the secreted mature peptides usually ranging in size from 20 to 60 amino acids, and possessing bactericidal activity (Joerger & Klaenhammer, 1986). Almost all bacteriocin peptides have a net charge at neutral or slightly acidic pH, and they usually contain stretches of the molecular sequence that are hydrophobic and/or amphiphilic (Eijsink *et al.*, 2002). As the list of bacteriocins continues to grow, a significant diversity in their structure and activity is evident, which has meant that their classification continues to be updated. Current classification divides bacteriocins into three main classes, which will be discussed in this section (see also Tamime, 2005).

Class I bacteriocins – lantibiotics

Class I bacteriocins, known as the lantibiotics, are defined as: (a) small peptides (21–38 amino acids; <5 kDa), (b) posttranslationally modified, (c) heat-stable peptides and (d) containing the unusual amino acids lanthionine or β-methyllanthionine, which form characteristic intramolecular (thioether) ring structures (McAuliffe *et al.*, 2000; Twomey *et al.*, 2002; Nes *et al.*, 2007; Rea *et al.*, 2011). Many lantibiotics also possess other modified residues, such as dehydro amino acids, D-alanine residues (Skaugen *et al.*, 1994; Ryan *et al.*, 1999), N-terminal α-keto amines (Kellner *et al.*, 1991) and oxidative decarboxylation of C-terminal cysteine, to yield a C-terminal S-aminovinylcysteine (Schnell *et al.*, 1988; Bierbaum *et al.*, 1996). Lantibiotics were initially broadly grouped according to structure: type A is elongated amphiphilic peptides, and type B is more compact and globular (Kellner *et al.*, 1991). Twomey *et al.* (2002) further subdivided lantibiotics into six subgroups based on primary sequence comparisons. Most of the characterised lantibiotics appear to have a common mode of action in that they dissipate the proton motive force in target organisms through the formation of pores in the cytoplasmic membrane (Garcia-Garcera *et al.*, 1993; Montville & Bruno, 1994; Abee *et al.*, 1995; Moll *et al.*, 1996). The genetic determinants of several lantibiotics from LAB have now been fully determined, including nisin (Kaletta & Entian, 1989; Kuipers *et al.*, 1993), lacticin 3147 (Dougherty *et al.*, 1998; Ryan *et al.*, 1999), cytolysin (Gilmore *et al.*, 1994), lactocin S (Skaugen *et al.*, 1994), lacticin 481 (van den Hooven *et al.*, 1996) and plantaracin C (Holo *et al.*, 2001).

The most representative lantibiotics are nisin A, lacticins (produced by *Lac. lactis* subsp. *lactis*), plantaricins (produced by *Lb. plantarum*) and lactocin S (produced by *Lb. sakei*) (Mortvedt *et al.*, 1991; Gonzalez *et al.*, 1994; Holo *et al.*, 2001; Nes *et al.*, 2007; Rea *et al.*, 2011). Plantaricin C, a broad-spectrum bacteriocin produced by *Lb. plantarum* isolated from ripening cheese (Gonzalez *et al.*, 1994), is a 27 amino acid peptide with a linear N-terminal end and a globular C-terminus. Structure similarity with lacticin 481 resulted in its designation as a member of subgroup II lacticin 481 (Twomey *et al.*, 2002). In contrast, plantaricin W is a two-component bacteriocin, with both peptides displaying inherent antimicrobial activity (Holo *et al.*, 2001). Lactocin S produced by *Lb. sakei* L45 is not sub-grouped with any other lantibiotic group as its primary sequence shares little similarity with any other known lantibiotics. Interestingly, molecular characterisation of this bacteriocin demonstrated that it contained D-alanine, which the authors suggest is derived from serine (Skaugen *et al.*, 1994).

From a probiotic application point of view, the lantibiotic class are possibly the most interesting group of bacteriocins, particularly since their inhibition spectra vary from medium to very broad.

Class II bacteriocins

The class II bacteriocins constitute a large and diverse group of intermediate peptides, (generally 30–60 amino acids, <15 kDa), non-posttranslationally modified, heat stable and consisting of standard amino acid residues (Rea *et al.*, 2011). The group includes three subclasses:

- *Subclass IIa*: The members of this subclass are characterised by showing high anti-listerial activity. These bacteriocins include 37–48 amino acid residues in their molecular structure, their N-terminal portion has a pleated sheet configuration and the C-terminal portion contains one or two α-helices. Regarding their mechanism of action, they reach the cell membrane of the target cell via the C-terminus promoting pore formation and consequent dissipation of proton motive force that causes high adenosine triphosphate (ATP) consumption and consequently death (Güllüce *et al.*, 2013). Pediocin, enterocin and sakacin are the most representative examples of subclass IIa bacteriocins. Bifidocin B, a member of this subclass that is produced by *B. bifidum* NCFB 1454, is one of the few bifidobacteria bacteriocins characterised fully to date (Yildirim & Johnson, 1998; Yildirim *et al.*, 1999). Bifidocin B displays activity against a number of foodborne pathogens and food spoilage bacteria, such as *Listeria*, *Enterococcus*, *Bacillus*, *Lactobacillus*, *Leuconostoc* and *Pediococcus* species. Curing experiments indicated that the genetic determinants of bifidocin B are associated with an 8 Kb plasmid.

- *Subclass IIb*: This subclass includes heterodimeric bacteriocins that consist of two peptides. Members of this subclass meet three criteria: (a) full antimicrobial activity needs both peptides – the individual peptides show little or no activity; (b) one immunity protein is sufficient to get immunity; and (c) the genetic organisation of the bacteriocin system includes two sequential bacteriocin structural genes encoding the individual peptides and a single immunity gene. Lactococcin G was the first bacteriocin of this group that was discovered; its antimicrobial activity depends on both α- and β-peptides. Plantaricin and lactacin F are also other important representative examples (Rea *et al.*, 2011; Güllüce *et al.*, 2013). In addition, lactacin F, produced by *Lb. johnsonii* VPI 11088 (previously classified as *Lb. acidophilus* 11088), has a relatively narrow spectrum of inhibition, being bactericidal against only closely related bacteria (Abee *et al.*, 1994). Plantaricin NC8 is produced by *Lb. plantarum* NC8 only when co-cultured with other Gram-positive bacteria, such as *Pediococcus*, *Lactococcus* and *Listeria* species (Maldonado *et al.*, 2004a, 2004b). The ABP-118, a novel two-component bacteriocin produced by the probiotic bacterium *Lactobacillus salivarius* subsp. *salivarius* UCC 118, was purified and characterised (Flynn *et al.*, 2002). *Lactobacillus salivarius* subsp. *salivarius* UCC 118 is a human intestinal isolate chromosomally encoding the genetic determinants for ABP-118. This bacteriocin exhibits a broad spectrum of inhibition, and it is capable of inhibiting medically significant pathogens, such as *Bacillus*, *Listeria*, *Enterococcus* and *Staphylococcus* species. Another bacteriocin produced by *Lb. salivarius* subsp. *salivarius* is salivaricin P (Rea *et al.*, 2011). Although the health-contributory effects of this bacteriocin-producing probiotic are yet to be assessed, in view of its spectrum, it should give the bacterium a competitive advantage in the complex microbial environment of the gut.
- *Subclass IIc*: Bacteriocins of this subclass are unique in having a circular structure associated with a covalent bond between C- and N-terminals that causes a head to tail cyclic shape of the peptides. The main representatives of this subclass are gassericin A (from *Lb. gasseri* LA39), acidocin B (from *Lb. acidophilus* M46), lactocyclicin Q (from *Lactococcus* spp. QU12) and reutericin 6 (from *Lb. reuteri* LA6) (Rea *et al.*, 2011).

Class III bacteriocins

Class III bacteriocins are large (>30 kDa), heat-labile protein peptides of which very few have been described as being produced by *Lactobacillus* and *Bifidobacterium*. An important criterion for members of this group is that class III bacteriocins have complex activity and a protein structure that provides a totally different mechanism action from other bacteriocins, in which they induce lysis of the cell wall of the target micro-organism. In the mode of action process, the N-terminal portion of the molecule acts as an endopeptidase and the C-terminal portion recognises the target cell (Nes *et al.*, 2007; Güllüce *et al.*, 2013). These bacteriocins are less well characterised; examples include helveticin J produced by *Lb. helveticus* 481 (Joerger & Klaenhammer, 1986), helveticin V produced by *Lb. helveticus* 1829 (Vaughan *et al.*, 1992) and lactacin B produced by *Lb. acidophilus* N2 (Barefoot & Klaenhammer, 1984). They all share a narrow inhibitory spectrum, only antagonistic to closely related species.

9.4.4 *Antimicrobial potential of* Lactobacillus *spp.*

The application of antimicrobial agents produced by *Lactobacillus* spp. has been demonstrated in many food systems, which in many cases demonstrates the effectiveness of these potent inhibitors to control undesirable bacteria (Table 9.1). The most prominent bacteriocin-producer species are *Lb. plantarum* (some strains can produce plantaricin C, plantaricin S, plantaricin ER or plantaricin JK), '*Lb. sakei*' [presumed to be *Lb. sakei* subsp. *sakei*; (can produce different types of sakacins or lactocin S)], *Lb. acidophilus* (can synthesise acidocin), *Lb. gasseri* (producing different types of gassericins) and *Lb. casei* (caseicin 80 and lactocin 705). To a lesser extent, production of bacteriocins has also been described in other *Lactobacillus* species (e.g. *Lactobacillus curvatus*, *Lactobacillus crispatus*, *Lb. rhamnosus*, *Lb. helveticus*, *Lb. johnsonii*, *Lb. reuteri*, *Lb. salivarius* subsp. *salivarius*, *Lactobacillus amylovorus* and *Lactobacillus bavaricus*) (Table 9.1).

With regard to bacteriocin activity, lactocin 705, for example, reduced numbers of the foodborne pathogen *Listeria monocytogenes* in a ground meat slurry from 1.4×10^4 to 1×10^2 colony-forming units (cfu) mL^{-1} after 24 h (Vignolo *et al.*, 1996; Palacios *et al.*, 1999). In addition, the same food pathogen was inhibited by the bacteriocin producer *Lactobacillus bavaricus* in three different beef systems at refrigeration temperatures (Winkowski *et al.*, 1993). Acidocin A, which was produced by the starter culture (*Lb. acidophilus* TK9201) for the production of fermented milk, inhibited food spoilage bacteria and food pathogens including *Propionibacterium* spp. and *Enterococcus faecalis* (Kanatani *et al.*, 1995). Another example is plantaricin C, which was isolated from matured cheese, and which was active against a wide range of Gram-positive bacteria including *Clostridium* spp. (Gonzalez *et al.*, 1994). Furthermore, gassericin A, produced by *Lb. gasseri* LA39 and isolated from human infant faeces, showed bactericidal activity against pathogens, such as *Listeria monocytogenes*, *Bacillus cereus* and *Staphylococcus aureus* (Kawai *et al.*, 1994). This intestinal strain is permitted in fermented products to promote health and inhibit pathogens, even in Japan, whose Food Hygiene Law prohibits the use of pure bacteriocins (including nisin) as bio-preservatives to prolong the shelf life of food (Saito, 2004). In addition, lactobacilli strains, such as *Lb. johnsonii* LA-1, *Lb. gasseri* K7 and *Lb. acidophilus* IBB 801, exhibited bacteriocins production when grown in milk supplemented with yeast extract (Majhenic *et al.*, 2003; Avonts *et al.*, 2004).

Several *in vitro* studies have also examined the antimicrobial potential of *Lactobacillus* spp. towards clinically significant pathogens and infections. For example, *Lactobacillus*

Table 9.1 Some examples of bacteriocins produced by lactic acid bacteria, propionibacteria and bifidobacteria.

Micro-organism and strain identification	Bacteriocin	References
Lactococcus lactis subsp. *lactis*[1]	Nisin A and Nisin Z	De Vuyst & Vandamme (1994)
Lac. lactis subsp. *lactis* ADRIA 85L030	Lactococcin DR	Dufour *et al.* (1991)
CNRZ 481	Lacticin 481	Piard *et al.* (1992)
DPC 3147	Lacticin 3147 (LtnA1 and LtnA2)	Dougherty *et al.* (1998)
IPLA 972	Lactococcin 972	Martínez *et al.* (1999)
9B4, LMG 2130, WM 4	Lactococcins A and B	Stoddard *et al.* (1992)
UL720	Diacetin B	Ali *et al.* (1995)
9B4	Lactococcin MN (LcnM and LcnN)	Van Belkmun *et al.* (1991)
LMG 2081	Lactococcin G (LcnGα and LcnGβ)	Nissen-Meyer *et al.* (1992)
Lactococcus spp. QU 12	Lactocyclicin Q	Rea *et al.* (2011)
Lactobacillus plantarum LCPO 10	Plantaricin S (Plsα and Plsβ)	Jimenez-Diaz *et al.* (1995)
C 11	Plantaricin EF (PlnE and PlnF)	Anderssen *el al.* (1998)
C 11	Plantaricin JK (PlnJ and PlnK)	
LL 441	Plantaricin C	Gonzalez *et al.* (1994)
Lactobacillus sakei LB 706	Sakacin A	Holck *et al.* (1992)
LTH 673	Sakacin P	Huhne *et al.* (1996)
LB 674	Sakacin 674	Tichaczek *et al.* (1993)
L 45	Lactocin S	Skaugen *et al.* (1994)
Lactobacillus acidophilus TK 8192	Acidocin 8912	Kanatani *et al.* (1995)
TK 9201	Acidocin A	
M 46	Acidocin B	Leer *et al.* (1995)
LF 221	Peptide A/B	Bogovic-Matijasic *et al.* (1998)
Lactobacillus casei CRL 705	Lactocin 705	Cuozzo *et al.* (2000)
Lb. casei B 80	Caseicin 80	Yang *et al.* (2014)
Lactobacillus curvatus LTH 1174	Curvacin A	Tichaczek *et al.* (1993)
Lactobacillus fermentum CCRC 14018	Fermentcin B	Yan & Lee (1997)
Lactobacillus crispatus JCM 2009	Crispacin A	Tahara & Kanatani (1997)
Lactobacillus rhamnosus 68	Rhamnosin A	Dimitrijević *et al.* (2009)
Lactobacillus helveticus 481	Helveticin J	Joerger & Klaenhammer (1986)
Lactobacillus gasseri HCM 2124	Gassericin B3	Tahara *et al.* (1997)
LA 39	Gassericin A	Kawai *et al.* (1994)
LF 221	Acidocin LF221B (Gassericin K7 B)	Majhenic *et al.* (2004)

(Continued)

Table 9.1 (Continued)

Micro-organism and strain identification	Bacteriocin	References
Lactobacillus johnsonii VPI 11088	Lactacin F (LafA and LafX)	Güllüce *et al.* (2013)
Lactobacillus reuteri LA 6	Reutericin 6	Nes *et al.* (2007)
Lactobacillus bavaricus MI 401	Bavaricin A	Larsen *et al.* (1993)
Lactobacillus salivarius UCC 118	ABP-118	Flynn *et al.* (2002)
DPC 6005	Salivaricin P	Barrett *et al.* (2007)
Lactobacillus amylovorus LMG P-13139	Lactobin A	Contreras *et al.* (1997)
Streptococcus salivarius 20P3	Salivaricin A	Ross *et al.* (1993)
Streptococcus thermophilus SPi13	Thermophilin 13 (ThmA/ ThmB)	Güllüce *et al.* (2013)
SBT 1277	Thermophilin 1277	Kabuki *et al.* (2007)
Leuconostoc spp. MF 215B	Leucocin H (α and β)	Blom *et al.* (1999)
Leuconostoc gelidum UAL 187	Leucocin	Savadogo *et al.* (2006)
Leuconostoc mesenteroides spp. TA 33a	Leucocin B-TA33a	Papathanasopoulos *et al.* (1998)
Y 105	Mesentericin Y105	Héchard *et al.* (1992)
FR 52	Mesentericin 52B	Héchard *et al.* (1999)
B 105	Mesentericin B105	Revol-Junelles *et al.* (1996)
Pediococcus pentosaceus FBB 61/L-7230	Pediocin A	Savadogo *et al.* (2006)
Pediococcus damnosus NCFB 1832	Pediocin PD-1	Nes *et al.* (2007)
Pediococcus acidilactici PAC-1.0	Pediocin PA1	Henderson *et al.* (1992)
H	Pediocin AcH	Bhunia *et al.* (1988)
Propionibacterium freudenreichii LMGT 2946[2]	Propionicin F	Nes *et al.* (2007)
Bifidobacterium bifidum NCFB 1454	Bifidocin B	Yildirim *et al.* (1998)
Bifidobacterium infantis BCRC 14602[3]	Bifidin I	Cheikhyoussef *et al.* (2010)
Bifidobacterium lactis BB-12[4]	Bifilact BB-12	
Bifidobacterium longum BB-46[5]	Bifilong BB-46	
Bifidobacterium thermophilum RBL 67	Thermophilicin B67	

[1] Strain identification was not reported.

[2] '*Propionibacterium freudenreichii* LMGT 2946' (presumed to be *Propionibacterium freudenreichii* subsp. *freudenreichii* LMGT 2946).

[3] '*Bifidobacterium infantis* BCRC 14602' (presumed to be *Bifidobacterium longum* subsp. *infantis* BCRC 14602).

[4] '*Bifidobacterium lactis* BB-12' (presumed to be *Bifidobacterium animalis* subsp. *lactis* BB-12).

[5] '*Bifidobacterium longum* BB-46' (presumed to be *Bifidobacterium longum* subsp. *longum* BB-46).

spent culture supernatants significantly inhibited proliferation of a protozoan (*Giardia intestinalis*) that causes diarrhoeal disease worldwide (Perez *et al.*, 2001). In addition, the human *Lb. acidophilus* LB strain displayed antagonistic activity against both Gram-positive and Gram-negative pathogens, including antagonistic activity against *H. pylori* and *Salmonella* Typhimurium infection *in vitro* and *in vivo* (Coconnier *et al.*, 1997, 1998, 2000). Interestingly, certain *Lactobacillus* strains have the ability to interfere with the adherence and growth of uropathogenic bacteria (McGroarty & Reid, 1988). This interaction is believed to be important in the maintenance of a normal urogenital microbiota and in the prevention of infection in females. For instance, vaginal *Lactobacillus* isolates displaying bacteriocin-like substances inhibitory towards *Ent. faecalis*, *Ent. faecium* and *Neisseria gonorrhoeae* (Ocana *et al.*, 1999) as well as *Gardnerella vaginalis* (Aroutcheva *et al.*, 2001; Alpay-Karaoglu, 2003) show potential application in probiotic products to prevent urogenital infections. Combinations of *Lactobacillus* strains were selected for preparation of vaginal tablets to treat such infections. The performance of the formulation was optimised *in vitro*; however, *in vivo* studies still need to be performed. Interestingly, among the characteristics of selected *Lactobacillus* strains was growth inhibition of *G. vaginalis* (Mastromarino *et al.*, 2002).

From clinical studies, the role of *Lactobacillus* antimicrobial agents as one of the desirable properties of a probiotic is becoming apparent. For instance, *Lb. acidophilus* LA-1 culture supernatant – shown to be effective *in vitro* – has a partial, acid-independent, long-term suppressive effect on *H. pylori* in clinical trials (Michetti *et al.*, 1999). More recently, it was reported that there was evidence of bactericidal activity and significantly reduced density of *H. pylori* in a study where humans were fed *Lb. johnsonii* LA-1 fermented milk (Felley *et al.*, 2001). These results suggest that consuming *Lb. johnsonii* acidified milk can downregulate *H. pylori* infection and gastritis.

9.4.5 *Antimicrobial potential of* Bifidobacterium *spp.*

Unlike antimicrobial agents produced by *Lactobacillus* spp., only a limited number of studies have been performed to demonstrate production of antimicrobial compounds or bacteriocins among bifidobacterial strains (Table 9.1). Recently, specific antagonistic activities against both Gram-positive and Gram-negative bacteria have been observed with bifidobacteria; however, as the mechanism for this activity has not been completely defined, authors often refer to production of bacteriocin-like substances (Poltavska *et al.*, 2012). In this respect, bifidobacteria are generally inhibitory to a wide range of micro-organisms due to their intense production of lactic and acetic acids as part of their normal metabolic processes (Eklund, 1983). However, Gibson and Wang (1994) showed that antagonistic activity of several species of bifidobacteria towards both Gram-positive and Gram-negative pathogens was not solely attributed to the presence of acids. They demonstrated that eight strains of bifidobacteria were able to excrete bactericidal or bacteriostatic substances. A few bifidobacterial bacteriocins have been reported, such as bifidocin B (produced by *Bif. bifidum*), bifidin I (produced by '*Bif. infantis*' – presumed to be *Bif. longum* subsp. *infantis*), bifilong BB-46 (produced by '*Bif. longum*' – presumed to be *Bif. longum* subsp. *longum*), bibilact BB-12 (produced by '*Bif.*

lactis' – presumed to be *Bif. animalis* subsp. *lactis* BB-12) and thermophilicin B67 (produced by *Bifidobacterium thermophilum*) (Table 9.1). Nowadays, bifidocin B is the only bifidobacterial bacteriocin characterised. It is a class IIa bacteriocin, homologous to pediocin-like bacteriocins, which strongly inhibits foodborne pathogens by a proposed mechanism involving its binding to specific cell wall receptors, such as lipoteichoic acid, to form pores (Yildirim *et al.*, 1999). Christopher *et al.* (2004) reported antimicrobial activity by *Bif. animalis* subsp. *animalis* BB-12 and *Bif. bifidum* DSM 20456 against *Shigella dysenteriae*, *Escherichia coli*, *Salmonella* Typhimurium and *Yersinia enterocolitica* using an agar well assay technique; they suggested that these strains could be used as dietary adjuncts in fermented milk products. In another study, Saleh and El-Sayed (2004) reported the activity of two bacteriocins produced by *Bif. animalis* subsp. *lactis* BB-12 (known as bifilact BB-12) and '*Bif. longum* BB-46' (presumed to be *Bif. longum* subsp. *longum* BB-46) (known as bifilong BB-46); sodium dodecyl sulphate–polyacrylamide gel electrophoresis (SDS-PAGE) profiles indicated a heterogeneous composition for both bacteriocins. Bifilact BB-12 consisted of four bands with molecular weights ranging between 25 and 89 kDa, whilst 13 bands were obtained for bifilong BB-46 with molecular weights ranging between 25 and 127 kDa. Both bacteriocins were sensitive to pepsin and trypsin, but were resistant to α-amylase or lipase when tested by agar diffusion assay or by reverse-phase high-pressure liquid chromatography (RP-HPLC).

9.4.6 *Other lactic acid bacteria species with antimicrobial potential*

Apart from *Lactobacillus* spp. and *Bifidobacterium* spp., other bacterial genera relevant to food fermentation have been reported to produce bacteriocins. Some *Lac. lactis* subsp. *lactis* strains can produce nisin, the first antibacterial peptide found in LAB (Rogers & Whittier, 1928). Nisin was first introduced commercially as a food preservative in the United Kingdom about 50 years ago. Nowadays, it is commercially used as a food preservative against contamination by micro-organisms, and is marketed as Nisaplin®. It is the only bacteriocin approved for use as a food preservative by the FDA, and it is approved as a food additive in over 45 countries (Yang *et al.*, 2014). Other bacteriocins, such as lactococcin, lacticin, diacetin or lactocyclicin Q, are also produced by *Lac. lactis* spp.

Some species of pediococci (mainly *Pediococcus pentosaceus*, *Pediococcus damnosus* and *Pediococcus acidilactici*) can produce pediocins. Other species, such as *Leuconostoc gelidum* and *Leuconostoc mesenteroides* spp., have the potential to produce different types of leucocins and mesentericins. A few strains of *Str. thermophilus* and *Streptococcus salivarius* can biosynthesise antimicrobial peptides (salivaricin and thermophilin). Finally, *P. freudenreichii* subsp. *freudenreichii* LMGT 2946 can synthesise propionicin F (Table 9.1).

9.5 Overall conclusions

For millennia, milk has been preserved by fermentation using LAB, whose primary role is to convert lactose to lactic and other organic acids, thus lowering the pH. As well as this mechanism of activity, many of these food cultures can produce a range of secondary metabolites during fermentation, which may influence not only the quality, safety, flavour and texture of the fermented food, but also its nutritional and health status. Lactobacilli, bifidobacteria and, to a lesser extent, propionibacteria are being increasingly exploited in probiotic dairy products such as cheese, yoghurt and milk drinks, due to the accumulating clinical evidence for their human health-promoting activities. Such probiotic-containing dairy foods are associated with a range of health claims, including alleviation of symptoms of lactose intolerance, treatment of infectious diarrhoea, cancer suppression and reduction of blood cholesterol. A number of mechanisms have been proposed to explain these health benefits, and these include the secretion by probiotic bacteria of beneficial nutrients, such as short-chain fatty acids, vitamins, bioactive peptides and fatty acids, bacterial–host signalling molecules and antimicrobial substances (Linares *et al.*, 2016).

This review presented some examples whereby the bioactivities of LAB, particularly probiotic bacteria, can be exploited and/or accentuated to produce complex biomolecules. For example, the production of vitamins by LAB provides a very attractive approach to improve the nutritional profile of fermented foods, while EPS production is associated with enhanced rheological properties, in addition to the biogenic effects associated with their ingestion. The ability of probiotic LAB to produce antimicrobial substances, such as bacteriocins, may provide them an advantage in terms of their survival and proliferation in the gut, and may have a controlling influence on the composition of the gut microbiota. Furthermore, production of these antimicrobials offers a type of self-preservation of foods containing bacteriocin-producing micro-organisms, by protecting the food against certain pathogenic and spoilage micro-organisms. It is important to emphasise that while production of these bioactive molecules may have important nutritional and safety implications for the food itself, what may be even more important is the production by probiotics of bioactive molecules *in vivo* in the gut, where these may have a greater impact on human health. In addition, such endeavours will be greatly facilitated by the recent increment in the number of genomes available, which is contributing and will continue to contribute to our greater understanding of the mechanisms underlying the health-promoting effects of these bioactive molecules.

Acknowledgements

This work was funded by the LONGLIFE Project (Joint Programming Initiative - JPI; A Healthy Diet for a Healthy Life) and the APC Microbiome Institute, a Centre for Science and Technology (CSET) funded by the Science Foundation Ireland (SFI), through the Irish Government's National Development Plan.

References

Abee, T., Klaenhammer, T.R. & Letellier, L. (1994) Kinetic studies of the action of lactacin F, a bacteriocin produced by *Lactobacillus johnsonii* that forms poration complexes in the cytoplasmic membrane. *Applied and Environmental Microbiology*, **60**, 1006–1013.

Abee, T., Krockel, L. & Hill, C. (1995) Bacteriocins: modes of action and potentials in food preservation and control of food poisoning. *International Journal of Food Microbiology*, **28**, 169–185.

Ale, E.C., Perezlindo, M.J., Burns, P., Tabacman, E., Reinheimer, J.A. & Binetti, A.G. (2016) Exopolysaccharide from *Lactobacillus fermentum* Lf2 and its functional characterization as a yogurt additive. *Journal of Dairy Research*, **83**, 487–492.

Alegría, A., Delgado, S., Guadamuro, L., Belén Flórez, A., Felis, G.E., Torriani, S. & Mayo, B. (2014) The genome of *Bifidobacterium pseudocatenulatum* IPLA 36007, a human intestinal strain with isoflavone-activation activity. *Gut Pathogens*, **6**, 31. http://www.gutpathogens.com/content/6/1/31

Ali, D., Lacroix, C., Thuault, D., Bourgeois, C.M. & Simard, R.E. (1995) Characterization of diacetin B, a bacteriocin from *Lactococcus lactis* subsp. *lactis* bv. *diacetylactis* UL720. *Canadian Journal of Microbiology*, **41**, 832–841.

Alpay-Karaoglu, S., Aydin, F., Kilic, S.S. & Kilic, A.O. (2003) Antimicrobial activity and characteristics of bacteriocins produced by vaginal lactobacilli. *Turk Journal of Medical Science*, **33**, 7–13.

Anderssen, E.L., Diep, D.B., Nes, I.F., Eijsink, V.G. & Nissen-Meyer, J. (1998) Antagonistic activity of *Lactobacillus plantarum* C11: two new two-peptide bacteriocins, plantaricins EF and JK and the induction factor plantaricin A. *Applied and Environmental Microbiology*, **64**, 2269–3372.

Arena, M.P., Russo, P., Capozzi, V., López, P., Fiocco, D. & Spano, G. (2014) Probiotic abilities of riboflavin-overproducing *Lactobacillus* strains: a novel promising application of probiotics. *Applied Microbiology and Biotechnology*, **98**, 7569–7581.

Aroutcheva, A.A., Simoes, J.A. & Faro, S. (2001) Antimicrobial protein produced by vaginal *Lactobacillus acidophilus* that inhibits *Gardnerella vaginalis*. *Infectious Diseases in Obstetrics and Gynecology*, **9**, 33–39.

Avonts, L., van Uytven, E. & De Vuyst, L. (2004) Cell growth and bacteriocin production of probiotic *Lactobacillus* strains in different media. *International Dairy Journal*, **14**, 947–955.

Balzaretti, S., Taverniti, V., Guglielmetti, S., Fiore, W., Minuzzo, M., Ngo, H.N., Ngere, J.B., Sadiq, S., Humphreys, P.N. & Laws, A.P. (2016) A novel rhamnose-rich hetero-exopolysaccharide isolated from *Lactobacillus paracasei* DG activates THP-1 human monocytic cells. *Applied Environmental Microbiology*, 83:e02702-16. https://doi.org/10.1128/AEM.02702-16.

Barefoot, S.F. & Klaenhammer, T.R. (1984) Purification and characterization of the *Lactobacillus acidophilus* bacteriocin lactacin B. *Antimicrobial Agents and Chemotherapy*, **26**, 328–334.

Barrett, E., Hayes, M., O'Connor, P., Gardiner, G., Fitzgerald, G.F., Stanton, C., Ross, R.P. & Hill, C. (2007) Salivaricin P, one of a family of two-component antilisterial bacteriocins produced by intestinal isolates of *Lactobacillus salivarius*. *Applied Environmental Microbiology*, **73**, 3719–3723.

Basavanna, G. & Prapulla, S.G. (2013) Evaluation of functional aspects of *Lactobacillus fermentum* CFR 2195 isolated from breast fed healthy infants' fecal matter. *Journal of Food Science and Technology*, **50**, 360–366.

Bhunia, A.K., Johnson, M.C. & Ray, B. (1988) Purification, characterization and antimicrobial spectrum of a bacteriocin produced by *Pediococcus acidilactici*. *Journal of Applied Bacteriology*, **65**, 261–268.

Bierbaum, G., Gotz, F., Peschel, A., Kupke, T., van de Kamp, M. & Sahl, H.G. (1996) The biosynthesis of the lantibiotics epidermin, gallidermin, Pep5 and epilancin K7. *Antonie van Leeuwenhoek*, **69**, 119–127.

Blom, H., Katla, T., Holck, A., Sletten, K., Axelsson, L. & Holo, H. (1999) Characterization, production and purification of leucocin H, a two-peptide bacteriocin from *Leuconostoc* MF215B. *Current Microbiology*, **39**, 43–48.

Bodana, A.R. & Rao, D.R. (1990) Antimutagenic activity of milk fermented by *Streptococcus thermophilus* and *Lactobacillus bulgaricus*. *Journal of Dairy Science*, **73**, 3379–3384.

Bogovic-Matijasic, B., Rogelj, I., Nes, I.F. & Holo, H. (1998) Isolation and characterization of two bacteriocins of *Lactobacillus acidophilus* LF221. *Applied Microbiology and Biotechnology*, **49**, 606–612.

Burgess, C., O'Connell-Motherway, M., Sybesma, W., Hugenholtz, J. & van Sinderen, D. (2004) Riboflavin production in *Lactococcus lactis*: potential for *in situ* production of vitamin-enriched foods. *Applied and Environmental Microbiology*, **70**, 5796–5777.

Burgess, C.M., Smid, E.J. & van Sinderen, D. (2009) Bacterial vitamin B2, B11 and B12 overproduction: an overview. *International Journal of Food Microbiology*, **133**, 1–7.

Cárdenas, N., Laiño, J.E., Delgado, S., Jiménez, E., Juárez del Valle, M., Savoy de Giori, G., Sesma, F., Mayo, B., Fernández, L., LeBlanc, J.G. & Rodríguez, J.M. (2015) Relationships between the genome and some phenotypical properties of *Lactobacillus fermentum* CECT 5716, a probiotic strain isolated from human milk. *Applied Microbiology and Biotechnology*, **99**, 4343–4353.

Cheikhyoussef, A., Cheikhyoussef, N., Chen, H., Zhao, J., Tang, J., Zhang, H. & Chen, W. (2010) Bifidin I – a new bacteriocin produced by *Bifidobacterium infantis* BCRC 14602: purification and partial amino acid sequence. *Food Control*, **21**, 746–753.

Christopher, M.D., Reddy, V.P., Mrudula, N. & Venkateswarulu, K. (2004) Assessment of antimicrobial activity of *Bifidobacterium* ssp. *Cheiron*, **32**, 132–134.

Ciszek-Lenda, M., Nowak, B., Sróttek, M., Gamian, A. & Marcinkiewicz, J. (2011) Immunoregulatory potential of exopolysaccharide from *Lactobacillus rhamnosus* KL37: effects on the production of inflammatory mediators by mouse macrophages. *International Journal of Experimental Pathology*, **92**, 382–391.

Coconnier, M.H., Lievin, V., Bernet-Camard, M.F., Hudault, S. & Servin, A.L. (1997) Antibacterial effect of the adhering human *Lactobacillus acidophilus* strain LB. *Antimicrobial Agents and Chemotherapy*, **41**, 1046–1052.

Coconnier, M.H., Lievin, V., Hemery, E. & Servin, A.L. (1998) Antagonistic activity against *Helicobacter* infection *in vitro* and *in vivo* by the human *Lactobacillus acidophilus* strain LB. *Applied and Environmental Microbiology*, **64**, 4573–4580.

Coconnier, M.H., Lievin, V., Lorrot, M. & Servin, A.L. (2000) Antagonistic activity of *Lactobacillus acidophilus* LB against intracellular *Salmonella enterica* serovar Typhimurium infecting human enterocyte-like Caco-2/TC-7 cells. *Applied and Environmental Microbiology*, **66**, 1152–1157.

Contreras, B.G.L., de Vuyst, L., Devreese, B., Busanyova, K., Raymaeckers, J., Bosman, F., Sablon, E. & Vandamme, E.J. (1997) Isolation, purification and amino acid sequence of lactobin A, one of the two bacteriocins produced by *Lactobacillus amylovorus* LMG P-13139. *Applied and Environmental Microbiology*, **63**, 13–20.

Crittenden, R.G., Martinez, N.R. & Playne, M.J. (2003) Synthesis and utilisation of folate by yoghurt starter cultures and probiotic bacteria. *International Journal of Food Microbiology*, **80**, 217–222.

Crociani, F., Alessandrini, A., Mucci, M.M. & Biavati, B. (1994) Degradation of complex carbohydrates by *Bifidobacterium* spp. *International Journal of Food Microbiology*, **24**, 199–210.

Cuozzo, S.A., Sesma, F., Palacios, J.M., de Ruiz Holgado, A.P. & Raya, R.R. (2000) Identification and nucleotide sequence of genes involved in the synthesis of lactocin 705, a two-peptide bacteriocin from *Lactobacillus casei* CRL 705. *FEMS Microbiology Letters*, **185**, 157–161.

De Angelis, M., Bottacini, F., Fosso, B., Kelleher, P., Calasso, M., Di Cagno, R., Ventura, M., Picardi, E., van Sinderen, D. & Gobbetti, M. (2014) *Lactobacillus rossiae*, a vitamin B12 producer, represents a metabolically versatile species within the genus *Lactobacillus*. *PLoS One*, **9**, e107232.

De Vuyst, L. & Vandamme, E.J. (1994) Nisin, a lantibiotic produced by *Lactococcus lactis* subsp. *lactis*: properties, biosynthesis, fermentation and applications. In *Bacteriocins of Lactic Acid Bacteria* (eds. L. De Vuyst and E.J. Vandamme), 151–221. Blackie Academic & Professional, London.

De Vuyst, L., De Vin, F., Vaningelgem, F. & Degeest, B. (2001) Recent developments in the biosynthesis and applications of heteropolysaccharides from lactic acid bacteria. *International Journal of Food Microbiology*, **11**, 687–707.

Deepak, V., Ramachandran, S., Balahmar, R.M., Pandian, S.R., Sivasubramaniam, S.D., Nellaiah, H. & Sundar, K. (2016) *In vitro* evaluation of anticancer properties of exopolysaccharides from *Lactobacillus acidophilus* in colon cancer cell lines. *In Vitro Cellular and Developmental Biology Animal*, **52**, 163–173.

Deguchi, Y., Morishita, T. & Mutai, M. (1985) Comparative studies on synthesis of water-soluble vitamins among human species of bifidobacteria. *Agricultural and Biological Chemistry*, **49**, 13–19.

Dimitrijević, R., Stojanović, M., Živković, I., Petersen, A., Jankov, R.M., Dimitrijević, L. & Gavrović-Jankulović, M. (2009) The identification of a low molecular mass bacteriocin, rhamnosin A, produced by *Lactobacillus rhamnosus* strain 68. *Journal of Applied Microbiology*, **107**, 2108–2115.

Divya, J.B. & Nampoothiri, K.M. (2015) Encapsulated *Lactococcus lactis* with enhanced gastrointestinal survival for the development of folate enriched functional foods. *Bioresource Technology*, **188**, 226–230.

Dougherty, B.A., Hill, C., Weidman, J.F., Richardson, D.R., Venter, J.C. & Ross, R.P. (1998) Sequence and analysis of the 60 kb conjugative, bacteriocin-producing plasmid pMRC01 from *Lactococcus lactis* DPC3147. *Molecular Microbiology*, **29**, 1029–1038.

Duboc, P. & Mollet, B. (2001) Applications of exopolysaccharides in the dairy industry. *International Dairy Journal*, **11**, 759–768.

Dufour, A., Thuault, D., Boulliou, A., Bourgeois, C.M. & Le Pennec, J.P. (1991) Plasmid-encoded determinants for bacteriocin production and immunity in a *Lactococcus lactis* strain and purification of the inhibitory peptide. *Journal of General Microbiology*, **137**, 2423–2429.

Eijsink, V.G., Axelsson, L., Diep, D.B., Havarstein, L.S., Holo, H. & Nes, I.F. (2002) Production of class II bacteriocins by lactic acid bacteria; an example of biological warfare and communication. *Antonie van Leeuwenhoek*, **81**, 639–654.

Eklund, T. (1983) The antimicrobial effect of dissociated and undissociated sorbic acid at different pH levels. *Journal of Applied Bacteriology*, **54**, 383–389.

Fanning, S., Hall, L.J. & van Sinderen, D. (2012) *Bifidobacterium breve* UCC2003 surface exopolysaccharide production is a beneficial trait mediating commensal-host interaction through immune modulation and pathogen protection. *Gut Microbes*, **3**, 420–425.

FAO/WHO (2001) *Evaluation of Health and Nutritional Properties of Probiotics in Food Including Powder Milk with Live Lactic Acid Bacteria*. Report of a Joint FAO/WHO Expert Consultation. Food and Agriculture Organization of the United Nations (FAO), Rome.

Felley, C.P., Corthesy-Theulaz, I., Rivero, J.L., Sipponen, P., Kaufmann, M., Bauerfeind, P., Wiesel, P.H., Brassart, D., Pfeifer, A., Blum, A.L. & Michetti, P. (2001) Favourable effect of an acidified milk (LC-1) on *Helicobacter pylori* gastritis in man. *European Journal of Gastroenterology and Hepatology*, **13**, 25–29.

Flynn, S., van Sinderen, D., Thornton, G.M., Holo, H., Nes, I.F. & Collins, J.K. (2002) Characterization of the genetic locus responsible for the production of ABP-118, a novel bacteriocin produced by the probiotic bacterium *Lactobacillus salivarius* ssp. *salivarius* UCC118. *Microbiology*, **148**, 973–984.

Garcia-Garcera, M.J., Elferink, M.G., Driessen, A.J. & Konings, W.N. (1993) In vitro pore-forming activity of the lantibiotic nisin: role of proton motive force and lipid composition. *European Journal of Biochemistry*, **212**, 417–422.

Gardiner, G.E., Casey, P.G., Casey, G., Lynch, P.B., Lawlor, P.G., Hill, C., Fitzgerald, G.F., Stanton, C. & Ross, R.P. (2004) Relative ability of orally administered *Lactobacillus murinus* to predominate and persist in the porcine gastrointestinal tract. *Applied and Environmental Microbiology*, **70**, 1895–1906.

Gibson, G.R. & Wang, X. (1994) Regulatory effects of bifidobacteria on the growth of other colonic bacteria. *Journal of Applied Bacteriology*, **77**, 412–420.

Gibson, G.R., Hutkins, R., Sanders, M.E., Prescott, S.L., Reimer, R.A., Salminen, S.J., Scott, K., Stanton, C., Swanson, K.S., Cani, P.D., Verbeke, K. & Reid, G. (2017). The International Scientific Association for Probiotics and Prebiotics (ISAPP) consensus statement on the definition and scope of prebiotics, *Nature Reviews, Gastroenterology & Hepatology*, PUBMED publication, which is online, but not yet in print,

Gilmore, M.S., Segarra, R.A., Booth, M.C., Bogie, C.P., Hall, L.R. & Clewell, D.B. (1994) Genetic structure of the *Enterococcus faecalis* plasmid pAD1-encoded cytolytic toxin system and its relationship to lantibiotic determinants. *Journal of Bacteriology*, **176**, 7335–7344.

Goldin, B.R. (1998) Health benefits of probiotics. *British Journal of Nutrition*, **80**, S203–S207.

Gonzalez, B., Arca, P., Mayo, B. & Suarez, J.E. (1994) Detection, purification, and partial characterization of plantaricin C, a bacteriocin produced by a *Lactobacillus plantarum* strain of dairy origin. *Applied and Environmental Microbiology*, **60**, 2158–2163.

Güllüce, M., Karaday, M. & Barış, Ö. (2013) Bacteriocins: promising natural antimicrobials. In *Microbial Pathogens and Strategies for Combating Them: Science, Technology and Education* (ed. A. Méndez-Vilas), Vol. **2**, 1016–1027. Formatex Research Centre, Extremadura.

Hakim, A.M. & Pappius, H.M. (1981) The effect of thiamine deficiency on local cerebral glucose utilization. *Annals of Neurology*, **9**, 334–339.

Harutoshi, T. (2013) Exopolysaccharides of lactic acid bacteria for food and colon health applications. In *Lactic Acid Bacteria – R&D for Food, Health and Livestock Purposes* (ed. M. Kongo), 515–537. Intech, Rijeka, Croatia.

Héchard, Y., Derijard, B., Letellier, F. & Cenatiempo, Y. (1992) Characterization and purification of mesentericin Y105, an anti-Listeria bacteriocin from *Leuconostoc mesenteroides*. *Journal of General Microbiology*, **138**, 2725–2731.

Héchard, Y., Berjeaud, J.M. & Cenatiempo, Y. (1999) Characterization of the *mesB* gene and expression of bacteriocins by *Leuconostoc mesenteroides* Y105. *Current Microbiology*, **39**, 265–269.

Hegarty, J.W., Guinane, C.M., Ross, R.P., Hill, C. & Cotter, P.D. (2016) Bacteriocin production: a relatively unharnessed probiotic trait? *F1000Res*, **5**, 2587. doi:10.12688/f1000research.9615.1

Henderson, J.T., Chopko, A.L. & van Wassenaar, P.D. (1992) Purification and primary structure of pediocin PA-1 produced by *Pediococcus acidilactici* PAC-1.0. *Archives in Biochemistry and Biophysics*, **295**, 5–12.

Hidalgo-Cantabrana, C., López, P., Gueimonde, M., de Los Reyes-Gavilán, C.G., Suárez, A., Margolles, A. & Ruas-Madiedo, P. (2012) Immune modulation capability of exopolysaccharides synthesised by lactic acid Bacteria and Bifidobacteria. *Probiotics Antimicrob Proteins*, **4**, 227–237.

Hidalgo-Cantabrana, C., Nikolic, M., López, P., Suárez, A., Miljkovic, M., Kojic, M., Margolles, A., Golic, N. & Ruas-Madiedo, P. (2014) Exopolysaccharide-producing *Bifidobacterium animalis* subsp. *lactis* strains and their polymers elicit different responses on immune cells from blood and gut associated lymphoid tissue. *Anaerobe*, **26**, 24–30.

Hidalgo-Cantabrana, C., Algieri, F., Rodriguez-Nogales, A., Vezza, T., Martínez-Camblor, P., Margolles, A., Ruas-Madiedo, P. & Gálvez, J. (2016) Effect of a ropy exopolysaccharide-producing *Bifidobacterium animalis* subsp. *lactis* strain orally administered on DSS-Induced colitis mice model. *Frontiers in Microbiology*, **7**, 868. doi:10.3389/fmicb.2016.00868

Hill, C., Guarner F., Reid, G., Gibson, G.R., Merenstein, D.J., Pot, B., Morelli, L., Canani, R.B., Flint, H.J., Salminen, S., Calder, P.C. & Sanders, M.E. (2014) The International Scientific Association for Probiotics and Prebiotics consensus statement on the scope and appropriate use of the term probiotic (Expert consensus document). *Nature Reviews Gastroenterology and Hepatology*, **11**, 506–514.

Hillman, J.D. (2002) Genetically modified *Streptococcus mutans* for the prevention of dental caries. *Antoine van Leeuwenhoek*, **82**, 361–366.
Holck, A., Axelsson, L., Birkeland, S.E., Aukrust, T. & Blom, H. (1992) Purification and amino acid sequence of sakacin A, a bacteriocin from *Lactobacillus sake* Lb706. *Journal of General Microbiology*, **138**, 2715–2720.
Holo, H., Jeknic, Z., Daeschel, M., Stevanovic, S. & Nes, I.F. (2001) Plantaricin W from *Lactobacillus plantarum* belongs to a new family of two-peptide lantibiotics. *Microbiology*, **147**, 643–651.
Holzapfel, W.H. & Schillinger, U. (2002) Introduction to pre- and probiotics. *Food Research International*, **35**, 109–116.
Holzapfel, W.H., Haberer, P., Snel, J., Schillinger, U. & Huis in't Veld, J.H. (1998) Overview of gut flora and probiotics. *International Journal of Food Microbiology*, **41**, 85–101.
Hosono, A., Lee, J., Ametani, A., Natsume, M., Hirayama, M., Adachi, T. & Kaminogawa, S. (1997) Characterisation of a water soluble polysaccharide fraction with immunopotentiating activity from *Bifidobacterium adolescentis* M101-4. *Bioscience, Biotechnology and Biochemistry*, **61**, 312–316.
Hou, J.W., Yu, R.C. & Chou, C.C. (2000) Changes in some components of soymilk during fermentation with bifidobacteria. *Food Research International*, **33**, 393–397.
Hugenholtz, J., Hunik, J., Santos, H. & Smid, E.J. (2002a) Neutraceutical production by propionibacteria. *Lait*, **82**, 103–112.
Hugenholtz, J., Sybesma, W., Groot, M.N., Wisselink, W., Ladero, V., Burgess, K., van Dinderen, D., Piard, J.-C., Eggink, G., Smid, E.J., Savoy, G., Sesma, F., Jansen, T., Hols, P. & Kleerebezem, M. (2002b) Metabolic engineering of lactic acid bacteria for the production of nutraceuticals. *Antonie van Leeuwenhoek*, **82**, 217–235.
Huhne, K., Axelsson, L., Holck, A. & Krockel, L. (1996) Analysis of the sakacin P gene cluster from *Lactobacillus sake* Lb674 and its expression in sakacin-negative *Lb. sake* strains. *Microbiology*, **142**, 1437–1448.
Hunik, J. (2002) Process for the production of vitamin B12. USA Patent No. 6492141B1.
Jiang, J., Bjorck, L. & Fonden, R. (1998) Production of conjugated linoleic acid by dairy starter cultures. *Journal of Applied Microbiology*, **85**, 95–102.
Jimenez-Diaz, R., Ruiz-Barba, J.L., Cathcart, D.P., Holo, H., Nes, I.F., Sletten, K.H. & Warner, P.J. (1995) Purification and partial amino acid sequence of plantaricin S, a bacteriocin produced by *Lactobacillus plantarum* LPCO10, the activity of which depends on the complementary action of two peptides. *Applied and Environmental Microbiology*, **61**, 4459–4463.
Joerger, M.C. & Klaenhammer, T.R. (1986) Characterization and purification of helveticin J and evidence for a chromosomally determined bacteriocin produced by *Lactobacillus helveticus* 481. *Journal of Bacteriology*, **167**, 439–446.
Jolly, L., Vincent, S.J., Duboc, P. & Neeser, J.R. (2002) Exploiting expolysaccharides from lactic acid bacteria. *Antonie van Leeuwenhoek*, **82**, 367–374.
Kabuki, T., Uenishi, H., Watanabe, M., Seto, Y. & Nakajima, H. (2007) Characterization of a bacteriocin, Thermophilin 1277, produced by *Streptococcus thermophilus* SBT1277. *Journal of Applied Microbiology*, **102**, 971–980.
Kaletta, C. & Entian, K.D. (1989) Nisin, a peptide antibiotic: cloning and sequencing of the nisA gene and posttranslational processing of its peptide product. *Journal of Bacteriology*, **171**, 1597–1601.
Kanatani, K., Oshimura, M. & Sano, K. (1995) Isolation and characterization of acidocin A and cloning of the bacteriocin gene from *Lactobacillus acidophilus*. *Applied and Environmental Microbiology*, **61**, 1061–1067.
Kang, H., Choi, H.S., Kim, J.E. & Han, N.S. (2011) Exopolysaccharide-overproducing *Lactobacillus paracasei* KB28 induces cytokines in mouse peritoneal macrophages via modulation of NF-κβ and MAPKs. *Journal of Microbiology and Biotechnology*, **21**, 1174–1178.

Kawai, Y., Saito, T., Toba, T., Samant, S.K. & Itoh, T. (1994) Isolation and characterization of a highly hydrophobic new bacteriocin (gassericin A) from *Lactobacillus gasseri* LA39. *Bioscience, Biotechnology and Biochemistry*, **58**, 1218–1221.

Kellner, R., Jung, G. & Sahl, H.G. (1991) Nisin and novel lantibiotics. In *Nisin and Novel Lantibiotics* (ed. G. Jung & H.G. Sahl), 141–158. ESCOM Scientific Publishers BV, Leiden.

Kitazawa, H., Ishii, Y., Uemura, J., Kawai, Y., Saito, T., Kaneko, T., Noda, K. & Itoh, T. (2000) Augmentation of macrophage functions by an extracellular phosphopolysaccharide from *Lactobacillus delbrueckii* ssp. *bulgaricus*. *Food Microbiology*, **17**, 109–118.

Korakli, M., Ganzle, M.G. & Vogel, R.F. (2002) Metabolism by bifidobacteria and lactic acid bacteria of polysaccharides from wheat and rye, and exopolysaccharides produced by *Lactobacillus sanfranciscensis*. *Journal of Applied Microbiology*, **92**, 958–965.

Krause, L.J., Forsberg, C.W. & O'Connor, D.L. (1996) Feeding human milk to rats increases *Bifidobacterium* in the cecum and colon which correlates with enhanced folate status. *Journal of Nutrition*, **126**, 1505–1511.

Kuipers, O.P., Beerthuyzen, M.M., Siezen, R.J. & de Vos, W.M. (1993) Characterization of the nisin gene cluster *nisABTCIPR* of *Lactococcus lactis*. Requirement of expression of the nisA and nisI genes for development of immunity. *European Journal of Biochemistry*, **216**, 281–291.

Kumar, H., Salminen, S., Verhagen, H., Rowland, I., Heimbach, J., Bañares, S., Young, T., Nomoto, K. & Lalonde, M. (2015) Novel probiotics and prebiotics: road to the market. *Current Opinion in Biotechnology*, **32**, 99–103.

Larsen, A.G., Vogensen, F.K. & Josephsen, J. (1993) Antimicrobial activity of lactic acid bacteria isolated from sour doughs: purification and characterization of bavaricin A, a bacteriocin produced by *Lactobacillus bavaricus* MI401. *Journal of Applied Bacteriology*, **75**, 113–122.

Laws, A., Gu, Y. & Marshall, V. (2001) Biosynthesis, characterisation, and design of bacterial exopolysaccharides from lactic acid bacteria. *Biotechnology Advances*, **19**, 1–28.

LeBlanc, J.G., Laiño, J.E., del Valle, M.J., Vannini, V., van Sinderen, D., Taranto, M.P., de Valdez, G.F., de Giori, G.S. & Sesma, F. (2011) B-group vitamin production by lactic acid bacteria: current knowledge and potential applications. *Journal of Applied Microbiology*, **111**, 1297–1309.

Leer, R.J., van der Vossen, J.M., van Giezen, M., van Noort, J.M. & Pouwels, P.H. (1995) Genetic analysis of acidocin B, a novel bacteriocin produced by *Lactobacillus acidophilus*. *Microbiology*, **141**, 1629–1635.

Li, S., Huang, R., Shah, N.P., Tao, X., Xiong, Y. & Wei, H. (2014) Antioxidant and antibacterial activities of exopolysaccharides from *Bifidobacterium bifidum* WBIN03 and *Lactobacillus plantarum* R315. *Journal of Dairy Science*, **97**, 7334–7343.

Li, P., Zhou, Q. & Gu, Q. (2016) Complete genome sequence of *Lactobacillus plantarum* LZ227, a potential probiotic strain producing B-group vitamins. *Journal of Biotechnology*, **234**, 66–70.

Li, W., Xia, X., Tang, W., Ji, J., Rui, X., Chen, X., Jiang, M., Zhou, J., Zhang, Q. & Dong, M. (2015) Structural characterization and anticancer activity of cell-bound exopolysaccharide from *Lactobacillus helveticus* MB2-1. *Journal of Agricultural and Food Chemistry*, **63**, 3454–3463.

Lin, M.Y. & Young, C.M. (2000) Folate levels in cultures of lactic acid bacteria. *International Dairy Journal*, **10**, 409–413.

Linares, D.M., Ross, P. & Stanton, C. (2016) Beneficial microbes: the pharmacy in the gut. *Bioengineered*, **7**, 11–20.

London, L.E., Kumar, A.H., Wall, R., Casey, P.G., O'Sullivan, O., Shanahan, F., Hill, C., Cotter, P.D., Fitzgerald, G.F., Ross, R.P., Caplice, N.M. & Stanton, C. (2014) Exopolysaccharide-producing probiotic lactobacilli reduce serum cholesterol and modify enteric microbiota in ApoE-deficient mice. *Journal of Nutrition*, **144**, 1956–1962.

Looijesteijn, P.J., Trapet, L., de Vries, E., Abee, T. & Hugenholtz, J. (2001) Physiological function of exopolysaccharides produced by *Lactococcus lactis*. *International Journal of Food Microbiology*, **64**, 71–80.

Majhenic, A.C., Matijasic, B.B. & Rogelj, I. (2003) Chromosomal location of the genetic determinanta for bacteriocins produced by *Lactobacillus gasseri* K7. *Journal of Dairy Research*, **70**, 199–203.
Majhenic, A.C., Venema, K., Allison, G.E., Matijasic, B.B., Rogelj, I. & Klaenhammer, T.R. (2004) DNA analysis of the genes encoding acidocin LF221 A and acidocin LF221 B, two bacteriocins produced by *Lactobacillus gasseri* LF221. *Applied Microbiology and Biotechnology*, **63**, 705–714.
Makino, S., Sato, A., Goto, A., Nakamura, M., Ogawa, M., Chiba, Y., Hemmi, J., Kano, H., Takeda, K., Okumura, K. & Asami, Y. (2016) Enhanced natural killer cell activation by exopolysaccharides derived from yogurt fermented with *Lactobacillus delbrueckii* ssp. *bulgaricus* OLL1073R-1. *Journal of Dairy Science*, **99**, 915–923.
Maldonado, A., Jimenez-Diaz, R. & Ruiz-Barba, J.L. (2004a) Induction of plantaricin production in *Lactobacillus plantarum* NC8 after coculture with specific gram-positive bacteria is mediated by an autoinduction mechanism. *Journal of Bacteriology*, **186**, 1556–1564.
Maldonado, A., Ruiz-Barba, J.L. & Jimenez-Diaz, R. (2004b) Production of plantaricin NC8 by *Lactobacillus plantarum* NC8 is induced in the presence of different types of gram-positive bacteria. *Archives of Microbiology*, **181**, 8–16.
Marin, M.L., Tejada-Simon, M.V., Lee, J.H., Murtha, J., Ustunol, Z. & Pestka, J.J. (1998) Stimulation of cytokine production in clonal macrophage and T-cell models by *Streptococcus thermophilus*: comparison with *Bifidobacterium* sp. and *Lactobacillus bulgaricus*. *Journal of Food Protection*, **61**, 859–864.
Martínez, B., Fernández, M., Suárez, J.E. & Rodríguez, A. (1999) Synthesis of lactococcin 972, a bacteriocin produced by *Lactococcus lactis* IPLA972, depends on the expression of a plasmid-encoded bicistronic operon. *Microbiology*, **145**, 3155–3161.
Mastromarino, P., Brigidi, P., Macchia, S., Maggi, L., Pirovano, F., Trinchieri, V., Conte, U. & Matteuzzi, D. (2002) Characterization and selection of vaginal *Lactobacillus* strains for the preparation of vaginal tablets. *Journal of Applied Microbiology*, **93**, 884–893.
Matsuzaki, C., Kamishima, K., Matsumoto, K., Koga, H., Katayama, T., Yamamoto, K. & Hisa, K. (2014) Immunomodulating activity of exopolysaccharide-producing *Leuconostoc mesenteroides* strain NTM048 from green peas. *Journal of Applied Microbiology*, **116**, 980–989.
Mattarelli, P., Bonaparte, C., Pot, B. & Biavati, B. (2008) Proposal to reclassify the three biotypes of *Bifidobacterium longum* as three subspecies: *Bifidobacterium longum* subsp. *longum* subsp. nov., *Bifidobacterium longum* subsp. *infantis* comb. nov. and *Bifidobacterium longum* subsp. *suis* comb. nov. *International Journal of Systematic and Evolutionary Microbiology*, **58**, 767–772.
McAuliffe, O., Hill, C. & Ross, R.P. (2000) Identification and overexpression of *ltnl*, a novel gene which confers immunity to the two-component lantibiotic lacticin 3147. *Microbiology*, **146**, 129–138.
McGroarty, J.A. & Reid, G. (1988) Detection of a *Lactobacillus* substance that inhibits *Escherichia coli*. *Canadian Journal of Microbiology*, **34**, 974–978.
Michetti, P., Dorta, G., Wiesel, P.H., Brassart, D., Verdu, E., Herranz, M., Felley, C., Porta, N., Rouvet, M., Blum, A.L. & Corthesy-Theulaz, I. (1999) Effect of whey-based culture supernatant of *Lactobacillus acidophilus* (johnsonii) La1 on *Helicobacter pylori* infection in humans. *Digestion*, **60**, 203–209.
Moll, G., Ubbink-Kok, T., Hildeng-Hauge, H., Nissen-Meyer, J., Nes, I.F., Konings, W.N. & Driessen, A.J. (1996) Lactococcin G is a potassium ion-conducting, two-component bacteriocin. *Journal of Bacteriology*, **178**, 600–605.
Montville, T.J. & Bruno, M.E. (1994) Evidence that dissipation of proton motive force is a common mechanism of action for bacteriocins and other antimicrobial proteins. *International Journal of Food Microbiology*, **24**, 53–74.
Morishita, T., Tamura, N., Makino, T. & Kudo, S. (1999) Production of menaquinones by lactic acid bacteria. *Journal of Dairy Science*, **82**, 1897–1903.
Morrison, H.I., Schaubel, D., Desmeules, M. & Wigle, D.T. (1996) Serum folate and risk of fatal coronary heart disease. *Journal of American Medical Association*, **275**, 1893–1896.

Mortvedt, C.I., Nissen-Meyer, J., Sletten, K. & Nes, I.F. (1991) Purification and amino acid sequence of lactocin S, a bacteriocin produced by *Lactobacillus sake* L45. *Applied and Environmental Microbiology*, **57**, 1829–1834.

Nagai, T., Makino, S., Ikegami, S., Itoh, H. & Yamada, H. (2011) Effects of oral administration of yogurt fermented with *Lactobacillus delbrueckii* ssp. *bulgaricus* OLL1073R-1 and its exopolysaccharides against influenza virus infection in mice. *International Immunopharmacology*, **11**, 2246–2250.

Nakajima, H., Suzuki, Y., Kaizu, H. & Hirota, T. (1992) Cholesterol lowering activity in ropy fermented milk. *Journal of Food Science*, **57**, 1327–1329.

Nes, I.F., Yoon, S.S. & Diep, D.B. (2007) Ribosomally synthesized antimicrobial peptides (bacteriocins) in lactic acid bacteria: a review. *Food Science Biotechnology*, **16**, 675–690.

Nissen-Meyer, J., Holo, H., Håvarstein, L.S., Sletten, K. & Nes, I.F. (1992) A novel lactococcal bacteriocin whose activity depends on the complementary action of two peptides. *Journal of Bacteriology*, **174**, 5686–5692.

Ocana, V.S., Pesce, de Ruiz Holgado, A.A. & Nader-Macias, M.E. (1999) Characterization of a bacteriocin-like substance produced by a vaginal *Lactobacillus salivarius* strain. *Applied and Environmental Microbiology*, **65**, 5631–5635.

Palacios, J., Vignolo, G., Farias, M.E., de Ruiz Holgado, A.P., Oliver, G. & Sesma, F. (1999) Purification and amino acid sequence of lactocin 705, a bacteriocin produced by *Lactobacillus casei* CRL 705. *Microbiological Research*, **154**, 199–204.

Papathanasopoulos, M.A., Dykes, G.A., Revol-Junelles, A.M., Delfour, A., von Holy, A. & Hastings, J.W. (1998) Sequence and structural relationships of leucocins A-, B- and C-TA33a from *Leuconostoc mesenteroides* TA33a. *Microbiology*, **144**, 1343–1348.

Patel, S., Majumder, A. & Goyal, A. (2012) Potentials of exopolysaccharides from lactic acid bacteria. *Indian Journal of Microbiology*, **52**, 3–12.

Perdigon, G., Valdez, G.F. & Rachid, M. (1998) Antitumour activity of yogurt: study of possible immune mechanisms. *Journal of Dairy Research*, **65**, 129–138.

Perez, P.F., Minnaard, J., Rouvet, M., Knabenhans, C., Brassart, D., De Antoni, G.L. & Schiffrin, E.J. (2001) Inhibition of *Giardia intestinalis* by extracellular factors from lactobacilli: an *in vitro* study. *Applied and Environmental Microbiology*, **67**, 5037–5042.

Piard, J.C., Muriana, P.M., Desmazaud, M.J. & Klaenhammer, T.R. (1992) Purification and partial characterization of lacticin 481, a lanthionine-containing bacteriocin produced by *Lactococcus lactis* subsp. *lactis* CNRZ 481. *Applied and Environmental Microbiology*, **58**, 279–284.

Poltavska, O.A. & Kovalenko, N.K. (2012) Antimicrobial activity of bifidobacterial bacteriocin-like substances. *Mikrobiolohichnyi zhurnal*, **74**, 32–42.

Presti, I., D'Orazio, G., Labra, M., La Ferla, B., Mezzasalma, V., Bizzaro, G., Giardina, S., Michelotti, A., Tursi, F., Vassallo, M. & Di Gennaro, P. (2015) Evaluation of the probiotic properties of new *Lactobacillus* and *Bifidobacterium* strains and their in vitro effect. *Applied Microbiology Biotechnology*, **99**, 5613–5626.

Rea, M.C., Ross, R.P., Cotter, P.D. & Hill C. (2011) Classification of bacteriocins from Gram-positive bacteria. In *Prokaryotic Antimicrobial Peptides* (ed. D. Drider & S. Rebuffat), 29–53. Springer-Verlag, New York.

Revol-Junelles, A.M., Mathis, R., Krier, F., Fleury, Y., Delfour, A. & Lefebvre, G. (1996) *Leuconostoc mesenteroides* subsp. *mesenteroides* synthesizes two distinct bacteriocins. *Letters in Applied Microbiology*, **23**, 120–124.

Ricciardi, A. & Clementi, F. (2000) Exopolysaccharides from lactic acid bacteria: structure production and technological applications. *Italian Journal of Food Science*, **12**, 23–45.

Rodríguez, C., Medici, M., Rodríguez, A.V., Mozzi, F. & Font de Valdez, G. (2009) Prevention of chronic gastritis by fermented milks made with exopolysaccharide-producing *Streptococcus thermophilus* strains. *Journal of Dairy Science*, **92**, 2423–2434.

Rogers, L.A. & Whittier, E.D. (1928) Limiting factors in lactic fermentation. *Journal of Bacteriology*, **16**, 211–229.

Ross, K.F., Ronson, C.W. & Tagg, J.R. (1993) Isolation and characterization of the lantibiotic salivaricin A and its structural gene *salA* from *Streptococcus salivarius* 20P3. *Applied Environmental Microbiology*, **59**, 2014–2021.

Rossi, M., Amaretti, A. & Raimondi, S. (2011) Folate production by probiotic bacteria. *Nutrients*, **3**, 118–134.

Ruas-Madiedo, P., Hugenholtz, J. & Zoon, P. (2002) An overview of the functionality of exopolysaccharides produced by lactic acid bacteria. *International Dairy Journal*, **12**, 163–171.

Ruas-Madiedo, P., Salazar, N. & de los Reyes-Gavilán, C.G. (2009) Biosynthesis and chemical composition of exopolysaccharides produced by lactic acid bacteria. In *Bacterial Polysaccharides: Current Innovations and Future Trends* (ed. M. Ullrich), 279–310. Caister Academic Press, Poole.

Ruijssenaars, H.J., Stingele, F. & Hartmans, S. (2000) Biodegradability of food-associated extracellular polysaccharides. *Current Microbiology*, **40**, 194–199.

Ruiz-Barba, J.L. & Jimenez-Diaz, R. (1994) Vitamin and amino acid requirements of *Lactobacillus plantarum* strains isolated from green olive fermentations. *Journal of Applied Bacteriology*, **76**, 350–355.

Ruiz-Barba, J.L., Cathcart, D.P., Warner P.J. & Jimenez-Diaz, R. (1994) Use of *Lactobacillus plantarum* LPCO10, a bacteriocin producer, as a starter culture in Spanish-style green olive fermentations. *Applied and Environmental Microbiology*, **60**, 2059–2064.

Russo, P., Capozzi, V., Arena, M.P., Spadaccino, G., Dueñas, M.T., López, P., Fiocco, D. & Spano, G. (2014) Riboflavin-overproducing strains of *Lactobacillus fermentum* for riboflavin-enriched bread. *Applied Microbiology and Biotechnology*, **98**, 3691–3700.

Ryan, M.P., Rea, M.C., Hill, C. & Ross, R.P. (1996) An application in Cheddar cheese manufacture for a strain of *Lactococcus lactis* producing a novel broad-spectrum bacteriocin, lacticin 3147. *Applied and Environmental Microbiology*, **62**, 612–619.

Ryan, M.P., Jack, R.W., Josten, M., Sahl, H.G., Jung, G., Ross, R.P. & Hill, C. (1999) Extensive post-translational modification, including serine to D-alanine conversion, in the two-component lantibiotic, lacticin 3147. *Journal of Biological Chemistry*, **274**, 37544–37550.

Ryan, P.M., Guinane, C.M., London, L.E., Kelleher, P.R., Fitzgerald, G.F., Caplice, N.M., Ross, R.P. & Stanton, C. (2015) Genome sequence of the heteropolysaccharide-producing strain *Lactobacillus mucosae* DPC 6426. *Genome Announcements* **3**, pii: e01350-14.

Saito, T. (2004) Selection of useful probiotic lactic acid bacteria from the *Lactobacillus acidophilus* group and their applications to functional foods. *Animal Science Journal*, **75**, 1–13.

Salazar, N., Gueimonde, M., de Los Reyes-Gavilán, C.G. & Ruas-Madiedo, P. (2016) Exopolysaccharides produced by lactic acid bacteria and bifidobacteria as fermentable substrates by the intestinal microbiota. *Critical Reviews in Food Science and Nutrition*, **56**, 1440–1453.

Saleh, F.A. & El-Sayed, E.M. (2004) Isolation and characterization of bacteriocins produced by *Bifidobacterium lactis* BB-12 and *Bifidobacterium longum* BB-46. In *9th Egyptian Conference for Dairy Science and Technology, Cairo, 9–11 October 2004, Research Papers I*, 323–337.

Savadogo, A., Ouattara, C., Bassole, I.H. & Traore, S.A. (2006) Bacteriocins and lactic acid bacteria – a minireview. *African Journal of Biotechnology*, **5**, 678–683.

Schnell, N., Entian, K.D., Schneider, U., Gotz, F., Zahner, H., Kellner, R. & Jung, G. (1988) Prepeptide sequence of epidermin, a ribosomally synthesized antibiotic with four sulphide-rings. *Nature*, **333**, 276–278.

Shao, L., Wu, Z., Zhang, H., Chen, W., Ai, L. & Guo, B. (2014) Partial characterization and immunostimulatory activity of exopolysaccharides from *Lactobacillus rhamnosus* KF5. *Carbohydrate Polymers*, **107**, 51–56.

Shin, J.S., Jung, J.Y., Lee, S.G., Shin, K.S., Rhee, Y.K., Lee, M.K., Hong, H.D. & Lee, K.T. (2016) Exopolysaccharide fraction from *Pediococcus pentosaceus* KFT18 induces immunostimulatory activity in macrophages and immunosuppressed mice. *Journal of Applied Microbiology*, **120**, 1390–1402.

Skaugen, M., Nissen-Meyer, J., Jung, G., Stevanovic, S., Sletten, K., Inger, C., Abildgaard, M. & Nes, I.F. (1994) *In vivo* conversion of L-serine to D-alanine in a ribosomally synthesized polypeptide. *Journal of Biological Chemistry*, **269**, 27183–27185.

Smid, E.J., Starrenburg, M., Mireau, I., Sybesma, W. & Hugenholtz, J. (2001) Increase of folate levels in fermented foods. *Innovations in Food Technology*, **February/March**, 13–15.

Sreekumar, O. & Hosono, A. (1998) The antimutagenic properties of a polysaccharide produced by *Bifidobacterium longum* and its cultured milk against some heterocyclic amines. *Canadian Journal of Microbiology*, **44**, 1029–1036.

Stoddard, G., Petzel, J.P., van Belkum, M.J., Kok, J. & McKay, L.L. (1992) Molecular analyses of the lactococcin A gene cluster from *Lactococcus lactis* subsp. *lactis* biovar. diacetylactis WM4. *Applied and Environmental Microbiology*, **58**, 1952–1961.

Sybesma, W., Starrenburg, M., Kleerebezem, M., Mierau, I., de Vos, W.M. & Hugenholtz, J. (2003a) Increased production of folate by metabolic engineering of *Lactococcus lactis*. *Applied and Environmental Microbiology*, **69**, 3069–3076.

Sybesma, W., Starrenburg, M., Tijsseling, L., Hoefnagel, M.H. & Hugenholtz, J. (2003b) Effects of cultivation conditions on folate production by lactic acid bacteria. *Applied and Environmental Microbiology*, **69**, 4542–4548.

Sybesma, W., van Den Born, E., Starrenburg, M., Mierau, I., Kleerebezem, M., de Vos, W.M. & Hugenholtz, J. (2003c) Controlled modulation of folate polyglutamyl tail length by metabolic engineering of *Lactococcus lactis*. *Applied and Environmental Microbiology*, **69**, 7101–7107.

Sybesma, W., Burgess, C., Starrenburg, M., van Sinderen, D. & Hugenholtz, J. (2004) Multivitamin production in *Lactococcus lactis* using metabolic engineering. *Metabolic Engineering*, **6**, 109–115.

Tahara, T. & Kanatani, K. (1997) Isolation and partial characterization of crispacin A, a cell-associated bacteriocin produced by *Lactobacillus crispatus* JCM 2009. *FEMS Microbiology Letters*, **148**, 97–100.

Tahara, T., Yoshioka, S., Utsumi, R. & Kanatani, K. (1997) Isolation and partial characterization of bacteriocins produced by *Lactobacillus gasseri* JCM2124. *FEMS Microbiology Letters*, **148**, 97–100.

Tamime, A.Y. (ed.) (2005) *Fermented Milks*. Blackwell Publishing, Oxford.

Taranto, M.P., Medici, M., Perdigon, G., Ruiz Holgado, A.P. & Valdez, G.F. (2000) Effect of *Lactobacillus reuteri* on the prevention of hypercholesterolemia in mice. *Journal of Dairy Science*, **83**, 401–403.

Taranto, M.P., Vera, J.L., Hugenholtz, J., de Valdez, G.F. & Sesma, F. (2003) *Lactobacillus reuteri* CRL1098 produces cobalamin. *Journal of Bacteriology*, **185**, 5643–5647.

Tichaczek, P.S., Vogel, R.F. & Hammes, W.P. (1993) Cloning and sequencing of curA encoding curvacin A, the bacteriocin produced by *Lactobacillus curvatus* LTH1174. *Archives in Microbiology*, **160**, 279–283.

Timmerman, H.M., Koning, C.J.M., Mulder, L., Rombouts, F.M. & Beynen, A.C. (2004) Monostrain, multistrain and multispecies probiotics: a comparison of functionality and efficacy. *International Journal of Food Microbiology*, **96**, 219–233.

Twomey, D., Ross, R.P., Ryan, M., Meaney, B. & Hill, C. (2002) Lantibiotics produced by lactic acid bacteria: structure, function and applications. *Antonie van Leeuwenhoek*, **82**, 165–185.

Umu, Ö.C., Bäuerl, C., Oostindjer, M., Pope, P.B., Hernández, P.E., Pérez-Martínez, G. & Diep, D.B. (2016) The potential of Class II bacteriocins to modify gut microbiota to improve host health. *PLoS One*, **211**, e0164036.

Van Belkmun, M.J., Hayema, B.J., Jeeninga, R.E., Kok, J. & Venema, G. (1991) Organization and nucleotide sequence of two lactococcal bacteriocin operons. *Applied and Environmental Microbiology*, **57**, 492–498.

van den Hooven, H.W., Lagerwerf, F.M., Heerma, W., Haverkamp, J., Piard, J.C., Hilbers, C.W., Siezen, R.J., Kuipers, O.P. & Rollema, H.S. (1996) The structure of the lantibiotic lacticin 481 produced by *Lactococcus lactis*: location of the thioether bridges. *FEBS Letters*, **391**, 317–322.

Vaughan, E.E., Daly, C. & Fitzgerald, G.F. (1992) Identification and characterization of helveticin V-1829, a bacteriocin produced by *Lactobacillus helveticus* 1829. *Journal of Applied Bacteriology*, **73**, 299–308.

Vignolo, G., Fadda, S., de Kairuz, M.N., de Ruiz Holgado, A.A. & Oliver, G. (1996) Control of *Listeria monocytogenes* in ground beef by 'Lactocin 705', a bacteriocin produced by *Lactobacillus casei* CRL 705). *International Journal of Food Microbiology*, **29**, 397–402.

Wang, K., Li, W., Rui, X., Chen, X., Jiang, M. & Dong, M. (2014) Characterization of a novel exopolysaccharide with antitumor activity from *Lactobacillus plantarum* 70810. *International Journal of Biological Macromolecules*, **63**, 133–139.

Wegkamp, A., Starrenburg, M., de Vos, W.M., Hugenholtz, J. & Sybesma, W. (2004) Transformation of folate-consuming *Lactobacillus gasseri* into a folate producer. *Applied and Environmental Microbiology*, **70**, 3146–3148.

Winkowski, K., Crandall, A.D. & Montville, T.J. (1993) Inhibition of *Listeria monocytogenes* by *Lactobacillus bavaricus* MN in beef systems at refrigeration temperatures. *Applied and Environmental Microbiology*, **59**, 2552–2557.

Yan, T. & Lee, C. (1997) Characterization of a partially purified bacteriocin, Fermentcin B, from *Lactobacillus fermentum*. *Biotechnology Letters*, **19**, 741.

Yang, S.C., Lin, C.H., Sung, C.T. & Fang, J.Y. (2014) Antibacterial activities of bacteriocins: application in foods and pharmaceuticals. *Frontiers in Microbiology*, **5**, 241. doi:10.3389/fmicb.2014.00241

Yildirim, Z. & Johnson, M.G. (1998) Characterization and antimicrobial spectrum of bifidocin B, a bacteriocin produced by *Bifidobacterium bifidum* NCFB 1454. *Journal of Food Protection*, **61**, 47–51.

Yildirim, Z., Winters, D.K. & Johnson, M.G. (1999) Purification, amino acid sequence and mode of action of bifidocin B produced by *Bifidobacterium bifidum* NCFB 1454. *Journal of Applied Microbiology*, **86**, 45–54.

Zarate, G., Gonzalez, S. & Perez Chaia, A.P. (2004) Assessing survival of dairy propionibacteria in gastrointestinal conditions and adherence to intestinal epithelia. *Methods in Molecular Biology*, **268**, 423–432.

Zarate, G., Morata, D., Chaia, A.P. & Gonzalez, S.N. (2002a) Adhesion of dairy propionibacteria to intestinal epithelial tissue in vitro and in vivo. *Journal of Food Protection*, **65**, 534–539.

Zarate, G., Morata de Ambrosini, V., Perez Chaia, A. & Gonzalez, S. (2002b) Some factors affecting the adherence of probiotic *Propionibacterium acidipropionici* CRL 1198 to intestinal epithelial cells. *Canadian Journal of Microbiology*, **48**, 449–457.

Zhang, Z., Zhou, Z., Li, Y., Zhou, L., Ding, Q. & Xu, L. (2016) Isolated exopolysaccharides from *Lactobacillus rhamnosus* GG alleviated adipogenesis mediated by TLR2 in mice. *Scientific Reports*, **6**, 36083.

Živković, M., Miljković, M.S., Ruas-Madiedo, P., Markelić, M.B., Veljović, K., Tolinački, M., Soković, S., Korać, A. & Golić, N. (2016) EPS-SJ Exopolisaccharide produced by the strain *Lactobacillus paracasei* subsp. *paracasei* BGSJ2-8 is involved in adhesion to epithelial intestinal cells and decrease on *E. coli* association to Caco-2 cells. *Frontiers in Microbiology*, **7**, 286. doi: 10.3389/fmicb.2016.00286

10 Future Development of Probiotic Dairy Products

M. Saarela

10.1 Developments in the probiotic field in the European Union (EU)

Since the first edition of this book was published in 2005, a lot has happened in the probiotic area. In particular, two factors have had a major impact on this field in the EU: diminishing public funding for probiotic research and the failure of probiotics in obtaining any approved health claims. In 2005, my colleagues and I were just finishing the last big EU projects on pro- and prebiotics (the FP5 PRO-EUHEALTH cluster of eight EU projects) without realising that these would be the last dedicated EU framework research projects in this area and that the future would be much more challenging. Health claims regulation in the EU came into force in 2006, and the assessment of health claims started soon after. Although initially there were high hopes regarding the health claims, the outcome regarding probiotics was a huge disappointment. The only accepted claim (so far) has been for starter culture bacteria (*Lactobacillus delbrueckii* subsp. *bulgaricus* and *Streptococcus thermophilus*) and improved lactose digestion (http://www.efsa.europa.eu/fr/efsajournal/pub/1763). The frustration and disappointment resulted in heavy criticism of the European Food Safety Authority (EFSA) and its competence in evaluating health claim applications. On the other hand, the probiotic industry was also criticised for submitting applications of poor quality (Katan, 2012). Since in the EU, like many other regions, medicinal claims (for prevention, treatment or cure of diseases) are not allowed on foods labels, it is not surprising that many probiotics have failed to achieve positive assessment outcomes. After all, the efficacy of probiotics has largely been shown in diseased populations. Difficulties in identifying suitable target populations ('representing the general population') and validated biomarkers, especially in the gut health area, have proved to be difficult, although not impossible. As of this writing (November 2016), a frustrating number of over 300 probiotic health claim applications have been rejected by EFSA (http://www.nutraingredients.com/Regulation-Policy/Probiotic-health-claim-for-iron-absorption-rejected-by-EFSA). Although the EU and EFSA have been the focus of heated discussions on the assessment of the health claims, there also have been criticisms of the current regulatory system in the United States of America (USA) (Sanders *et al.*, 2016), whereas the regulatory system of Canada is

Probiotic Dairy Products, Second Edition. Edited by Adnan Y. Tamime and Linda V. Thomas.

considered to use a better model for the assessment of the health claims (Hoffmann *et al.*, 2013). Chapter 5 reviews in detail the current legislation of probiotic dairy products worldwide. In the EU, it would be possible to apply for a medicinal claim for probiotics from the European Medicines Agency (EMA), but this approach is usually not feasible for the food industry due to the prohibitive cost of the process. The failure of probiotics to obtain any approved health claims has led some people to think that probiotics do not have any efficacy at all, and that they are just a fad (http://www.dailymail.co.uk/health/article-3602382/We-bombarded-health-giving-foods-coconut-water-probiotics-FAD-FACT.html). A recent systemic review by Kristensen *et al.* (2016) on the lack of effect of probiotics on the faecal microbiota in healthy adults has evoked further comments in the media that 'probiotics are a waste of money for healthy adults' (http://www.nutraingredients-usa.com/Markets/Experts-react-to-waste-of-money-headlines-against-probiotics). This interpretation of results is on the wrong track, however, since already early culture-based studies performed up to 10–15 years ago showed that probiotics do not have any lasting effect on the resident gut microbiota of healthy individuals (Berggren *et al.*, 2003; Mättö *et al.*, 2006). This was later verified in a multitude of studies in healthy populations. Sanchez *et al.* (2016) actually suggest that changes in the metabolic activity of the microbiota are more important for probiotic efficacy than changes in microbiota composition. It can even be argued, however, that the manipulation of the composition of a healthy person's microbiota could actually be harmful and should not even be the aim of such probiotic studies.

A recent study by van den Nieuwboer *et al.* (2016) identified the main barriers in the probiotic innovation process, based on key-opinion-leader analysis. The most important innovation barriers were the following (in descending order):

- Difficulty in demonstrating efficacy;
- Competition with marketed probiotics with no evidence base;
- Regulatory approval (of health claims);
- Competition between the food and pharma industries;
- High cost of the clinical trials;
- Poor investment in probiotic research and development (R&D);
- Lack of scientific knowledge (e.g. on the mechanisms of action);
- Poor collaboration between industries (food and pharma) and academia;
- Lack of professional R&D;
- Negative perception of probiotics (due to failed health claims); and
- Small return on investment.

An important barrier related to the efficacy is the poor translation from preclinical animal models to human clinical studies, and the inability to generalise the outcomes. This case is well-known among researchers and largely due to the much larger variability of the diets and gut microbiota of humans compared to those of rats or mice. Another noteworthy barrier is the lack of investment in R&D. This has resulted in a situation where clinical studies are often heterogeneous and under-powered; such small projects just do not enable a sufficient evidence base to be acquired to successfully obtain a health claim in the EU. Changes in regulations are needed. Currently, probiotics fall somewhere between food and medicine, and therefore a simplified process for the medicinal claims of probiotic strains that are generally recognised as safe (GRAS) in the USA or have a qualified presumption of safety (QPS) in the EU (GRAS, http://www.fda.gov/Food/IngredientsPackagingLabeling/GRAS/; QPS, https://www.efsa.europa.eu/en/topics/topic/qps) would be needed (Sanders *et al.*, 2016; van den Nieuwboer *et al.*, 2016).

10.2 The current probiotic market and its trends

Although the situation on the probiotic research front has been challenging due to health claim issues and declining EU financing, the probiotic market is still growing and probiotic food products are becoming increasingly diverse. A wide range of probiotic dairy products is available in different markets, including (flavoured) milks, fermented milks, ice-cream, cheeses and infant formulas (see Chapter 4). The global sales of probiotic dairy-based foods were around US\$22 400 million in 2013, and it is anticipated that this market will continue to grow (an estimate for 2018 is US\$33 500 million). Probiotic yoghurt represents about 35% of the products, cultured drinks about 25%, Kefir and cheeses both ~10% and infant formula ~8% (BCC Research, 2014). In Northern Europe (Finland and Sweden), where fermented dairy products are commonly consumed, the total consumption of cultured milks (yoghurt and drinkable and spoonable cultured milks) is around 35 kg per year per capita (http://statistik.sjv.se/PXWeb/pxweb/sv/Jordbruksverkets%20statistikdatabas/?rxid=5adf4929-f548-4f27-9bc9-78e127837625; http://www.maataloustilastot.fi/ravintotase). In Finland, the consumption of yoghurt has been constantly increasing (it doubled to 21 kg in 14 years), and that of more traditional products is decreasing. Elsewhere in Europe, the consumption of yoghurt varies between 11 kg per capita in the UK up to 20 kg per capita in France (http://www.agr.gc.ca/eng/industry-markets-and-trade/statistics-and-market-information/agriculture-and-food-market-information-by-region/europe/market-intelligence/dairy-products-western-europe/?id=1420644518049#c). Dairy products high in protein (e.g. Skyr) seem to be more rapidly increasing in popularity compared to other products. No figures are available on the per capita consumption of probiotic fermented milks in different countries.

Compared to the situation in 2005, there has been a noticeable decrease in the variety of probiotic strains used in dairy products in the EU (see Chapter 4). Many dairies that used to produce many different probiotic products in the EU market have clearly reduced the variety of strains they use. This is probably due to the uniform failure of probiotics in securing any health claims in the EU; companies have instead started to use the same, technologically robust probiotic strains and discarded their 'own' strains. Ten years ago, several *Lactobacillus* and *Bifidobacterium* strains were used, whereas today in the more traditional products, typically the 'classical AB combination' of *Lactobacillus acidophilus* LA-5 and *Bifidobacterium animalis* subsp. *lactis* BB-12 is common, and in others only a couple of species and strains dominate (*Lactobacillus casei* Danone/Shirota/F19, *Lactobacillus johnsonii* and *Lactobacillus rhamnosus* GG) (http://www.arla.dk/produkter/arla-cultura/; http://www.valio.com/gefilus/). These species have been used by the dairy industry for a long time now and, due to their good availability and optimised technological properties, dairies will find them easier to use than many other species and strains.

10.3 Recent developments in the probiotic research

Over the past 10–15 years, the gut microbiota has been associated with almost all aspects of human health and disease. Thus, today it seems to be more difficult to identify a condition where the gut microbiota does *not* play a role than one where it does. Concomitant with this, the potential application fields of probiotics have also become broader. For example, paediatric diseases, such as necrotising enterocolitis, infantile colic, asthma, atopic disease, diabetes, malnutrition, mood/anxiety disorders and autism spectrum disorders, have become associated with microbiome alterations. Probiotic efficacy has been studied in many of these – with variable results – as well as other conditions such as antibiotic-associated diarrhoea (AAD), irritable bowel syndrome (IBS), inflammable bowel disease (IBD), *Helicobacter pylori* infection, arthritis, cardiovascular diseases and central nervous system related conditions, such as mood symptoms (Ebel *et al.*, 2014; Tillisch, 2014; Bravo-Blas *et al.*, 2016; Sanchez *et al.*, 2016; Slattery *et al.*, 2016). However, since it is hard to show causality between microbiota changes and a disease in humans (often there is only correlation; it is difficult to prove whether a certain condition results from microbiota changes or *vice versa*) (Wu *et al.*, 2016), it is probable that in many cases probiotics do not show consistent performance. Bacteria–gut–brain signalling has been a very active research area lately. One new application of probiotics has arisen from these studies – probiotics conferring mental health benefits (Tillisch, 2014; Sarkar *et al.*, 2016). These studies have mainly utilised various rodent models on stress and behaviour. In humans, so far little evidence on positive outcomes has been obtained (Sarkar *et al.*, 2016). Thus it remains to be seen whether in this case, results from animal models can be translated to human clinical studies. Another active research area has been the study of interactions between probiotics and the host. Probiotics interact with the host gut on three levels: the mucus layer, the epithelial layer and the gut-associated lymphoid tissue (GALT). Bacterial surface molecules are thought to play an important role in these interactions (Venema & do Carmo, 2015; Sanchez *et al.*, 2016).

The idea of using heat-killed 'probiotic' cells to activate the human immune system has been revisited recently (Iwasaki *et al.*, 2016; Maruyama *et al.*, 2016). Heat-killed 'probiotics' are an attractive option compared to the live probiotic products since problems with storage stability and challenges generated by the food matrix formulation can be avoided. Also, killed bacteria may be safer for immunocompromised patients even though possible adverse immunological effects cannot be ruled out. However, it should be noted that probiotics are, strictly speaking, live micro-organisms according to the widely accepted definition.

In the probiotic application, issues with the quality (mainly viability and stability) of probiotic products persist. Normally probiotics need to be viable in the product and in the human gastrointestinal (GI) tract, and some probiotic products struggle to achieve this. The well-established probiotic products are usually of consistent quality (http://www.dailymail.co.uk/health/article-2752798/The-probiotic-drinks-don-t-bring-benefits-Study-finds-good-bacteria-products-does-not-reach-small-intestine.html; Drago *et al.*, 2010; Fredua-Agyeman & Gaisford, 2015). Microencapsulation technologies have been widely studied to solve problems with viability and stability. Today, the focus is on technologies that would allow more cost-efficient production of high-quality probiotic products (Martin *et al.*, 2015). The challenges with probiotic product quality become even larger when we consider introducing novel bacterial genera into the products. To date, mainly certain robust lactobacilli and bifidobacterial strains have been used as probiotics; however, as a result of gut microbiota research, there is increasing interest in using other bacteria recently recognised to be important for human health. These include *Faecalibacterium prausnitzii* and other members of *Ruminococcaceae*, *Clostridium* XIVa cluster bacteria and *Akkermansia* spp. (Sanchez *et al.*, 2016).

10.4 Future target areas for research and conclusion

Since the first edition of this book was published, the research on probiotics continued to be active. One of the main challenges stated 10 years ago, namely achieving recognised scientific creditability for the probiotic health claims, is still an acute problem, at least in the EU (http://www.nutraingredients.com/Regulation-Policy/6-years-of-hurt-Probiotic-heavyweights-debate-the-EU-s-health-claim-blockade). Within the current regulatory framework, it is unlikely that any changes will occur in the near future. There have been attempts to persuade the EU to modify the regulation, but so far only a relevant guidance document was updated in 2015 by EFSA (http://www.efsa.europa.eu/en/efsajournal/pub/4369). In spite of this, probiotic research is moving on. There are several active research lines, such as: (a) 'unconventional' new probiotics, which will probably never enter the food chain but instead could be used as pharmaceuticals, (b) new probiotic targets, such as manipulation of the gut–brain axis (probiotics for mood, or 'psychobiotics'), (c) elucidating the mechanism of action of probiotics, which is still largely unclear (including their interaction with the host), (d) identifying the molecules responsible for probiotic health benefits, and (e) the dose dependency of any probiotic health effect. From the application side, issues with quality (viability and stability) still remain. This topic, as well as safety, will become even more important with new probiotic genera which have never previously been used in the human diet and which will tend to be technologically very challenging.

The study of van den Nieuwboer *et al.* (2016) that identified the main barriers in the probiotic innovation process also indicated what could be future targets for the probiotic research field. In addition to the above-mentioned aspects, improved animal models would allow better translation from preclinical animal studies to human clinical studies, and better quality clinical studies. Overcoming the innovation barriers would require increasing co-operation and communication between industry, academia and regulatory bodies, as well as increasing scientific research efforts (e.g. fundamental research on the mode of action and multi-centre clinical trials).

References

BCC Research (2014) *The Probiotics Market: Ingredients, Supplements, Foods*. FOD035D. BCC Research, Wellesley, MA.

Berggren, A., Söderberg, L., Önning, G., Johansson Hagslätt M.-L. & Axelsson, I. (2003) Intestinal function, microflora and nutrient intake of children after administration of a fermented oat product containing *Lactobacillus plantarum* DSM 9843 (299v). *Microbial Ecology in Health and Disease*, **15**, 160–168.

Bravo-Blas, A., Wessel, H. & Milling, S. (2016) Microbiota and arthritis: correlations or cause? *Current Opinion in Rheumatology*, **28**(2), 161–167.

Drago, L., Ridighiero, V., Celeste, T., Rovetto, L. & De Vecchi, E. (2010) Microbiological evaluation of commercial probiotic products available in the USA in 2009. *Journal of Chemotherapy*, **22**(6), 373–377.

Ebel, B., Lemetais. G., Beney. L., Cachon, R., Sokol, H., Langella, P. & Gervais, P. (2014) Impact of probiotics on risk factors for cardiovascular diseases: a review. *Critical Reviews in Food Science and Nutrition*, **54**, 175–189.

Fredua-Agyeman, M. & Gaisford, S. (2015) Comparative survival of commercial probiotic formulations: tests in biorelevant gastric fluids and real-time measurements using microcalorimetry. *Beneficial Microbes*, **6**(1), 141–151.

Hoffmann, D.E., Fraser, C.M., Palumobo, F.B., Ravel, J., Rothenberg, K., Rowthorn, V. & Schwartz, J. (2013) Probiotics: finding the right regulatory balance. *Science*, **342**, 314–315.

Iwasaki, K., Maeda, K., Hidaka, K., Nemoto, K., Hirose, Y. & Deguchi, S. (2016) Daily intake of heat-killed *Lactobacillus plantarum* L-137 decreases the probing depth in patients undergoing supportive periodontal therapy. *Oral Health and Preventive Dentistry*, **14**(3), 207–214. doi:10.3290/j.ohpd.a36099

Katan, M.B. (2012) Why the European Food Safety Authority was right to reject health claims for probiotics. *Beneficial Microbes*, **3**, 85–89.

Kristensen, N.B., Byrup, T., Allin, K.H., Nielsen, T., Hansen T.H. & Pedersen O. (2016) Alterations in fecal microbiota composition by probiotic supplementation in healthy adults: a systemic review of randomized controlled trials. *Genome Medicine*, **8**, 52.

Martin, M.J., Lara-Villoslada, F., Ruiz, M.A. & Morales, M.E. (2015) Microencapsulation of bacteria: a review of different technologies and their impact on the probiotic effects. *Innovative Food Science and Emerging Technologies*, **27**, 15–25.

Maruyama, M., Abe, R., Shimono, T., Iwabuchi, N., Abe, F. and Xiao, J.Z. (2016) The effects of non-viable *Lactobacillus* on immune function in the elderly: a randomised, double-blind, placebo-controlled study. *International Journal of Food Science and Nutrition*, **67**(1), 67–73. doi :10.3109/09637486.2015.1126564

Mättö, J., Fonden, R., Mikkola, T., von Wright, A., Vilpponen-Salmela, T., Satokari, R. & Saarela, M. (2006) Intestinal survival and persistence of probiotic *Lactobacillus* and *Bifidobacterium* strains administered in triple-strain yoghurt. *International Dairy Journal*, **16**, 1174–1180.

Sanchez, B., Delgado, S., Blanco-Miguel, A., Lourenco, A., Guiemonde, M. & Margolles, A. (2016) Probiotics, gut microbiota, and their influence on host health and disease. *Molecular Nutrition and Food Research* (published online). doi:10.1002/mnfr.201600240

Sanders, M.E., Shane, A.L. & Merenstein, D.J. (2016) Advancing probiotic research in humans in the United States: challenges and strategies. *Gut Microbes*, **7**, 97–100.

Sarkar, A., Lehto, S.M., Hart, S., Dinan, T.G., Cryan, J.F. & Burnet, P.W.J. (2016) Psychobiotics and the manipulation of bacteria–gut–brain signals. *Trends in Neurosciences*, **39**(11), 763–781.

Slattery, J., MacFabe, D.F & Frye, R.E. (2016) The significance of the enteric microbiome on the development of childhood disease: a review of prebiotic and probiotic therapies in disorders of childhood. *Clinical Medicine Insights: Pediatrics*, **10**, 91–107. doi:10.4137/CMPed.S38338

Tillisch, K. (2014) The effects of gut microbiota on CNS function in humans. *Gut Microbes*, **5**(3), 404–410. doi:10.4161/gmic.29232

van den Nieuwboer, M., van de Burgwal, L.H.M. & Claassen, E. (2016) A quantitative key-opinion-leader analysis of innovation barriers in probiotic research and development: valorisation and improving the tech transfer cycle. *PharmaNutrition*, **4**, 9–18.

Venema, K. & do Carmo, A.P. (eds.) (2015) *Probiotics and Prebiotics: Current Research and Future Trends*. Caister Academic Press, Poole.

Wu, H., Tremaroli, V. & Bäckhed, F. (2015) Linking microbiota to human diseases: a systems biology perspective. *Trends in Endocrinology and Metabolism*, **26**, 758–770.

Index

Probiotic Dairy Products, Second Edition. Edited by Adnan Y. Tamime and Linda V. Thomas.